D1620132

PLANETARY
ECOLOGY

Planetary Ecology

Edited by

Douglas E. Caldwell
University of Saskatchewan

James A. Brierley
Advanced Mineral Technologies, Inc.

Corale L. Brierley
Advanced Mineral Technologies, Inc.

VNR VAN NOSTRAND REINHOLD COMPANY
————————————————————— *New York*

Manufactured in the United States of America.

Published by Van Nostrand Reinhold Company Inc.
135 West 50th Street
New York, New York 10020

Van Nostrand Reinhold Company Limited
Molly Millars Lane
Wokingham, Berkshire RG11 2PY, England

Van Nostrand Reinhold
480 Latrobe Street
Melbourne, Victoria 3000, Australia

Macmillan of Canada
Division of Gage Publishing Limited
164 Commander Boulevard
Agincourt, Ontario MIS 3C7, Canada

15   14   13   12   11   10   9   8   7   6   5   4   3   2   1

Library of Congress Cataloging in Publication Data
Main entry under title:
Planetary ecology.
  Selected papers from the Sixth International Symposium on
Environmental Biogeochemistry, held Oct. 10-14, 1983, in Santa Fe, N.M.
  Includes index.
  1. Ecology—Congresses.   2. Biogeochemistry—Congresses.
I. Caldwell, Douglas E.   II. Brierley, James A.   III. Brierley, Corale L.
IV. International Symposium on Environmental Biogeochemistry (6th :
1983 : Santa Fe, N.M.)
QH540.P57   1985   574.5′222   84-27061
ISBN 0-442-24007-4

Contents

Contents

PART III: TRANSPORT, DEPOSITION, AND WEATHERING

Preface

The sixth International Symposium on Environmental Biogeochemistry (ISEB), held October 10-14, 1983, in Santa Fe, New Mexico, had as its theme, "Planetary Ecology." As in the previous five symposia, the 1983 ISEB provided an opportunity for scientists in paleoecology, ecosystem ecology, geochemistry, geomicrobiology, atmospheric chemistry, and other diverse fields to achieve a broader biogeochemical perspective. The first scientific session, "Biological Evolution and Planetary Chemistry," and the National Aeronautics and Space Administration (NASA) roundtable, "Biogeochemistry through Time and Space," provided a planetary view. They emphasized that individual biogeochemical events are linked by oceans and atmosphere and that present biogeochemistry is the consequence of a geologic and biologic heritage of which human beings have only recently become an influential part. Although an interdisciplinary approach is needed to study these relationships most effectively, research remains largely within traditional areas of specialization. As a result, the remaining sessions focused on specific environments, nutrient cycles, or organisms, relating these to the development of Earth chemistry.

Like the sixth ISEB Symposium, this book is interdisciplinary and international in scope. Reflected in these collected works are the symposium theme, "Planetary Ecology," and the principal objectives of ISEB:

Promotion of scientific knowledge of biogeochemistry and related subjects;

Stimulation of scientific investigation;

Advancement of education in biogeochemistry and related subjects.

The cochairmen of the sixth ISEB acknowledge the support of the U.S. Geological Survey, the Department of Energy, the National Science Foundation, NASA, Advanced Mineral Technologies, Inc., New Mexico Institute of Mining and Technology, University of New Mexico, Sandia National Laboratories, and Van Nostrand Reinhold Company, Inc. We also thank the enthusiastic participants for their stimulating and informative contributions, which made the meeting a success. We are especially grateful for the efficient and skillful contributions of our organizational team: Ms. Betsy Wilson assisted proficiently with the operational details of the meeting and performed word processing and typing; Ms. Shirley Whyte, with exceptional competence and patience, processed the manuscripts for publication; Ms. Pat Cataldo expedited the review of manuscripts and preparations for production of the volume.

DOUGLAS E. CALDWELL
JAMES A. BRIERLEY
CORALE L. BRIERLEY

Contributors*

Richard Aguilar
Colorado State University
Fort Collins, Colorado

P. G. Arcara
Istituto per lo Studio e la Difesa del
  Suolo
Firenze, Italy

Gert Asmund
The Geological Survey of Greenland
Copenhagen, Denmark

L. E. Bågander
University of Stockholm
Stockholm, Sweden

Lawrence A. Baker
University of Minnesota
Minneapolis, Minnesota

David L. Balkwill
Florida State University
Tallahassee, Florida

Jill Baron
Colorado State University
Fort Collins, Colorado

Terry J. Beveridge
University of Guelph
Guelph, Canada

Jerry M. Bigham
Ohio State University
Columbus, Ohio

Alex E. Blum
University of Wyoming
Laramie, Wyoming

James G. Bockheim
University of Wisconsin
Madison, Wisconsin

Anne-Marie Bodergat
Université Claude Bernard
Villeurbanne, France

Simon C. Brassell
University of Bristol
Bristol, England

Patrick L. Brezonick
University of Minnesota
Minneapolis, Minnesota

Fabio Caporali
Tuscia University
Viterbo, Italy

Yiu-Kwok Chan
Research Branch, Agriculture Canada
Ottawa, Canada

Ralf Conrad
Max-Planck-Institut für Chemie
Mainz, Federal Republic of Germany

Charles W. Culbertson
U. S. Geological Survey
Menlo Park, California

*The most recent known addresses are given.

Contributors

Brian E. Davies
University College of Wales
Penglais, Aberystwyth, Dyfed, Wales

Liesbeth W. de Jong
Univesity of Leiden
Leiden, The Netherlands

Hans P. M. de Vrind
University of Leiden
Leiden, The Netherlands

James I. Drever
University of Wyoming
Laramie, Wyoming

Brian J. Eadie
Great Lakes Environmental Research
    Laboratory
Ann Arbor, Michigan

Martin Eccleston
University of Warwick
Coventry, England

Henry L. Ehrlich
Rensselaer Polytechnic Institute
Troy, New York

Steven Emerson
University of Washington
Seattle, Washington

W. Fabig
Universitat Hohenheim
Stuttgart (Hohenheim), Federal
    Republic of Germany

Timothy J. Fahey
Cornell University
Ithaca, New York

S. S. D. Foster
Institute of Geological Sciences
Wallingford (Oxford), England

James J. Germida
University of Saskatchewan
Saskatoon, Canada

William C. Ghiorse
Cornell University
Ithaca, New York

Stjepko Golubic
Boston University
Boston, Massachusetts

Rolf O. Hallberg
University of Stockholm
Stockholm, Sweden

Robert D. Heil
Colorado State University
Fort Collins, Colorado

Mark E. Hines
University of New Hampshire
Durham, New Hampshire

Sally G. Horner
Virginia Polytechnic Institute and
    State University
Blacksburg, Virginia

Stephen R. Hutchins
Advanced Mineral Technologies, Inc.
Socorro, New Mexico

Richard James
University of East Anglia
Norwich (Norfolk), England

Edward A. Jepsen
University of Wisconsin
Madison, Wisconsin

Timothy Jickells
Bermuda Biological Station for
    Research
Ferry Reach, Bermuda

Galen E. Jones
University of New Hampshire
Durham, New Hampshire

Kevin C. Jones
Chelsea College, University of London
London, England

Don P. Kelly
University of Warwick
Coventry, England

Eugene F. Kelly
Colorado State University
Fort Collins, Colorado

Paul E. Kepkay
Bedford Institute of Oceanography
Dartmouth, Nova Scotia

Donald A. Klein
Colorado State University
Fort Collins, Colorado

Anthony Knap
Bermuda Biological Station for
    Research
Ferry Reach, Bermuda

Roger Knowles
Macdonald Campus of McGill
    University
Ste. Anne de Bellevue, Canada

Steven W. Leavitt
University of Wisconsin-Parkside
Kenosha, Wisconsin

Austin Long
University of Arizona
Tucson, Arizona

Douglas H. Loring
Bedford Institute of Oceanography
Dartmouth, Nova Scotia

W. Berry Lyons
University of New Hampshire
Durham, New Hampshire

S. O. McDonald
Utah State University
Logan, Utah

Peter R. Marshall
Research Branch, Agriculture Canada
Ottawa, Canada

Georg Matthess
Kiel University
Kiel, Federal Republic of Germany

R. R. Meglen
University of Colorado at Denver
Denver, Colorado

Philip A. Meyers
University of Michigan
Ann Arbor, Michigan

Christine L. Miller
Colorado School of Mines
Golden, Colorado

Leo J. Miller
Miller Exploration Co.
Evergreen, Colorado

Alan W. Morris
Institute for Marine Environmental
    Research
Plymouth, England

Arvin R. Mosier
U.S. Department of Agriculture—ARS
Fort Collins, Colorado

Monir Naguib
Max-Planck-Institut für Limnologie
Plon/Holstein, Federal Republic of
    Germany

Paolo Nannipieri
Istituto Chemica Terreno
Pisa, Italy

Kenneth H. Nealson
Scripps Institution of Oceanography
La Jolla, California

Arie Nissenbaum
The Weizmann Institute of Science
Rehovot, Israel

Ronald S. Oremland
U.S. Geological Survey
Menlo Park, California

J. C. G. Ottow
Universitat Hohenheim
Stuttgart (Hohenheim), Federal
  Republic of Germany

Nancy L. Parduhn
U.S. Geological Survey
Denver, Colorado

William J. Parton
Colorado State University
Fort Collins, Colorado

Jorge Pérez
New Mexico Institute of Mining and
  Technology
Socorro, New Mexico

Peter J. Peterson
Chelsea College, University of London
London, England

Reijo T. T. Rantala
Bedford Institute of Oceanography
Dartmouth, Nova Scotia

Clifford P. Rice
University of Michigan
Ann Arbor, Michigan

John A. Robbins
Great Lakes Environmental Research
  Laboratory
Ann Arbor, Michigan

Robert A. Sanford
Colorado State University
Fort Collins, Colorado

David S. Schimel
Colorado State University
Fort Collins, Colorado

Morris Schnitzer
Research Branch, Agriculture Canada
Ottawa, Canada

O. Tacheeni Scott
Northern Arizona University
Flagstaff, Arizona

Wolfgang Seiler
Max-Planck-Institut für Chemie
Mainz, Federal Republic of Germany

Andrei Serban
The Weizmann Institute of Science
Rehovot, Israel

J. A. Kent Simmons
Bermuda Biological Station for
  Research
Ferry Reach, Bermuda

John Skujiņš
Utah State University
Logan, Utah

Richard L. Smith
U.S. Geological Survey
Menlo Park, California

Mary Jo Spencer
University of New Hampshire
Durham, New Hampshire

Karl O. Stetter
Universität Regensburg
Regensburg, Federal Republic of
  Germany

John W. B. Stewart
University of Saskatchewan
Saskatoon, Canada

Janneke Tanke-Visser
University of Leiden
Leiden, The Netherlands

Craig D. Taylor
Woods Hole Oceanographic Institution
Woods Hole, Massachusetts

Bradley M. Tebo
Scripps Institution of Oceanography
La Jolla, California

Holm Tiessin
University of Saskatchewan
Sakatoon, Canada

Mason B. Tomson
Rice University
Houston, Texas

Arpad E. Torma
New Mexico Institute of Mining and
    Technology
Socorro, New Mexico

Joyce B. Tugel
University of New Hampshire
Durham, New Hampshire

Olli H. Tuovinen
Ohio State University
Columbus, Ohio

David M. Updegraff
Colorado School of Mines
Golden, Colorado

P. Mark Walthall
Colorado State University
Fort Collins, Colorado

Calvin H. Ward
Rice University
Houston, Texas

John R. Watterson
U.S. Geological Survey
Denver, Colorado

Peter Westbroek
University of Leiden
Leiden, The Netherlands

Bo Wikner
Chalmers University of Technology and
    University of Göteborg
Göteborg, Sweden

Ann P. Wood
University of Warwick
Coventry, England

Joseph B. Yavitt
University of Wyoming
Laramie, Wyoming

Caroline Yonker
Colorado State University
Fort Collins, Colorado

Sung Ok Yoon
Virginia Polytechnic Institute and
    State University
Blacksburg, Virginia

Part I

BIOLOGICAL EVOLUTION AND PLANETARY CHEMISTRY

1

Microbial Mats and Modern Stromatolites in Shark Bay, Western Australia

Stjepko Golubic

Boston University

ABSTRACT: Distribution, external morphology, texture, and microbial composition of microbial mats in Hamelin Pool, Shark Bay, Western Australia, have been studied and reviewed along a composite representative profile starting from the permanently submerged zone, across the zones of periodic flooding, toward permanently emerged land and coastal dunes. The following nine types of algal mats have been recognized: colloform, gelatinous, smooth, pincushion, tufted, mamillate, film, reticulate, and blister. Solar ponds represent a particular environment. The mat types represent microbial communities that are characterized by one or more dominant microorganisms. The colonization and stabilization of loose sediment is carried out by a microbial assemblage of generalists that prepare the ground for later replacement and succession by specialized microflora. Lithification of microbial mats takes place periodically, mainly during the austral summer. This process is destructive for the microbial community but increases the preservation potential of the stromatolitic structures.

INTRODUCTION

The hypersaline embayment Hamelin Pool of Shark Bay (Pl. 1-1, fig. 1) harbors a unique set of microbial communities that form algal or microbial mats and stromatolites. These organo-sedimentary formations are considered the closest modern analogs of ancient stromatolites known from Precambrian strata (Logan, 1961; Logan, Hoffman, and Gebelein, 1974; Hoffman, 1976; Playford, 1979, 1980).

PLATE 1-1. Environmental setting and subtidal and lower intertidal algal mats in Hamelin Pool, Shark Bay, Western Australia.

Fig. 1. Satellite image of two hypersaline basins of Shark Bay: Lharidon Bight (left) and Hamelin Pool (right).

Fig. 2. Aerial view of the research site north of Flagpole Landing station. Intertidal algal mats are seen as dark areas.

Fig. 3. Subtidal stromatolites covered by colloform mat in a 4-5 m depth at Carbla Point.

Fig. 4. Smooth mat and shallow pools, lower intertidal zone.

Fig. 5. Fenestrate margins of smooth mat.

Fig. 6. Columnar stromatolites of the lower intertidal zone covered by smooth mat.

Fig. 7. Surface view of the pincushion *Gardnerula* mat.

Fig. 8. Tufts of *Scytonema* filaments on the surface of tufted mat.

The present study is based on field work and collection of algal mats at several wave-exposed and protected sites carried out between July 20 and August 10, 1973 (austral winter), and between November 15 and 30, 1980 (the onset of austral summer).

The purpose of the present paper is to summarize the existing information, reconcile conflicting views, and bring new insight into ecological interactions with special reference to biological influences in carbonate precipitation. The types of algal mats as microbial communities are defined in terms of their morphology, microbial composition, and distribution with respect to principal environmental determinants. Biogenic influences on carbonate precipitation in this setting are discussed. The names of algal mat types published previously are used in the present paper whenever possible, although some renaming and redefining of algal mat types has been necessary.

Only a few of the microorganisms that build stromatolites have previously been studied and formally described; thus many of the microbial taxa encountered are new to science. Since formal description of these taxa is beyond the scope of the present paper, they will be identified to the nearest higher taxonomic level. Species identifications that have been published in the past using Drouet's system of cyanophytes have to be considered invalid (see Stam and Venema, 1977; Golubic and Focke, 1978; Lewin, 1981).

DISTRIBUTION AND MICROBIAL COMPOSITION OF ALGAL MAT TYPES

The various types of algal mats and stromatolites in Hamelin Pool are arranged in zones lying parallel to the coastline. Zonation of mats reflects the zonal arrangement of average conditions of wetting and drying (Golubic, 1976). The steepness of the coastal slope determines the horizontal distance covered by oscillation of the sea level, which in turn determines the width of each algal mat zone (Pl. 1-1, fig. 2, dark areas). This general pattern is complicated by topographic irregularities, patterns of sediment accumulation, and water drainage—which itself is modified by algal mat growth. The following nine mat types can be recognized along a composite representative profile, from the permanently submerged habitats landward, traversing zones with progressively decreasing frequency of wetting (Fig. 1-1): colloform, gelatinous, smooth, pincushion, tufted, mamillate, film, reticulate, and blister.

Colloform Mat

This mat covers subtidal stromatolites and indurated pavements that occur in Hamelin Pool under permanently submerged conditions, from one to several meters depth. These are autochthonous structures that are actively growing. The subtidal stromatolites are large (up to 60-100 cm high), internally lithified structures of distinct morphology (Fig. 1-1. [1]; Pl. 1-1, fig. 3).

The convexities and protrusions of the colloform mat are beige-translucent when sectioned, becoming light blue-green in the interior. This mat is built by a coccoid cyanophyte (cyanobacterium) that produces extracellular gelatinous matter of firm but elastic consistency. The gelatinous envelopes are deposited asymmetrically around the cells, with thicker layers oriented toward the outside of the colony. The cells are spherical, 3.5-5.5 μm in diameter. They divide in three planes and may elongate in the division process up to 7 μm. The extracellular envelopes are always hyaline and colorless. This microorganism is new to science at the species and genus levels. It belongs to the family Entophysalidaceae of the order Chroococcales. In addition to this principal formative microorganism, the colloform mat harbors a number of other prokaryotes and eukaryotes such as *Schizothrix* sp. and diatoms of the genera *Orthoneis* and *Mastogloia*.

The colloform mat appears to be vertically differentiated when viewed in a cross section. Beneath the translucent colloform mat there is a porous encrusted layer, about 1.5 mm thick, composed of intertwined calcareous serpulid tubes. At the time of collection most of these tubes were empty and lined by a chasmolithic (fissure-inhabiting) vegetation dominated by a rod-shaped cyanophyte (cells 2.5 x 5.0 μm isodiametric 2.5 x 2.5 μm upon division). These cells are inserted into a thin but firm gel, from which they are easily removed by pressure, leaving cup-shaped envelope bases similar to the pseudovagina of the Chamaesiphonales. The orientation of the rods is perpendicular to the surface, and the cell division is transversal. This microorganism represents a new genus and species of cyanophyte.

The interior of these stromatolites is composed of white, firmly-cemented,

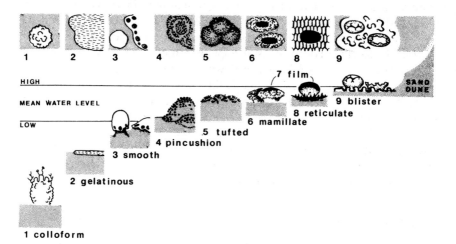

Figure 1-1.   Algal mat types in Hamelin Pool, arranged along a representative profile from submerged habitats (left) to coastal dunes (right).

carbonate deposit. Aragonite precipitation has been observed locally within colonies of the entophysalidacean builder of the colloform mat, as well as within scattered colonies of *Schizothrix* (sectio *Chromosiphon*). These concretions appear to be species-specific and morphologically distiguishable. The presence of the serpulid level possibly marks a single temporal event in which the growth and accretion of subtidal stromatolites were interrupted in the recent past. However, whether the conditions and microbial composition of these stromatolites prior to that event were the same as today, or different, cannot be ascertained at this time.

Gelatinous Mat

The name "gelatinous mat" has been applied by Logan, Hoffman, and Gebelein (1974) to a prostrate mat in the upper intertidal zone. They did not identify the microorganisms that build it. A mat that corresponds well to this name is distributed mainly in permanently submerged habitats up to the mean low water level (Fig. 1-1 [2], dashed). This mat is bright orange in color, very soft gelatinous to slimy, though not as cartilaginous as the colloform mat. The bulk of this mat is comprised of several species of the gel-producing, rod-shaped, coccoid cyanophyte *Aphanothece,* while the surface of the mat is dominated by stalk-forming diatoms. Diatoms in this mat produce polar, dichotomously-branched, gelatinous stalks that contribute to the entrapment of carbonate grains and serve as nuclei for carbonate precipitation. The internal structure of these diatom mats is laminated. Stalked diatoms can be considered eukaryotic stromatolite-building microorganisms (see also Golubic, 1976). Gelatinous mat dominated by *Aphanothece,* but lacking the stalked diatom component, has also been found in larger ponds in the upper intertidal zone.

Smooth Mat

This mat occurs in a wide range of intertidal habitats, but it dominates in the lowermost intertidal zone. It is characterized by a beige-pink color and a smooth, leathery surface (Fig. 1-1 [3]; Pl. 1-1, figs. 4-6). In contrast to the translucent colloform and gelatinous mats, it is always opaque. It incorporates substantial amounts of entrapped sediment (mostly carbonate skeletal fragments) that are arranged in laminae. Precipitation of $CaCO_3$ occurs periodically in distinct horizons as cement between entrapped particles and forms calcareous crusts. Vertical section through the mat shows it to consist of two layers. The surface layer is composed of a small *Schizothrix* sp. (section *Hypheothrix*) and a dense population of filamentous flexibacteria. A lower, light blue-green layer (2-3 mm below the surface) is dominated by *Microcoleus chthonoplastes* Thuret.

In the main range of its distribution, the smooth mat forms a contiguous cover (Pl. 1-1, fig. 4). Toward the lower end of its distribution, this cover becomes perforated and fenestrated (Pl. 1-1, fig. 5). The fenestrae are round to oval openings in the mat with rounded edges, 25-50 mm in diameter. They often line the margins of the smooth mat around depressions and ponds. Headlike and domal stromatolites within that range are covered by a contiguous smooth mat on their tops, while the fenestrate margins droop over the stromatolite slopes (Pl. 1-1, fig. 6). The rounded edges of the fenestrae give these margins a wavy appearance. In the upper ranges of its distribution, the smooth mat seems most competitive in colonizing and stabilizing loose sediment but later gives way to more specialized microorganisms that form new mat types (pincushion, tufted, mamillate, and reticulate) in the process of biological succession.

Pincushion Mat

In the lower intertidal zone, within the range of the smooth mat dominance, flat cushions of the branched brown filaments of *Gardnerula* sp. are commonly found (Fig. 1-1 [4]). This rivulariacean cyanophyte grows in a narrowly defined zone and is not present in all transect censuses because all the conditions required for its growth are not always fulfilled. The *Gardnerula* thalli occur most often embedded in the sediment with only the tips of filaments protruding. Thus they resemble dark pinheads stuck in a white pincushion (Pl. 1-1, fig. 7). On wave-exposed coasts with high sediment flux, *Gardnerula* grows through the sediment in an upward-radiating fashion, traps and consolidates sediment grains, and forms large (up to 60 cm high) stromatolites. No carbonate precipitation could be detected within these structures. Occasionally, *Gardnerula* is found free of sediment on hard, rocky substrates.

Tufted Mat

The name "tufted mat" has been suggested by Logan, Hoffman and Gebelein (1974) to describe mats characterized by upright tufts of algal filaments. However, they did not specify to which mat-forming microorganism this name should refer. The term describes well the mats formed by a marine *Scytonema* (Fig. 1-1 [5]). "Tufted mat" will be used here consistently in reference to *Scytonema* mats. Mats formed by *Lyngbya aestuarii*, which is also characterized by upright-growing filaments, will be referred to as reticulate mat.

The tufted mat occurs on moderately exposed, gently sloping coasts having good drainage. It occupies a zone between the lower intertidal smooth and pincushion mats and the upper intertidal mamillate and reticulate mats. It is characterized by distinctive, evenly distributed tufts of falsely branched *Scytonema* filaments that protrude out of the sediment (Pl. 1-1, fig. 8). The microorganism was discovered in 1858 by Harvey in Key West, Florida, and described under the name *Calothrix pilosa*. However, this organism cannot be considered *Calothrix* because it has a subterminal meristematic zone and does not differentiate a cellular hair. Reassignment of this taxon from the genus *Calothrix* to the genus *Scytonema* is being prepared for a separate publication.

Mamillate Mat

This is one of the most common microbial communities in the Hamelin Pool. It occurs in areas of good drainage over a wide spectrum of environmental energies. Along wave-exposed coasts, it forms domal and headlike, often elongated, stromatolites that later harden and lithify (Fig. 1-1 [6]; Pl. 1-2, figs. 1, 2). In protected environments, it forms prostrate covers. The mamillate mat is light brown (greenish-brown when wet) at the surface and translucent pale green in the interior. The surface of this mat is mamillate (warty), beset with small (0.08-8.0 mm) rounded or billowy protrusions (Pl. 1-2, fig. 4). The texture is gelatinous and elastic but crumbles easily. The mamillate mat (Golubic, 1973) has also been called "cinder zone" (Kendall and Skipwith, 1968) and "pustular mat" (Logan, Hoffman, and Gebelein, 1974).

The mamillate mat is built by the coccoid cyanophyte *Entophysalis major* Ercegovic. Cells are 3-6 μm in diameter and are surrounded by multiple gelatinous envelopes with brown pigmentation at the surface of the colony (Golubic, 1976). In the interior parts of the colony the envelopes are colorless, and the cells appear to be aligned in vertical rows. The gelatinous envelopes have a high preservation potential and can be recognized in very old buried mats. The envelopes preserve as rounded or ellipsoid bodies slightly reduced in size, while the cells shrivel into little granules and specks or disappear completely. *Entophysalis major* is the modern counterpart of the Precambrian microfossil *Eoentophysalis belcherensis* Hofmann, which built similar stromatolites in hypersaline environments about 2 billion years (2 Ga) ago (Golubic and Hofmann, 1976).

Mamillate mat stabilizes sediment to form extensive platforms. Erosion by waves damages the mat and subdivides platforms into elongated ridges that run roughly perpendicular to the coast (Logan, 1961), although this configuration may change over time (see Pl. 1-2, fig. 1). Once a ridge or a mound is carved out of a platform, the mat covers and stabilizes the slopes. The structures are subsequently hardened by carbonate precipitation that occurs within the mat. Hardened mat areas act as obstacles to currents and cause eddies that deepen the areas surrounding the stromatolites, narrow the stromatolite mounds, and carve out narrow necks at their bases (Pl. 1-2, fig. 3).

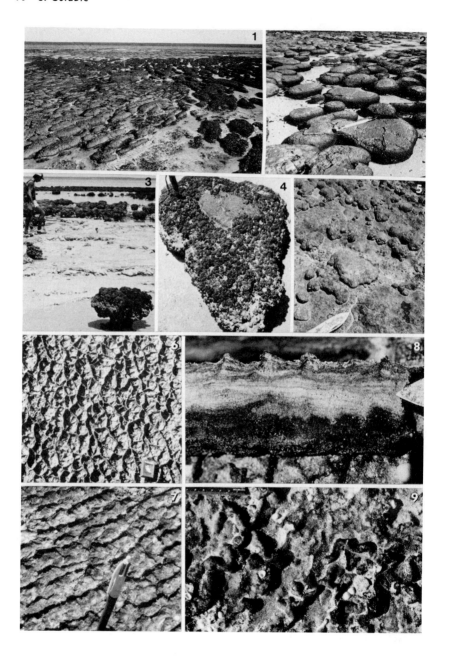

The result is formation of fields of isolated stromatolite domes and heads (Golubic, 1983).

Lithification occurs during the austral summer, starting as isolated carbonate concretions within the gelatinous envelopes of the mamillate mat, which then spread rapidly throughout the entire mat. The cells shrivel and are destroyed in the process of carbonate precipitation, which permeates the mat until the entire structure becomes firmly cemented. The carbonate is initially amorphous (Golubic and Campbell, 1981; Golubic, 1983) but later crystallizes into aragonite needles. Hard, lithified stromatolite surfaces reach the consistency of a dense carbonate crust and are subsequently colonized by endolithic, carbonate-destroying microflora called film mat.

Film Mat

The name "film mat" (Logan, Hoffman, and Gebelein, 1974) refers to a thin veneer of epilithic-endolithic colonizers, mostly *Hormathonema violaceonigrum* Ercegovic, which replaces the mamillate (or any other) mat upon lithification (Fig. 1-1 [7]; Pl. 1-2, fig. 5). Film mat does not form over loose sediment. The role of this mat is boring into and destroying stromatolites rather than building them (Golubic, 1976; Golubic, 1983, Fig. 6).

Lithified stromatolites in the highest ranges of their distribution are inactive fossil structures (Golubic, 1983) that are exposed to gradual erosion. Their surfaces are beige to rusty red in color. These surfaces are bare of any

PLATE 1-2. Stromatolites and algal mats in the upper intertidal ranges of Hamelin Pool.

Fig. 1. Lithified stromatolites (left) and nonlithified growing stromatolites formed by mamillate mat (right). Note the change in direction of the wave scour that shaped these stromatolites at different times.

Fig. 2. Lithified "fossil" stromatolites in the uppermost intertidal and supratidal zones.

Fig. 3. Solitary stromatolite head in the center of a pool, surrounded by unconsolidated sediment.

Fig. 4. Stromatolite head overgrown by mamillate mat and lithified on top.

Fig. 5. Lithified stromatolite top (detail from Fig. 4) covered by (endolithic) film mat.

Fig. 6. Reticulate mat in its typical development with sharp ridges outlining small polygonal depressions. Matchbox is for scale.

Fig. 7. Parallel ridges of the reticulate mat in a fast-draining channel lie roughly perpendicular to the direction of water flow. A ballpoint pen provides scale.

Fig. 8. Vertical section through reticulate mat, close-up view. Surface elevations are ridges in cross section. A pattern similar to that of the surface profile is seen in subsurface laminae, where preserved sheaths of *Lyngbya aestuarii* (Cyanophyta) can be found.

Fig. 9. Crenulate surface of blister mat at the upper boundary of the intertidal zone.

vegetation and reflect the color of mineral substances, such as carbonate stained by ferric iron (Pl. 1-2, fig. 2).

Reticulate Mat

This mat covers depressions and shallow ponds having poor drainage. It has a firm and leathery consistency and a sculptured surface that forms a network (reticulum) of interconnecting ridges. They outline small polygonal depressions 2-5 cm in diameter (Fig. 1-1 [8]; Pl. 1-2, fig. 6). The mesh size of this network is directly correlated with the water supply. In channels and fast-draining ponds, the ridges are arranged perpendicularly to the direction of the prevailing current, and the mesh shape is elongated (Pl. 1-2, fig. 7). This polygonal pattern of elevated ridges should be distinguished from desiccation polygons that have an opposite profile, being outlined by cracks. During low tides, each polygon acts as a micropond that facilitates water retention by the mat.

The reticulate mat is dominated and shaped by the oscillatoriacean filamentous cyanophyte *Lyngbya aestuarii* Liebman. Its trichomes are composed of short cells arranged like stacked coins (9-12 μm wide, 1-2 μm long, and surrounded by a thick brown sheath) having parallel or funnel-shaped layers.

Initially, *Lyngbya* filaments invade areas colonized by the smooth mat. They first appear as small round inocula (colonies) spaced at regular distances of 2-5 cm apart. They later interconnect to form thin ridges composed of *Lyngbya* filaments. When flooded, the filaments are upright, whereas during low tide they collapse and lean on each other, forming an A-profile in cross section. A vertical section through this mat often exposes laminae of preserved *Lyngbya* sheaths. They mark the positions of old mat surfaces and are often underlain by black anaerobic sediment (Pl. 1-2, fig. 8).

Blister Mat

Blistering of a mat is a characteristic, morphologically conspicuous feature involving crenulation and convolution of the mat, leaving the interior of the formed curvatures hollow (Fig. 1-1 [9]; Pl. 1-2, fig. 9). Blistering can occur in any mat type; thus it does not seem appropriate to assign the name to any particular microbial assemblage. However, this feature primarily afflicts those mats in elevated positions and in the uppermost intertidal ranges. Smooth mat in these areas often shows extensive areas of blistering.

Solar Ponds

On the flat, sabkha-like embayment bank on the west side of Hamelin Pool, there is a series of hypersaline ponds that are occasionally recharged by the water from Hamelin Pool brought in during high tides or prolonged on-shore winds. The ponds are circular, collapsed structures associated with dissolution of fossil gypsum deposits underneath and similar to karstic sinkholes. They range from a few meters to about 30m in diameter. At the time of study (November–December 1980), there was no pronounced thermal or chemical stratification. The entire benthos was covered by one type of mat, salmon pink in color at the surface and pale blue-green at its base. This mat forms fine pinnacles 1–3 cm high, similar to those described in the epilimnion of Solar Lake, Sinai (Krumbein, Cohen, and Shilo, 1977), and to a lesser extent to those found in Yellowstone themal springs (Walter, Bould, and Brock, 1976). The microbial composition of this mat is very diverse; it contains a large number of various metabolic types of microorganisms. However, two dominant cyanophytes, a thin *Phormidium* sp. and an *Aphanothece* sp., are responsible for the architecture of the mat.

DISCUSSION AND CONCLUSIONS

Since the discovery and first reports on modern stromatolites in Hamelin Pool, Shark Bay, Western Australia (Logan, 1961), these structures have been studied as a model for interpretation of ancient stromatolites known from Precambrian strata. However, some generalizations based on early observations are now outdated and need to be reassessed in view of new information.

Stratiform and domal stromatolites are formed by a number of different algal (microbial) mat communities in subtidal and intertidal coastal ranges. They are not a strictly intertidal phenomenon as initially claimed (see Logan, 1961; Logan, Rezak, and Ginsburg, 1964).

Stromatolitic structures are formed by active microbial communities that are restricted to a few-millimeters-thick layer or mat at the sediment-water interface (Golubic, 1973). This microbial activity is drastically reduced in the underlying layers of stromatolitic structures, which may be considered the sedimentary record of past algal mat activities. These lower, or interior, layers of stromatolites may be altered further by bacterial activity as a part of their early diagenesis.

The principal formative element of each stromatolite type studied is photosynthetic microorganisms, mostly cyanophytes (cyanobacteria), but some eukaryotic algae such as diatoms are important. In this sense the term "algal mat" is appropriate. These primary producers of algal mats provide the bulk of the mat's biomass and serve as a trophic base for heterotrophs, mostly bacteria, and to a lesser extent for animals and fungi that degrade and consume the organic matter of the mat. In this sense the term "microbial mat" is appropriate.

The persistance and preservation of stromatolites as structures depends on the balance between constructive and destructive elements and the rates of their activities. Burial under anaerobic (euxinic) conditions promotes preservaton of organic matter because the rate of microbial degradation under these conditions is slow and often incomplete. Under aerobic conditions degradation usually keeps pace with the production of organic matter.

Stromatolites form as a result of interaction between microorganisms and the sediment. This interaction includes stabilizing, trapping, and binding of sediment particles as well as mineral precipitation (Golubic, 1976). Lithification of stromatolites caused by precipitation of minerals such as $CaCO_3$ is a significant factor in promoting stromatolite consolidation and preservation. Some evidence of carbonate precipitation is found in all of the algal mat types studied, but in only two (the colloform mat in the subtidal zones and the mamillate mat in the intertidal zones) does carbonate precipitation lead to extensive lithification and induration of entire stromatolites. Under the conditions prevailing in Hamelin Pool, carbonate precipitates mostly during the austral summer.

Microorganisms influence carbonate lithification by providing sites (organic templates) for mineral nucleation; this may lead to formation of species-specific morphologies of mineral concretions that differ from surrounding inorganic carbonate cement (Golubic and Campbell, 1981; Golubic, 1983). The extent of lithification, however, is not under organismal control and often results in complete obliteration of the mat. Lithified stromatolites may then be colonized by carbonate-penetrating endolithic microflora or by a new generation of stromatolite-building mats.

The morphology and internal texture of living, nonlithified stromatolites result from the growth, orientation, and movement patterns of the stromatolite-forming microorganisms as well as from the direct and indirect effects of environmental energies. The degree of involvement of biological versus environmental factors in the shaping of stromatolites varies from case to case. Attempts to treat stromatolites as organisms, or as strictly sedimentary structures, represent extreme views that do not provide explanations for the observed phenomena.

Most of the mat morphologies can be correlated with dominating, formative microorganisms; however, the presence of subdominant taxa and the overall microbial diversity can serve as a sensitive indicator of ecological conditions. The highest organismal diversity is observed in subtidal mats, while the diversity declines in zones where mats have prolonged exposure to air and undergo periodic desiccation.

The colonization of bare sediment is carried out by a pioneer assemblage of microorganisms, dominated by *Microcoleus chthonoplastes*. It is clear that this assemblage tolerates the widest range of environmental conditions as it occurs in the widest range of habitats across the intertidal zone. These early mats gradually become more complex in their microbial composition and show more pronounced vertical differentiation. They are eventually replaced by more specialized assemblages that form narrowly zonated mats that meet in sharp boundaries.

The establishment and activity of microbial mats modifies physical, chemical, and biological properties in the sedimentary environments they inhabit. In addition to sediment trapping, stabilization, and lithification, these modifications include water retention and drainage, Eh and pH change, change in the solubility and mobility of metals and nutrients, and accumulation and recycling of metabolic products. Additional trophic niches are thus provided, contributing further to the complexity and differentiation of microenvironments within stromatolitic mats.

ACKNOWLEDGEMENTS

Thanks are due to all colleagues who carried out the organization of the field work, particularly to W. Wiebe and M. Walter. Field research was supported by the Bureau of Mineral Resources and Baas Becking Institute, Canberra, and Roche Pharmaceutical Co., Sidney. Research was supported by the following grants: NSF GA43391, OCE12999, EAR7684233, 7911200, 8107686, and NASA NSG7588, NAGW141.

REFERENCES

Golubic, S., 1973, The relationship between blue-green algae and carbonate deposits, in *The Biology of Blue-Green Algae,* N. Carr and B. A. Whitton, eds., Blackwell, Oxford, pp. 434-472.

Golubic, S., 1976, Organisms that build stromatolites, in *Stromatolites, Developments in Sedimentology,* vol. 20, M. R. Walter, ed., Elsevier, Amsterdam, pp. 113-126.

Golubic, S., 1983, Stromatolites, fossil and recent: A case history, in *Biomineralization and Biological Metal Accumulations,* P. Westbroek and E. W. deJong, eds., D. Reidel, Dordrecht, pp. 313-326.

Golubic, S., and S. E. Campbell, 1981, Biogenically formed aragonite concretions in marine *Rivularia,* in *Phanerozoic Stromatolites,* C. Monty, ed., Springer-Verlag, Berlin, pp. 204-224.

Golubic, S., and J. W. Focke, 1978, *Phromidium hendersonii* Howe: Identity and significance of a modern stromatolite building microorganism, *J. Sediment. Petrol.* **48:**751-764.

Golubic, S., and H. J. Hofmann, 1976, Comparison of modern and mid-Precambrian Entophysalidaceae (Cyanophyta) in stromatolitic algal mats: Cell division and degradation, *J. Paleontol.* **50:**1074-1082.

Harvey, W. H., 1858, Nereis Boreali-Americana. Contribution to the history of the marine algae of North America, v. III: Smithsonian Contribution.

Hoffman, P., 1976, Stromatolite morphogenesis in Shark Bay, Western Australia, in *Stromatolites, Developments in Sedimentology,* vol. 20, M. R. Walter, ed., Elsevier, Amsterdam, pp. 261-271.

Kendall, C. G. St. C., and P. A. d'E. Skipwith, 1968, Recent algal mats of a Persian Gulf lagoon, *J. Sediment. Petrol.* **38:**1040-1050.

Krumbein, W. E., Y. Cohen, and M. Shilo, 1977, Solar Lake (Sinai). 4. Stromatolitic cyanobacterial mats, *Limnol. Oceanogr.* **22:**635-656.

Lewin, R., 1981, Introduction and guide to the marine blue-green algae, by H. J. Humm and S. R. Wicks. Book Reviews. *Phycologia* **20:**216-217.

Logan, B. W., 1961, *Cryptozoon* and associate stromatolites from the Recent of Shark Bay, Western Australia, *J. Geol.* **69:**517-533.

Logan, B. W., R. Rezak, and R. N. Ginsburg, 1964, Classification and environmental significance of algal stromatolites, *J. Geol.* **72:**68-83.

Logan, B. W., P. Hoffman, and C. D. Gebelein, 1974, Algal mats, cryptalgal fabrics and structures, Hamelin Pool, Western Australia, *Am. Assoc. Pet. Geol. Mem.* **22:**140-194.

Playford, P. E., 1979, Stromatolite research in Western Australia, in *Aspects of Science in Western Australia 1829-1979,* A. E. Cockbain, ed., *Royal Soc. West. Australia J.* **62:**13-20.

Playford, P. E., 1980, Environmental controls on the morphology of modern stromatolites at Hamelin Pool, Western Australia, *West. Aust. Geol. Survey Ann. Rept. for 1979,* pp. 73-77.

Stam, W. T. and G. Venema, 1977, The use of DNA-DNA hybridization for determination of the relationship between some blue-green algae (Cyanophyceae), *Acta Bot. Neerl.* **26:**327-342.

Walter, M. R., J. Bould, and T. D. Brock, 1976, Microbiology and morphogenesis of columnar stromatolites (Conophyton, Vacerilla) from hot springs in Yellowstone National Park, in *Stromatolites, Developments in Sedimentology,* vol. 20, M. R. Walter, ed., Elsevier, Amsterdam, pp. 273-310.

2

Biomimetic Catalysis Mediated by Humic Substances and Melanoidins: "Geo-Enzyme" Activity?

Andrei Serban and Arie Nissenbaum

The Weizmann Institute of Science

ABSTRACT: Laboratory experiments show that humic substances and synthetic melanoidins exhibit catalytic activities usually associated with enzymes such as peroxidase, diaphorase, NADH dehydrogenase, and esterase. The extent of catalytic activity appears to be dependent on the particular composition of the polymer. The addition of coenzyme enhances the catalytic activity of humic substances, suggesting the formation of an enzymelike active association similar to the coenzyme-apoenzyme type of complex. The measured catalytic activities are small (ca milliunits/mg). However, when considering the time scale of geochemical processes, this catalysis can indeed be relevant. In the absence of biological activity, it can be speculated that such polymers might play a diagenetic role usually assigned to organisms or cell-free enzymes.

INTRODUCTION

It is generally assumed that biological activity is the main factor in electron transfer reactions occurring in the recent sedimentary environments. The evidence to support this assumption is more than ample and requires no further documentation. It is interesting to note, however, that very often such biogeochemical processes are identified by analyzing for the reactants and the products of the reaction (e.g., the sulfate-sulfide couple, the fermentation of organic matter, and so on). Knowledge of the actual processes involved is far more scanty. The very important questions of how many organisms occur in the sediments and how their numbers change with depth are not often addressed. These questions go unanswered because experimental devices that give accurate and precise answers do not yet exist. There is considerable indirect evidence that suggests biological activity at depths of hundreds of meters in unlithified sediments. However, when estimates of the magnitude of biological activity are made, they are often in conflict with the paradigm that the number of active microorganisms decreases very rapidly below the water-sediment interface.

Recent work (Maggioni and Cacco, 1977; Suflita and Bollag, 1980) has indicated the occurrence of extracellular enzymes in soils. Neither the mode of occurrence nor the biogeochemical activity of the enzymes is known. If the enzymes are active, biological-like reactions may occur where living systems are not usually operative.

The present report speculates on the possibility that sedimentary organic matter may mimic biotic reactions. This report is based on laboratory studies, and it is not known whether such reactions occur in the natural environment. The proposed processes may have been of more importance in a prebiotic milieu because microorganisms would not be competitive with the pseudo-enzymes.

A geologically plausible scenario for the chemical evolution of enzymatic-like catalysts that could be involved in redox reactions in the prebiotic environment has been proposed by Nissenbaum, Kenyon, and Oro (1975). It suggests that the simple organic molecules formed in the prebiotic environment would normally not reach a high enough concentration assumed for the primordial soup (Nissenbaum, 1976). In present-day oceans one of the mechanisms for scavenging of dissolved organic matter is the one leading to the formation of humic substances. It seems reasonable to assume that a similar process, based on the Maillard type of condensation between abiotically formed reducing sugars (carbonyl compounds) and amino acids (amino compounds) leading to melanoidin polymers, was already operational in the prebiotic hydrosphere. Nissenbaum, Kenyon, and Oro, (1975, p. 261) suggested that

under prebiotic conditions, melanoidins may have acted as primitive coenzymes, where the apoenzymes could be either amino acids or peptides adsorbed on the polymer, or even some of the amino acid residues of the polymer itself. In the latter case, the melanoidin itself, or certain regions of it, would be the holoenzyme.

By analogy, the term "geoenzyme" is suggested for these polymers.

MATERIALS AND METHODS

The humic substances used in this study originated from widely different sources: peat, organic rich soil, and marine sediments (Table 2-1). The

Table 2–1. Sources of Fulvic Acids Used in This Study

Sample	Source Material	Sampling Location
FA#1, 2, 3	soil	Hula Basin, Israel
FA#4-7	peat	Hula Basin, Israel
FA#8	marine sediment	Tanner Basin, Pacific Ocean
FA#9, 10	marine sediment	East Cortez Basin, Pacific Ocean
FA#11	marine sediment	Black Sea

extraction technique consists of three successive cycles of ultrasonic dispersion of the sample in 0.1 M sodium pyrophosphate at pH 7.5, followed by centrifugation, dialysis against distilled water using dialysis membranes with a molecular cut-off of 6000-8000 daltons, and lyophylization. The humic acids were separated from the fulvic acids by precipitation with HCl at pH 3.0. The present study is limited to the fulvic acid fraction of the humic substances, due to experimental difficulties caused by the insolubility of humic acids.

The melanoidin compounds were synthesized by either: (a) heating the 30% (w/w) water mixture of reactants (see below) in well-stoppered flasks at 120°C for 72 hr and maintaining the pH slightly alkaline by using solid calcium carbonate (samples 1, 2, and 3), or (b) heating at 95°C aqueous solutions of the reactants under reflux for 72 hr and maintaining an alkaline pH (9-9.5) by addition of sodium hydroxide pellets when necessary (samples 4, 5, and 6).

The compositions of the mixtures were as follows:

1. D-glucose (2×10^{-2} moles); L-aspartic acid and L-glutamic acid (10^{-2} moles each);
2. the same composition as (1) plus ferrous iron (2.5×10^{-3} moles, as ferrous sulfate);
3. D-glucose (3×10^{-2} moles), D,L-phenylalanine, L-tyrosine, L-tryptophane (10^{-2} moles each); and ferrous iron (3.5×10^{-3} moles, as ferrous sulfate);
4. D-glucose (8×10^{-2} moles) and L-lysine, L-aspartic acid, L-glutamic acid, D, L-phenylalanine, L-tyrosine, glycine, L-cysteine, and L-histidine (10^{-2} moles each);
5. the same composition as (4) but the lyophylized powder obtained was dialyzed against a solution of ferrous sulfate for 48 hr and then dialyzed again against water and re-lyophylized;
6. D-glucose (2.5×10^{-1} moles), L-histidine (2×10^{-2} moles), and L-lysine, D,L-phenylalanine and L-tyrosine (10^{-2} moles each).

The peroxidasic activity was assayed by measuring the rate of decomposition of hydrogen peroxide with o-dianisidine as hydrogen donor. One unit of peroxidase activity is that amount of catalyst that decomposes 1 micromole of hydrogen peroxide per minute at 25°C (Maehly and Chance, 1954). The oxidation of o-dianisidine was monitored at 460 nm. In the calculation, a molar absorbancy of 1.13×10^4/cm was used for the hydrogen peroxide.

The diaphorasic activity (the oxidation of $NADPH_2$) was determined at 25°C by measuring spectrophotometrically the reduction of 2,6-dichloro-phenolindophenol (DCPIP) at 600 nm. One unit of diaphorasic activity equals a decrease in absorbance of one unit A per minute (Mahler et al., 1952).

The esterasic activity was measured using as a substrate p-nitrophenyl acetate dissolved in 0.1 M phosphate buffer pH 6.2, at 25°C, and monitoring the concentration of released p-nitrophenol at 400 nm (Rohlfing and Fox, 1967).

The NADH dehydrogenase activity was measured in 0.2 M-Tris HCl buffer, pH 7.5, by following the decrease in absorbance at 340 nm due to the oxidation of NADH to NAD by FAD.

RESULTS

Fulvic acids originating from both soils and marine sediments, as well as some of the synthetic melanoidins (samples 3 and 6), exhibit some diaphorasic-like activity (from 10^{-2} units/mg to 10^{-4} units/mg) (Table 2-2). In the case of some fulvic acid samples (FA #3, FA #8), there is a delay of 5-25 min from

Table 2-2. Diaphorasiclike Activity of Fulvic Acids and Melanoidins

Sample	Activity (milliunits/mg)
Fulvic acid	
FA#1	1.6
FA#2	10
FA#3	4
FA#4	0.4
FA#5	1
FA#6	0.4
FA#7	0.2
FA#8	0.2
FA#9	1
FA#10	1
FA#11	0.2
Melanoidin	
#3	10
#6	1

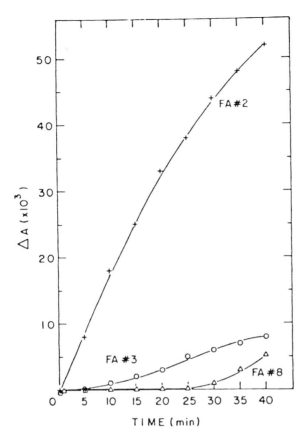

Figure 2-1. Diaphorasic activity (as a change in absorbance A) of fulvic acids FA#2, FA#3, and FA#8 as a function of time.

the moment the reactants are mixed in the cuvette of the spectrophotometer until the absorbance at 600 nm begins to decrease as a result of DCPIP reduction (Fig. 2-1). When an excess of substrate is present and the fulvic acid (catalyst) concentration is high, an anomalous reaction rate of nonlinear character is exhibited (Fig. 2-2).

To assess the influence of trace-metal ions present in the structure of the fulvic acid on the diaphorasic activity, three different fulvic acid samples were dissolved in 10^{-3} M HCl, extensively dialyzed against EDTA, then against water, and finally lyophylized. After that treatment, the diaphorasic-like activity of the samples decreased one order of magnitude (Table 2-3).

The potential catalytic effect of fulvic acids on the direct redox reaction between NADH and FAD (flavin-dependent NADH dehydrogenase) was also

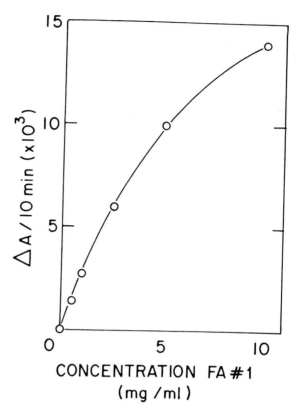

Figure 2-2.   Dependence of the diaphorasic activity (as the rate of change in absorbance A/min) of fulvic acid FA#1 on fulvic acid concentration.

investigated. The net increase in the rate of reaction mediated by fulvic acids was estimated after making the correction for the two noncatalytic NADH oxidation processes by FAD (control C_1) and fulvic acid (control C_2), respectively. The results are listed in Table 2-4; the reaction rate is expressed as change in absorbance per minute.

The analogy with enzyme-catalyzed reactions is also evident in the melanoidin-catalyzed hydrogen peroxide decomposition by a hydrogen acceptor-peroxidasic activity (Table 2-5). Verification of the nature of the reaction (catalytic or stoichiometric) for the most active compound (#3) was attempted. After a 60 min reaction time, when the rate of reaction was practically negligible, a second charge of substrate and donor was added to the reaction mixture, and the optical density change due to the donor oxidation was recorded for another 60 min. The rate of reaction remained at the same order of magnitude: 17.5×10^{-3} units/mg for the standard

Table 2–3. The Effect of Incubation with EDTA on Diaphorasiclike Activity of Fulvic Acids

Sample	Initial Activity (milliunits/mg)	Activity after EDTA Treatment (milliunits/mg)
FA#1	1.6	0.2
FA#5	1	0.1
FA#10	1	0.1

Table 2–4. Dehydrogenase-like Activity of Fulvic Acids

	$\Delta A/min \times 10^3$			
Sample	Control C_1	Control C_2	Test	Test$-C_1-C_2$
FA#1	1.8	0.4	3.9	1.7
FA#3	1.5	0.5	2.4	0.4
FA#5	1.7	0.5	3.4	1.2
FA#8	1.8	0.4	2.6	0.4
FA#10	1.8	0.7	4.0	1.5

Table 2–5. Peroxidase-like Activity of Melanoidins

Sample	Peroxidase Activity (milliunits/mg)
#1	0
#2	trace
#3	17.5
#4	~1
#5	5
#6	~1

condition and 12.5×10^{-3} units/mg after the addition of the second charge of substrate. This result implied that the reaction was catalytic. The effect of heat treatment on the activity of melanoidin #3 was also assessed. In the case of an enzymatic protein, heat treatment usually leads to a decrease of catalytic activity as a result of the denaturation of the macromolecular structure of the protein. Unlike the case with enzymes, 10 min incubation of the aqueous melanoidin suspension at 100°C caused an increase of about 40% in the peroxidasic activity (17.5-25 milliunits/mg).

The analogy between humic substances and melanoidins and some genuine enzymes is not limited to the group of redox enzymes. For example, melanoidins also exhibit some esterasic activity. The rate constant K_1 for the

Table 2-6. Esterasic Activity of Melanoidin

Melanoidin Concentration (mg/ml)	Rate Constant K_1 $1/min \times 10^3$
0 (autohydrolysis)	0.47
0.1	0.49
0.2	0.54
0.5	0.62

first order of reaction of hydrolysis of a 2×10^{-3} M solution of p-nitro-phenylacetate for the first 10 min of reaction was measured at three different concentrations of melanoidin: 0.1, 0.2, and 0.5 mg/ml reaction mixture. The results given in Table 2-6 show the increase in the rate constant of hydrolysis in the presence of the melanoidin (esterasic activity) as compared to the rate constant for the spontaneous, uncatalyzed hydrolysis of the ester.

A cautionary note has to be made with respect to the magnitude of the catalytic activities reported in this work. There are indications that the lyophylization procedure used to obtain humic compounds and melanoidin preparations that were clean and less susceptible to bacterial contamination had the undesirable effect of partially removing structural water—that is, water molecules playing a stabilizing role in the macromolecular structure of the polymers. Such partially denatured and collapsed polymeric structures are difficult to rehydrate, and as a result, parts of the active sites or of the cooperative sites remained unavailable for substrate binding. As an indication of this hypothesis, a 43% increase in the peroxidasic activity of melanoidin #3 subjected to 10 min incubation at 100°C was noted when compared to the test in which the activity was measured after the dissolution of the freeze-dried polymer in buffer at 25°C. Consequently, the enzymaticlike activities of the humic compounds might be even higher than those reported in this study.

DISCUSSION

The presence of an enzymelike flavoprotein moiety in the humic compound's macromolecular structure was proposed to explain some electron transfer reactions mediated by humic acids (Schindler, Williams, and Zimmerman, 1976; Zimmerman, 1981).

Humic compounds and their synthetic analogs, melanoidins, act as catalysts in a variety of chemical reactions, mainly of the redox type. In oxidation and reduction reactions conducted in aqueous media, humic compounds are capable of catalyzing electron-transfer processes over a large range of redox potentials, from the reduction of ferric ion ($E_o' = +0.77$ V) (Szilagyi, 1971) to the photodecomposition of water to molecular hydrogen ($E_o' = -0.827$ V) (A. Serban and A. Nissenbaum, unpub.).

Enzymes are characterized by very high substrate specificity. An indication that the catalytic activity of melanoidin also depends on the particular composition of the polymer can be inferred from the fact that of the six different melanoidin polymers investigated, only two had diaphorasiclike activity. This suggests that, depending upon the initial composition of the reactants, particular structures confer specific catalytic properties to the melanoidin. This may also explain the significant variations observed in the peroxidasic activity among catalytically active melanoidins. Since naturally occurring humic substances originate from a very diverse range of source materials, the possibility exists that some type of sedimentary humates may catalyze large spectra of enzyme-like reactions.

The analogy with enzymes is also apparent in the effect of the EDTA treatment on the diaphorasic activity of the fulvic acid. Massey and Veeger (1961) report that incubation of pig heart preparation with copper ions increases the native diaphorasic activity. NADH dehydrogenase, similar to diaphorase in its substrate specificity, is known to contain eight atoms of iron in a non-heme protein; this structure is essential for the enzymatic activity (Lehninger, 1970). The decrease in the diaphorasic activity subsequent to the dialysis against EDTA could be attributed to a change in the catalytically active structure of the fulvic acid, which is possibly stabilized by metal ions.

A process similar to the cooperative effect played by the macromolecular structure of the holoenzyme on the catalytic activity of the coenzyme can be inferred from the increase in the oxidation rate of the NADH by FAD in the presence of fulvic acid. In the case of the reaction between NADH and FAD, the net difference between the standard redox potentials of the initial and final states is small ($\triangle E_o' = E_o'$ FAD/FADH $- E_o'$ NAD/NADH $= +0.101$ V). This suggests that the reaction, even if thermodynamically possible, should have a very low reaction rate in the absence of a catalyst system. The presence of both fulvic acid and FAD leads to an increase of 18%-77% in the reaction rate as compared to the simple arithmetic summation of the reaction rates in the presence of either fulvic acid or FAD alone. A plausible explanation for the synergistic effect on the NADH oxidation rate is the formation of a coenzyme-apoenzyme type of a complex between the FAD and the fulvic acid.

The catalytic activities associated with humic substances and melanoidins are orders of magnitude lower than the corresponding enzymes that exhibit the same activities. However, from a geochemical point of view, even activities of the order of milliunits/mg could be significant.

ACKNOWLEDGEMENTS

The advice provided by Dr. A. Yaron is gratefully acknowledged. We would also like to thank Dr. S. Weiner for critically reviewing the manuscript.

REFERENCES

Lehninger, A. L., 1970, *Biochemistry,* Worth, New York.

Maehly, A. C. and B. Chance, 1954, The assay of catalases and peroxidases, in *Methods in Biochemical Analysis,* Vol. 1, D. Glick, ed., Interscience, New York.

Maggioni, A., and G. Cacco, 1977, Acetyl-naphtyl-esterase activity in humus-enzyme complexes of different molecular sizes, *Soil Sci.* **123:**122-125.

Mahler, H. R., N. K. Sarker, L. P. Vernon, and R. A. Alberty, 1952, Studies on diphosphopyridine nucleotide-cytochrome c reductase. II. Purification and properties, *J. Biol. Chem.* **199:**585-597.

Massey, V., and C. Veeger, 1961, Reaction mechanism of lipoyl dehydrogenase, *Biochim. Biophys. Acta.* **48:**33-47.

Nissenbaum, A., 1976, Scavenging of soluble organic matter from the prebiotic oceans, *Origins of Life* **7:**413-416.

Nissenbaum, A., D. H. Kenyon, and J. Oro, 1975, On the possible role of organic melanoidin polymers as matrices for prebiotic activity, *J. Mol. Evol.* **6:**253-270.

Rohlfing, D. L., and S. W. Fox, 1967, The catalytic activity of thermal polyanhydro alpha-amino acids for the hydrolysis of p-nitrophenyl acetate, *Arch. Biochem. Biophys.* **118:**122-126.

Schindler, J. E., and D. J. Williams, and A. P. Zimmerman, 1976, Investigations of extracellular electron transport by humic acids, in *Environmental Biochemistry,* Vol. 1, J. O. Nriagu, ed., Ann Arbor Science, Ann Arbor, Michigan, pp. 109-115.

Suflita, J. M., and J. M. Bollag, 1980, Oxidative coupling activity in soil extracts, *Soil Biol. Biochem.* **12:**177-183.

Szilagyi, M., 1971, Reduction of Fe^{+3} ion by a humic acid preparation, *Soil Sci.* **114:**233-235.

Zimmerman, A. P., 1981, Electron intensity, the role of humic acids in extracellular electron transport and chemical determination of pE in natural waters, *Hydrobiologia* **78:**259-265.

3

Peptidoglycan Envelope in the Cyanelles of *Glaucocystis nostochinearum*

O. Tacheeni Scott
Fort Lewis College

ABSTRACT: *Glaucocystis nostochinearum* is an alga of indeterminate phylogenetic identity. Various structural and pigment characters have been used by investigators to suggest its taxonomic affinity to the cyanobacteria, red algae, green algae, dinoflagellates, and the cryptomonads. With the advent of the endosymbiotic theory for the evolution of chloroplasts, other researchers suggested that *Glaucocystis* represents a symbiotic association involving a colorless green-algal host cell and intracellular cyanobacteria, or cyanelles.

The cyanelles found in *Glaucocystis* have been investigated in several ultrastructural and biochemical studies in unsuccessful attempts to detect the presence of a peptidoglycan envelope, a basal layer of non-Archaebacterial prokaryotic cell walls.

Disruption of intact *Glaucocystis* cells with a glass tissue homogenizer permitted the isolation of the uniquely shaped cyanelles. Preservation of intactness and shape of the cyanelles in a hypotonic isolation medium was terminated by lysozyme at final concentrations of less that $50\,\mu g/ml$. At $5\,\mu g/ml$, lysozyme achieves lysis within a few seconds, and this was found to be the optimal concentration.

Lysozyme-mediated lysis was inhibited by N-acetyl-glucosamine-2, a known competitive inhibitor. Diaminopimelic acid was detected by standard amino acid-analysis methods, prompting the conclusion that the cyanelles of *Glaucocystis* are cyanobacterial endosymbionts.

INTRODUCTION

Glaucocystis nostochinearum is a taxonomically enigmatic freshwater alga. During the past 117 years, *Glaucocystis* has been assigned to prokaryotic and eukaryotic taxa including the cyanobacteria, red algae, green algae, dinoflagellates, and the cryptomonads. *Glaucocystis* cells are ovoid shaped and characterized by a cellulosic cell wall, nucleus, mitochondria, dictyosomes, vacuoles, internal (rudimentary) flagella, blue-green-pigmented chloroplast-like bodies, extra chloroplastidic starch, and autospore formation as a means of asexual reproduction. In recognition of this combination of prokaryotic and eukaryotic characters in *Glaucocystis,* several investigators have considered it to be a unique eukaryotic organism that could be accommodated only by special taxonomic groupings, including Glaucocystideae (West, 1904) and Glauycophyta (Skuja, 1954; Lee, 1980).

An alternative to a taxonomic remedy for *Glaucocystis* was first offered by Geitler (1923), who suggested that *Glaucocystis* constituted a symbiotic association between a colorless green-algal host and several cyanobacterial endosymbionts ("cyanelles" after Pascher, 1914). Geitler's endosymbiotic explanation, however, received critical reviews from Fritsch (West and Fritsch, 1927) and Korschikov (1930), who argued that confirmation of the symbiotic explanation rested on the capacities of the cyanelles to tolerate isolation from the host cytoplasm and to demonstrate growth while in the isolated state. Fritsch and Korschikov concluded that the apparent symbiotic nature of *Glaucocystis* was not proved.

In response to the contention that independent culture of the symbionts is necessary as validation of an endosymbiotic association in *Glaucocystis,* Skuja (1948) concluded that the establishment of such a symbiosis by the dissolution of the host cell wall was impossible since no extracellular cellulases were known to be produced by cyanobacteria. Skula argued that *Glaucocystis* represented an endosymbiotic association that occurred very early in the geologic record, resulting in the complete mutual dependence of host and cyanobacteria. With this direct application of the endosymbiotic theory for the origin of chloroplasts (Mereschkowsky, 1905; Pascher, 1914), Skuja felt the need to create a new taxonomic group: the phylum Glaucophyta (Skuja, 1954). Subsequently, however, many investigators have argued that Glaucophyta does not represent a group of organisms sharing phylogenetic affinities, but represents an artificial taxon.

A direct result of the unsettled issue regarding the phylogeny of *Glaucocystis* was the search for a prokarytic cell wall surrounding the cyanelles. Evidence for a remnant cyanobacterial cell wall became the subject of several extensive ultrastructural studies (Bourdu and Lefort, 1967; Echlin, 1967; Hall and Claus, 1967; Lefort, 1965; Lefort and Pouphile, 1967; Schnepf, 1965; Schnepf, Koch, and Deichgräber, 1966). Only Schnepf and his coworkers (1966) reported evidence for a remnant cell wall. They stated that only under highly favorable conditions involving fixation with OsO_4 and plasmolysis of the cyanobacterial cells were they able to detect the 6-nm thick, osmiophilic innermost layer of the cell walls. They concluded that the plasmalemma and cell wall of the cyanelles are so intimately appressed to one another that they appear as a single line in electron micrographs.

The cryptic nature of the phylogenetic issue was reemphasized when nonstructural evidence regarding the biochemistry of the cyanelle cell wall and the spectral analysis of the photosynthetic pigments was considered. Holm-Hansen, Prasad, and Lewin (1965) failed to detect in the cyanelles diaminopimelic acid, an exotic amino acid found only in the walls of prokaryotes (not, however, in the Archaebacteria [Fox et al., 1980]). Although their finding supported the conclusion that the cyanelles lacked this similarity with the cyanobacteria, Holm-Hansen and coworkers concluded that it may be attributable to the loss of the cell wall.

Chapman (1966) found evidence for the following photosynthetic pigments in the cyanelles: chlorophyll a, β-carotene, zeaxanthin, cyanophycocyanin (the prefix is necessary to distinguish it from the related pigment, rhodophycocyanin), and allophycocyanin. The two common cyanobacterial xanthophylls, echinenone and myxoxanthophyll, could not be detected, however. Chapman (1966, p.2) concluded that "as the limit of detection with thin-layer chromatography approaches 1 microgram or less, these two xanthophylls, should they be present, would be there in very small amounts." Otherwise, he considered the pigments to be characteristic of the cyanobacteria. More recently, Schmidt, Kies, and Weber (1979) not only substantiated Chapman's findings but found β-cryptoxanthin in *Glaucocystis* cyanelles.

Schenk (1970) reported his findings of a lysozyme-sensitive remnant cell wall in cyanelles isolated from *Cyanophora,* an apparent endosymbiotic association involving a cyanobacterium and a flagellated host. In his application of the same experimental protocol to *Glaucocystis* cyanelles, Schenk (1971) was unable to demonstrate lysozyme-mediated lysis as demonstrated by phycocyanin release. Schenk concluded (Schenk and Hofer, 1972, p. 2095) that the cyanelles have only a "lipid-containing envelope membrane similar to that of the chloroplasts."

The present taxonomic status of *Glaucocystis* remains enigmatic. In a statement apparently made out of frustration—and certainly resignation— Lewin (1974, p. 26) concluded that "not even with the help of the electron microscope have we been able to establish unequivocally whether its pigmented bodies are blue-green algae or plastids." It is evident that Lewin was referring to the lack of substantial evidence for the presence of a peptidoglycan envelope in the cyanelles of *Glaucocystis.*

The purpose of this report, therefore, is to disseminate the results of a recent study on the biochemical nature of the envelope surrounding the cyanelles of *Glaucocystis.* Lysozyme-sensitivity and related biochemical manifestations of the envelope are examined in order to qualify the envelope as consisting of peptidoglycan, a component of non-Archaebacterial, prokaryotic cell walls. The results are intended to add to the knowledge and appreciation of the cyanelles as endosymbiotic cyanobacteria that have become obligately, and thus irreversibly, dependent on an equally dependent eukaryotic partner.

MATERIALS AND METHODS

Culture Conditions

Axenic cultures of *Glaucocystis nostochinearum* Itzigsohn (#229/1) were obtained from the Cambridge Collection of Algae and Protozoa, Cambridge University, Cambridge, United Kingdom. Cultures were grown axenically in cotton-plugged 500-ml bubbler tubes containing 450-ml volumes of culture medium bubbled with sterile air at 24°C under 1,700 lx of continuous, cool-white fluorescent illumination. The growth medium, devised by Gustave Görtlind, University of Uppsala, Uppsala, Sweden, was of the following composition (given per liter): 0.085 g $NaNO_3$, 0.015 g KCl, 0.027 g KH_2PO_4, 0.174 g K_2HPO_4, 0.049 g $MgSO_4$, 0.047 g $Ca(NO_3)_2 \cdot 4H_2O$, 0.5 g N-Z-Case (Sheffield Chemical, Union, New Jersey), 0.005 mg vitamin B_{12}, and 1.0 ml of a micronutrient solution containing the following (given as mg in 50-ml stock): 300 mg nitrilotriacetic acid dissolved in 4 ml of 1 N NaOH, 11 mg $ZnSO_4 \cdot 7H_2O$, 18 mg $MnCl_2 \cdot 4H_2O$, 48 mg $FeCl_3 \cdot 6H_2O$, 17.5 mg H_3PO_3, 6.3 mg $Na_2MoO_4 \cdot 2H_2O$, 0.37 mg $CuSO_4 \cdot 5H_2O$, 0.020 mg $CoCl_2 \cdot 6H_2O$. Double-distilled water (DD-water) was used in the preparation of all media and experimental stock solutions. The initial pH of the culture medium was 7.1.

Cell Counts

Cell numbers were determined with a Bright-line Improved Neubauer hemacytometer (American Optical Corp.).

Aseptic Isolation of Cyanelles from *Glaucocystis*

All routine transfers, inoculations of experimental cultures, samplings, and homogenization procedures were performed in an EdgeGARD laminar flow transfer hood (The Baker Co., Sanford, Maine). The suspension cultures were permitted to achieve mid-log growth (1-2 × 10^4 cells/ml) before they were pelleted by centrifugation, 1,000 × gravity, 2 min, in sterile polycarbonate centrifuge tubes with caps fashioned from aluminum foil (caps utilized in all subsequent centrifugations). Approximately 2 ml of packed *Glaucocystis* cells were resuspended in 40 ml of chilled growth medium and homogenized aseptically in a 40-ml-capacity glass tissue homogenizer resting in an ice bath. The brei was centrifuged, 800 × gravity, 1 min, 6°C (all subsequent centrifugations at 6°C) before the supernatant (crude cyanelle suspension) was poured into sterile centrifuge tubes and centrifuged, 1,000 × gravity, 1 min. The supernatants from the 1,000 × gravity centrifugations were poured into a sterile collecting tube fitted at one end with microfilament cloth (Nitex; Tobler, Ernst, and Traber, Inc., New York) having a mesh size of 10 μm. The filtrate was passed through a second Nitex cloth (8-μm mesh) and centrifuged, 1,000 × gravity, 20 min. The pellet contained the isolated cyanelles. Several pellets were resuspended in small volumes of culture medium, consolidated, and centrifuged, 1,000 × gravity, 20 min.

For lysozyme-sensitivity experiments or microscopy, the pellets were resuspended in 12-15 ml of culture medium and permitted to warm to room temperature (24°C) before use. Pellets used for diaminopimelic acid analysis were frozen immediately.

Lysozyme Treatment and Phycocyanin Assay

Hen egg white (HEW) lysozyme (M.W. 13,930: muramidase, N-acetylmuramide glycanhydrolase, EC 3.2.1.17, Sigma Chemical Co.) was used in all lysozyme-mediated lysis experiments. HEW lysozyme (75 μg/ml) was prepared in water before each experiment and did not contain EDTA. The pH of the lysozyme solution was 7.1.

The release of the water-soluble phycocyanin was monitored at 620 nm, the maximum absorption peak of phycocyanin from *Glaucocystis* cyanelles. All lysozyme-mediated digestions were performed in 5-ml test tubes. The final volume of the incubation medium was 1.5 ml and contained known volumes of the following components added in the following order: (1) suspension of freshly isolated cyanelles; (2) for appropriate experiments, either a solution of a competitive inhibitor of lysozyme for experimental samples, or DD-water for the controls; and (3) either lysozyme for experimental samples or DD-water for the controls. Each tube was immediately capped with a small square of parafilm, mixed by inversion, and immediately transferred to a filtration apparatus, which consisted of a 10-ml disposable syringe attached to a Swinnex-25 membrane filter holder (Millipore Corp.) loaded with a 25-mm Nuclepore filter (polycarbonate, 0.22 μm). Mixing and filtration required 10 sec. The filtrates were collected in 10-ml screw-cap centrifuge tubes, capped, immediately placed in an ice bath, and analyzed for absorbance at 620 nm with a Beckman DU monochromator (Beckman Instruments) equipped with a Gilford Model 2220 adapter. The 620-nm values were corrected for turbidity by subtracting values obtained at 720 nm.

Competitive Inhibition

N-acetyl-glucosamine-2 (NAG-2) was used as a competitive inhibitor for the hydrolytic activity of lysozyme. The NAG-2 was isolated from a partial acid hydrolysis of chitin.

Isolation of Peptidoglycan

The method of Golecki (1977) was used for the isolation of peptidoglycan from *Anacystis nidulans,* a small, unicellular cyanobacterium used as a positive control, and *Glaucocystis* cyanelles. Golecki's procedure was modified only in the initial homogenization step. *Anacystis* cells (25-30 mg, wet wt) were harvested from agar cultures grown at 28°C, resuspended in 20 ml DD-water containing 10 mM $MgCl_2$ and 1 mg DNAase, homogenized in a French pressure cell (Aminco, American Instruments Co.) at 20,000 lbs/sq in and centrifuged, 20,000 × gravity, 10 min.

The previously frozen *Glaucocystis* cyanelle samples were not subjected to the homogenization step but were thawed and centrifuged, 1,000 × gravity, 10 min. The pellets were resuspended in 5 ml of the $MgCl_2$-DNAase solution, vortexed for several minutes and centrifuged, 20,000 × gravity, 10 min.

Subsequent to the homogenization step, both *Anacystis* and *Glaucocystis* were treated according to Golecki (1977). The final pellets were resuspended in 5 ml DD-water and centrifuged, 150,000 × gravity, 30 min. The pellets were lyophilized and stored at 4°C.

Amino Acid Analysis

Lyophilized peptidoglycan preparations from *Anacystis* and *Glaucocystis* cyanelles were hydrolyzed with 6 M HCl at 110°C, 24 hr. Hydrolyzed *Glaucocystis* peptidoglycan material was subjected to ascending paper chromatography as an intermediate step for separating the amino acids methionine (Met) and diaminopimelic acid (DAP) that were found to elute together in standard amino acid-analysis methods.

Hydrolyzed peptidoglycan preparations were equilibrated with loading buffer at pH 5.28, loaded onto a short (0.5 cm x 6 cm) ion-exchange resin (type A chrombeads, Technicon) column, and analyzed. Identical samples were resuspended in loading buffer at pH 3.25 and loaded onto long columns (0.5 cm x 24 cm). For long-column samples, the pH was increased to pH 4.25 after the elution of proline and allowed to run until the complete elution of the phenylalanine peak. All analyses were performed in a Technicon TSM Sequential Multisample Amino Acid Analyzer at 60°C at a wavelength of 540 nm.

Ascending Paper Chromatography

One-gallon glass jars with screw lids were used as chromatography vessels into which 150 ml of the following solvent (4:1:1, N-butyl alcohol:glacial acetic acid:DD-water, respectively) were introduced and equilibrated, 2 hr, 24°C. Cylinders of Whatman 3MM chromatography paper (16 cm x 21.5 cm) were fashioned by stapling the 21.5-cm sides together after a horizontal line (origin) was drawn 2 cm from the edge of one of the 16-cm sides.

Samples (25 ng) of authentic amino acid standards (Met and DAP) and a sample of resuspended (DD-water), hydrolyzed, *Glaucocystis* cyanelle peptidoglycan were spotted on the line, dried with a hand-held hair dryer, and placed in the chromatography vessel for 4 hr.

The chromatogram was air dried and vertical strips containing the authentic standards separated from the cyanelle peptidoglycan sample. The authentic standards were developed with ninhydrin (0.2%, aerosol) and heat-dried (80°C, 8 min), and the R_f values were determined for each amino acid. A horizontal zone of the undeveloped portion of the chromatogram characterized by R_f values approximating the DAP R_f value was cut out and subsectioned into 0.5-sq-cm pieces. The pieces were eluted into 10 ml of DD-water, filtered through a 0.22-μm Nuclepore filter, lyophilized, and subjected to amino acid analysis.

Competitive Inhibition of Lysozyme

Because only a limited amount of NAG-2 was available for the determination of its competitive inhibitory properties, preliminary studies were evaluated by the cell-count method. Lysozyme (5 μg/ml) was added to a suspension of cyanelles containing NAG-2 (8-20 mM). Samples were removed at 1 min and 5.5 min and lysed, and intact cyanelles were counted (minimum, 500 cyanelles); the percent of inhibition of lysis was determined.

The final studies of NAG-2 inhibition were performed in accordance with the phycocyanin-release method. After exposure to lysozyme for 2.5 min (2-min incubation period and 0.5 min for transfer and filtration), the samples were placed on ice and subsequently evaluated for the presence of phycocyanin.

RESULTS

Lysozyme-Mediated Lysis

Following isolation from the host in an hypotonic culture medium, *Glaucocystis* cyanelles retained their integrity and unique shape. The gross morphology of isolated *Glaucocystis* cyanelles was markedly affected, however, by lyso-zyme at concentrations of less than 50 μg/ml at 24°C (Fig. 3-1). At concentrations below 50 μg/ml, lysozyme was especially effective at 5 μg/ml.

A time-course examination of lysozyme-mediated lysis under optimal conditions (lysozyme at 5 μg/ml, 24°C) indicated that lysis was essentially complete within the first minute of exposure to lysozyme (Fig. 3-2).

In an effort to examine more closely the apparently rapid enzymic reaction, the effect of decreased incubation temperature upon the time-course of lysozyme-mediated lysis was investigated. At an incubation tem-perature of 6°C, the hydrolytic activity of lysozyme was retarded at incubation times of up to 20 min (Fig. 3-3).

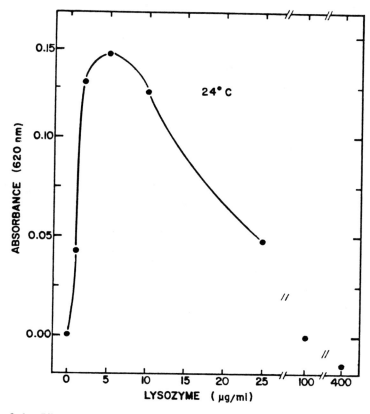

Figure 3-1. Effect of lysozyme concentration on the release of phycocyanin in *Glaucocystis* cyanelles at 24°C. Lysis acheived by various concentrations of lysozyme was determined by the release of phycocyanin into the incubation medium. Phycocyanin was quantified by photometric determination of absorbance at 620 nm following filtration.

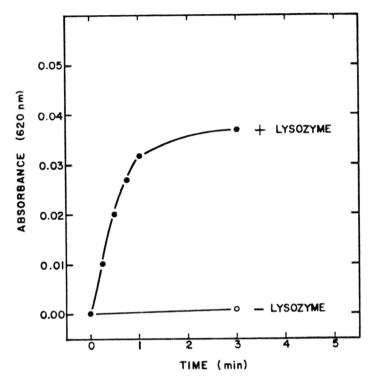

Figure 3-2. Release of phycocyanin from isolated *Glaucocystis* cyanelles treated with lysozyme (5 μg/ml) at 24°C.

Competitive Inhibition by N-acetyl-glucosamine-2

The cell-count method for evaluating the inhibition of lysozyme activity by NAG-2 indicated that at 4.5 min after the addition of lysozyme (5 μg/ml), NAG-2 (8-20 mM) was capable of achieving 8%-41% inhibition (Table 3-1).

Competitive inhibition of lysozyme activity, as determined by the phycocyanin-assay method, indicated that NAG-2 (10 mM and 30 mM) was capable of achieving 35% and 62% inhibition, respectively.

Diaminopimelic Acid Analysis

The R_f values obtained for DAP and Met, chromatographed simultaneously with hydrolyzed *Glaucocystis* cyanelle peptidoglycan, were 0.06 and 0.52, respectively. *Glaucocystis* cyanelle peptidoglycan was subjected to amino acid analysis following the removal from the chromatogram of amino acids

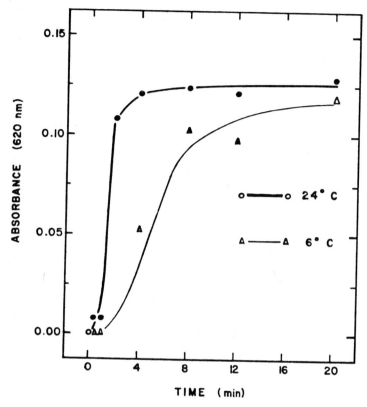

Figure 3-3. Effect of temperature on lysozyme treatment of isolated *Glaucocystis* cyanelles. The cyanelles were incubated in lysozyme-containing medium (5 μg/ml) at temperatures of 6°C and 24°C (control). Phycocyanin was extracted with 0.22-μm Nuclepore filters. Absorption values for phycocyanin were obtained at 620 nm.

Table 3-1. Competitive Inhibition of Lysozyme by NAG-2 at 24°C

min*	% Inhibition by NAG-2		
	8 mM	12 mM	20 mM
4.5	8	10	41
9	6	7	19

Note: *Samples were taken at 1 and 5.5 min; 3.5 min were required for counting a minimum of 500 cyanelles.

characterized by R_f values greater than 0.20. *Glaucocystis* cyanelle DAP was increased by authentic DAP (Fig. 3-4).

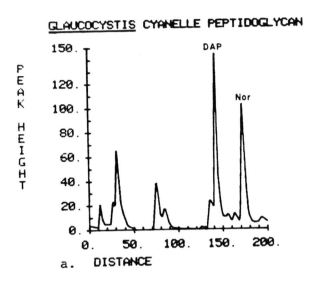

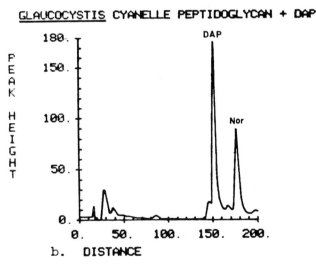

Figure 3-4. Amino acid analysis of chromatographed peptidoglycan from the cyanelles of *Glaucocystis*. a. *Glaucocystis* cyanelle peptidoglycan. Elution distance: 15, Thr; 30, Ser; 35, Glu; 45, Pro; 75, Gly; 85, Ala; 135, Val; 145, DAP; 155, Ile; 165, Leu; 175, Nor (internal standard, 5 nM); 195, Phe. b. *Glaucocystis* cyanelle peptidoglycan + authentic diaminopimelic acid. Elution distance: 150, DAP; 180, Nor (internal standard, 5 nM).

DISCUSSION

All living organisms fall into two major categories, prokaryotes and eukaryotes. Prokaryotes are generally considered more primitive because they lack the well-defined compartmentalization of the internal structure of the eukaryotes; DNA-containing and double-membraned organelles (nucleus, mitochondria, chloroplasts) are characteristically absent in the prokaryotes. The great dichotomy of cellular organization is further exemplified by the absence in eukaryotes of peptidoglycan, a major prokaryote character. Peptidoglycan, a compound unique to the eubacteria, is characterized by the amino sugar, N-acetylmuramic acid, and the amino acid, diaminopimelic acid.

The involvement of prokaryotes in endosymbiotic associations is presently enjoying considerable attention (Margulis, 1981; Wallace,1982; Whately and Whatley, 1981). The impetus for the current interest in the topic of prokaryotic origins of eukaryotic organelles, however, was provided by what now is known as the serial endosymbiotic hypothesis (Margulis, 1970). Margulis's model proposed that nonphotosynthetic bacteria were phagocytically taken up by a primitive amoebaflagellate, a heterotrophic cell purported to be ancestral to all eukaryotes. According to the model, the endosymbiotic, nonphotosynthetic bacteria gave rise to the mitochondria presently found in all eukaryotes. In addition, cyanobacteria were taken up by ancestral plant cells, resulting in the organelles presently known as chloroplasts.

Critics of the serial endosymbiotic hypothesis have maintained that an independent origin of organelles is untested and difficult to subject to experimental scrutiny, but most have recognized that the origin of chloroplasts is perhaps the strongest aspect of Margulis's model.

While cyanobacteria have long been argued as precursors for red algal chloroplasts, cyanobacterial endosymbionts of clearly definable red-algal cells (nonflagellated, eukaryotic algae) have never been found. An extant, achlorotic plant cell permanently endowed with intracellular, photosynthetic prokaryotes also has never been found.

The discovery of a peptidoglycan envelope in the cyanelles of *Glaucocystis* provides the first, unequivocal substantiation for Geitler's endosymbiotic explanation for *Glaucocystis*. Furthermore, *Glaucocystis* represents a model system for experimentally examining some aspects of the strongest support for Margulis's serial endosymbiotic hypothesis. Caution must be exercised, however, because the cyanelles are photosynthetic prokaryotes and not organelles.

ACKNOWLEDGEMENTS

This report represents a portion of the research conducted for the Ph.D. (Scott, 1981) earned at the University of Oregon, Eugene, Oregon, under the cosponsorship of Dr. Richard W. Castenholz and Dr. Howard T. Bonnett.

During the course of the study, the author was the recipient of financial support from the Ford Foundation, Pre-Doctoral Biomedical Research Traineeship (University of Oregon), American Indian Graduate Fellowship Program, and the Navajo Nation.

REFERENCES

Bourdu, R., and M. Lefort, 1967, Structure fine, observée en cryodécapage, des lamelles photosynthetiques des Cyanophycées endosymbiotiques: *Glaucocystis nostochinearum* et *Cyanophora paradoxa, C. r. Hedb. Seanc. Acad. Sci.* **265**:37-40.

Chapman, D.J., 1966, The pigments of the symbiotic algae (cyanomes) of *Cyanophora paradoxa* and *Glaucocystis nostochinearum* and two Rhodophyceae, *Porphyridium aerugineum* and *Asterocystis ramosa, Arch. Mikrobiol.* **55**:17-25.

Echlin, P., 1967, The biology of *Glaucocystis nostochinearum*. I. The morphology and fine structure, *Br. Phycol. Bull.* **3**:225-239.

Fox, G. E., E. Stackebrandt, R. B. Hespell, J. Gibson, J. Maniloff, T. A. Dryer, R. S. Wolfe, W. E. Balch, R. S. Tanner, L. J. Magrum, L. B. Zablen, R. Blakemore, R. Gupta, L. Bonen, B. J. Lewis, D. A. Stahl, K. R. Luehrsen, K. N. Chen, and C. R. Woese, 1980, The phylogeny of prokaryotes, *Science* **209**:457-463.

Geitler, L., 1923, Der Zellbau von *Glaucocystis nostochinearum* und *Gleochaete wittrockiana* und die Chromatophoren-Symbiosetheorie von Mereschkowsky, *Arch. Protist.* **47**:1-24.

Golecki, J. R., 1977, Studies on ultrastructure and composition of cell walls of the cyanobacterium *Anacystis nidulans, Arch. Microbiol.* **114**:35-41.

Hall, W. T., and G. Claus, 1967, Ultrastructural studies on the cyanelles of *Glaucocystis nostochinearum* Itzigsohn, *J. Phycol.* **3**:37-51.

Holm-Hansen, O., R. Prasad, and R. A. Lewin, 1965, Occurrence of α, ϵ-diaminopimelic acid in algae and flexibacteria. *Phycologia* **5**:1-14.

Korschikov, A. A., 1930, *Glaucosphaera vacuolata,* a new member of Glaucophyceae, *Arch. Protist.* **70**:217-222.

Lee, R. E., 1980, *Phycology,* Cambridge University Press, Cambridge.

Lefort, M., 1965, Sur de chromatoplasma d'une Cyanophyceae endosymbiotique: *Glaucocystis nostochinearum* Itzigs, *C. r. Hedb. Seanc. Acad. Sci.* **261**:233-236.

Lefort, M., and M. Pouphile, 1967, Données cytochimiques sur l'organisation structurale du chromaplasm de *Glaucocystis nostochinearum, C. r. Soc. Biol.* **161**:992-994.

Lewin, R. A., 1974, Biochemical taxonomy, in *Algal Physiology and Biochemistry,* W. D. P. Stewart, ed., University of California Press, Berkeley.

Margulis, L., 1970, *Origin of Eukaryotic Cells,* Yale University Press, New Haven.

Margulis, L., 1981, *Symbiosis in Cell Evolution,* W. H. Freeman and Co., San Francisco.

Mereschkowsky, C., 1905, Uber Natur und Ursprung der Chromatophoren in Pflanzenreiche, *Biol. Centralblatt.* **25**:593-604.

Pascher, A., 1914, Uber Symbiosen von Spaltpilzen und Flagellaten, *Ber. Deutsch. Bot. Ges.* **32**:339-351.

Schenk, H. E. A., 1970, Nach einer lysozymempfindlichen Stützmembran Endocyanellen von *Cyanophora paradoxa* Korschikoff, Z. Naturforsch. **25**:656.

Schenk, H. E. A., 1971, Isolierung der Endocyanellen aus *Glaucocystis nosto-chinearum* Itz. mit hilfe eines quirschnittvariablen Durchlaufhomogenisators und chemische Charakterisierung der Hüllmembran, *Hoppe-Seyler's Z. Physiol. Chem.* **352**:321-324.

Schenk, H. E. A., and I. Hofer, 1972, About the light and dark fixation of CO_2 in the cyanoms *Cyanophora paradoxa* and *Glaucocystis nostochinearum* and their endocyanelles, in *Proceedings of the Second International Congress on Photosynthesis Research*, Vol. 3, G. Forti, M. Avron, and A. Melandri, eds., Dr. W. Junk NV, The Hague, pp. 2095-2099.

Schmidt, V. B., L. Kies, and A. Weber, 1979, Die Pigmente von *Cyanophora paradoxa, Gloeochaete wittrockiana* und *Glaucocystis nostochinearum, Arch. Protist.* **122**:164-170.

Schnepf, E., 1965, Struktur der Zellwande und Cellulosefibrillen bei *Glaucocystis, Planta* **67**:213-224.

Schnepf, E., W. Koch, and G. Diechgraber, 1966, Zur Cytologie und taxonomischen Einordnung von *Glaucocystis, Arch. Microbiol.* **55**:149-174.

Scott, O. T., 1981, *Evidence for a Peptidoglycan Envelope in the Cyanelles of* Glaucocystis nostochinearum *Itzigsohn,* PhD. dissertation, University of Oregon, Eugene.

Skuja, H., 1948, Taxonomic des Phytoplanktons einiger seen in Uppland, Sweden, *Symb. Bot. Upsal.* **9**:1-400.

Skuja, H., 1954, Glaucophyta, in *Die Naturlichen Pflanzenfamilien Syllabus*, A. Engler and K. Prantl, eds., Borntraeger, Berlin, pp. 56-57.

Wallace, D. C., 1982, Structure and evolution of organelle genomes, *Microbiol. Rev.* **46**:208-240.

West, G. S., 1904, *The British Freshwater Algae*, Cambridge University Press, Cambridge.

West, G. S., and F. E. Fritsch, 1927, *A Treatise on the British Freshwater Algae,* Cambridge University Press, Cambridge.

Whatley, J. M., and F. R. Whatley, 1981, Chloroplast evolution, *New Phytol.* **87**:233-247.

4

Recovery of Viable *Thermoactinomyces vulgaris* and Other Aerobic Heterotrophic Thermophiles from a Varved Sequence of Ancient Lake Sediment

Nancy L. Parduhn and John R. Watterson

U. S. Geological Survey

ABSTRACT: A varved sequence of lake sediment from Elk Lake, Minnesota, contains a 10,284-year record of sedimentation determined on the basis of ^{14}C analysis and varve chronology. Fifty core samples were studied to determine the viability of thermophilic bacterial endospores. Recovery of viable *Thermoactinomyces vulgaris* and other aerobic heterotrophic thermophiles shows that the spores of these organisms can remain viable for more than 7000 years. In a similar study, viable endospores of *Bacillus stearothermophilus* were recovered from 5800-year-old sediments off the coast of Catalina Island, California (Bartholomew and Paik, 1966). Elk Lake bacterial counts correlate with paleolimnological data. In particular, maximum bacterial counts occur in sediments deposited during the mid-Holocene hypsothermal (warm) period.

INTRODUCTION

The plausibility of culturing viable endospore-forming bacteria from ancient sediments is currently being investigated by several workers. Early studies by Kieffer (1923) and Bulloch (1928) showed that *Bacillus anthracis* and *Clostridium tetani* can remain viable for 30 or 40 years. Wilson and Shipp (1938) isolated viable endospores of thermophilic *Bacillus* sp. in a tin of meat 118 years old. Sneath (1962) enumerated *Bacillus* from soil adhering to roots of plants collected 300 years earlier. Cross and Attwell (1971) estimated that thermoactinomycetes and *Bacillus* endospores recovered from lake sediments remain viable over periods in excess of 1000 years, whereas spores of non-endospore-forming actinomycete genera survive for periods of only 200-300 years. In ocean basin cores, viable *Bacillus stearothermophilus* spores have been recovered and identified from sediments greater than 5800 years old (Bartholomew and Paik, 1966). Accounts of viable bacteria isolated from Permian salt (Reiser and Tasch, 1960) and from Precambrian rocks and Permian coal (Lipman, 1966) are questionable. Possible contamination from outside sources must be considered when evaluating the validity of these findings. However, because of the limited number of viability studies there is much to be learned about the viability of bacterial endospores in ancient sediments and rocks.

A continuous 10,284-year sequence of varved sediment from Elk Lake, Minnesota, was sampled and examined for viable endospores. Thermophilic organisms were chosen for investigation of viable spore count because of their long-lived character (Cross and Attwell, 1971; Bartholomew and Paik, 1966). *Thermoactinomyces vulgaris* is an organism that inhabits high-temperature environments such as overheated grain (Festenstein et al., 1965), compost (Waksman, Umbriet, and Cordon, 1939), and dung (Henssen, 1957). It also has been isolated from peat (Kuster and Locci, 1963) and from the air (Lacey and Lacey, 1964). In most surface waters, *T. vulgaris* is considered to be a classic example of a wash-in organism—that is, an organism indicative of waters originating from surface runoff (Cross and Johnson, 1971)—and may have applications in the interpretation of water geochemistry. Two experiments were conducted to determine the number of viable thermophilic organisms present throughout the core, which ranges in age from 493 years to 10,284 years before present. The first experiment involved culturing 17 samples for both *T. vulgaris* and other thermophilic organisms. A second series of 33 samples was cultured for *T. vulgaris* only to investigate whether the first results could be replicated and to compare total counts (both *T. vulgaris* and other thermophiles) with *T. vulgaris* counts. Previous paleolimnological studies of the core made possible a comparison between paleolimnological data and bacteria counts.

MATERIALS AND METHODS

Collection and Sampling

A 20-m core was collected in December 1978 in 1-m and 3-m sections from the southern basin of Elk Lake in a water depth of 29.5 m. By mistake there was a 1-m coring gap, and in 1982 the lake was resampled to fill in this gap. Samples marked 24 and 58 (Table 4-2) were collected from this 1982 core. Varve years, depth below water surface, and sub-bottom depth for the samples used in experiments 1 and 2 are given in Tables 4-1 and 4-2. The core

Table 4-1. Varve Years Before Present, Depth Below Water Surface, and Sub-Bottom Depth for All Samples Used in Experiment 1

Sample Number	Varve Years B.P.	Depth Below Water Surface (m)	Sub-Bottom Depth (m)
JH3000415	2001	33.36	3.86
JI3001050	2498	34.00	4.50
JJ3000850	3047	35.02	5.52
JK1000390	3492	35.96	6.46
JL1000595	4260	38.04	8.54
KA1000580	4392	39.04	9.54
KB2001065	4587	39.39	9.89
KC3500540	5144	40.65	11.15
KD2000530	5582	41.45	11.95
KE2500263	5908	42.47	12.97
KF2000575	6341	43.64	14.14
KG2000995	6710	44.70	15.20
KH2000250	7173	45.68	16.18
KI2000815	7518	46.42	16.92
KJ4500545	8091	47.35	17.85
KK3000330	9067	48.57	19.07
KL1000015	10,284	49.90	20.50

contains more than 10,000 annual laminations throughout its entire length, allowing accurate dating of the lake sediment. Dating was also done by ^{14}C analysis, which closely correlates with the ages determined by varve chronology.

Core sections were extruded, wrapped in plastic wrap and aluminum foil, and frozen. Frozen core sections were sampled in 1979 for paleomagnetic, geochemical, pigment, and micropaleontological analyses. The core remained frozen until March 1983, when it was sampled for viable thermophiles. A sterile scalpel was used to scrape away the surface of the section sampled. The undisturbed laminae were then sampled on a fresh surface and weighed directly into screw-cap tubes containing sterile one-quarter strength Ringers solution (see below).

Medium

The basal medium used for all studies was CYC agar containing: 33.4 g Czapek-Dox Broth (Difco), 2.0 g yeast extract (Difco), 6.0 g vitamin-free casamino acids (Difco), and 16.0 g agar (Kodak) per liter deionized water. A selective medium for *T. vulgaris* was prepared by adding 25 mg novobiocin (U.S. Biochem.) and 50 mg cycloheximide per liter (U.S. Biochem.) after Cross and Attwell (1971). Antibiotics were sterilized by 0.45-μ membrane filtration and added to the autoclaved agar upon cooling to 47°C. Antibiotics were left out of the first experiment, resulting in growth of other

Table 4–2. Varve Years Before Present, Depth Below Water Surface, and Sub-Bottom Depth for All Samples Used in Experiment 2

Sample Number	Varve Years B.P.	Depth Below Water Surface (m)	Sub-Bottom Depth (m)
JA1500600	493	30.04	.54
JA1500060	520	30.09	.89
JB3000695	783	30.63	1.13
JB2000350	900	30.93	1.43
JC3000515	1243	31.45	1.95
JC2500265	1306	31.77	2.27
JD1500670	1585	32.23	2.73
JD0500275	1705	32.47	2.97
JH3500925	1926	33.31	3.81
JI3000480	2527	34.05	4.55
JJ3000270	3077	35.07	5.57
JK2000000	3400	35.70	6.20
240000000	3428	36.10	6.60
580000000	4129	37.50	8.00
JL1000890	4250	38.01	8.51
KA2000300	4298	38.60	9.10
KA1001735	4347	38.93	9.43
KB2000315	4612	39.47	9.97
KB1000560	4703	40.00	10.50
KC6000280	4903	40.50	11.00
KC5000720	4987	40.60	11.10
KC3500860	5132	40.61	11.11
KC2000820	5284	41.00	11.50
KD2001530	5546	41.35	11.85
KD1001200	5658	41.90	12.40
KE2500565	5898	42.44	12.94
KF2000270	6352	43.67	14.17
KG2001090	6707	44.69	15.19
KH2000320	7170	45.67	16.17
KI2500563	7643	46.44	16.94
KJ5000500	8058	47.50	18.00
KK2000025	9198	48.70	19.20
KL2500590	10,060	49.60	20.10

thermophilic bacteria in addition to *T. vulgaris*. The second experiment contained both novobiocin and cycloheximide, allowing for *T. vulgaris* growth only.

Culturing Procedures

Filtering techniques were used for concentrating *T. vulgaris* and other thermophilic organisms in all Elk Lake samples because of earlier unsuccessful results using standard spread-plate techniques. Using a sterile scalpel flamed in 70% ethynol, 0.5 g of sample was weighed into 16 x 150-mm screw-cap culture tubes containing 4.5 ml sterile quarter-strength Ringers solution (pH 7.0) consisting of 2.25 g NaCl, 0.105 g KCl, 0.12 g $CaCl_2$, and 0.05 g $NaHCO_3$ per liter (Cross and Johnson, 1971). Tubes were shaken for 10 min in a mechanical shaker and centrifuged for 3 min at 56.5 × gravity. Supernatent liquid (2.5-4.0 ml) from each sample was vacuum-filtered using 0.45-μ Black Gelman membranes (Gelman part no. 66378). Filters were then placed right side up on selective CYC agar. All samples were incubated at 50°C for at least 48 hr. Colony-forming units were counted with the aid of a stereo binocular microscope at 9X.

RESULTS

Culture results for the first experiment are shown in Figure 4-1*A*. The agar used contained no antibiotics to suppress *Bacillus* and fungal growth. No fungal growth was observed, however, and the colony-forming units included typical white colonies of *T. vulgaris* and other aerobic heterotrophic thermophiles. Sample KI2000815 (7518 varve years) contained the oldest significant number of viable thermophilic bacteria. There was considerable fluctuation in the number of organisms present throughout the length of the core, with a major peak at 5144 varve years B.P. Other significant peaks occurred at 4392, 6341, and 7173 varve years B.P. Colonies of *T. vulgaris* and other thermophiles from samples 4392 and 6341 varve years are illustrated in Figure 4-2*A* and *B*.

Colony counts for the second experiment using selective CYC agar with antibiotics are shown in Figure 4-1*B*. Although antibiotics were added, two colony types were observed: flat white colonies with aerial mycelia (*T. vulgaris*) and transparent colonies of uncertain identity (Fig. 4-3). Both types were included in the colony counts (see following discussion). Sample JB2000350 (900 varve years) contained the highest counts of colony-forming units. Significant bacterial counts were mainly in samples from between 493 and 1705 varve years B.P. and 4,129 and 7170 varve years B.P. The oldest significant counts occurred at 7170 varve years B.P. Colony-forming units/g were considerably less than those observed in the first experiment when no antibiotics were used.

DISCUSSION

Cycloheximide and novobiocin retard the growth of fungi and thermophilic *Bacillus* sp. and permit optimum selection for *T. vulgaris* (Cross, 1968). No antibiotics were used in the first experiment, resulting in growth of *T. vulgaris* and other thermophilic organisms. Cross (1968) has shown that the

EXPERIMENT 1   EXPERIMENT 2

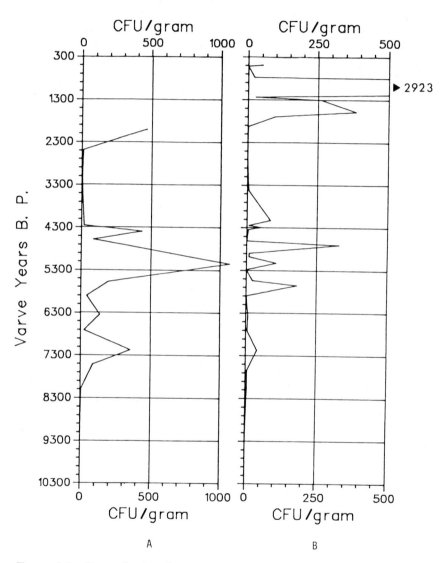

Figure 4-1. Plots of colony-forming units per gram versus varve years B.P. for experiment 1 (*A*) and experiment 2 (*B*).

A

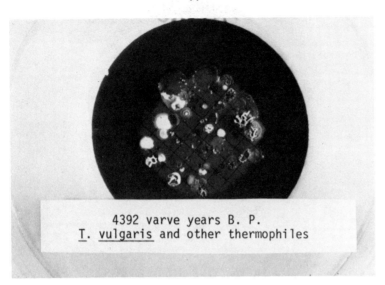

B

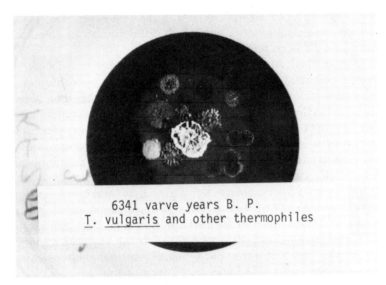

Figure 4-2. Colonies of *T. vulgaris* and other aerobic heterotrophic thermophiles in samples from 4392 varve years B.P. and 6341 varve years B.P. (unmagnified).

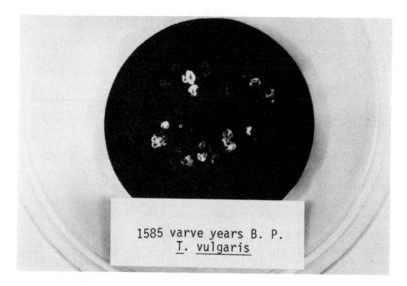

Figure 4–3. White colonies of *T. vulgaris* with aerial mycelia and transparent wrinkled colonies of unknown identity (1585 varve years B. P., unmagnified).

simultaneous growth of thermophilic *Bacillus* sp. and fungi can in many cases overgrow the plate and suppress the growth of actinomycetes such as *T. vulgaris*. Therefore, *T. vulgaris* colonies in the first experiment may have been suppressed, resulting in a smaller observed number of colonies than may have actually been present.

On the other hand, Attwell and Colwell (1982) showed that the antibiotics novobiocin and nalidixic acid may retard *T. vulgaris* colony development when dormant spores (as in the case with Elk Lake), rather than vegetative cells, are used as the inoculum. The retarding effect of the antibiotics, along with the fact that other thermophiles besides *T. vulgaris* were present in the first experiment, may explain the larger numbers of colonies in the first experiment (no antibiotics) compared to the second (with antibiotics).

Paleolimnological studies of Elk Lake have been done by Boucherle (1982) and Stark (1976). Extensive chemical analyses also have been done on Elk Lake cores. Peaks of *T. vulgaris* and other thermophilic organisms correlate with (1) increases in sedimentation rate, (2) increases in organic carbon, (3) a period from 4000 to 8500 years B.P. (hypsothermal) when conditions were dry and warm and semiarid prairie conditions expanded over the drainage basin of Elk Lake, and (4) increases of *Daphnia longispina,* a zooplankter, which indicates eutrophic conditions (Brugman, 1978; Boucherle and Zullig, in press). Figure 4-4 shows plots of sediment mass accumulation rate (mg/sq cm/yr) and percent organic carbon in comparison with colony-

POLLEN ZONES

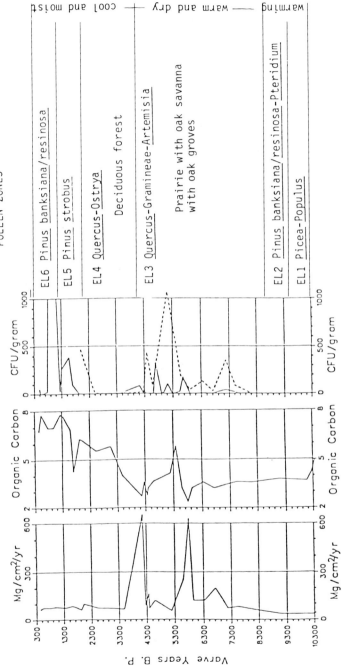

Figure 4–4. Plots of sediment mass accumulation rate (mg/cm²/yr), percent organic carbon, and colony-forming units per gram versus varve years B.P. Dashed line indicates results for experiment 1 (*T. vulgaris* and other thermophiles), and solid line indicates results for experiment 2 (*T. vulgaris* only). *(Pollen zones from Boucherle, 1982.)*

forming units (CFU) per gram, which includes *T. vulgaris* and other thermophiles (dashed line) in experiment 1, and only *T. vulgaris* (solid line) in experiment 2. Pollen zones also are shown (Boucherle, 1983). It is apparent from these data that peaks in counts of viable thermophilic bacteria correspond to a time when conditions were warmer and drier. Proliferation of *T. vulgaris* and other thermophiles during this altithermal period, when prairie grasslands ideally suited for *T. vulgaris* growth existed, would account for this increase. Increases of the zooplankter *D. longispina,* indicative of eutrophic conditions, also correspond to this period. These short periods of increased eutrophic conditions in the lake water produced significant oxygen depletion during the altithermal period. The low dissolved oxygen would then provide an environment suitable for endospore preservation as the spore settled through the water column. This phenomenon, in conjunction with an increase in thermophilic bacteria at this time, would account for the pattern of bacterial peaks observed.

The highest bacterial counts for the second experiment correspond to the elevated number of spores recovered in the first experiment (altithermal period, 4000 to 8500 varve years B.P.) and occur more recently between 493 to 1705 varve years B.P. Increased numbers of *T. vulgaris* near the surface (493–1705 varve years) do not correspond to increases in sedimentation or a warm, dry period. Conditions for growth of *T. vulgaris* may have existed for limited times during the summer and may account for the increased numbers of *T. vulgaris* near the surface. It has been shown that *T. vulgaris* can grow and sporulate within four days at 32°C and may grow at even lower temperatures with extended incubation (Foerster, 1975).

Although selective CYC agar was used to enumerate *T. vulgaris* in the second experiment, considerable growth of colonies atypical of *T. vulgaris* morphology occurred. Instead of exhibiting a flat, white appearance, these colonies appeared transparent and were of uncertain identity. Upon microscopic examination of both colony types, it was found that both contained typical *T. vulgaris* aerial mycelia, similar in appearance to those studied by Attwell and Colwell (1982). Photomicrographs of both the flat, white colonies of *T. vulgaris* and those of the transparent colonies are shown in Figure 4-5*A* and Figure 4-5*B*.

It appears that organisms cultured from Elk Lake must be of the same age as the sediments from which they were cultured. It is evident from ^{14}C dating and varve chronology that Elk Lake cores extend as far back as 10,284 years. The coincidence of warmer, drier conditions and high counts of thermophilic organisms suggests that the expansion of prairie over the Elk Lake area during the altithermal period resulted in windier conditions. Because airborne spores can be carried long distances (Hirst, Stedman, and Hogg, 1967; Hirst, Stedman, and Hurst, 1967), it may be that the recovered spores were transported by wind. If this is true, then other temperate lakes at this

A

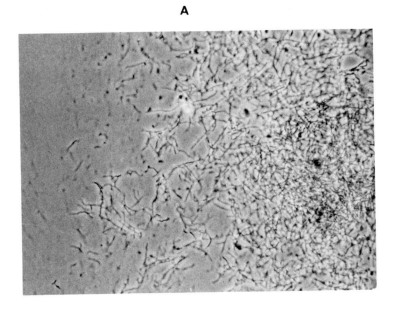

B

Figure 4-5. Photomicrographs of typical *T. vulgaris* aerial mycelia (A) and colonies of unknown identity also showing aerial mycelia typical of *T. vulgaris* (B), 400X (bar = 10 μ).

latitude also should have received larger numbers of spores during the warm, dry, altithermal period.

The oldest stratum from which a significant number of colonies were isolated is approximately 7518 years old. Contamination is possible by displacement of younger spores downward through older sediment, mainly by bioturbation and by coring and sampling methods. The first theory is highly unlikely because there is no evidence of bioturbation in the finely laminated varved sequence. The highest counts are found not in the deepest (oldest) parts of the core, but at shallower depths, which would not be expected if uniform contamination occurred. Samples were selected from the core center rather than from the core surface. The fact that numerous colonies were isolated from several samples, in addition to the many samples in which no bacteria could be isolated, supports the argument against substantial contamination. Furthermore, correlations with paleolimnological studies suggest that the patterns observed in bacterial distributions are not random but correspond to periods that were warmer and drier than the present.

ACKNOWLEDGEMENTS

We thank J. P. Bradbury and W. E. Dean for critical reviews of the manuscript and for providing samples. D. M. Updegraff provided valuable guidance and assistance, and Sarah Broadbent provided samples used in preliminary studies.

REFERENCES

Attwell, R. W., and R. R. Colwell, 1982, Use of epifluorescence microscopy in studies of the germination and recovery of thermoactinomycetes, *Appl. Envir. Microbiol.* **43:**478-482.

Bartholomew, J. W., and G. Paik, 1966, Isolation of obligate thermophilic sporeforming bacilli from ocean basin cores, *J. Bacteriol.* **92:**635-638.

Boucherle, M. M., 1982, An ecological history of Elk Lake, Clearwater Co., MN, based on cladocera remains, Ph.D. thesis, Indiana University, Bloomington, Indiana, pp. 1-38.

Boucherle, M. M., and H. Zullig, in press, Cladoceran remains as evidence of change in trophic state in three Swiss lakes, *Hydrobiologica.*

Brugman, R. B., 1978, Human disturbance and the historical development of Linsley Pond, *Ecology* **59:**19-36.

Bulloch, W., 1928, The viability of bacteria in antiseptic solutions, *Zbl. I Abt. Orig.* **106:**21.

Cross, T., 1968, Thermophilic actinomyces, *J. Appl. Bacteriol.* **31:**36-53.

Cross, T., and R. W. Attwell, 1971, Recovery of viable thermoactinomycete endospores from deep mud cores, in *Spore Research,* A. N. Barker, G. W. Gould, and J. Wolf, eds., Academic Press, London, pp. 11-20.

Cross, T. and D. W. Johnson, 1971, *Thermoactinomyces vulgaris* II. Distribution in natural habitats, in *Spore Research,* A. N. Barker, G. W. Gould, and J. Wolf, eds., Academic Press, London, pp. 315-330.

Festenstein, G. N., J. Lacey, F. A. Skinner, P. A. Jenkins, and K. C. Pepys, 1965, Self-heating of hay and grain in Dewar flasks and the development of farmers' lung antigens, *J. Gen. Microbiol.* **41:**38.

Foerster, H. F., 1975, Germination characteristics of some thermophilic actinomycete spores, in *Spores VI,* P. Gerhardt, ed., American Soc. for Microbiol., Washington, D.C., pp. 36-43.

Henssen, A., 1957, Beitrage zur Morphologie und Systenatik der thermophilen Actinomyceten, *Arch. Mikrobiol.* **26:**373.

Hirst, J. M., O. J. Stedman, and W. H. Hogg, 1967, Long-distance spore transport: Methods of measurement, vertical spore profiles and the detection of immigrant spores, *J. Gen. Microbiol.* **48:**329-355.

Hirst, J. M., O. J. Stedman, and G. W. Hurst, 1967, Long-distance spore transport: Vertical sections of spore clouds over the sea, *J. Gen. Microbiol.* **48:**357-377.

Kieffer, K. H., 1923, Ein beitrag zur Lebensfahigkeit dur Bakterien, *Zbl. Bakt. J. Abt. Orig.* **90:**1.

Kuster, E., and R. Locci, 1963, Studies on peat and peat microorganisms I. Taxonomic studies on thermophilic acinomycetes isolated from peat, *Arch. Mikrobiol.* **45:**88.

Lacey, J., and M. E. Lacey, 1964, Spore concentrations in the air of farm buildings, *Trans. British Mycol. Soc.* **47:**547.

Lipman, C. B., 1966, Living microorganisms in ancient rocks, *J. Bacteriol.* **12**(3): 183-198.

Reiser, R., and P. Tasch, 1960, Investigation of the variability of osmophile bacteria of great geologic age, *Kansas Acad. Sci. Trans.* **63:**31-34.

Sneath, P. H., 1962, Longevity of micro-organisms, *Nature* **195:**643.

Stark, D. M., 1976, Paleolimnology of Elk Lake, Itasca State Park, northwestern Minnesota, *Arch. Hydrol. Suppl.* **50:**208-274.

Waksman, S. A., W. W. Umbriet, and T. C. Cordon, 1939, Thermophilic acinomycetes in solids and composts, *Soil Sci.* **47:**37.

Wilson, G. S., and H. L. Shipp, 1938, The examination of some tin foods of historical interest, *Chem. and Ind.* **57:**834-836.

5

Immunological Studies on the Organic Matrix of Recent and Fossil Invertebrate Shells

Peter Westbroek, Hans P. M. de Vrind, Janneke Tanke-Visser, and Liesbeth W. de Jong

University of Leiden

ABSTRACT: Antibodies directed against soluble macromolecules of the cephalopod *Nautilus* were allowed to react with shell fragments of a number of invertebrates. An adaptation of the Enzyme Linked Immuno Sorbent Assay (ELISA) was used to measure the reaction intensities. Significant reactions were obtained with most mollusks and also with a few representatives of other invertebrate phyla. Some fossil mollusks, including the 70-million-year-old (70 Ma) belemnite *Gonioteuthis*, also gave positive reactions. More specific antibody preparations were obtained by adsorption on etched powders of selected shell materials. With such preparations it could be demonstrated that no differences are detected between the matrices of *Nautilus* nacreous and prismatic layers (which does not mean that such differences do not exist); aspects of the existing systematic classification of the mollusks were confirmed. In light of these findings the potential of the immunological approach for the study of calcification mechanisms and evolution and for geology is discussed.

INTRODUCTION

Biological calcification is a process of major biogeochemical interest; virtually all limestone on earth has been formed by living systems. The degree of biological control over the crystallization of calcium carbonate may vary, however. Lowenstam and Weiner (1983) have distinguished between biologically induced and organic matrix-mediated mineralization, which represent any members of an intergrading spectrum in which the organism exercises increasing control over the mineral to be deposited. Biologically induced mineral deposits are similar to those formed by inorganic processes; the biological system merely facilitates their formation. In contrast, in organic matrix-mediated systems crystallization takes place within a macromolecular framework—the organic matrix. The crystals are arranged in more or less ordered arrays, and as a rule their size, shape, and crystallographic orientation are well defined. The formation of a gastropod shell, a coral skeleton, or a coccolith is not a random physico-chemical process; rather, crystallization takes place within a complex biological medium and is modulated in subtle ways so as to fulfill specific tasks in the living organization.

The idea of a matrix function for the organic macromolecules that are invariably found in close association with the crystalline phase of the well-organized biominerals is generally accepted; in other words, these materials are believed to play a key regulatory role in crystallization. But even in biologically induced mineralization, biopolymers may have a definite effect on crystallization (Westbroek, 1983).

Biochemical analysis of the organic matrix is a first step towards understanding the mechanisms of biological calcification. Moreover, comparative studies on the structure of matrices produced by different taxa may give important information on the evolution of the calcification process. Westbroek et al. reported earlier (1983) that immunology is a useful tool in these two approaches, and in this paper further evidence is presented in support of this contention. Moreover, the earlier finding that antigenic determinants of matrix components may be preserved over geological time is further corroborated.

Matrix Function and Evolution

The calcified tissues of animals are examples of matrix-mediated mineralization (Weiner, Traub, and Lowenstam, 1983). After demineralization with EDTA (pH 7.4-8.0), a soluble and an insoluble macromolecular fraction are generally liberated. The soluble material may be a very complex mixture—up to 40 fractions could be separated in *Mytilus* (Weiner, in press)—but as a rule it contains a high proportion of polyanions. The insoluble residue is of a more hydrophobic nature. In mollusks it may contain chitin and silk-fibroin-like fractions. The insoluble matrix is believed to act as a framework, supporting and orienting the soluble material. The latter would then assist in the nucleation, growth, and termination of the crystallization process (Weiner and Traub, 1984). The available evidence suggests that the presence of a soluble polyanionic and an insoluble, more supporting, phase is widespread among the matrix-mediated biomineralizations. Models similar to those for the mollusks have been proposed for dentin (Veis and Sebsay, 1983), bone (Glimcher, 1984), and coccoliths (Westbroek et al., 1983).

In many phyla of the animal kingdom, matrix-mediated calcification has originated in a relatively short period of time around the beginning of the Cambrian (Lowenstam, 1981). The important question has been raised as to whether these appearances were all independent innovations, or whether they originated by common descent from a biologically mediated ancestral system (Weiner, Traub, and Lowenstam, 1983).

Preservation of Matrix Components in Fossils

Because of their intimate association with the crystalline phase, some of the organic constituents of calcified tissues may be preserved over extended periods of time. Weiner, Lowenstam, and Hood (1976) reported preserved characteristic amino acid sequences of shell protein 80 Ma old. Most significantly, Krampitz and his coworkers have recently discovered that a sequence comprising 15 amino acids in a calcium-binding anionic polypeptide is preserved in 150-Ma-old mollusks (G. Krampitz, pers. comm.).

Antigenic determinants, the macromolecular domains recognized by antibodies, generally comprise between four and eight amino acids (or a nonproteinaceous structure of similar size). This is large enough to endow the immune system with its great specificity; on the other hand such small domains have a much better chance than entire macromolecules to be preserved in an appropriate environment such as a mineralized tissue. Thus, immunological techniques may be well suited for the study of macromolecular fossils. So far, only a few exploratory investigations have been conducted to study the applicability of immunology to geological materials.

In our laboratory, macromolecular structures still containing antigenic determinants were detected in three independent series of experiments in mollusk shells up to 70 Ma old (de Jong et al., 1974; Westbroek et al., 1979; Westbroek et al., 1983). Lowenstein and others have studied collagens and albumins in Pleistocene bone with a radioimmunoassay and have obtained species-specific reactions *inter alia,* with human fossils as old as 1.9 Ma (Lowenstein, 1980, 1981). An EDTA extract of bone powder, heated at 850°C prior to demineralization until the nitrogen content was reduced to a negligible amount, consistently gave radioimmunoassay evidence of small but definite amounts of surviving collagen and albumin.

NEW IMMUNOLOGICAL EVIDENCE

Experimental

Here are reported some results obtained with Recent and fossil invertebrate shells and antibodies elicited in a rabbit against EDTA-soluble macromolecules from the shell of the Recent cephalopod mollusk *Nautilus pompilius.*

In previously reported double-diffusion experiments, this antiserum was shown to give immunological reactions with EDTA-soluble macromolecules from shells of *Nautilus pompilius, Sepia officinalis* (another Recent cephalopod), and with a fraction of the Upper Cretaceous belemnite *Gonioteuthis* sp. (Cephalopoda), obtained by successive EDTA and acid extraction. A striking similarity was observed in the immunodiffusion patterns produced by the *Nautilus* and the *Gonioteuthis* extracts, although the reactive material of the latter was found to be present in very low concentrations in the fossil shell (Westbroek et al., 1979).

In the present experiments an adaptation of the Enzyme-Linked Immuno Sorbent Assay (ELISA), as described by Clark and Adams (1977), was used (Westbroek et al., 1983). The IgG-antibodies of the antiserum and of a preimmune serum of the same rabbit were purified by ammonium sulfate precipitation, followed by passage through a column of DE22 cellulose (Whatman Ltd.). Small, mechanically cleaned fragments of shell material (about 3 mm in diameter) were etched in 10% EDTA (pH 8.0) for 15 min, washed, and placed in the wells of microtiter plates (flat-bottom microelisa plates, M 129A, Cynatech). They were incubated for 1 hr at 37°C with a solution of antibodies (0.01 mg/ml), and thoroughly washed. Then the fragments were incubated in the presence of goat antirabbit IgG alkaline phosphatase conjugate (Sigma) (1 hr, 37°C; dilution 1:1000) and subsequently rinsed. The preparations were incubated with a p-nitrophenyl phosphate solution as a substrate. To arrest the phosphatase reaction, 3 N NaOH was added; the shell fragments were removed from the wells, and the adsorption of the remaining stained solution was measured automatically at 405 nm with a Titertek Multiskan photometer (Flow Laboratories). Data were obtained on the reactivity of EDTA-etched shell fragments from different sources with IgG extracts of both the anti-*Nautilus* and the preimmune serum, the latter serving as a blank. With this technique more than 100 reactions could be carried out per day. Duplicate experiments were performed in the same run, and many duplicates were repeated several times.

Reactions Between Antibodies Against
Nautilus and a Variety of
Invertebrate Shells

In Figure 5-1*a*, the intensity of the reactions is expressed as the percentage of the mean value obtained for *Nautilus*. Mean blank values (obtained with IgG from the same rabbit before immunization) are drawn from the baseline down. The bars running up from the baseline represent the mean values of the respective species with the standard deviation minus the mean values for the blanks. Note that the blank reactions are generally weak. Significantly stronger reactions obtained with the anti-*Nautilus* preparation are taken to

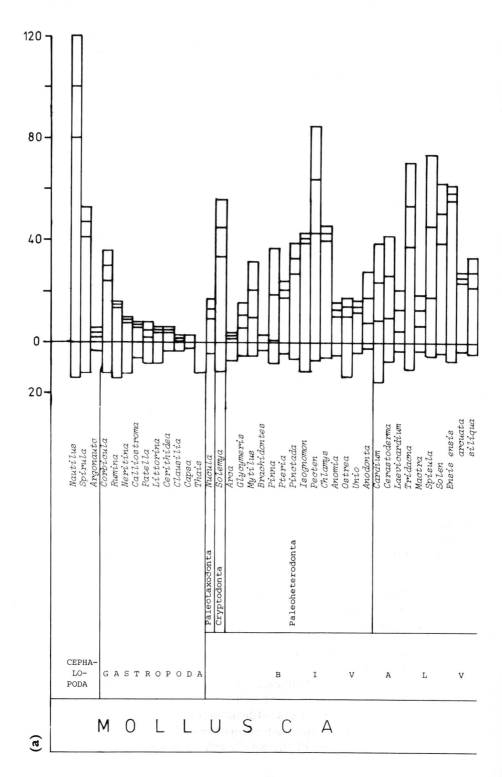

120

80

40

0

20

Nautilus
Spirula
Argonauta
Corbicula
Rumina
Neritina
Calliostroma
Patella
Littorina
Cerithidea
Clausilia
Capsa
Thais
Nucula
Solemya
Arca
Glycymeris
Mytilus
Brachidontes
Pinna
Pteria
Pinctada
Isognomon
Pecten
Chlamys
Anomia
Ostrea
Unio
Anodonta
Cardium
Cerastoderma
Laevicardium
Tridacna
Mactra
Spisula
Solen
Ensis ensis
arcuata
siliqua

Paleotaxodonta
Cryptodonta
Paleoheterodonta

CEPHA-
LO-
PODA

G A S T R O P O D A

B   I   V   A   L   V

M O L L U S C A

(a)

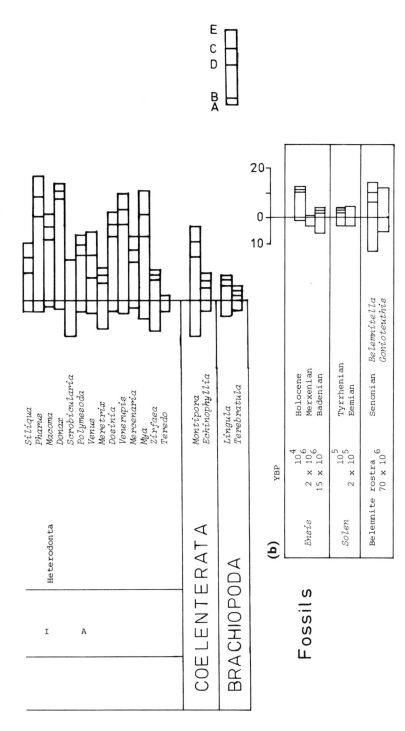

Figure 5-1. Intensity of immunological reactions between antibodies directed against soluble macromolecules of *Nautilus* shell and etched shell fragments of Recent (a) and fossil (b) invertebrates. For experimental details, see text. Insert: AB, blank reactions with antibodies isolated from pre-immune serum; BC, mean intensity of reaction with anti-*Nautilus* antibodies minus blank reaction, expressed as percentage of mean reaction with *Nautilus*; CD and CE, standard deviation.

represent specific immunological interactions. From Figure 5-1a it is evident that the antibodies elicited with the *Nautilus* extract react with a broad variety of Recent mollusk species. Particularly striking are the reactions with more than 50% intensity obtained with the bivalves *Pecten, Tridacna, Solen,* and *Ensis.* Interestingly, similar experiments conducted with antibodies raised against an extract of *Ensis* shell gave a strong reaction with *Nautilus* (more than 20%) (Westbroek et al., 1983). The reaction between anti-*Nautilus* and the cephalopod *Spirula* was of about the same strength as with the aforementioned bivalves (47%), while *Argonauta,* in which shell formation seems to have evolved independently, gives a very low response (4%). The gastropods studied gave, in general, weaker reactions than the bivalves. The strongest reaction was obtained with *Nautilus* itself, probably owing to the presence of determinants with a very narrow systematic range.

These results suggest that certain determinants are common to many, if not all, mollusks. The calcifying systems leading to shell formation in the different taxa would be derived from a common ancestor, with *Argonauta* as a probable exception.

Of special interest are the reactions obtained with skeletal material of corals (especially *Montipora* and of brachiopods (especially the calcium-phosphate-producing *Lingula*). In principle, this too may result from common ancestry of the calcifying systems, although these phyla diverged in evolution before they produced skeletons (see, however, Lowenstam and Weiner, 1983). But at this stage the possibility cannot be ruled out that the correspondence is due to convergent evolution or to low specificity of some of the antibodies contained in the anti-*Nautilus* preparation.

In some cases significant reactions were obtained with fossils (Fig. 5-1b). The reaction with the Upper Cretaceous *Gonioteuthis* confirms the earlier results.

Similar results were obtained with antibodies elicited against an extract from shells of the Recent bivalve *Ensis ensis,* in particular with preparation aEl, as reported by Westbroek et al. (1983). In the latter experiments the strongest reactions were obtained with *Ensis* and related species. Among the mollusk classes the cephalopods had a higher affinity than gastropods with aEl (not shown), and significant values were obtained with brachiopods (especially *Lingula*) and with several fossils.

However, these data should not be overinterpreted. The reaction intensities depend on the number of determinants recognized by the antibodies on the surface of the shell fragments, and this is not related only to phyletic distances or even to structural similarities of shell matrices. In the first place, the procedure for fragmentation of the shells was crude, and the exposed surfaces were not identical in size even after careful selection. Moreover, minor variations in the experimental conditions were found in some cases to have a considerable effect on the outcome of the experiments. However, when the genera used were listed according to reaction intensity, their sequence was virtually identical in successive experiments. Thus, although the experiments were essentially reproducible, the method can only be considered as semi-quantitative at this stage. Second, dependence of reaction intensity on shell structure must be considered. This problem is studied in the following experiment.

Shell Structure and Reaction Intensity

Isolated fragments of the (inner) nacreous layer of a *Nautilus* shell gave much stronger reactions than similarly sized parts of the (outer) prismatic layer from the same specimen (Fig. 5-2a). This could be due either to the recognition of different determinants in the two layers or to the fact that the same determinants were exposed in different concentrations. To decide between these possibilities the following experiment was carried out. Isolated nacreous and prismatic materials were powdered, and the powders were etched for a few minutes with 10% EDTA (pH 8). They were then incubated for 1 hr with aliquots of dilute antibody solution. The suspensions were centrifuged, and the supernatants were subjected to the treatment with fresh amounts of the same etched powders. The final supernatants, from which were extracted the antibodies that were directed against the nacreous and the prismatic material, respectively, were allowed to react with etched shell fragments of *Nautilus* containing both nacreous and prismatic material and with fragments of isolated nacreous and prismatic shell. Significant reactions were not obtained with either of the two supernatants (Fig. 5-2b). From this experiment it was concluded that the same determinants were detected in the two layers with these antibodies and that the difference in reactivity between these materials was due to the fact that they were exposed in different concentrations upon etching. The fact alone that no qualitative differences were recognized between the determinants of the two layers does not imply that their macromolecular contents are identical.

Comparative Studies Avoiding the Effect of Shell Structure

Similar experiments allowed the reactivity of shell fragments from different taxa to be compared independent of shell structure. This led to more meaningful systematic conclusions than could be drawn from the results of Fig. 5-a. Aliquots of the *Nautilus* antibodies were incubated with EDTA-etched shell powders of the bivalves *Ensis* and *Pecten,* and the supernatants were allowed to react with a variety of shell fragments (Figs. 5-3a and 5-3b). No significant reactions were obtained with *Ensis* and *Pecten,* respectively, indicating that all antibodies capable of recognizing determinants in these species were adsorbed to the powders. Strong reactions were obtained with the cephalopod representatives tested, while the other bivalves showed no or only a weak affinity for the supernatants. *Tridacna* gave some response with both supernatants, and *Pecten* reacted slightly only with the *Ensis* extract. In general terms, these results adequately reflect the existing systematic

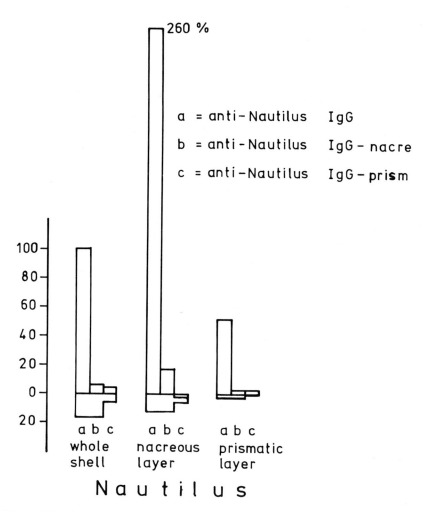

Figure 5–2. Intensity of immunological reactions between (*a*) antibody preparation used in experiment of Figure 5–1; (*b*) supernatant obtained after incubation of the preparation used in Figure 5–1 with excess EDTA-etched powder of *Nautilus pompilius* nacre and centrifugation; and (*c*) idem, but extracted with prismatic material of same species and etched fragments of the whole shell (left), the nacreous layer (middle), and the prismatic layer (right) of *Nautilus pompilius* shell. Note that virtually all antibodies are extracted by the two powders, suggesting that with unextracted antibody preparation no qualitative differences can be detected between the matrices of the two shell layers.

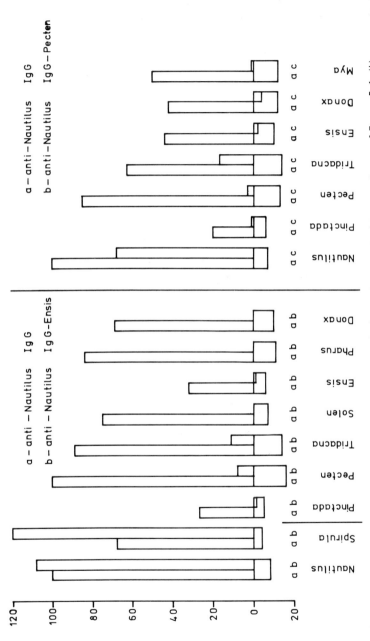

Figure 5-3. Intensity of immunological reactions between (*a*) antibody preparation used in experiment of Figure 5–1; (*b*) supernatant obtained after incubation of preparation used in Figure 5–1 with excess EDTA-etched powder of *Ensis* shell; and (*c*) idem, but with *Pecten* shell powder and etched shell fragments of some mollusks. Note that these results are in agreement with existing systematics.

classification; the original antibody preparation recognized similar populations of determinants in the bivalves tested, but these represented only a fraction of the total number of determinants against which the antibodies were elicited and of those common to *Nautilus* and *Spirula*. *Pecten* and *Tridacna* appear to have retained some determinants that also occur in *Nautilus* but that may have been lost during the evolution of the other bivalves.

CONCLUSIONS

The results presented in this paper clearly demonstrate that immunology is a powerful tool in the study of the calcification process and its evolution and of fossil macromolecules from biominerals. They suggest that comparative studies of the biochemical structure of matrix macromolecules may eventually lead to new insights into the origins of calcification in the animal kingdom. Of particular importance is the observation that adsorption of antibodies with selected, etched shell powders leads to highly specific preparations that are eminently suited for systematic and comparative purposes.

The applicability of other immunological techniques in the study of Recent and fossil biominerals, such as immunohistochemical localization of determinants, isolation of antigenic matrix components with affinity chromatography, and the production of monoclonal antibodies, are presently being tested in our laboratory. Preliminary experiments suggest that important applications are forthcoming, relevant both to biomineralization and geology.

ACKNOWLEDGEMENTS

We thank Dr. J. W. Bruning for valuable suggestions. F. Leupe and his staff (Department of Medical Microbiology, University of Leiden) were responsible for the biotechnical work. Shell materials were kindly provided by P. J. Felder (Museum of Natural History, Maastricht), Drs. E. Gittenberger, J. C. den Hartog, and N. R. R. B. Wijsman (State Museum of Natural History, Leiden), Dr. A. W. Janssen (State Museum of Geology and Mineralogy, Leiden), and Dr. Howarth (British Museum [Natural History], London). This research was supported by the Netherlands Foundation for Earth Science Research (AWON) with financial aid from the Netherlands Organization for the Advancement of Pure Research (ZWO).

REFERENCES

Clark, N. F., and A. N. Adams, 1977, Characteristics of the microplate method of enzyme-linked immunosorbent assay for the detection of plant viruses, *J. Gen. Virol.* **34:**475-483.

De Jong, E. W., P. Westbroek, J. E. Westbroek, and J. W. Bruning, 1974, Preservation of antigenic properties in macromolecules over 70 Myr old, *Nature* **252**:63.

Glimcher, N. J., 1984, Recent studies of the mineral phase in bone and its possible linkage to the organic matrix by protein-bound phosphate bonds, *R. Soc. Lond. Phil. Trans.,* **B304**:479-508.

Lowenstam, H. A., 1981, Mineral formed by organisms, *Science* **211**:1126-1131.

Lowenstam, H. A., and S. Weiner, 1983, Mineralization by organisms and the evolution of biomineralization, in *Biomineralization and Biological Metal Accumulation,* P. Westbroek and E. W. de Jong, eds., Reidel, Dordrecht, Netherlands, pp. 191-203.

Lowenstein, J. M., 1980, Species-specific proteins in fossils, *Naturwissenshaften* **67**:343-346.

Lowenstein, J. M., 1981, Immunological reactions from fossil material, *R. Soc. Lond. Phil. Trans.* **B292**:143-149.

Veis, A., and B. Sabsay, 1983, Bone and tooth formation. Insights into mineralization strategies, in *Biomineraliation and Biological Metal Accumulation,* P. Westbroek and E. W. de Jong, eds., Reidel, Dordrecht, Netherlands, pp. 273-284.

Weiner, S., in press, Mollusk shell formation: Isolation of two organic matrix proteins associated with calcite deposition in the bivalve *Mytilus californianus, Biochemistry.*

Weiner, S., and W. Traub, 1984, Macromolecules in mollusk shells and their function in biomineralization, *R. Soc. Lond. Phil. Trans.,* **B304**:425-434.

Weiner, S., H. A. Lowenstam, and L. Hood, 1976, Characterization of 80-million-year-old mollusk shell proteins, *Natl. Acad. Sci. Proc.* **73**:2514-2545.

Weiner, S., W. Traub, and H. A. Lowenstam, 1983, Organic matrix in calcified exoskeletons, in *Biomineralization and Biological Metal Accumulation,* P. Westbroek and E. W. de Jong, eds., Reidel, Dordrecht, Netherlands, pp. 205-224.

Westbroek, P., 1983, Biological metal accumulation and biomineralization in a geological perspective, in *Biomineralization and Biological Metal Accumulation,* P. Westbroek and E. W. de Jong, eds., Reidel, Dordrecht, Netherlands, pp. 1-11.

Westbroek, P., P. H. van der Meide, J. S. van der Wey-Kloppers, R. J. van der Sluis, J. W. de Leeuw, and E. W. de Jong, 1979, Fossil macromolecules from cephalopod shells: Characterization, immunological response and diagenesis, *Paleobiology* **5**:151-167.

Westbroek, P., J. Tanke-Visser, J. P. M. de Vrind, R. Spuy, W. van der Pol, and E. W. de Jong, 1983, Immunological studies on macromolecules from invertebrate shells— recent and fossil, in *Biomineralization and Biological Metal Accumulation,* P. Westbroek and E. W. de Jong, eds., Reidel, Dordrecht, Netherlands, pp. 249-253.

Westbroek, P., E. W. de Jong, P. van der Wal, A. H. Borman, J. P. M. de Vrind, D. Kok, W. C. de Bruijn, and S. B. Parker, 1984, Mechanism of calcification in the marine alga *Emiliania huxleyi, R. Soc. Lond. Phil. Trans.* **B304**:435-444.

Part II

C-1 COMPOUNDS

6

Biogenic Gases in Sediments Deposited Since Miocene Times on the Walvis Ridge, South Atlantic Ocean

Philip A. Meyers
University of Michigan

Simon C. Brassell
University of Bristol

ABSTRACT: Hydraulic piston coring done during Deep Sea Drilling Project Leg 75 recovered, from under the Benguela upwelling off Namibia, biogenic sediments of Late Miocene to Holocene age containing up to 6% organic carbon. The highest concentrations of organic carbon occur in Late Pliocene sediments at about 100 m subbottom, and concentrations become progressively less down to 290 m. Abundant interstitial gases were found in sediments from depths between 130 m and 200 m. Methane was by far the dominant component of all gas samples, and the composition of C_2 to C_5 hydrocarbons showed little variation. Hydrogen sulfide was obnoxiously present. The gases evidently were formed by *in situ* microbial activity and did not migrate from deeper sediments. Their presence suggests the possibility of viable microbial populations to depths as great as several hundred meters in deep-sea sediments.

INTRODUCTION

The formation of gases by microbial activity in sediments of lakes and oceans is a common occurrence. These biogenic gases include methane, carbon dioxide, hydrogen sulfide, and other compounds, and their zone of production usually begins where interstitial waters no longer contain dissolved oxygen. The downward extent of microbial activity and accompanying gas production is not generally considered to be great, yet Hunt (1979) suggests that in some fine-grained sediments, bacterial methanogenesis may continue to depths of several hundred meters. Belyaev, Lein, and Ivanov (1980) report methane production at low rates to sediment depths of 90 m in the Caspian Sea. Hathaway et al. (1979) describe profiles of sulfate depletion and methane appearance to depths of 300 m in sediments from the North American continental shelf and suggest that bacteria may have been active in these sediments since Miocene times. Viable methanogenic bacteria have been found as deep as 12 m in some sediments from the Gulf of California, but not at all sampling locations (Oremland, Culbertson, and Simoneit, 1982). The nonuniformity of bacterial activity, as well as lack of activity below 12-m depth, may be due to patchiness of microbial populations, or it may be due to sampling procedures and possible oxygen contamination of samples. Difficulties in maintaining viable methanogenic bacteria in sediments from deep in the sea floor are also mentioned by Ivanov, Belyaev, and Laurinavichus (1980), who find no methanogenic activity from subbottom depths ranging from 6 m to 1550 m in sediment samples from Deep Sea Drilling Project (DSDP) Leg 50 in the Moroccan Basin. This lack of activity may be due to oxygen contamination.

The results of gas analyses conducted aboard D/V *Glomar Challenger* during DSDP Leg 75 in the eastern South Atlantic Ocean in the region of the Benguela Current are described. Earlier results from DSDP Leg 40 suggested the presence of active bacteria in the sediments underlying these highly productive surface waters. Bolli et al. (1978) report high concentrations of biogenic gases in sediment cores from DSDP Site 362 on the Walvis Ridge. Gas pockets had formed quickly as the cores expanded on deck, causing the polypropylene end caps to bulge and sometimes forcing them from the core liner. The major constituents of the gas pockets, excluding air, were methane and carbon dioxide, with traces of hydrogen sulfide and free nitrogen (Foresman, 1978). In view of these results from Site 362, an analysis program involving gas sampling was conducted during DSDP Leg 75.

MATERIALS AND METHODS

Sampling

D/V *Glomar Challenger* sampling occupied two sites under the oceanic edge of the Benguela upwelling in August 1980 during DSDP Leg 75. Sediments were sampled by continuous hydraulic piston coring to a subbottom depth of 180 m at Site 530 in the Angola Basin and to 291 m at Site 532 on the Walvis Ridge (Fig. 6-1). Water depths at these locations are 4639 m and 1341 m, respectively. Use of the hydraulic piston corer minimizes sediment disturbance, thus allowing observation and study of closely spaced variations in sediment character.

A description of the Pleistocene to Miocene sediments from these two sites is given by Hay et al. (1982). Different types of sedimentary sequences are found at the two sites shown in Figure 6-1. At Site 530 in the Angola Basin, sediments consist of turbidites and debris-flow deposits, presumably originating from biogenic oozes originally laid down on the Walvis Ridge. The sediments at Site 532 on the Walvis Ridge are made up of alternating sequences of light and dark pelagic oozes and are heavily bioturbated.

Organic Carbon Analysis

Analyses were done on board the ship using a Hewlett-Packard 185-B CHN (Carbon-Hydrogen-Nitrogen) Analyzer. Portions of samples selected for carbonate measurements were treated with 1 N HCl to remove carbonate, washed with deionized water, and dried at 110°C. A Cahn Electrobalance was used to weigh 20-mg samples of acid-treated sediment for CHN analysis. Samples were combusted at 1050°C in the presence of an oxidant, and the volumes of the evolved gases were determined as measures of the C, H,

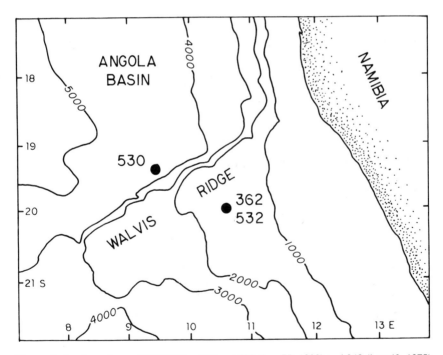

Figure 6-1. Locations of DSDP Sites 530 and 532 (Leg 75, 1980) and 362 (Leg 40, 1975) where gas samples were collected from sediment cores.

and N contents of sediment organic matter. Total organic carbon concentrations and atomic C/N ratios were calculated using response factors determined from standards and were corrected for the small blank of the complete procedure.

Gas Analysis

Gas samples were collected from sediment cores by piercing the plastic core liners with a needle to allow gases to pass through a stopcock into an evacuated glass tube (Vacutainer). Such sampling was performed as soon as the cores were brought onto the deck. Where possible, samples were taken from gas pockets visible through the core liners.

Two gas chromatographs (GC) were used for analysis of gaseous hydrocarbons. Samples were directly injected into a Carle Model 8000 GC equipped with a thermal conductivity detector and fitted with a QS column (1.5 m x 3.1 mm OD) that was operated isothermally at 45°C. Samples analyzed on a Hewlett-Packard Model 5711A GC first passed through an alumina-filled loop (20 cm x 3.1 mm OD packed with 60/80 mesh Al_2O_3) precooled to approximately $-70°C$ (refrigerated propan-2-ol bath). After 90 sec, to allow methane stripping, the trapping loop was closed and heated for 60 sec in a hot-water bath (90°-100°C); then the sample was flushed into the GC column (1.8 m x 3.1 mm OD 40-100 mesh Spherosil linked to 3.6 m x 3.1 mm OD 20% OV-101 on 100/110 mesh Anakrom AS) by the helium carrier gas. The column was programmed from 60°-200°C at 8°C/min. The use of the two GC systems allowed analysis of methane and ethane (Carle) and of ethane through pentane (Hewlett-Packard).

RESULTS AND DISCUSSION

Organic Carbon and C/N Ratios

Although lithologically different, the sediments at Sites 530 and 532 share similar organic matter patterns and compositions. Figure 6-2 shows individual values for organic carbon concentrations of samples from these sites. The percentages of organic carbon are relatively high, particularly in view of the amount of bioturbation evident in sediments from both Leg 75 locations (Hay et al., 1982). Maximum values of 5% to 6% total organic carbon (TOC) are found at subbottom depths of 40 m to 110 m at Site 530 and at 50 m to 150 m at Site 532. Below this level, values tend to decrease with depth, yet are still approximately 2% in the upper Miocene layers of these sediments. There is considerable variability in TOC of individual samples, with light-colored oozes having lower organic carbon content than the darker, olive-colored samples.

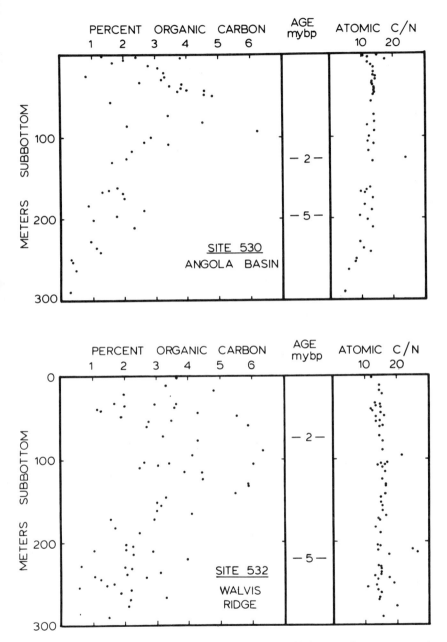

Figure 6-2. Organic carbon concentrations and atomic C/N ratios of organic matter in sediment samples from DSDP Sites 530 and 532. Approximate subbottom depths of the Pleistocene-Pliocene (2 MaBP) and Pliocene-Miocene (5 MaBP) boundaries are shown. Concentrations are given as percentages of whole sediment dry weight.

In contrast to the variability and depth changes in the organic carbon contents of Leg 75 sediments, the atomic C/N ratios change little from a value of 15 throughout the sediment sequence (Fig. 6-2). This monotonous depth trend shows that the processes altering the character of the organic matter have not changed significantly since the initiation of the Benguela upwelling system. Because C/N values of particulate matter average 7.3 in the upper water column in this area of the Atlantic Ocean (Bishop, Ketten, and Edmond, 1978), a selective loss of nitrogen may occur after particles sink out of the photic zone, but prior to their incorporation into the bottom sediments, as suggested by Suess and Müller (1980).

Pyrolysis of sediments from Sites 530 and 532 suggest that the sedimentary organic matter is of predominantly marine origin and was deposited under slightly oxic conditions (Meyers et al., 1983; Rullkötter, Mukhopadhyay, and Welte, 1984). In addition, the lipid contents in sediments from Site 362 (Boon et al., 1978), as well as their carbon isotope ratios (Erdman and Schorno, 1978), indicate predominantly marine sources. Thus, organic geochemical studies suggest that the high aquatic productivity associated with the Benguela Current is the primary origin of most of the organic matter in the sediments of this part of the southeastern Atlantic Ocean.

Biogenic Gases

The close proximity of Site 532 to Site 362 meant that significant quantities of gas would almost certainly be encountered during DSDP Leg 75. However, few gas pockets were observed in the first 30 cores recovered at Site 532 (down to 130 m subbottom depth). Methane first appeared at 123 m subbottom depth and was present to a sediment depth of 232 m in gas samples taken when the cores first arrived on deck. Considerable degassing of cores within this depth range subsequently occurred as they gradually warmed to ambient temperatures while awaiting opening. In several instances, end caps were punctured to release the pressure build-up, resulting in sediment extrusion or spurting through the puncture. This abundance of gas in Site 532 sediments contrasted with the relatively little gas present in samples from Site 530.

The total concentrations of methane and of C_2 to C_5 hydrocarbons present in gas pockets in cores from Site 532 are plotted against depth in Figure 6-3A. Three values of methane concentrations from Site 362 are included (Foresman, 1978). The compositions of the C_2 to C_5 hydrocarbons for each sample are shown in Figure 6-3B. Variations with depth of selected ratios of gas concentrations are given in Figure 6-4. Significant quantities of carbon dioxide were found in addition to methane. The concentrations of carbon dioxide and methane for Site 532 and the values for Site 362 normalized on an air-free basis (Bolli et al., 1978) are shown against depth in Figure 6-5. Attempts to measure hydrogen sulfide using the Carle GC were unsuccessful, indicating that the levels were below detection limits, but no hydrogen sulfide standard was available to calibrate the instrumental response. However, the strength of the hydrogen sulfide odor was sufficiently prominent to indicate that significant quantities of the gas were present. No attempts were made to quantify the intensity of the hydrogen sulfide odor versus core depth, although the smell in the core lab increased from faintly unpleasant to thoroughly obnoxious as successively deeper cores were opened.

Comparison of the gas data for Site 532 with those for Site 362 (Fig. 6-5) shows similar concentration levels of methane and carbon dioxide. The values for Site 532 show greater fluctuation, probably because they are not normalized on an air-free basis. The apparent direct relationship between methane and carbon dioxide concentrations reflects the association of sulfate reduction and methanogenesis in these sediments. These microbial

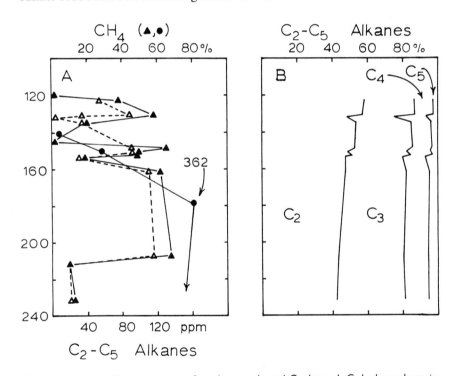

Figure 6-3. (A) Concentrations of methane and total C_2 through C_5 hydrocarbons in interstitial gas in sediments from DSDP Sites 532 and 362. *(Source: Data for Site 362 from Foresman, 1978.)* (B) Composition of C_2 through C_5 hydrocarbons in interstitial gas in sediments from DSDP Site 532.

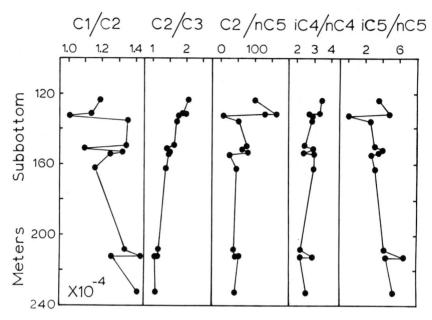

Figure 6-4. Ratios of hydrocarbon components in interstitial gas in sediments from DSDP Site 532.

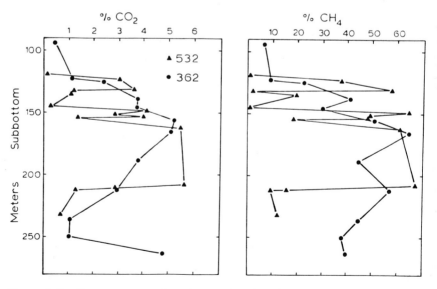

Figure 6-5. Concentrations of methane and carbon dioxide in interstitial gases in sediments from DSDP Sites 532 and 362. *(Source: Data from Site 362 from Bolli et al., 1978.)*

processes liberate hydrogen sulfide and carbon dioxide, and methane and carbon dioxide, respectively (Mechalas, 1974). Bacterial activity in the sediments of Site 362 is indicated by their lipid composition (Boon et al., 1978), which includes biological markers of *Desulfovibrio desulfuricans,* and by $\delta^{13}C$ values of -72/mil of methane in Miocene sediments (Foresman, 1978), which is characteristic of biogenic methane (Kvenvolden and Claypool, 1980). In addition, the lipids of surficial diatomaceous oozes from nearby Walvis Bay include hydrocarbons characteristic of methanogenic bacteria (Whelan, Hunt, and Berman, 1980; Brassell et al., 1981). There exist, therefore, various forms of indirect evidence for the presence and activity of bacteria in the sediments underlying the highly productive areas associated with the Benguela Current. Primary among the manifestations of their existence are the high concentrations of biogenic gas, including hydrogen sulfide, at Sites 532 and 362.

The composition of gaseous hydrocarbons at Site 532 remains almost constant with depth (Fig. 6-3*B*). Such "in-phase" behavior has been previously reported in sediments from the Black Sea and the Moroccan Basin (Hunt and Whelan, 1978; Whelan, 1979) and indicates that the gases are formed *in situ* and have not migrated from depth. However, without knowledge of the microbial processes operating within the sediment, it is uncertain whether *in situ* formation of gaseous hydrocarbons occurs in a discrete shallow zone of the sediment that subsequently becomes buried or whether it continues to depth. The depth trend for the downhole decrease in the ethane/propane ratio (Fig. 6-4) might reflect a difference in the composition of the products of microbial activity at depth. It is also possible that it reflects a gradual diffusive loss with increasing depth of ethane relative to propane, yet Schaefer and Leythaeuser (1984) conclude that little migration of light hydrocarbons has occurred in these sediments.

The methane/ethane ratio of gas samples from Site 532 is consistently greater than one to ten thousand and appears to increase between 120 m and 240 m sediment depth (Fig. 6-4). These high values result in high ratios of methane-to-ethane-plus-propane characteristic of biogenic gases (Kvenvolden and Claypool, 1980). The tendency to increase with depth may be a migrational effect, since methane in shallower sediments may more easily escape to seawater, or it may be related to differences in microbial activity.

An additional factor important to the abundance and composition of gaseous hydrocarbons in these deep-ocean sediments is the possible existence of gas hydrates. The combination of low temperatures and high pressures found in the upper 500-1000 m of sediments in the deep ocean is suitable for the formation of gas hydrates (Hunt, 1979; Kvenvolden and McMenamin, 1980). The occurrence of these hydrates has been reviewed by Kvenvolden and McMenamin (1980) and includes several DSDP sites. Although no direct evidence of hydrates was observed during DSDP Leg 75, it is possible that they were originally present and had decomposed with release of pressure as the cores were brought to the sea surface.

SUMMARY AND CONCLUSIONS

The abundance of gases dissolved in interstitial waters of sediments from DSDP Site 532 on the Walvis Ridge serves as a record of their *in situ* microbial generation. The dominance of methane, along with important amounts of carbon dioxide and hydrogen sulfide, relative to higher-molecular-weight gaseous hydrocarbons rules out thermogenic production of the gases in deeper strata and migration to the upper layers of sediments. As observed by Schaefer and Leythaeuser (1984), concentrations of gaseous hydrocarbons are directly related to concentrations of organic carbon in these sediments. This relationship indicates that microbial production of gases is dependent upon the amount of organic matter available in sediments and is further evidence against the probability of gas migration.

Although sediments from Site 530 in the Angola Basin contain amounts of organic carbon similar to those found at Site 532, their gas contents are much less. This difference in abundances of interstitial gases may be the result of higher water pressures at the Angola Basin location (approximately 460 atm) than at the Walvis Ridge site (approximately 130 atm), which may act to retard rates of microbial activity, or it may be due to the downslope debris-flow relocation of sediments found at Site 530, which may have released interstitial gases present in the sediments at their original site of deposition.

There remains an important question concerning the biogenic gases found in the Walvis Ridge sediments. In the absence of any direct evidence of the presence or absence of viable microbial populations in sediments several hundreds of meters below the seabottom, it is impossible to resolve whether the gases are the products of recent biosynthesis or are the molecular fossils of microbes that lived in Pliocene and Miocene times. Both possibilities have exciting implications.

ACKNOWLEDGEMENTS

We thank the National Science Foundation and the National Environmental Research Council (United Kingdom) for providing the opportunities for us to participate in Leg 75 of the Deep Sea Drilling Project. Portions of this project were funded by NSF Grants OCE 8214605 EAR 7822432.

REFERENCES

Belyaev, S. S., A. V. Lein, and M. V. Ivanov, 1980, Role of methane-producing and sulfate-reducing bacteria in the process of organic matter destruction, in *Biogeo-*

chemistry of Ancient and Modern Environments, P. A. Trudinger and M. R. Walter, eds., Proc. 4th International Symposium on Environmental Biogeochemistry, Australian Academy of Science, Canberra, pp. 235-242.

Bishop, J. K. B., D. R. Ketten, and J. M. Edmond, 1978, The chemistry, biology and vertical flux of particulate matter from the upper 400 m of the Cape Basin in the southeast Atlantic Ocean, *Deep Sea Res.* **25:**1121-1161.

Bolli, H. J., W. B. F. Ryan, and Leg 40 Shipboard Party, 1978, Walvis Ridge-Sites 362 and 363, in *Initial Reports DSDP,* Vol. 40, H. M. Bolli et al., eds., U.S. Govt. Printing Office, Washington, D.C., pp. 183-356.

Boon, J. J., F. W. von der Meer, P. J. W. Schuyl, J. W. de Leeuw, P. A. Schenck, and A. L. Burlingame, 1978, Organic geochemical analysis of core samples from Site 362, Walvis Ridge, DSDP Leg 40, in *Initial Reports DSDP,* Vol. 40, H. M. Bolli et al., eds., U.S. Govt. Printing Office, Washington, D.C., pp. 627-637.

Brassell, S. C., A. M. K. Wardroper, I. D. Thompson, J. R. Maxwell, and G. Eglinton, 1981, Specific acyclic isoprenoids as biological markers of methanogenic bacteria in marine sediments, *Nature* **290:**693-696.

Erdman, J. G., and K. S. Schorno, 1978, Geochemistry of carbon: Deep Sea Drilling Project Leg 40, in *Initial Reports DSDP,* Vol. 40, H. M. Bolli et al., eds., U.S. Govt. Printing Office, Washington, D.C., pp. 651-658.

Foresman, J. B., 1978, Organic geochemistry DSDP, Leg 40, Continental Rise of Southwest Africa, in *Initial Reports DSDP,* Vol. 40, H. M. Bolli et al., eds. U.S. Govt. Printing Office, Washington, D.C., pp. 557-567.

Hathaway, J. C., C. W. Poag, P. C. Valentine, R. E. Miller, D. M. Schultz, F. T. Manheim, F. A. Kohout, M. H. Bothner, and D. A. Sangrey, 1979, U.S. Geological Survey, core drilling on the Atlantic Shelf, *Science* **206:**515-527.

Hay, W. W., J. C. Sibuet, and Leg 75 Shipboard Party, 1982, Sedimentation and accumulation of organic carbon in the Angola Basin and on Walvis Ridge: Preliminary results of Deep Sea Drilling Project Leg 75, *Geol. Soc. Am. Bull.* **93:**1038-1050.

Hunt, J. M., 1979, *Petroleum Geochemistry and Geology,* W. H. Freeman and Co., San Francisco, Calif.

Hunt, J. M., and J. K. Whelan, 1978, Dissolved gases in Black Sea sediments, in *Initial Reports DSDP,* Vol. 42, D. A. Ross et al., eds., U.S. Govt. Printing Office, Washington, D.C., pp. 661-668.

Ivanov, M. V., S. S. Belyaev, and K. S. Laurinavichus, 1980, A short report on microbiology of sediments from Deep Sea Drilling Project Holes 415, 415A, and 416A, in *Initial Reports DSDP,* Vol. 50, Y. Lancelot et al., eds., U.S. Govt. Printing Office, Washington, D.C., pp. 656-666.

Kvenvolden, K. K., and G. E. Claypool, 1980, Origin of gasoline-range hydrocarbons and their migration by solution in carbon dioxide in Norton Basin, Alaska, *AAPG Bull.* **64:**1078-1086.

Kvenvolden, K. K., and M. A. McMenamin, 1980, *Hydrates of Natural Gas: A Review of Their Geological Occurrence,* U.S. Geological Survey Circular **825,** Arlington, Va.

Mechalas, B. J., 1974, Pathways and environmental requirements for biogenic gas production in the ocean, in *Natural Gases in Marine Sediments,* I. R. Kaplan, ed., Plenum Press, New York, pp. 11-25.

Meyers, P. A., A. V. Huc, S. C. Brassell, and DSDP Leg 75 Shipboard Party, 1983,

Organic matter patterns in South Atlantic sediment deposited since the late Miocene beneath the Benguela upwelling system, in *Advances in Organic Geochemistry 1981,* M. Bjorøy, ed., John Wiley and Sons, Chichester, pp. 465-470.

Oremland, R. S., C. Culbertson, and B. R. T. Simoneit, 1982, Methanogenic activity in sediment from Leg 64, Gulf of California, in *Initial Reports DSDP,* Vol. 64, J. R. Curray, et al., eds., U.S. Govt. Printing Office, Washington, D.C., pp. 759-762.

Rullkötter, J., P. K. Mukhopadhyay, and D. K. Welte, 1984, Geochemistry and petrography of organic matter in sediments from Deep Sea Drilling Project Holes 530A, Angola Basin, and 532, Walvis Ridge, in *Initial Reports DSDP,* Vol. 75, W. W. Hay et al., eds., U.S. Govt. Printing Office, Washington, D.C., pp.1069-1087.

Schaefer, R. G., and D. Leythaeuser, 1984, C_2-C_8 hydrocarbons in sediments from Deep Sea Drilling Project Leg 75, Hole 530A (Angola Basin) and Hole 532 (Walvis Ridge), South Atlantic, in *Initial Reports DSDP,* Vol. 75, W. W. Hay et al., eds., U.S. Govt. Printing Office, Washington, D.C., pp. 1055-1067.

Suess, E., and P. J. Müller, 1980, Productivity, sedimentation rate and sedimentary organic matter in the oceans II. Elemental fractionation, *Collog. int. du C.N.R.S.* **293:**17-26.

Whelan, J. K., 1979, C_1 to C_7 hydrocarbons from IPOD Holes 397 and 397A, in *Initial Reports DSDP,* Vol. 47, U. von Rad et al., eds., U.S. Govt. Printing Office, Washington, D.C., pp. 531-539.

Whelan, J. K., J. M. Hunt, and J. Berman, 1980, Volatile C_1-C_7 organic compounds in surface sediments from Walvis Bay, *Geochim. Cosmochim. Acta* **44:**1767-1785.

7

Aspects of the Biogeochemistry of Big Soda Lake, Nevada

Ronald S. Oremland, Richard L. Smith, and
Charles W. Culbertson

U.S. Geological Survey

INTRODUCTION

Despite interest in the physiology and biochemistry of alkalophilic microorganisms (Horikoshi and Akiba, 1982), surprisingly little research has been devoted to the study of the microbial biogeochemistry of alkaline lakes. Furthermore, there is geologic evidence that the sediments of alkaline, meromictic lakes may have provided the source material for certain lacustrine deposits of petroleum (Demaison and Moore, 1980). In order to better understand the biogeochemistry of these aquatic environments, an interdisciplinary study of Big Soda Lake, Nevada, was begun in October 1980. The limnological, hydrochemical, and microbiological features of the lake have been examined. Various aspects of the study are still in progress at the time of this writing. The purpose of this paper is to summarize the major scientific findings made thus far and to point out current avenues of research regarding the biogeochemistry of alkaline lakes.

SITE DESCRIPTION

Big Soda Lake is located in the Carson Desert in western Nevada near the town of Fallon (about 560 km east of San Francisco). The lake occupies a volcanic crater last active during the Holocene era. During the Pleistocene this region was part of Lake Lahontan, a large inland water body. Big Soda Lake lies at an altitude of approximately 1220 m above sea level and has a surface area of 156 ha and a maximum depth of approximately 62 m. At the turn of the century the lake's water level was about 18 m lower. However, diversion of irrigation waters from the Truckee River in 1906 recharged the local ground waters and caused the lake level to rise. Because the water entering the lake was of lower density than the waters already present, a condition of meromixis was established by 1930. At present the lake's water column can be divided into three zones: (1) the aerobic mixolimnion (0-20 m, spring-fall; 0-29 m, winter); (2) the anoxic mixolimnion (approximately 20-35 m, spring-fall; 29-35 m, winter); and (3) the anoxic monimolimnion (35-62 m). Details on the lake's bathymetry and hydrologic history have been summarized by Rush (1972). The lake's meromictic status was discussed by Kimmel et al. (1978).

MATERIALS AND METHODS

Methods for chemical analysis were reported by Kharaka et al. (1984) and limnological properties by Axler, Gersberg, and Paulson (1978) and Cloern, Cole, and Oremland (1983a). Productivity studies were made by Cloern, Cole, and Oremland (1983b) and Priscu et al. (1982). Extraction and quantification of hydrocarbon gases and determination of $\delta^{13}C[CH_4]$ are delineated by Oremland and Des Marais (1983). Water samples (500 ml) were filtered (0.2 μm; glass fiber) and extracted for ATP (Holm-Hansen and Booth, 1966) and cell protein (Lowry et al., 1951). Sulfides were determined by the method of Smith and Klug (1981).

RESULTS AND DISCUSSION

Water Column Chemistry

The major chemical components of the lake's water column are shown in Table 7-1. The pH of both the mixolimnion and monimolimnion is 9.7 and is well buffered by bicarbonate ions. Details on the hydrochemistry of the lake have been published (Priscu et al., 1982; Kharaka et al., 1984). The waters are rich in sulfate (5.6-6.7 g/l) but deficient in iron (<0.1 mg/l). Nutrient concentrations of the aerobic mixolimnion are highest in winter due to turnover (Cloern, Cole, and Oremland, 1983a). Surface (1-m) concentrations of dissolved inorganic nitrogen (mainly ammonia) range from <0.1 μM (spring) to 14 μM (winter), while silica levels range from 9 μM (spring) to 49 μM (winter). Levels of phosphate are always high (>100 μM) in the surface waters. Monimolimnion waters have a distinct straw-yellow color, probably caused in part by the presence of high levels (400 mg/l) of reduced sulfur compounds (Kharaka et al., 1984).

Table 7-1. Important chemical constituents of the mixolimnion and monimolimnion of Big Soda Lake. Units are in mg/l unless otherwise stated

	Mixolimnion	*Monimolimnion*
Na	8000	28,000
K	310	1100
Cl	6500	27,000
SO$_4$	5600	6700
Mg	145	6
Fe	<0.1	<0.1
Total dissolved solids	26,000	88,000
Dissolved organic carbon	20	60
Conductivity (mmhos/cm)	28	87

Source: Data from Kharaka et al. (1984) and Priscu et al. (1982).

Table 7-2. Depth profiles of temperature (T°C), oxygen (D.O.), light transmittance (L.T.), cell protein, and cell ATP taken in October 1982

Depth (m)	*T (°C)*	*D.O. (mg/l)*	*L.T. (%)*	*Protein (mg/l)*	*ATP (μg/l)*
1	15.4	7.2	100	1.61	0.08
5	14.8	7.2	95	0.65	0.16
10	14.6	7.1	95	0.49	0.04
15	12.3	3.2	99	0.77	0.01
16	11.1	1.5	96	—	—
18	9.4	0.9	75	—	—
19	8.9	0	66	1.23	0.44
20	8.5	0	64	2.54	0.71
21	8.1	0	6.6	10.2	7.9
22	7.7	0	7.5	7.45	0.41
25	6.5	0	39	2.77	0.63
30	6.6	0	54	1.85	0.15
35	11.3	0	14	3.00	0.39
40	11.7	0	27	2.69	0.25
50	11.7	0	27	2.66	1.73

Dissolved Gases and Sulfide

Table 7-2 lists a typical fall profile for dissolved oxygen and temperature. Surface heat flux governs mixing of the mixolimnion, while the depth of oxygen penetration is governed by light abundance (Cloern, Cole and Oremland, 1983a). During summer, surface temperatures rise to as high as approximately 24°C, and for most of the year (spring-fall), dissolved oxygen

Table 7-3. Depth profiles of dissolved methane and sulfide during October 1982

Depth (m)	Sulfide (mM)	Methane (mM)
0	0	0.06
6	0	0.11
14	0	0.11
18	0	0.54
20	0	1.10
22	0.13	2.3
24	0.34	3.2
30	0.73	3.8
34	7.2	52
40	6.7	55
48	6.1	49
56	6.7	49

disappears at a depth of 18-20 m (just beneath the thermocline). During winter, however, cold surface water (4°C) sinks to a depth of approximately 29 m, thereby oxygenating the water column to a depth of about 28 m. The temperature of the monimolimnion is always approximately 12°C.

Both methane and sulfide increase with depth in the anoxic mixolimnion and are abundant in the monimolimnion (Table 7-3). In addition, ethane (approximately 260 μM), propane (approximately 80 μM), and butanes (approximately 45 μM) are present in the monimolimnion (Oremland and Des Marais, 1983) and may originate from bacterial decomposition of organic matter (Davis and Squires, 1954; Oremland, 1981; Vogel, Oremland, and Kvenvolden, 1982). Values of δ^{13}C[CH$_4$] exhibit progressive enrichment in ^{13}C with vertical distance moving up from the monimolimnion sediments and through the anoxic water column. Values of δ^{13}C[CH$_4$] as light as -74/mil were found at a depth of 1.8 m in monimolimnion sediments (where methane bubbles were also evident), but reached values as heavy as -55/mil at the surface (0.1 m) of the sediment core. Monimolimnion values of δ^{13}C[CH$_4$] ranged from -55/mil to -60/mil; however, a further enrichment in ^{13}C was evident in the anoxic mixolimnion. The heaviest value of δ^{13}C[CH$_4$] encountered in the anoxic mixolimnion was -20.3/mil at 32 m (Oremland and Des Marais, 1983). This progressive enrichment of δ^{13}C[CH$_4$] in ^{13}C may have been caused by bacterial activities including anaerobic methane oxidation and methanogenesis from isotopically heavy substrates.

Productivity

Annual water column productivity of the lake has been estimated at 500 g C/m^2 of which 60% is due to phytoplankton (Cloern, Cole, and Oremland,

1983*b*). Nearly all of the phytoplankton productivity occurs during a winter bloom composed primarily of pennate diatoms (Cloern, Cole, and Oremland, 1983*a*, 1983*b*). The bloom is apparently caused by a release of nutrients as a consequence of winter turnover of the mixolimnion and perhaps by reduced zooplankton grazing in winter. Priscu et al. (1982) found that additions of dissolved inorganic nitrogen or iron stimulated algal growth in water samples taken in May, when the mixolimnion was thermally stratified.

A dense plate of purple photosynthetic bacteria identified as *Ectothiorhodospira vacuolata* (H. Truper, personal communication) is evident in the water column (summer through fall) at a depth of 21 m (just below the depth of dissolved oxygen disappearance). The presence of the plate is reflected in the high values of ATP and protein and low values of light transmittance evident at 21 m (Table 7-2). Integrated values of water column bacterial chlorophyll a (1,050 mg/m^2 during November 1981) can exceed maximum phytoplankton chlorophyll a values (950 mg/m^2 during February 1982) occurring in the winter bloom (Cloern, Cole, and Oremland, 1983*a*). However, photosynthetic bacteria account for only 10% of annual productivity (Cloern, Cole, and Oremland, 1983*b*). Bacterial chemoautotrophy occurring within the photosynthetic bacterial plate accounts for 30% of annual productivity, and experiments with inhibitors (nitrapyrin and acetylene) suggest that 40%–80% of dark CO_2 fixation is due to nitrification (Cloern, Cole, and Oremland, 1983*b*).

Bacterial Processes

Sediments taken from the monimolimnion were found to have methanogenic activity (Oremland, Marsh, and Des Marais, 1982). Methanogenesis in these sediments was unusual in two respects: activity was maximal at a pH of 9.7, and activity was stimulated by methanol, trimethylamine, and methionine, but not by hydrogen, acetate, or formate. Methylamines, methanol, and methionine were subsequently shown to be noncompetitive substrates of methanogenic bacteria in estuarine (high-sulfate) sediments, whereas acetate and hydrogen were substrates for which methanogens were outcompeted by sulfate-reducers (Oremland, Marsh, and Polcin, 1982; Oremland and Polcin, 1982). Thus, methanogenesis in the high-sulfate (5.6–6.5 g/l) environments of Big Soda Lake (e.g., anoxic water column and sulfate-containing monimolimnion sediments) may be restricted to metabolism of noncompetitive substrates. An enrichment culture of a small, methanogenic coccus was recovered from the monimolimnion sediments, which grew maximally on methanol at pH 9.7 with a stable carbon isotope fractionation factor of about 1.070 (Oremland, Marsh, and Des Marais, 1982).

The water column of the lake was studied with respect to nitrogen fixation, denitrification, methane oxidation, sulfate reduction, and methanogenesis.

No detectable nitrogen fixation or denitrification was found when the acetylene-reduction or the acetylene-blockage assays were applied to water samples recovered from the upper 30 m during all four seasons (R. S. Oremland, unpublished data). The absence of nitrogen fixation is perplexing because the system is severely nitrogen-limited from spring through fall, and high rates of nitrogen fixation occur in the littoral zone (see below). This may be a consequence of zooplankton grazing, which prevents a summer bloom of diazotrophic cyanobacteria, or perhaps trace-element limitation.

Experiments with ^{14}C-methane demonstrated the occurrence of anaerobic methane oxidation in the water column. Rates observed in the monimolimnion (49–85 nmol CH_4 oxidized/l/day) were about 10-fold higher than rates observed in the anoxic mixolimnion (N. Iversen, pers. comm.). Sulfate reduction was measured by using ^{35}S-sulfate. Rates in the monimolimnion were about 3000 nmol sulfate reduced/l/day, while those in the anaerobic mixolimnion were 7–600 nmol sulfate reduced/l/day (R. L. Smith, unpub. data). Preliminary results indicate that methanogenesis occurs in both the anoxic mixolimnion and the monimolimnion at rates equivalent to 0.1 to 1 nmol/l/d and 1.6 to 12 nmol/l/d, respectively. (R. S. Oremland, unpub. data).

Dense stocks of the rooted macrophyte *Ruppia* sp. grow in the lake's littoral zone from late spring through early winter. By midsummer the plant is heavily colonized by the nitrogen-fixing cyanobacterium *Anabaena* sp. The areal rate of N_2 fixation by these communities was estimated to be 103 μmol fixed/m^2/hr. These cyanobacterial aggregates also evolve hydrogen as a consequence of bacterial fermentations occurring in the dark but not in light. Areal rates of dark H_2 production were estimated to be 6.8 μmol H_2/m^2/hr (Oremland, 1983).

The *Ruppia* beds are denuded in the fall, when all above-sediment biomass senesces. This biomass contains approximately 0.1% oxalic acid by dry weight (Smith and Oremland, 1983), which presumably enters the sediments as detritus. Anaerobic degradation of oxalate occurs in both the littoral-zone and pelagic-zone (monimolimnion) sediments. However, littoral-zone sediments metabolized the compound at 50-fold higher rates, despite similar oxalate pool sizes (approximately 100 μmol/l sediment). Turnover times were 4.55 days in the littoral zone vs. 250 days for the monimolimnion (Smith and Oremland, 1983).

CONCLUSIONS

Big Soda Lake represents an alkaline, meromictic environment that is being studied with regard to its important microbial biogeochemical processes. To date, the lake has been characterized with regard to its pelagic productivity, rates of anaerobic mineralization (sulfate reduction, methane oxidation, methanogenesis), inorganic chemistry, seasonality, and littoral-zone processes. Considerably more work is needed in these areas. In addition, it is hoped that new avenues of investigation will be undertaken, such as characterizing the water column and sediments with regard to their important organic geochemical features, and isolation of and physiological studies on strictly anaerobic, alkalophilic bacteria. Studies along these lines should make important contributions to understanding the ecological role of microorganisms in highly alkaline environments.

REFERENCES

Axler, R. P., R. M. Gersberg, and L. J. Paulson, 1978, Primary productivity in meromictic Big Soda Lake, Nevada, *Great Basin Nat.* **38:**187-192.

Cloern, J. E., B. E. Cole, and R. S. Oremland, 1983*a,* Seasonal changes in the chemical and biological nature of a meromictic lake (Big Soda Lake, Nevada, U.S.A.), *Hydrobiologia* **105:**195-206.

Cloern, J. E., B. E. Cole, and R. S. Oremland, 1983*b,* Autotrophic processes in meromictic Big Soda Lake, Nevada, *Limnol. Oceanog.* **28:**104-106.

Davis, J. B., and R. M. Squires, 1954, Detection of microbially produced gaseous hydrocarbons other than methane, *Science* **119:**381-382.

Demaison, G. J., and G. T. Moore, 1980, Anoxic environments and oil source bed genesis, *Am. Assoc. Petrol. Geo. Bull.* **64:** 1179-1209.

Holm-Hansen, O., and C. Booth, 1966, The measurement of adenosine triphosphate in the ocean and its ecological significance, *Limnol. Oceanog.* **11:**510-519.

Horikoshi, K., and T. Akiba, 1982, *Alkalophilic Microorganisms: A New Microbial World,* Japan Scientific Societies Press, Tokyo: Springer-Verlag, Berlin.

Kharaka, Y. K., S. W. Robinson, L. M. Law, and W. W. Carothers, 1984, Hydrogeochemistry of Big Soda Lake, Nevada: An alkaline meromictic desert lake, *Geochim. Cosmochim. Acta* **48:**823-835.

Kimmel, B. C., R. M. Gersberg, L. J. Paulson, R. P. Axler, and C. R. Goldman, 1978, Recent changes in the meromictic status of Big Soda Lake, Nevada, *Limnol. Oceanog.* **23:**1021-1025.

Lowry, O. H., N. J. Rosebrough, A. L. Farr, and R. J. Randall, 1951, Protein measurement with the Folin phenol reagent, *J. Biol. Chem.* **193:** 265-275.

Oremland, R. S., 1981, Microbial formation of ethane in anoxic estuarine sediments, *Appl. Environ. Microbiol.* **42:**122-129.

Oremland, R. S., 1983, Hydrogen metabolism by decomposing cyanobacterial aggregates in Big Soda Lake, Nevada, *Appl. Environ. Microbiol.* **45:**1519-1525.

Oremland, R. S., and D. J. Des Marais, 1983, Distribution, abundance and carbon isotopic composition of gaseous hydrocarbons in Big Soda Lake, Nevada: An alkaline, meromictic lake, *Geochim. Cosmochim. Acta* **47:**2107-2114.

Oremland, R. S., and S. P. Polcin, 1982, Methanogenesis and sulfate reduction: Competitive and noncompetitive substrates in estuarine sediments, *Appl. Environ. Microbiol.* **44:**1270-1276.

Oremland, R. S., L. M. Marsh, and D. J. Des Marais, 1982, Methanogenesis in Big Soda Lake, Nevada: An alkaline, moderately hypersaline desert lake, *Appl. Environ. Microbiol.* **43:**462-468.

Oremland, R. S., L. M. Marsh, and S. P. Polcin, 1982, Methane production and simultaneous sulfate reduction in anoxic, salt-marsh sediments, *Nature* **296:**143-145.

Priscu, J. C., R. P. Axler, R. G. Carlton, J. E. Reuter, P. A. Arneson, and C. R. Goldman, 1982, Vertical profiles of primary productivity, biomass and physico-chemical properties in meromictic Big Soda Lake, Nevada, U.S.A., *Hydrobiologia* **96:**113-120.

Rush, E. F., 1972, Hydrologic reconnaissance of Big and Little Soda Lakes, Churchill County, Nevada, *Water Res. Info. Rep.* **11,** U.S. Geological Survey and Nevada Dept. of Conservation and National Resources, Carson City.

Smith, R. L., and M. J. Klug, 1981, Reduction of sulfur compounds in the sediments of a eutrophic lake basin, *Appl. Environ. Microbiol.* **41:**1230-1237.

Smith, R. L., and R. S. Oremland, 1983, Anaerobic oxalate degradation: Widespread natural occurrence in aquatic sediments, *Appl. Environ. Microbiol.* **46:**106-113.

Vogel, T. M., R. S. Oremland, and K. A. Kvenvolden, 1982, Low-temperature formation of hydrocarbon gases in San Francisco Bay sediment, *Chem. Geol.* **37:**289-298.

8

The Global Biosphere as Net CO_2 Source or Sink: Evidence from Carbon Isotopes in Tree Rings

Steven W. Leavitt and Austin Long

University of Arizona

INTRODUCTION

The well-documented growth in fossil-fuel usage since the Industrial Revolution (Rotty, 1981) contributes to increasing carbon dioxide levels in the atmosphere, causing concern over the global environmental consequences of a resulting "greenhouse effect." There is also evidence, however, that land-use changes may result in a substantial contribution of CO_2 to the atmosphere from forests and soils (Woodwell et al., 1978). There is disagreement as to the actual size of these biospheric inputs, and whether marine and other carbon sinks are sufficient to absorb this proposed large excess biospheric CO_2 input (Broecker et al., 1979; Keeling, 1983).

A possible method of resolving the controversy and determining the net historical activity of the biosphere (here to include soils) lies in the carbon-isotopic composition of tree rings. The tree rings record changes in the $^{13}C/^{12}C$ ratio of atmospheric CO_2 (offset by enzymatic fractionation) by fixing this carbon and assimilating it into the wood of annual growth rings. Because both the biosphere and fossil fuels are enriched, their additions to the atmosphere decrease the atmospheric $^{13}C/^{12}C$ ratio. A rough measure of the portion of this atmospheric $^{13}C/^{12}C$ change due only to fossil-fuel input may be estimated from known parallel changes in the $^{14}C/^{12}C$ ratio in the atmosphere as it is diluted by the "dead" carbon of fossil fuels (Stuiver, 1978). When this fossil-fuel contribution is subtracted from a tree-ring $\delta^{13}C$ chronology, the remaining trend should reflect the history of the biosphere as CO_2 source or sink.

Unfortunately, different approaches to $^{13}C/^{12}C$ analysis of tree rings have yielded contradictory reconstructions. Contributing to some of this divergence are site selection, the wood component chosen for analysis, environmental influences on fractionation, and natural intra-individual and intrasite isotopic variability. A recent study by Leavitt and Long (1983a) was aimed particularly at eliminating both climate effects on isotopic fractionation and the radial isotopic variations within individuals, as contained in a 50-year juniper (*Juniperus* spp.) tree-ring record from Arizona, U.S.A. The research reported herein examines a much longer set of $^{13}C/^{12}C$ measurements from pinyon pine trees growing in the American Southwest.

MATERIALS AND METHODS

Pinyon pine trees (*Pinus edulis* Engelm.) were harvested from nine sites in Arizona and New Mexico with a complete cross section from near each base collected for analysis. Samples were obtained where dendrochronological investigations had previously been carried out so that the specimens could be easily dated. The sites were far removed from urban centers, major roads, and point pollution sources such as smelters and power plants. Further, the trees were taken from open stands as opposed to closed-canopy forests. The trees were sampled at the end of the 1981 growing season, and all exceeded 180 years in age (Table 8-1). Staff of the Laboratory of Tree-Ring Research of the University of Arizona dendrochronologically dated the surfaced cross sections. Five-year groups of tree rings were split from the full circumference in order to avoid effects of radial variation (Tans and Mook, 1980).

Following the grinding of the wood to 20 mesh, a method modified after Green (1963) was used to isolate the relatively stable cellulose component. Leaching of the wood first with 2:1 toluene-ethanol and then with 100% ethanol in a Soxhlet extraction apparatus removed oils and resins. Treatment with hot (70°C), acidified sodium chlorite followed to decompose lignin, and a residual cellulose (holocellulose) resulted.

The cellulose samples were burned to CO_2 in a recirculating microcombustion line at 800°C. Cold ethanol traps (−80°C) removed the water combustion product, and liquid nitrogen traps collected the purified CO_2 for analysis on a VG Micromass 602C mass spectrometer. The isotopic compositon was calculated after the procedures of Craig (1957) and expressed as δ values in per mil units with respect to the PDB calcite standard:

$$\delta^{13}C/\text{mil} = \left[\frac{(^{13}C/^{12}C) \text{ sample}}{(^{13}C/^{12}C) \text{ PDB}} - 1 \right] \times 1000$$

Table 8-1. Pinyon Pine Sampling Sites and Tree-Ring Dating

Site	Date Sampled	Location	Elevation (m)	Dated Rings*	Percent Missing Rings
Red Mountain	22 Nov 81	35° 01′ 10″ N 112° 49′ 10″ W	1635	1788–1981	2½
Walnut Canyon	23 Nov 81	35° 10′ 50″ N 111° 29′ 15″ W	2005	1795–1981	1½
Stoneman	23 Nov 81	35° 45′ 50″ N 111° 38′ 40″ W	1660	1795–1981	1½
Hackberry	23 Nov 81	34° 27′ 10″ N 111° 41′ 50″ W	1195	1720–1981 (35)	8½
Nutrioso	5 Jan 82	34° 02′ 10″ N 109° 09′ 55″ W	2395	1690–1981	7½
Hay Hollow	6 Jan 82	34° 30′ 50″ N 109° 58′ 15″ W	1785	1800–1981 (25)	7½
Defiance	7 Jan 82	35° 43′ 15″ N 109° 21′ 45″ W	2180	1660–1981	9
Reserve	8 Jan 82	33° 36′ 35″ N 108° 45′ 00″ W	2020	1805–1981	8
Dry Creek	31 Mar 82	34° 53′ 45″ N 111° 49′ 30″ W	1380	1780–1981 (15)	2

*If the section was not datable to the center, the approximate number of rings prior to the first dated ring is given in parentheses.

The standard deviation for repeated combustions and analysis of a laboratory cellulose standard is ±0.05/mil. Random replicate analysis of different pinyon samples ($n = 34$) yielded a mean absolute difference of 0.08 ± 0.07/mil. Separate chemical extraction, combustion, and analysis of two splits of the ground wood from 22 different samples gave a mean absolute difference of 0.10 ± 0.08/mil.

Because local meteorological measurements are typically limited to 50-100 years, ring-width measurements were used as proxy climate indicators (Fritts, 1976). The ring widths of each cross section were measured at the Laboratory of Tree-Ring Research, and the ring areas were calculated. From the raw ring-width plots, each tree's standardized tree-ring indices were obtained by removing the growth trend. The removal of the growth trend also served to reduce the sensitivity of the indices to long-term ($\simeq$ the tree's lifetime) climatic shifts. From previous studies each locality already had site-standardized tree-ring indices (STRI), which are a linear combination of 10-20 individual STRIs.

RESULTS

The $\delta^{13}C$ curves for all sites are plotted in Figure 8-1a. Although there are different overall trends among them, they show some remarkably consistent short-term fluctuations. These include $\delta^{13}C$ maxima at 1820-1824, 1860-1864,

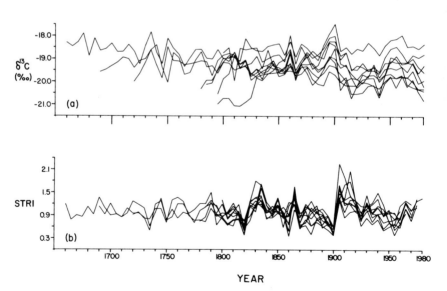

Figure 8-1.   Comparison of (a) the $\delta^{13}C$ trends of each of the pinyon trees with (b) the standardized tree-ring indices (STRI) for each of the sites.

1900-1904, 1910-1914, and 1970-1974, and minima at 1865-1869, 1905-1909, and 1940-1944. The environmental influence causing these similar trends must be at least regional in scope because some of these sites are several hundred kilometers apart. Figure 8-1b contains the five-year mean STRI of each site, and comparison with Figure 8-1a indicates at least a possible climatic influence on the $\delta^{13}C$ fluctuations.

Figure 8-2a exhibits the mean $\delta^{13}C$ curve for all sites. The first 40 years of each tree were first deleted to avoid "juvenile" effects, perhaps related to greater exposure of small trees to respired CO_2 from larger neighbors (Craig, 1954; Freyer, 1979). Vertical bars represent the ±1s scatter among trees at each interval. The numerals above the bars indicate the number of sites in the mean. The similar curve in Figure 8-2b was generated by first normalizing each $\delta^{13}C$ curve as deviations from its respective 1845-1980 mean $\delta^{13}C$. This allows for a more direct comparison of relative $\delta^{13}C$ trends among sites without regard to their different absolute $\delta^{13}C$ values. Figure 8-2b is then the mean of these normalized $\delta^{13}C$ curves. In most cases, the error bars in this mean curve are smaller than those in the Figure 8-2a absolute mean curve. The smoothing curves in Figures 8-2a and 8-2b are five-interval running means.

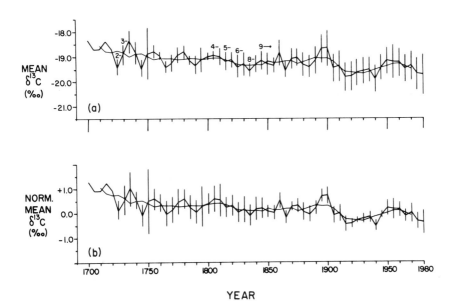

Figure 8-2. The mean $\delta^{13}C$ curves representing (a) the direct mean of curves in Figure 8-1a after dropping the first 40 years from each, and (b) a mean taken after each curve is first normalized as deviations from its respective 1845–1980 mean. Vertical bars are ±1s and numerals above the bars show the number of values in the mean. A five-interval running mean is plotted in both figures.

Table 8-2. Regression Coefficients and Statistical Significance for δ^{13}C with Environmental Parameters (See Text for Parameter Definitions)

Site	δ vs. STRI	δ vs. RRW	δ vs. AREA	δ vs. ISTRI	δ vs. STRI and RRW	
					STRI	RRW
Defiance	-0.41	1.45***	-4.19×10^{-4}	0.042	-0.89	2.08***
Nutrioso	-0.57*	-0.00	-8.39×10^{-5}	-0.0046	-0.69	0.36
Hackberry	-0.94***	0.05	-2.42×10^{-4}***	-0.69***	-1.03	0.25***
Red Mountain	-0.66**	-0.00	-9.89×10^{-5}	0.15	-0.90	0.37***
Stoneman	-1.07**	-0.50	-6.08×10^{-4}**	0.041	-1.11	-0.01**
Reserve	-0.34	-0.13	-9.93×10^{-5}	-0.054	-0.29	-0.09
Dry Creek	-0.20	0.37*	-2.28×10^{-4}	-0.15	-0.73	0.80***
Hay Hollow	-0.43	0.30	-2.20×10^{-4}	-0.0027	-1.05	1.05***
Walnut Canyon	-0.30	0.46*	-2.58×10^{-4}	-0.12	-0.93	0.80**

*P < .05.
**P < .01.
***P < .001 (F-test significance).

94

Environmental influences in these curves were tested (Table 8-2) by regressing the δ values of each tree with their respective site STRI, individual raw ring-width measurements (RRW), individual standardized ring-width measurements (ISTRI), and ring areas (AREA). In the broadest of terms, STRI could be considered a measure of local climate variations, with long-term trends removed (Fritts, 1976); RRW contains a growth curve superimposed on the individual's response to local climate and other environmental influences; ISTRI is RRW without the growth curve; and ring areas may be a measure of net carbon assimilation by the tree (Stuiver, 1982). Regarding ring areas, the plant carbon-fractionation model of Francey and Farquhar (1982) had indicated the $\delta^{13}C$ of a plant to be dependent on the rate of CO_2 assimilation, and Stuiver determined that for trees under certain conditions the ring areas may be a proxy indicator of net assimilation.

As expected from Figure 8-1, δ vs. STRI coefficients all have negative signs, as do all δ vs. AREA and most of the δ vs. ISTRI regressions. The δ vs. RRW coefficients have mixed signs, but the significant coefficients have positive signs. For δ vs. STRI, all sites except Reserve, Dry Creek, and Walnut Canyon are significant at 95% or close to it. Of these three, only the Reserve $\delta^{13}C$ curve appears unrelated to any of these measures. Multiple linear regression of $\delta^{13}C$ with more than one variable may yield substantially improved relationships, as exemplified in the δ vs. STRI, RRW coefficients in Table 8-2. Because STRI and RRW are not independent variables for most of the trees, however, these multiple regressions may not give the true specific response of $\delta^{13}C$ to each variable.

With these sets of regressions, it is possible to remove environmental effects represented by these variables. As an example, the $\delta^{13}C$ curve in Figure 8-3a is the result of using the δ vs. STRI, RRW coefficients from Table 8-2 to correct each curve to a value of STRI = 1 and a RRW value equal to its respective mean RRW. Figure 8-3a is the mean curve after first normalizing the individual climate-corrected curves to their respective 1845-1960 means (all sites had STRI values only through 1960-1964), and it may be compared directly to Figure 8-2b. The major effects are a flattening of the overall curve and reduction in magnitude of some maxima (e.g., 1900-1904) and minima (e.g., 1865-1869).

The curve in Figure 8-3a has actually been adjusted to the atmospheric $\delta^{13}C$ value of -7.34/mil as measured for 1978 (Keeling, Mook, and Tans, 1979). The fossil-fuel $\delta^{13}C$ contribution to the atmosphere is approximated in Figure 8-3b after the algorithm of Stuiver (1978) and the industrial source $\delta^{13}C$ values of Tans (1981). If one assumes Figure 8-3a is the actual atmospheric curve, this fossil-fuel contribution may be subtracted from the running mean curve with the resulting biospheric curve shown in Figure 8-3c. Where the curve rises, as after 1930, the biosphere would be considered a net sink.

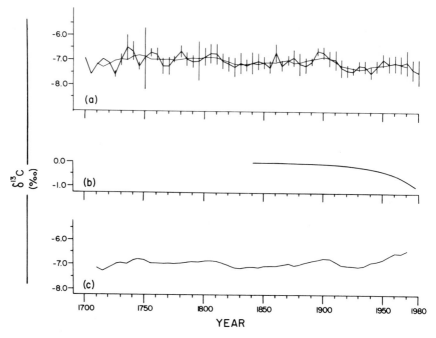

Figure 8-3.   Removal of fossil-fuel effects from a $\delta^{13}C$ curve. (*a*) A mean normalized $\delta^{13}C$ curve (and running mean) after each site has first been "corrected" with the δ vs. STRI, RRW coefficients from Table 8-2; the 1978 value has been set to an atmospheric value of $-7.34°$/mil (Keeling, Mook, and Tans, 1979). (*b*) The fossil-fuel $\delta^{13}C$ effect on the atmosphere after Stuiver (1978) and Tans (1981). (*c*) The "biospheric" curve resulting when (*b*) is subtracted from the running mean curve in (*a*).

DISCUSSION

The correspondence of $\delta^{13}C$ maxima and minima and maxima of the STRI curves in Figure 8-1 suggests an influence of climate on $\delta^{13}C$ of pinyon pine. In the American Southwest, narrow rings (small STRI) generally represent warm or dry years. An inverse relationship between $\delta^{13}C$ and STRI was also seen by Mazany, Lerman, and Long (1980) in rings of ancient white fir (*Abies concolor*) and ponderosa pine (*Pinus ponderosa*) from New Mexico. This δ-STRI inverse relationship could imply a positive relationship between $\delta^{13}C$ and temperature. In fact, for the leaf cellulose from these trees there is evidence of a positive temperature coefficient (Leavitt and Long, 1983*b*). For juniper tree rings in the study of Leavitt and Long (1983*a*), however, $\delta^{13}C$ was inversely related to temperature. Such contradictory results have also been observed in various *in vivo* and *in vitro* experiments (Long, 1982).

There was no single proxy environmental parameter that was strongly correlated to $\delta^{13}C$ in all sites. There were more statistically significant relationships for δ with STRI than with any of the other parameters. Because the other three parameters represent measurements of the individual's growth response, it appears that gross factors affecting the whole site (e.g., temperature and rainfall) generally seem to influence $\delta^{13}C$ of the individual more than do other specific individual influences (e.g., competition and heredity). Exceptions include the Dry Creek and Walnut Canyon trees, which show a significant relationship of $\delta^{13}C$ with individual ring thickness. The fact that these two relationships are positive suggests a correlation between the decreasing ring thickness of the natural growth curve with generally decreasing $\delta^{13}C$ values. This is supported by no significant correlations between δ with ISTRI (raw thickness with growth curve removed) for these two sites. On the other hand, Reserve $\delta^{13}C$ values seem to be independent of any of these parameters. Additional samples from the Reserve site are being obtained in an attempt to better understand this anomaly.

The mean $\delta^{13}C$ curves in Figure 8-2 show a generally decreasing trend. The drop from pre-Industrial (1800-1850) to present is only about one-half per mil, considerably less than the 1.5-2/mil drop found by Freyer (1979) for European trees or the 1.3/mil drop found in Arizona juniper trees for 1930-1979 alone (Leavitt and Long, 1983a). The pinyon curve, however, does show a $\delta^{13}C$ increase from about 1930 to 1960, approximately coincident with one observed by Freyer (1979), which he suggested may be related to hemispheric cooling at that time.

Another feature of the mean $\delta^{13}C$ curve is that the 1955-1980 period actually shows a drop of about 0.5/mil. Given the error bars of ~0.3/mil for the 1955-1980 points in Figure 8-2b, this drop is consistent with the measured atmospheric $\delta^{13}C$ shift from 1956 to 1978 of -0.65 ± 0.13/mil (Keeling, Mook, and Tans, 1979). Whereas the pinyon post-1955 trend is continuously downward, that measured in Arizona juniper rings was downward with a flattening after 1965.

When the $\delta^{13}C$ trends were corrected for environmental effects as in Figure 8-3a, the mean $\delta^{13}C$ curves were smoothed, and the drop from pre-Industrial to present was further reduced. This resembles the flat $\delta^{13}C$ trends observed in Tasmanian tree rings (Francey, 1981). If these then represent atmospheric $\delta^{13}C$ curves, and the fossil-fuel contribution is removed, a generally flat curve with a post-1930 upward trend results. This suggests a previously neutral biosphere that has become a net carbon sink over the last 50 years.

Given the magnitude of error bars in Figures 8-2 and 8-3, however, there is still room for alternative interpretation. Other methods of removing environmental influences from these $\delta^{13}C$ trends, such as regression coefficients derived from first differences, are being tested. Error may be further reduced if the natural variability of $\delta^{13}C$ in and among trees is known and considered in the sampling strategy.

CONCLUSIONS

The $\delta^{13}C$ trends in tree rings of sites widely scattered about the American Southwest show remarkably similar short-term features. On the basis of strong inverse $\delta^{13}C$ with site STRI serial correlations, good growth years (cool/wet) favor more negative $\delta^{13}C$ values and poor years (warm/dry) favor less negative values. If this is a temperature effect, then it may serve as the basis for a carbon-isotope tree-ring paleothermometer.

If a record of atmospheric $\delta^{13}C$ changes is contained in these tree rings, it is probably modified by local environmental conditions. Regression coefficients used to correct the individual $\delta^{13}C$ trends for these climate effects tend to flatten out the mean $\delta^{13}C$ trend. Because fossil fuels contibute to more negative atmospheric $\delta^{13}C$ values, the flat atmospheric $\delta^{13}C$ trend derived here would imply the biosphere has had to act as a net sink to counteract the fossil-fuel effect. However, the current thinking seems to be that the biosphere is either a net CO_2 source or is approximately in balance (Keeling, 1983). Sufficient errors remain in these $\delta^{13}C$ reconstructions as to allow these other interpretations, but results seem to be quite distinct from the very large $\delta^{13}C$ drop measured in European trees (Freyer, 1979). In addition to reducing scatter by accounting for climatic influences, scatter may be further reduced in the sampling process if the natural variability is known.

ACKNOWLEDGEMENTS

This research was supported under the U.S. Department of Energy Carbon Cycle Research Program through Oak Ridge-Union Carbide Subcontract No. 19X-22290C. We thank C. Sullivan and M. Leavitt for assistance in field sampling and L. Warneke for chemical preparation and analysis. J. Dean and D. Bowden from the Laboratory of Tree-Ring Research dated the sections, and E. Sutherland measured widths, separated rings, and standardized raw data.

REFERENCES

Broecker, W. S., T. Takahashi, H. J. Simpson, and T.-H. Peng, 1979, Fate of fossil-fuel carbon dioxide and the global carbon budget, *Science* **206:**409-418.

Craig, H., 1954, Carbon-13 variations in sequoia rings and the atmosphere, *Science* **119:**141-143.

Craig, H., 1957, Isotopic standards for carbon and oxygen and correction factors for mass-spectrometric analysis of CO_2, *Geochim. Cosmochim. Acta* **12:**133-149.

Francey, R. J., 1981, Tasmaian tree rings belie suggested anthropogenic $^{13}C/^{12}C$ trends, *Nature* **290:**232-235.

Francey, R. J., and G. D. Farquhar, 1982, An explanation of $^{13}C/^{12}C$ variations in tree rings, *Nature* **297:**28-31.

Freyer, H. D., 1979, On the ^{13}C record in tree rings. Part I. [13]C variations in northern hemispheric trees during the last 150 years, *Tellus* **31**:308-312.

Fritts, H. C., 1976, *Tree Rings and Climate,* Academic Press, New York.

Green, J. W., 1963, Wood cellulose, in *Methods of Carbohydrate Chemistry, Vol. III,* R. L. Whistler, ed., Academic Press, New York, pp. 9-21.

Keeling, C. D., 1983, The global carbon cycle: What we know and could know from atmospheric, biospheric, and oceanic observations, in *Proceedings: Carbon Dioxide Research Conference: Carbon Dioxide, Science and Concensus,* Sept. 19-23, 1982, Berkeley Springs, W. Virginia. U.S. Department of Energy: CONF-820970, pp. II.3-II.62.

Keeling, C. D., W. G. Mook, and P. P. Tans, 1979, Recent trends in the ^{13}C/^{12}C ratio of atmospheric carbon dioxide, *Nature* **277**:121-123.

Leavitt, S. W., and A. Long, 1983*a,* An atmospheric ^{13}C/^{12}C reconstruction generated through removal of climate effects from tree-ring ^{13}C/^{12}C measurements, *Tellus* **35B**:92-102.

Leavitt, S. W., and A. Long, 1983*b,* Possible climatic response of δ^{13}C leaf cellulose of pinyon pine in Arizona, U.S.A., *Isot. Geos.* **1**:169-180.

Long, A., 1982, Stable isotopes in tree rings, in *Climate from Tree Rings,* M. K. Hughes, P. M. Kelly, J. R. Pilcher, and V. C. LaMarche, Jr., eds., Cambridge University Press, Cambridge, pp. 13-18.

Mazany, T., J. C. Lerman, and A. Long, 1980, Carbon-13 in tree-ring cellulose as an indicator of past climates, *Nature* **287**:432-435.

Rotty, R. M., 1981, Data for global CO$_2$ production from fossil fuels and cement, in *Carbon Cycle Modelling,* B. Bolin, ed., Wiley and Sons, New York, pp. 121-125.

Stuiver, M., 1978, Atmospheric carbon dioxide and carbon reservoir changes, *Science* **199**:253-258.

Stuiver, M., 1982, The history of the atmosphere as recorded by carbon isotopes, in *Atmospheric Chemistry,* E. D. Goldberg, ed., Springer-Verlag, Berlin, pp. 159-179.

Tans, P. P., 1981, ^{13}C/^{12}C of industrial CO$_2$, in *Carbon Cycle Modelling,* B. Bolin, ed., Wiley and Sons, New York, pp. 127-129.

Tans, P. P., and W. G. Mook, 1980, past atmospheric CO$_2$ levels and the ^{13}C/^{12}C ratios in tree rings, *Tellus* **32**:268-283.

Woodwell, G. M., R. H. Whittaker, W. A. Reiners, G. E. Likens, C. C. Delwiche, and D. B. Botkin, 1978, The biota and the world carbon budget, *Science* **199**:141-146.

9

Methanogenesis in the Sediments of Inland Waters: Kinetics of Methane Production from Methanol

Monir Naguib

Max-Planck-Institut für Limnologie

ABSTRACT: The kinetic parameters estimated for methane production from methanol with and without peptone in the sediments of Lake Plussee showed a common K_m of 17 mM methanol end-concentration. The V_{max} values were different in fresh and aged sediments and were of the order 0.53 and 0.26 μmol/d/g wet wt, respectively. In the presence of peptone, the V_{max} increased in both types of sediments up to 0.83 μmol. Peptone not only caused the increase of V_{max} but also caused an increase in the total methane production and a shortening of the lag phase, especially with aged sediments. The type of kinetic curves obtained, $1/S$ vs. $1/V$, would suggest that the methanogenesis in the sediments might be subjected to some kind of noncompetitive inhibition that would be reduced or eliminated by the peptone effect.

An attempt was made to evaluate the possible contribution of methane produced from methanol to the total methane produced in the lake. Advantage was taken of the fact that Lake Plussee has been thoroughly studied and many parameters well estimated. Allowing for certain assumptions and considering some prevailing ecological factors of Lake Plussee, it was estimated that methane from the methanol conversion could contribute as much as 50% of the total methane produced in the lake.

INTRODUCTION

Anaerobic sediments are no doubt the site of intensive metabolic couplings subjected variously to competition and inhibition, with methane and CO_2 as the final end products if the electron flow is not directed to some noncarbon electron acceptors. These interactions are described in general reviews on methanogenesis (Zeikus, 1977; Mah et al., 1977; Jain, 1980; Smith, Zinder, and Mah, 1980; Sahm, 1981) and in specific reviews (MacGregor and Keeney, 1973; Zhilina and Zavarzin, 1973; Cappenberg, 1975; Winfrey et al., 1977; Strayer and Tiedje, 1978; Abram and Nedwell, 1978; Van Kessel, 1978; Kelly and Chynoweth, 1981; Lovley, Dwyer, and Klug, 1982; Oremland, Marsh, and Polcin, 1982; Oremland and Polcin, 1982; Lovley and Klug, 1983; Lynd and Zeikus, 1983). In most of these reports, acetate and CO_2/H_2 are examined as the major methane precursors. Methanol, on the other hand, is either not considered or found unimportant as a possible significant methane precursor. In only a few reports was methanol explicitly described as an important and probably most dominant methane precursor in sediments of the eutrophic Lake Plussee (Naguib, 1982a) and in salt-marsh sediments from San Francisco Bay (Oremland, Marsh, and Polcin, 1982). Contrary to these findings, Lovley and Klug (1983) found that methanol does not account for more than 5% of the produced methane in the sediments of the eutrophic Wintergreen Lake in Michigan.

According to the present data, it seems that *Methanosarcina* are among the dominant methanogens that would convert methanol to methane. Apparently the mode of attack on the methanol molecule is different among the different species. *M. barkeri,* isolated in pure culture and well described, is a methyloclastic methanogen acting on both methanol and acetate. *M. methanica,* presumably never isolated, is supposed to use methanol and acetate as H-donors (Mah et al., 1977, 1981; Schlegel, 1969). *Methanococcus mazei,* known as an acetoclastic methanogen, is shown to be able to grow on, and produce methane from, methanol (Oremland, Marsh, and Polcin, 1982).

Previous work with the sediments of the eutrophic Lake Plussee (Naguib, 1982a) revealed that methanol stimulated the methanogenetic activities remarkably more than acetate and CO_2/H_2. Moreover, the influence of inorganic nitrogen, in the form of ammonium ions, was found to cause a partial inhibition with methanol, acetate, and CO_2/H_2. The addition of organic nitrogen in the form of peptone, on the other hand, increased the methane production from methanol more than from acetate ad CO_2/H_2.

Inorganic nitrogen in the form of nitrate was found to cause a reversible inhibition on methanogenesis (Laskowski and Moraghan, 1967; Bell, 1969; Bollag and Czlonkowski, 1973; MacGregor and Keeney, 1973). Kusnezow (1959) found that a deficiency of readily assimilable proteinaceous material might cause limitation in the degradability of organic deposits in lake sediments. Strayer and Tiedje (1978) could not find any stimulating influence on sediment methanogenesis when H_2-donating amino acids such as valine and leucine were added to the sediments. Either these amino acids are not important H_2-donors, or they are already being turned over at maximum rates. Nagase and Matsuo (1982) examined the anaerobic degradation of amino acids in terms of the interactions between amino acid-degrading bacteria and methanogenic bacteria. They found that certain amino acids, such as alanine, valine, and leucine, are degraded oxidatively with methanogenic bacteria acting as H_2-acceptors. However, a coupled oxidation-reduction reaction between two amino acids, one acting as H_2-donor and the other acting as H_2-acceptor, as in the Stickland reaction, might occur, thus by-passing the methanogenic bacteria as H_2-acceptors. Hoogerheide and Kocholaty (1938) reported that glycine, proline, and other amino acids would undergo reductive deamination and would then compete with methanogenic bacteria for the H_2 as a reductant.

This paper reports on the methanogenic activities of fresh and aged sediments of the eutrophic Lake Plussee under the influence of added methanol in different concentrations in the presence and absence of peptone. The kinetic parameters K_m and V_{max} were simultaneously estimated for the different sediment conditions. Moreover, in order to have more insight into the importance of methanol as a methane precursor, an attempt was made to evaluate the contribution of the methane converted from methanol to the whole methane budget of the investigated lake.

MATERIALS AND METHODS

Sediments

The present investigation studied surface profundal sediments collected from the middle of the eutrophic Lake Plussee (29 m) in Ploen, Federal Republic of Germany. Sediments collected on November 23, 1982, were thoroughly mixed and divided into two portions. The first portion (designated as fresh sediments) was directly treated, and the other portion (designated as aged sediments) was kept incubated under He atmosphere for about three weeks to induce a deficiency of assimilable carbon. The aging period was determined by the termination of endogenous methane production. All sediment manipulations were performed in a constructed gas hood under He atmosphere.

Incubation

The technique used is principally the same as that previously described in detail (Naguib, 1982a). Incubations were carried out in series of 106-ml standard brown bottles sealed with a silicon injection stopper. The bottles were further sealed with a screw cap fitted with an extra silicon gasket. Twenty-gram sediment samples were placed in the bottles with a specially prepared 50-ml syringe. Methanol and peptone wer injected where needed, flushed additionally with He, sealed with the double seal as described, and incubated at 27°C. The methanol was injected in a series of end concentrations of 5, 7.5, 12.5, 25, 100 and 200mM. The peptone injected was 1 ml of a 1% solution per incubation. Incubations were run in triplicate, so that for one investigation of the above order 50 bottles would be needed. This necessitated at least 150 gas analyses per investigation.

Gas Chromatography

Methane was estimated by direct injections of 100 μl from the gas head space of each bottle into a Becker Multigraph Type 409 analyzer provided with an FID detector. Carrier gas was pure N_2 at a flow rate of 20 ml/min. The oven and detector temperatures were 120°C. A Porapak N (80/100 mesh) column was used. Integrations were performed with a standardized and computerized Shimadzu Chromatopac C-R2A. Methane analyses were carried out at least each second day.

Kinetic Analyses

The data obtained from the slopes of the time course of methane production for each methanol concentration, with and without peptone, were computed to obtain the correlations S vs. V and $1/S$ vs. $1/V$. The progress curves and the straights obtained indicated the validity of the application of Michaelis-Menten kinetics. The same was also established for methane precursors (e.g., acetate and H_2) in the sediments of other lakes (Strayer and Tiedje, 1978; Lovley, Dwyer, and Klug, 1982).

For estimation of the K_m and V_{max} values, the Lineweaver-Burk reciprocal plots, $1/S$ vs. $1/V$, were adopted. Although this representation is criticized for not being precise, a clear trend was obtained with allowance for appropriate interpretations.

RESULTS

The total methane produced during incubation of fresh and aged sediments with different methanol concentrations, with and without peptone, is illustrated in Figure 9-1. The percent conversion of methanol is given in each case at the top of the corresponding column. The time factor is not included in this representation. These results indicate that:

1. The methanogenic activities in fresh sediments with methanol alone were remarkably higher than in aged sediments. The methane production in fresh sediments had an average of 10.5 μmol CH_4/d/g wet wt, corresponding to a 17.3% conversion efficiency, compared to 2.2 μmol and an average conversion of 6.8% in aged sediments.
2. The addition of peptone consistently stimulated an increased methane

production, both in fresh and aged sediments. With fresh sediments the addition of peptone seemed to stabilize the methanol conversion efficiency to about 36% at different methanol concentrations. With aged sediments this phenomenon was variable, ranging from 4% to 70% methanol conversion.

3. The total methane produced in fresh sediments increased consistently from 10% to 31% conversion efficiency as the methanol concentration increased from 5 to 100 mM. No toxic effects were noted at the 100 mM methanol end concentration. In aged sediments the maximum methanol conversion efficiency was 16% with methanol concentrations of 25 mM. Increasing the methanol caused a decrease in conversion efficiency to a minimum of 0.4% at methanol end concentrations of 200 mM.

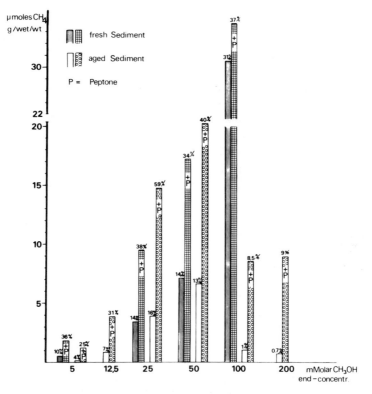

Figure 9-1. The cumulative methane production in incubations with fresh and aged sediments under the influence of different methanol concentrations with and without peptone. The percent conversion of methanol is given at the top of each column. Notice the interrupted scale on the ordinate.

The time course of methane formation during these incubations is shown in Figure 9-2 and Figure 9-3 for fresh and aged sediments, respectively. The effect of the peptone is not only to increase the methane production, but also to shorten the lag phase before methanogenesis sets in.

The common K_m value of about 17 mM for the fresh and aged sediments both with and without peptone is remarkable. The differences lie in the V_{max} values. Whereas the methane production, in the presence of the corresponding methanol concentration, would proceed in fresh sediments with a V_{max} of 0.53 μmol/d/g, in aged sediments this was reduced to about half this value, or 0.26 μmol. The presence of peptone, however, seemed to bring both aged

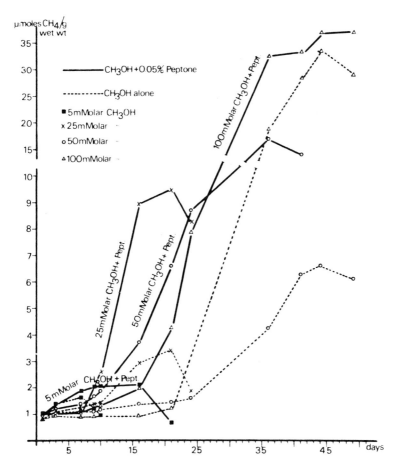

Figure 9-2. The time-course production of methane in incubations of fresh sediments with different methanol concentrations in the absence and presence of peptone. Notice the interrupted scale on the ordinate.

and fresh sediments to an increased V_{max} of 0.82 μmol. Reciprocal plots with a common K_m and different V_{max} suggest a noncompetitive type of inhibition or limitation. This possible inhibition will be discussed later on.

DISCUSSION

Methanol stimulated the methane production in the sediments of the eutrophic Lake Plussee; stimulation was more intense in the presence of peptone. The endogenous methane production in the sediments was very low and was exhausted within the first three days. From the behavior of aged sediments it can be concluded that if, for any reason, the continuous supply of methane precursors were interrupted, the sediment methanogenesis

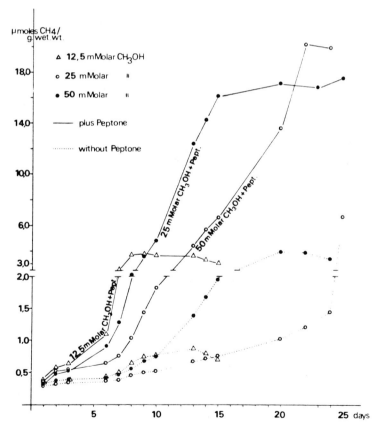

Figure 9-3. The time-course production of methane incubations of aged sediments with different methanol concentrations in the absence and presence of peptone. Notice the interrupted scale on the ordinate.

would react variably with a prolonged lag phase before active methane production set in (Fig. 9-3).

The peptone effect not only shortens the lag phase and increases the total methane production, but it also increases the V_{max}; the K_m remained constant in all examined cases. Peptone seems to neutralize some kind of noncompetitive inhibitor(s) to which aged sediments are more subject than are fresh sediments. This effect is suggested by the kinetic reciprocal curves of fresh and aged sediments, which showed a common K_m and a different V_{max}, indicating a noncompetitive type of inhibition being eliminated or reduced by the addition of peptone (Fig. 9-4). Aquatic microorganisms might, however, react to changes in the environmental conditions with a change in K_m (Witzel, 1980). Unlike the case of methanol, a competitive type of inhibition would be expected for acetate, since acetate is a main substrate for many metabolic routes in methanogens and nonmethanogens. Which amino acid or group of amino acids would induce such effects cannot be answered with certainty. Some methanogens seem to have a demand for organic nitrogen (Bryant et al., 1971). The oxidative degradation of some amino acids by dehydrogenation is a possibility for participation as H_2-donors in a Stickland-type reaction with methanogens as H_2-acceptors (Nagase and

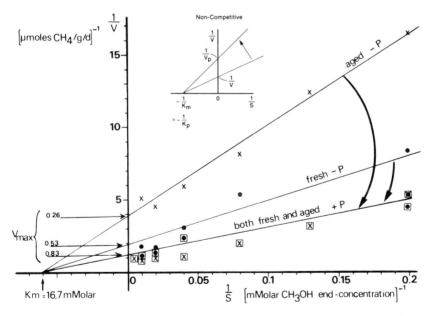

Figure 9-4. The Lineweaver-Burk reciprocal plots of the methane produced from different methanol concentrations in the absence and presence of peptone ($-P$ and $+P$). The straights obtained indicate a type of noncompetitive inhibition as shown at the top of the illustration.

Matsuo, 1982). However, the inability of the potential H_2-donors, valine and leucine, to stimulate sediment methanogenesis (Strayer and Tiedje, 1978) would not support this mechanism. Also, peptone alone did not markedly increase the methane production from examined sediments. Moreover, a common K_m would suggest that the methanogenic population, most probably belonging to the genus *Methanosarcina,* established under the different conditions should be almost identical. However, the possibility cannot be ruled out that methanol/peptone might have induced some side reactions causing methane production from other sediment ingredients, which did not show in the controls.

One of the major questions is how much the methane converted from methanol would contribute to total methane production in the lake. Recent investigations indicate contrary trends. Lovley and Klug (1983) emphasize that acetate and hydrogen are the major methane precursors in the sediments of the eutrophic Wintergreen Lake in Michigan and that methanol does not contribute more than 5% of the total methane produced. On the other hand, Oremland, Marsh, and Polcin (1982) and Naguib (1982*a*) indicate that methanol could be a dominant methane precursor, probably contributing to the bulk of the methane produced in two investigated ecosystems, a salt-marsh lake and a freshwater eutrophic lake.

Evaluation of the possible contribution of the methane obtained from methanol conversion to the total methane production in the sediments of Lake Plussee is attempted here. Lake Plussee has been thoroughly studied (Overbeck, 1979, 1981) and the different compartments of the carbon cycle have been quantified as annual budgets (Naguib, unpublished work). Of main concern here is the total methane production, which is estimated to be 6.15×10^3 kg C. Plussee is a forest lake with an autochthonous primary production of about 25×10^3 kg C/yr (Overbeck, 1981), and an allochthonous input on the order of 5000 kg C/yr (Naguib, 1982*b*) partially originating from the fall leaves from the surrounding forest. Methanol is now known to be an end product of aerobic and anaerobic pectin degradation and can reach concentrations on the order of 16 mM by some examined methanogenic species (Schink and Zeikus, 1980). Donelly and Dagley (1980) consider lignin, the most significant methoxylated polymer in nature, to be a potential source of methanol. Moreover, the peptone effect seems to be operable since sediments of eutrophic lakes seem to be rich in proteinaceous material, as indicated in the findings of Molongoski and Klug (1976) that proteolytic *Clostridia* are the dominant isolated heterotrophs in Wintergreen Lake.

The approach in this investigation is based on two assumptions. The first is that methanogenesis from methanol in the sediments of Lake Plussee would proceed with an average V_{max} of 0.52 μmol/d/g wet wt (Fig. 9-4). This would give a methane budget of 13×10^3 kg C/yr/lake at *in vitro* conditions of 27°C. The second assumption is that the reaction rate would be halved for every decrease of 10°C in temperature. This would give a methane budget of 3×10^3 kg C at 4°C, which is the prevailing sediment temperature during most of the year. With these assumptions, the methane contributed from methanol conversion would amount to 49% of the total methane estimated to be produced in Lake Plussee.

In conclusion, it seems apparent that the preference and domination of a particular methane precursor would be dependent on the magnitude of its availability, since it is itself dependent on the magnitude of sedimenting allochthonous and autochthonous materials and the degree of its incorporation in the prevailing metabolic routes.

ACKNOWLEDGEMENTS

I wish to thank Miss Karin Eckert for precision in the performance of the analytical work and for finishing the illustrations. I am also indebted to Dr. Karl-Paul Witzel for reviewing the manuscript and for valuable discussion.

REFERENCES

Abram, J. W., and D. B. Nedwell, 1978, Inhibition of methanogenesis by sulfate reducing bacteria competing for transferred hydrogen, *Arch. Microbiol.* **117**:89-92.

Bell, R. G., 1969, Studies on the decomposition of organic matter in flooded soil, *Soil Biol. Biochem.* **1**:105-116.

Bollag, J. M., and S. T. Czlonkowski, 1973, Inhibition of methane formation in soil by various nitrogen-containing compounds, *Soil Biol. Biochem.* **5**:673-678.

Bryant, M. P., S. F. Tzeng, I. M. Robinson, and A. E. Joyner, 1971, Nutrient requirements of methanogenic bacteria, in *Anaerobic Biological Treatment Processes,* F. G. Pohland, ed., Adv. Chem. Ser. 105, American Chemistry Society, Washington, D.C., pp. 23-40.

Cappenberg, T. E., 1975, A study of mixed continuous culture of sulfate reducing and methane producing bacteria, *Microb. Ecol.* **2**:60-72.

Donelly, M. I., and S. Dagley, 1980, Production of methanol from aromatic acids by *Pseudomonas putida, J. Bacteriol.* **142**:916-924.

Hoogerheide, J. C., and W. Kocholaty, 1938, *Biochem. J.* **32**:949. (See Nagase, M., and T. Matsuo, 1982.)

Jain, M. K., 1980, Microbial methanogenesis, *Int. J. Ecol. Environ. Sci.* **6**:5-17.

Kelly, C. A., and D. P. Chynoweth, 1981, The contributions of temperature and of the input of organic matter in controlling rates of sediment methanogenesis. *Limnol. Oceanog.* **26**:891-897.

Kusnezow, S. L., 1959, *Die Rolle der Mikroorganismen im Stoffkreislauf der Seen,* A. Pochmann, trans., VEB, Deutscher Verlag der Wissenschaften, Berlin.

Laskowski, D., and J. T. Moraghan, 1967, The effect of nitrate and nitrous oxide on hydrogen and methane accumulation in anaerobically icubated soils, *Plant Soil* **27**:357-368.

Lovley, D. R., and M. J. Klug, 1983, Methanogenesis from methanol and methyl-amines and acetogenesis from hydrogen and carbon dioxide in the sediments of a eutrophic lake, *Appl. Environ. Microbiol.* **45**:1310-1315.

Lovley, D. R., D. F. Dwyer, and M. J. Klug, 1982, Kinetic analysis of competition between sulfate reducers and methanogens in sediments, *Appl. Environ. Microbiol.* **43**:1373-1379.

Lynd, L. H., and J. G. Zeikus, 1983, Metabolism of H_2-CO_2, methanol, and glucose by *Butyrobacterium methylotrophicum, J. Bacteriol.* **153**:1415-1423.

MacGregor, A. N., and D. R. Keeney, 1973, Methane formation by lake sediments during in vitro incubation, *Water Resour. Bull.* **9**:1153-1158.

Mah, R. A., D. M. Ward, L. Baresi, and T. L. Glass, 1977, Biogenesis of methane, *Ann. Rev. Microbiol.* **31**:309-341.

Molongoski, J. J., and M. J. Klug, 1976, Characterization of anaerobic heterotrophic bacteria isolated from freshwater lake sediments, *Appl. Environ. Microbiol.* **31**:83-90.

Nagase, M., and T. Matsuo, 1982, Interactions between amino acid-degrading bacteria and methanogenic bacteria in anaerobic digestion, *Biotechnol. Bioeng.* **24**:2227-2239.

Naguib, M., 1982*a,* Methanogenesis in sediments of inland waters. 1. Methanol as the dominant methane-precursor in the sediments of an eutrophic lake, *Arch. Hydrobiol.* **95**:317-329.

Naguib, M., 1982*b,* Bilanzierung des Kohlenstoffs und die Möglichkeit der Ermittlung des allochthonen Eintrags in Binnengewässer. Kurzfassungen. 12. GFÖ *Jahrestagung in Bern,* p. 19.

Oremland, R. S., and S. Polcin, 1982, Methanogenesis and sulfate reduction: Competitive and noncompetitive substrates in estuarine sediments, *Appl. Environ. Microbiol.* **44**:1270-1276.

Oremland, R. S., L. M. Marsh, and S. Polcin, 1982, Methane production and simultaneous sulfate reduction in anoxic salt-marsh sediments, *Nature* **296**:143-145.

Overbeck, J., 1979, Studies on heterotrophic functions and glucose metabolism of microplankton in Plussee. *Arch. Hydrobiol. Beih. Ergebn. Limnol.* **13**:56-76.

Overbeck, J., 1981, A new approach for estimating the overall heterotrophic activity in aquatic ecosystems, *Verh. Internat. Verein. Limnol.* **21**:1355-1358.

Sahm, H., 1981, Biology of methane formation, *Chem. Ing. T.* **53**:884.

Schinck, B., and J. G. Zeikus, 1980, Microbial methanol formation: A major end-product of pectin metabolism, *Curr. Microbiol.* **4**:387-389.

Schlegel, H. G., 1969, *Allgemeine mikrobiologie,* Georg Thieme Verlag, Stuttgart.

Smith, M. R., S. H. Zinder, and R. A. Mah, 1980, Microbial methanogenesis from acetate, *Process Biochem.* **15**:34-39.

Strayer, R. F., and J. M. Tiedje, 1978, Kinetic parameters of conversion of methane precursors to methane in a hypereutrophic lake sediment, *Appl. Environ. Microbiol.* **36**:330-340.

Van Kessel, J. F., 1978, Gas production in aquatic sediments in the presence and absence of nitrate, *Water Res.* **12**:291-297.

Winfrey, M. R., D. R. Nelson, S. C. Klevickis, and J. G. Zeikus, 1977, Association of hydrogen metabolism with methanogenesis in Lake Mendota sediments, *Appl. Environ. Microbiol.* **33**:312-318.

Witzel, K. P., 1980, Temperature compensation of [U-^{14}C]-glucose incorporation by microbial communities in a river with a fluctuating thermal regime, *Appl. Environ. Microbiol.* **39:**790-796.

Zeikus, J. G., 1977, The biology of methanogenic bacteria, *Bacteriol. Rev.* **41:**514-541.

Zhilina, T. N., and G. A. Zavarzin, 1973, Trophic relations between *Methanosarcina* and its accompanying cultures, *Microbiology* **42:**266-273.

10

Destruction and Production Rates of Carbon Monoxide in Arid Soils Under Field Conditions

Ralf Conrad and Wolfgang Seiler
Max-Planck-Institut für Chemie

INTRODUCTION

Carbon monoxide destruction and production processes take place simultaneously in soils (Seiler, 1978; Conrad and Seiler, 1980, 1982). The resulting net effect of both processes is usually the destruction of atmospheric CO (Inman, Ingersoll, and Levy, 1971; Heichel, 1973; Ingersoll, Inman, and Fisher, 1974; Liebl and Seiler, 1976; Seiler, 1978). Under arid soil conditions, however, soils may act as a net source for atmospheric CO. Desert and semidesert soils in southern Africa have been found to act permanently as a source for atmospheric CO (Conrad and Seiler, 1982). On the other hand, German soils under arid conditions showed a net emission of CO during daytime, but a net deposition of atmospheric CO at night (Seiler, 1978; Conrad and Seiler, 1982). The observed change of soil activity is caused by the changing rates of the individual CO destruction, or by production processes in the upper soil layers, or both. However, the parameters influencing the production and destruction processes of CO in the soil under field conditions are unknown.

Field measurements were carried out on arid soils in Andalusia, Spain. The results indicate that the net flux of CO at the soil-air interface is determined mainly by changing CO production rates. These rates were found to be strongly dependent on the soil surface temperature.

MATERIALS AND METHODS

Field measurements were carried out in August and September 1982 on an unplanted field at the experimental station of the BASF Espanola, located near Utrera, approximately 30 km south of Sevilla, Andalusia. The soil is reddish brown, consisting of a loamy sand (pH 7.4) with 0.5% organic carbon. The soil had not received any water for more than one month and had a soil moisture content of approximately 0.5% in the upper 10-cm soil layer. The CO flux rates between soil and atmosphere were determined by applying the closed box method described by Seiler (1978). A rectangular stainless steel frame (area = 800 cm^2) with a groove on the top outside of the frame was driven into the soil to a depth of 10 cm. The soil plot was closed by putting a glass box (height = 10 cm) into the groove, which had been filled with boiled water. The glass box contained a vent consisting of a glass coil (length = 29 cm, inside diameter = 3 mm) to ensure equilibration of the air pressure inside and outside of the box. Gas samples (1-10 ml) were taken by means of gas-tight syringes and analyzed immediately in a CO-analyzer based on the HgO-to-Hg vapor conversion technique (Seiler, 1978; Seiler, Giehl, and Roggendorf, 1980). The soil temperature was recorded at approximately 3-5 mm depth inside the glass box by using a temperature probe (iron/constantan). The glass box was not shaded from sunlight. The soil surface temperature inside the glass box was identical to that outside.

The CO destruction and production rates were determined by measuring the temporal decrease of the CO mixing ratio, m, inside the glass box until the equilibrium value, m_e, was reached. The equilibrium value is the CO mixing ratio that establishes when destruction and production of CO are in steady state (Seiler, 1978; Conrad and Seiler, 1980). Whereas the CO destruction is an apparent first-order reaction with k_d as the rate constant

$$-dm/dt = k_d m, \qquad \text{(Eq. 10-1)}$$

the CO production is an apparent zero-order reaction

$$dm/dt = k_p. \qquad \text{(Eq. 10-2)}$$

Since destruction and production of CO occur simultaneously, the following equation is derived:

$$dm/dt = k_p - k_d m, \qquad \text{(Eq. 10-3)}$$

or in its integrated form with the boundary conditions $m = m_o$, for time $t = 0$, and $m = m_e$ when t is infinity:

$$k_d = (1/t) \left[ln \, (m_o - m_e) - ln \, (m - m_e) \right] \qquad (1/s). \qquad \text{(Eq. 10-4)}$$

The CO destruction rate, D, is calculated from its rate constant, k_d, by multiplication with the ambient CO mixing ratio, m_a, and the height, H, of the glass box (height = volume/soil surface area):

$$D = k_d \, m_a \, H \, (nl \, m^{-2} s^{-1}). \qquad \text{(Eq. 10-5)}$$

For steady-state conditions, CO destruction equals CO production so that the CO production rate, P, is given by:

$$P = k_d \, m_e \, H. \qquad \text{(Eq. 10-6)}$$

RESULTS

When the soil surface is covered by a box containing ambient CO mixing ratios, the CO inside the box changes with time until an equilibrium value is reached that is due to the simultaneous destruction and production of CO in soil. This equilibrium value changes with changing soil conditions and reaches high values when soil conditions are arid (Seiler, 1978; Conrad and Seiler, 1982). Figure 10-1A shows the CO equilibrium values measured as a function of time of day on an arid soil in Andalusia, Spain. The CO equilibrium values showed a strong diurnal variation, with maximum values in the afternoon at about 3 P.M. (European Summer Time), and minimum values in the early morning at about 6 A.M. The CO equilibrium values showed a positive correlation with the soil surface temperature, which is shown in the lower part of Figure 10-1.

At soil temperatures higher than 35°C, the CO equilibrium values exceeded the ambient CO mixing ratios. During this period (i.e., from 11:30 A.M. to 7:30 P.M.) the soil acted as a source for atmospheric CO. For the rest of the day the soil was a sink. It is of interest that similar diurnal patterns were observed in German soils, usually during summertime, when soil conditions may become arid after long periods without rain. Figure 10-1B gives an example of this situation as reported for Deuselbach in August 1969 by Seiler (1978) and shows a diurnal change of the CO equilibrium values very similar to that observed in Andalusian soils.

The magnitude of the CO equilibrium value and thus the influence of the soil on the atmospheric CO is a function of the rates of CO destruction and CO production in soil. These may change individually with changing soil conditions—for example, soil temperature, T:

$$m_e = k_p \, (T)/k_d \, (T). \qquad \text{(Eq. 10-7)}$$

It is, therefore, necessary to determine at least two of the three variables (m_e, k_p, k_d) in order to find out whether the magnitude and direction of the net CO flux between soil and atmosphere is dependent on variations of the CO production and the CO destruction.

Since the CO equilibrium values are often higher than ambient, the CO destruction rates were determined by injecting CO into the glass box, resulting in a CO mixing ratio significantly higher than equilibrium (Fig. 10-2). The CO destruction rate was then determined as described earlier in this paper. The change of the CO equilibrium values during the period of artificially increased CO mixing ratios due to CO injection was extrapolated by considering the change in soil surface temperature, as shown by the dotted line in Figure 10-2. This procedure had the advantage that the glass box remained in place and was not opened and closed for the flux determinations. The equilibrium values were not significantly influenced by

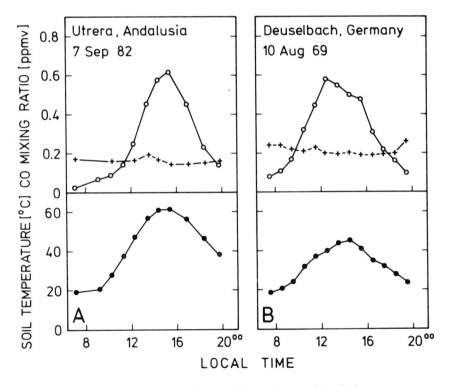

Figure 10-1. Variation of CO equilibrium values with time of day. Soil temperature was measured in 3–5mm depth (*A*) and in 1cm depth (*B*), respectively.
The CO equilibrium value is shown as (o); the ambient CO mixing ratio is shown as (+).

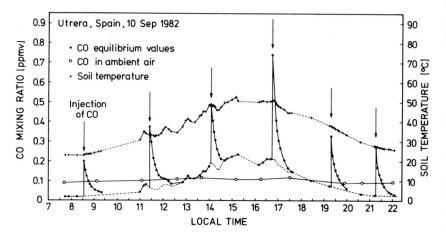

Figure 10-2. Measurements of CO equilibrium values and decrease of CO mixing ratios after injection of CO.

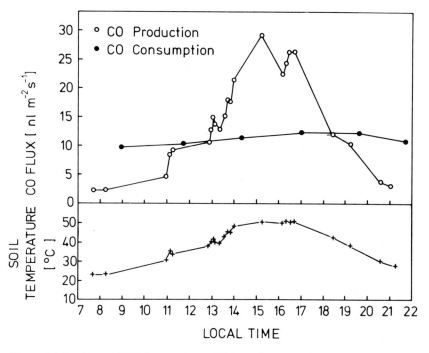

Figure 10-3. Rates of CO destruction and CO production as function of time of day in an arid soil in Andalusia.

possible changes in the chemical composition of the air inside the box (e.g., elevated carbon dioxide) during long-term closure of the box.

Figure 10-3 shows the CO destruction rates, D, that were determined from the CO injection experiments taken from Figure 10-2 by using the ambient CO mixing ratio of $m_a = 0.1$ ppmv in equation 10.5. It can be recognized that the CO destruction rates do not change significantly with time of day. The rates ranged between 10 and 13 $nl/m^2/sec$ and were not significantly correlated with the soil surface temperature.

The CO production rates, P were determined from the equilibrium values, m_e, measured between the individual CO injection experiments. Figure 10-3 shows that the CO production rates changed significantly with time of day and were positively correlated with the soil surface temperature. In the afternoon (1 P.M. to 6 P.M.) the CO production rates reached values of 30 $nl/m^2/sec$, thus exceeding the CO destruction rates so that the soil acted as a net source for atmospheric CO. During the rest of the day (i.e., morning and night) the CO production rates were smaller than the CO destruction rates, dropping to values of 2 $nl/m^2/sec$. During this time the soil acted as a net sink for atmospheric CO.

DISCUSSION

Observations in Andalusian soils confirm earlier observations that arid soils act as a source as well as a sink for atmospheric carbon monoxide (Seiler, 1978; Conrad and Seiler, 1982). The direction of the net flux of CO between soil and atmosphere seems to be caused exclusively by the rate of the CO production processes occurring in the soil. The rate of CO production is a function of the soil surface temperature, indicating that the CO production processes are located in the upper soil layers where soil temperature shows strong diurnal variations. Conrad and Seiler (1980, 1982) have shown that CO production in soils is mainly due to chemical processes, most probably oxidation of organic matter. Possible organic compounds are phenols, such as pyrogallol, gallic acid, hydrochinone, from which CO can be released during autoxidation (Loewus and Delwiche, 1963; Miyahara and Takahashi, 1971). The potential magnitude of CO production seems to be dependent on the content of soil organic carbon (Conrad and Seiler, 1982); however, other soil parameters (e.g., soil moisture) may influence the CO production as well. At the moment it is impossible to evaluate the influence of the various parameters on the CO production. It is of interest that soils in completely different climatic zones—Andalusia and Germany—show similar patterns in CO destruction and production.

In contrast to CO production, CO destruction rates did not change significantly with changing soil surface temperatures under field conditions. Similar observations were reported by Liebl and Seiler (1976), who also found only a slight dependence of CO destruction rates on soil surface temperature. Under laboratory conditions, however, they observed a significant correlation between CO destruction and soil temperature with a maximum at 35°C. Interestingly, they observed complete inactivation of the CO destruction process at temperatures higher than 41°C. This is in agreement with the microbiological nature of the CO destruction processes (Conrad and Seiler, 1980, 1982). Since surface temperatures in arid soils reach values of >60°C, it is assumed that microbial CO consumption is unlikely in surface soils. Hence, data suggest that in arid soils with hot surface temperatures, CO destruction processes are located in deeper soil layers where temperature is significantly lower and where diurnal variations of temperature are small. This interpretation would explain why the CO destruction rates did not change significantly with changing soil surface temperature under field conditions.

ACKNOWLEDGEMENTS

We thank the BASF Company (Dr. K. Schelberger) for giving permission to perform field measurements on the grounds of the experimental station in Utrera, Andalusia. We thank G. Walther (BASF Espanola) for his cooperation during the fieldwork and for providing the soil characteristics. This study was financially supported by the Bundesministerium für Forschung und Technologie, project FKW 18.

REFERENCES

Conrad, R., and W. Seiler, 1980, Role of microorganisms in the consumption and production of atmospheric carbon monoxide by soil, *Appl. Environ. Microbiol.* **40:**437-445.

Conrad, R., and W. Seiler, 1982, Arid soils as a source of atmospheric carbon monoxide, *Geophys. Res. Lett.* **9:**1353-1356.

Heichel, G. H., 1973, Removal of carbon monoxide by field and forest soils, *J. Environ. Qual.* **2:**419-423.

Ingersoll, R. B., R. E. Inman, and W. R. Fisher, 1974, Soil's potential as a sink for atmospheric carbon monoxide, *Tellus* **26:**151-159.

Inman, R. E., R. B. Ingersoll, and E. A. Levy, 1971, Soil: A natural sink for carbon monoxide, *Science* **172:**1229-1231.

Liebl, K. H., and W. Seiler, 1976, CO and H₂ destruction at the soil surface, in *Production and Utilization of Gases,* H. G. Schlegel, G. Gottschalk, and N. Pfenning, eds., E. Goltze KG, Göttingen, pp. 215-229.

Loewus, M. W., and C. C. Delwiche, 1963, Carbon monoxide production by algae, *Plant Physiol.* **38:**371-374.

Miyahara, S., and H. Takahashi, 1971, Biological CO evolution: Carbon monoxide evolution during auto- and enzymatic-oxidation of phenols, *J. Biochem.* **69:**231-233.

Seiler, W., 1978, The influence of the biosphere on the atmospheric CO and H$_2$ cycles, in *Environmental Biogeochemistry and Geomicrobiology,* Vol. 3, W. E. Krumbein, ed., Ann Arbor Science, Ann Arbor, Michigan, pp. 773-810.

Seiler, W., H. Giehl, and P. Roggendorf, 1980, Detection of carbon monoxide and hydrogen by conversion of mercury oxide to mercury vapor, *Atmos. Technol.* **12:**40-45.

Part III

TRANSPORT, DEPOSITION, AND WEATHERING

11

Sources, Transport, and Degradation of Organic Matter Components Associated with Sedimenting Particles in Lake Michigan

Clifford P. Rice and Philip A. Meyers
University of Michigan

Brian J. Eadie and John A. Robbins
Great Lakes Environmental Research Laboratory

ABSTRACT: Processes important to the dispersion and degradation of natural and anthropogenic organic matter have been studied in Lake Michigan. Solvent-extractable compounds have been obtained from samples of atmospheric material, rainfall, surface microlayers, sediment-trap contents, and bottom sediments in order to estimate the pathways of these organic compounds from input to eventual incorporation in the lake bottom. PCBs and land-plant n-alkanes have a major source in atmospheric deposition, and most of the PCBs are introduced on particles through washout or dryfall. Good correlations exist between PCBs and n-alkanes in the particulate phase of surface microlayers and atmospheric deposition. Fluxes of sediment-trap contents agree with atmospheric inputs of PCBs and show augmentation of hydrocarbons from aquatic and land runoff sources. Trap fluxes also agree well with atmospheric inputs of the natural radionuclides ^{7}Be and ^{137}Cs. Near-bottom sediment traps reveal considerable resuspension of bottom sediments and suggest the possibility of repartitioning of organic components on the basis of the size of their carrier sediment particles.

INTRODUCTION

Recent investigations have shown the importance and complexity of processes affecting organic matter components associated with sinking particles in bodies of water. Crisp et al. (1979) employed a number of molecular, elemental, and isotopic measurements to distinguish aquatic, terrigenous, and anthropogenic inputs of organic matter in particulate material collected in sediment traps at four locations offshore from southern California. A similar approach was used by Prahl, Bennett, and Carpenter (1980) to determine that aquatic organic matter is preferentially degraded during sinking and at the sediment-water boundary in Dabob Bay, Washington. Studies of fatty acid and hydrocarbon compositions of sediment-trap material from Lake Michigan (Meyers, Edwards, and Eadie, 1980; Meyers et al., 1984) suggest that a complex interaction of different organic inputs, alteration, and remineralization processes, as well as near-bottom sediment resuspension, occurs in this large lake.

A number of investigations of the processes important to the dispersion and degradation of organic materials on suspended and sinking particles have been conducted in the Great Lakes over the past decade. Many of these have employed sediment collection from lake bottoms or sediment traps suspended above the bottoms. Filtration of particulate matter from rainwater, lake water, and surface microlayers has also been performed. In this paper aspects of these studies, conducted by the authors and others, have been integrated with newly-made measurements of aliphatic and chlorinated hydrocarbon components that can be used as tracers of sources, transport, and degradation of organic materials in Lake Michigan.

MATERIALS AND METHODS

Sampling

Rain.   Samples of rainfall were collected on Beaver Island in Lake Michigan (see Fig. 11-1) in April and June of 1980. The rain collectors were simple bucket collectors that were carefully washed with soap and water and rinsed with pesticide-grade methylene chloride immediately prior to each use. For sample storage, the contents of the buckets were poured into precleaned, brown glass, 1-gal bottles. The interiors of the buckets were rinsed with methylene chloride, and these rinses were added to the sample bottles. Additional methylene chloride (total 200 ml/bottle) was added for stabilization in storage.

Rain samples were similarly collected in April 1980 at two urban locations in Rochester, New York, and a rural location near Hemlock Lake, New York (see Fig. 11-1) as contrasts to the Beaver Island site.

Air and Surface Microlayer.   Samples of particulate matter in air and in surface microlayers were collected as described by Meyers, Rice, and Owen (1983) at locations on Beaver Island and in Lake Michigan.

Sediment Traps.   Sediment traps consisting of 50 cm x 10 cm Plexiglas cylinders attached to 500-ml polyethylene bottles were deployed at sites in

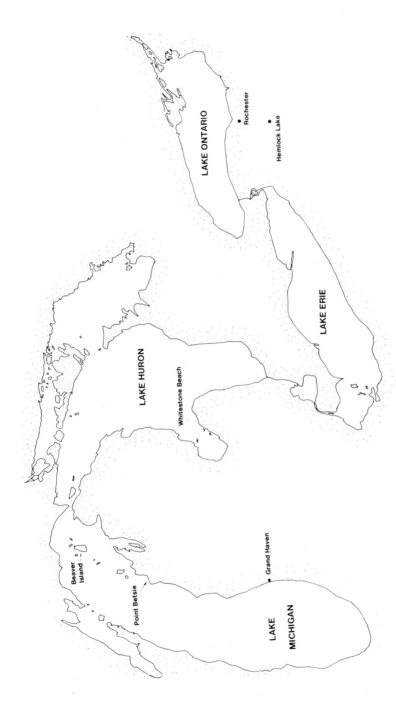

Figure 11-1. Locations of sample collections in the Great Lakes. Most sediment trap deployment was offshore of Grand Haven. This and other locations provided atmospheric samples.

Lake Michigan. Samples of settling matter were obtained from traps positioned at various depths in the water column and at 1 m above the lake bottom. Further information about sediment trap deployment and locations is given by Chambers and Eadie (1981) and Meyers et al. (1984).

Suspended Matter. Samples of suspended particulate matter were collected in July 1980 1 m below the water surface from near the sediment-trap sites. Water samples were filtered through glass fiber filters having a nominal retention size of 1 μm. The collection locations shown in Meyers et al. (1984) included one station within the plume of the Grand River, one near-shore station close to but not within this turbid plume, and one open-lake location 25 km offshore.

Analysis

Saturated Hydrocarbons. Rain samples were extracted with methylene chloride as described by Meyers and Hites (1982). Samples of microlayer particulate-phase material were obtained and extracted as explained by Meyers and Kawka (1982). Freeze-dried samples of sediment-trap material and suspended particulate matter were subjected to Soxhlet extraction with toluene/methanol as detailed by Leenheer (1981). Column chromatography on silica gel was used to isolate saturated hydrocarbons from the extracted material. Identification of hydrocarbon components was achieved by splitless injection of samples into a Hewlett-Packard 5830 FID gas chromatograph equipped with a 10-m SP2100 glass capillary column. Quantification was accomplished through the use of known amounts of internal standards added to each sample. Individual compounds were identified both by retention times and by GC-MS analysis of representative samples with a Finnigan 1015 mass spectrometer interfaced with a Varian 1400 gas chromatograph.

Chlorinated Hydrocarbons. Samples of particulate matter and water obtained from rainfall and from surface microlayers were extracted with dichloromethane as described by Rice, Eadie, and Erstfeld (1982). Sediment-trap samples were dried, wetted with extracted distilled water, and Soxhlet-extracted with acetone/hexane (1:1). The extracts were partitioned with water, and the resulting hexane phase was concentrated prior to column chromatography, first on alumina and then on silica gel, to isolate polychlorinated biphenyls and toxaphenes.

Gas chromatography was done on a Varian 3700 ECD instrument using both packed and capillary glass columns. Identification and quantification of sample components were based upon comparison to authentic standards of known concentration. Identifications were verified by combined gas chromatography-mass spectrometry on a Hewlett-Packard 5993C instrument.

Radionuclides. Sediment-trap samples were dried and ground, and weighed portions were placed in standardized counting vials. Amounts of ^{137}Cs and ^{7}Be were measured from gamma-radiation counting with a Nuclear Data, Inc., model 66 multichannel analyzer equipped with a lithium-drifted germanium crystal.

RESULTS AND DISCUSSION

Radionuclides and Particle Dynamics

A manmade radionuclide, ^{137}Cs, and ^{7}Be, produced by cosmic radiation in the atmosphere, provide different types of information about the sedimentation of particles in Lake Michigan. Production of ^{137}Cs maximized in 1963, and very little of this radioisotope is presently entering the lake. Strongly associated with clay minerals, ^{137}Cs consequently has a short residence time within the water column. Most of it now resides in the sediments of the lake bottom. Therefore, most of the ^{137}Cs present in the trap materials is likely to originate from sediments and is an indicator of the amount of resuspended materials. During nonstratified periods, the activity of the isotope is essentially uniform in trap materials throughout the water column. As stratification begins to develop in May, the activity in samples from the upper part of the water column vanishes. Particles from the epilimnion entering the traps do not carry ^{137}Cs, and the resuspended component is confined to bottom waters with the reduction in turbulent mixing that accompanies stratification of the lake. During November and December, the ^{137}Cs activity again reappears in near-surface samples, as surface waters turn over. Resuspension of bottom materials during times of no thermal stratification of lake waters is clearly occurring in Lake Michigan.

In contrast to ^{137}Cs, amounts of ^{7}Be in sediment traps in the upper water column reach maxima in June when thermal stratification is well developed and, at the same time, amounts in near-bottom traps decrease. Evidently, atmospheric inputs of ^{7}Be remain in the upper part of the stratified water column and do not sink into the deeper parts to replenish the short-half-life ^{7}Be in the lake bottom. During July and August, amounts of this radioisotope decrease markedly in upper traps, probably as a result of scavenging by plankton and of formation of authigenic calcite. Such separation into compartments in the water and scavenging can also influence organic-matter components of settling particles.

In the summertime in Lake Michigan, high concentrations of suspended materials are found in two depth zones, one in the region of the thermocline and another near the lake bottom (Chambers and Eadie, 1981; Rea, Owen, and Meyers, 1981). Comparison of sediment mass-accumulation rates in near-bottom sediment traps with long-term sedimentation rates commonly shows substantially more material accumulating in the traps than in the bottoms of the Great Lakes (Charlton, 1975; Chambers and Eadie, 1981), suggesting that considerable sediment resuspension occurs. This provides opportunities for sediment lateral transport before permanent deposition. During such transport, selective settling of organic particles is possible (cf. Davis and Brubaker, 1973).

Saturated Hydrocarbons

Differences in compositions suggest that the organic-matter contents of particulate matter in microlayer samples, at a water depth of 35 m and near the water-sediment boundary, are not the same (Meyers and Kawka, 1982; Meyers et al., 1984). Selective degradation probably contributes to this variance; however, different sources of organic matter to the respective depths in Lake Michigan seem to be the major cause of differences in the character of organic matter.

Organic matter associated with suspended particles collected from the surface microlayer has a strong terrigenous hydrocarbon character in river-mouth, near-shore, and open-lake locations (Meyers and Kawka, 1982). Hydrocarbon distributions contain large contributions of the C_{27}, C_{29} and C_{31} n-alkanes that dominate hydrocarbon compositions of Midwest rain and snow (Meyers and Hites, 1982). Airborne particles from land with associated land-plant organic material probably are an important fraction of the suspended particulate matter at the surface of Lake Michigan.

Deeper in the water, aquatic production becomes a more important source of organic matter (Meyers, Edwards, and Eadie, 1980). Hydrocarbon compositions of sediment traps at the base of the metalimnion indicate equal inputs of lake and land materials. Furthermore, these n-alkanes show a stronger aquatic character in the near-bottom trap material than in the contents of the shallower traps (Meyers et al., 1984). Hydrocarbon distributions in the shallower sediment traps contain evidence of weathered petroleum, but distributions in the near-bottom traps lack these components. Their absence indicates that the particles with which they are associated do not sink into the deeper waters of Lake Michigan during stratified periods. This is evidence for segregation of organic materials on the basis of density differences and selective sinking through the water column.

Hydrocarbons differ between the sediment-trap samples and the underlying sediment in Lake Michigan. In general, the organic-matter composition of surficial sediment has less of an aquatic character than does the content of any of the sediment traps (Meyers et al., 1984). The more aquatic character of hydrocarbons in the near-bottom sediment trap relative to those in underlying bottom sediments probably results from selective resuspension of smaller particles or from degradation of organic matter in the benthic boundary layer.

Rainfall samples collected as part of this study contained amounts of n-alkanes from land plants that made up a large proportion of their total hydrocarbon loadings. In addition, a sample from downtown Rochester, New York, had a strong petroleum hydrocarbon character. Atmospheric inputs may be quite important to the particulate hydrocarbon composition in the upper waters of the Great Lakes during summer stratification.

Chlorinated Hydrocarbons

Some results of analyses of rainfall and dryfall samples generated by studies of the authors and of other investigators are summarized in Table 11-1. While PCBs have long been recognized as Great Lakes contaminants, it is only recently that toxaphenes have been discovered to be widely present in fish in these lakes. Toxaphene appears to be more strongly associated with atmospheric inputs to bodies of water than is PCB. This is especially true for rainfall inputs, as first observed by Bidleman and Christensen (1979). The toxaphene rainfall flux is greater than the dryfall flux (Table 11-1). In contrast, the greater flux of PCBs appears to result from the settling of dry particles into Lake Michigan.

Observations of material collected in sediment traps arrayed in Lake Michigan for various intervals between 1977 and 1983 indicated that traps in the epilimnion appear to give a reasonably good measure of atmospheric deposition of chlorinated hydrocarbons. For example, the mean flux of PCBs from near-surface sediment traps was 130 ng/m^2/day during the stratified season of 1980. This flux value yields an annual atmospheric deposition of 2750 kg to Lake Michigan, which compares within a factor of two to the estimate of 4700 kg made by Murphy et al., 1982 and derived from shore-based rainfall and dryfall collectors (Table 11-1). The agreement between atmospheric PCB flux and that measured in upper-water-column sediment traps becomes more significant because Eadie, Rice, and Frez (1983) show that the flux of PCBs to surficial sediments is very similar to the sediment-trap value. These observations suggest that the principal source of PCBs that become buried in the sediments of the Great Lakes is from atmospheric sources.

An important factor in investigating processes involved in PCB distributions in water is their partitioning between particulate and dissolved phases. In studies of PCBs in surface microlayers on Lake Michigan, Rice, Meyers, and Brown (1983) found that a large fraction of the total PCBs is in the particulate phase. This fraction becomes higher in microlayer samples having higher PCB concentrations. Moreover, the particulate phases of these enriched samples contain proportionately more Aroclor 1242 than Aroclor 1254, contrary to solubility considerations. Both the higher particulate loading and the solubility disequilibrium agree in favor of recent additions of PCBs from atmospheric dryfall or particle washout to the microlayer

Table 11-1. Some Representative Values of Atmospheric Deposition of PCB and Toxaphene into Lake Huron and Lake Michigan (Number of Multiple Measurements Given in Parentheses)

Location	Date	Concentration (ng/l)		Flux (g/mol/km^2)		Reference
		Toxaphene	PCB	Toxaphene	PCB	
			Wet			
Beaver Island (Lake Michigan)	April 1981	70.2	64.8			This report
	Sept. 1981	31.6	46.1			This report
Point Betsie (Lake Michigan)	May 1976		138 (13)		8.5	Murphy and Rzeszutko, 1978
	1979–1980		15.6 (9)		0.75	Murphy et al., 1982
Whitestone Beach (Lake Huron)	1977–1978		35 (32)		0.91	Murphy et al., 1982
Lake Michigan	1981	68–136*		4.2–8.4*		Rice and Samson, 1983
Lake Michigan	1977–1980		22.9 (25)		1.46	Murphy et al., 1982
			Dry			
Lake Michigan	1981			0.53–1.0*		Rice and Samson, 1983
Lake Michigan	1977–1980				5.3 (19)	Murphy et al., 1982

*Values calculated from measured air concentrations, washout ratios, and particle deposition velocities.

Table 11-2. Summer and Winter Depth Profiles of PCB Measured at Two Closely Spaced Sediment Trap Locations in Southern Lake Michigan

Depth (m)	Particle Mass Flux (g/m²/day)	PCB Concentration ng/g	Percent of Total PCB Measured as Arochlor 1242
		Winter (1981)	
35	4.26	202	71.0
60	6.04	245	72.9
80	6.40	273	75.9
95	10.76	201	65.7
		Summer (1978)	
35	0.70	1300	77
73	14.0	517	50
82	28.0	216	56

samples having high PCB levels. Hence, study of phase partitioning points out again the probable importance of atmospheric inputs of contaminants, such as PCBs, to large bodies of water such as the Great Lakes.

Significant differences exist in amounts of PCBs associated with particles settling through water at different depths and during different seasons. The data in Table 11-2 illustrates some of this variability in material collected in sediment traps positioned in Lake Michigan off-shore from Grand Haven (see Chambers and Eadie, 1981). Concentrations of total PCBs are relatively uniform at all depths during the nonstratified winter collection period, and the contribution of Aroclor 1242 remains close to around 72% of the total PCBs except in the samples nearest the bottom. Turbulent water motion has distributed PCB-containing particles throughout the lake waters and has created high settling-flux values, although the presence of more suspended material nearer to the lake bottom is evident from the increase in particulate fluxes closer to the bottom.

During summer stratification, resuspension of bottom material is confined to the hypolimnion, and little of this material reaches the upper parts of the lake water column. As a result, particle flux rates are low in the 35-m trap but high in the deeper traps (Table 11-2). Concentrations of PCBs, undiluted by resuspended mineral matter, can become quite high in particles collected in the epilimnion. In addition to major depth-related differences in concentration, the composition of PCBs varies on particles collected by sediment traps at different depths. For both the winter and summer profiles, the proportion of Aroclor 1242 decreases closer to the lake bottom. This is especially true for the summer profile in Table 11-2, which is another example of the thermocline barrier isolating recent inputs of PCB from the bottom. These data agree with other observations in Lake Michigan made by the authors and with those made by Eisenreich, Capel, and Looney (1983) in Lake Superior.

The change with depth in Aroclor composition of PCBs that was found for the summer profile was further investigated by determining the isomeric constitution of the PCB mixtures. As described by Anderson (1980), the method involves expressing the amount of possible selective alteration of individual isomers via a ratio of the area of the isomer peak on a capillary gas chromatogram to that of another isomer determined to be a persistent PCB component. Anderson (1980) has shown that the 2,3,2',3'-tetrachlorobiphenyl isomer degrades minimally under natural sediment conditions; this compound is used as the basis of estimates of selective alteration. Another reason for selecting the 2,3,2',3' isomer as a reference compound is that it is a measurable component of Aroclor 1242 but not of Aroclor 1254. This choice avoids the complication of mutual overlap of a common isomer arising from different Aroclors. The peak area units of each peak in the capillary column recorder trace (see Fig. 11-2) are divided by the peak area units of 2,3,2',3'. The resulting ratio should be the same for the commercial mixture, Aroclor

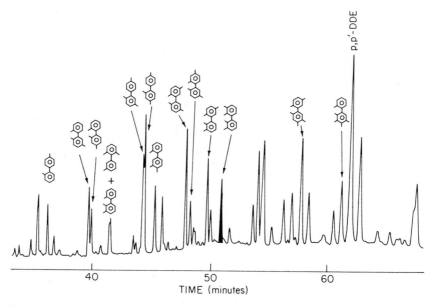

Figure 11-2. Capillary column gas chromatographic trace of chlorohydrocarbons from Lake Michigan summer sediment trap sample from a depth of 73m. Conditions: 30m x 0.25mm i.d. borosilicate glass capillary column coated with SP-2100, temperature programmed from 100°C to 200°C at 2°C/min; N2 carrier gas flowing at 1ml/min. Reference PCB isomer marked in black.

**Table 11-3. Relative Isomeric Composition of PCB in the
Summer Sediment Trap Depth Profile. Isomeric Content of Each Sample Was
Compared to 2, 3, 2′, 3′-Tetrachlorobiphenyl* (Figure 11-2) as Detailed in Text.**

PCB Isomer	$R_i(73\,meters)$ / $R_i(35\,meters)$	$R_i(82\,meters)$ / $R_i(35\,meters)$
2,5,2′	0.72	0.39
2,4,2′	0.85	0.37
2,5,4′	0.82	0.77
2,4,4′	0.83	0.84
3,4,2′	0.72	0.51
2,5,2′,4′	0.79	0.53
2,3,2′,5′	0.83	0.73
2,3,2′,3′	1.0	1.0
2,4,5,2′,5′	0.98	1.22
2,3,4,5,2′	1.25	1.07

$$*R_i = \frac{\text{area of peak } i}{\text{peak area of 2,3,2′,3′ isomer}}$$

1242, and the environmental sample, if no alteration has occurred. If alteration has occurred in the samples, the ratio of each peak (i) to 2,3,2′,3′ will be less than the ratio in Aroclor 1242.

The results of the capillary isomeric ratio technique are shown in Table 11-3. In applying this technique to the vertical profile results, the isomers in the two deeper samples are compared to the 35-m sample. The PCB isomeric composition in the extract of the sediment-trap sample from the 73-m depth is only slightly lower than that from the 35-m depth ($R_i = 0.72 - 0.98$). The 82-m sample, however, is quite different for a number of isomers. Isomers 2,5,2′- and 3,4,2′-chlorobiphenyl, in particular, are significantly lower ($R_i = 0.37 - 0.51$) than in the 35-m or 73-m samples. The other identified isomers in the 82-m trap sample are approximately the same as the other two depths.

A possible explanation for the shift in Aroclor composition near the bottom would be bacterial decomposition. It has been shown that both singly isolated bacteria (Furukawa and Matsumura, 1976) and bacteria in mixed culture isolated from Saginaw Bay, Michigan, in Lake Huron (Anderson, 1980) are capable of degrading chlorinated biphenyls. These investigations have also shown that the degree of degradation is inversely proportionate to the degree of chlorine substitution on the biphenyl molecule. PCB-degrading bacteria are probably not homogeneously distributed in the water column but may have areas of greater or lesser population density, and they are expected to have a higher affinity for particulate material (Paerl, 1975) and sediments. Consequently, it is reasonable to find a greater degree of decomposition within the benthic nepheloid layer, extending 1-10 m up from the lake bottom, and also within the sediments, than higher in the water column.

It must be emphasized, however, that the observed changes in PCB composition with depth may also be mediated by chemico-physical phenomena. Resuspension or differential adsorption phenomena alone can account for significant changes in Aroclor composition due to the different partition coefficients and chemical properties of the individual components (Halter and Johnson, 1977).

SUMMARY AND CONCLUSIONS

The Great Lakes are particularly susceptible to contamination by synthetic materials, primarily because of the heavy industrialization of the region and the slow flushing times of the lakes. For compounds characterized by slow decomposition rates, such as highly chlorinated PCB isomers, settling into the benthic boundary layer and burial are a prime removal mechanism.

Differences in the amount and type of a particulate organic matter found at various depths and locations in Lake Michigan suggest a complex interplay of organic-matter input, transport, and degradation that accompanies the settling of particles through the water. Some of the elements of this interplay are generalized as:

1. Particulate matter in the microlayer of open lake waters contains a major component of land-derived organic matter brought to these locations by eolian transport and atmospheric fallout.
2. Petroleum hydrocarbons are associated with particles collected in sediment traps in midwater depths but not with shallower or deeper particles, evidently as a result of density segregation of settling particles during summer stratification.
3. A significant resuspension of sediment-bound PCBs occurs during nonstratified periods. The net effect of resuspension is to increase the residence time for contaminants in the active ecosystem.
4. Microbial degradation of some PCB components in the lake bottom is indicated but is not yet nonequivocally established.

ACKNOWLEDGEMENTS

We thank Max Anderson and William Frez for their contributions to this paper. Portions of the investigations described in this paper have been variously supported by grants from the National Oceanic and Atmospheric Administration, the Michigan Sea Grant College Program, the National Science Foundation, the United States Environmental Protection Agency, and the Petroleum Research Fund administered by the American Chemical Society. This paper is contribution number 385 of the Great Lakes Environmental Research Laboratory.

REFERENCES

Anderson, M. L., 1980, *Degradation of Polychlorinated Biphenyls in Sediments of the Great Lakes,* Ph.D. thesis, University of Michigan, Ann Arbor.

Bidleman, T. F., and E. J. Christensen, 1979, Atmospheric removal processes of high molecular weight organochlorine compounds, *J. Geophys. Res.* **84:**7857-7862.

Chambers, R. L., and B. J. Eadie, 1981, Nepheloid and suspended particulate matter in southeastern Lake Michigan, *Sedimentology* **28:**439-447.

Charlton, M. N., 1975, Sedimentation: Measurements in experimental enclosures, *Verh. Int. Ver. Theor. Agnew. Limnol.* **19:**267-272.

Crisp, P. T., S. Brenner, M. I. Venkatesan, E. Ruth, and I. R. Kaplan, 1979, Organic geochemical characterization of sediment-trap particulates from San Nicolas, Santa Barbara, Santa Monica, and San Pedro Basins, California, *Geochim. Cosmochim. Acta* **43:**1791-1801.

Davis, M. B., and L. B. Brubaker, 1973, Differential sedimentation of pollen grains in lakes, *Limnol. Oceanogr.* **18:**635-646.

Eadie, B. J., C. P. Rice, and W. A. Frez, 1983, The role of the benthic boundary in the cycling of PCBs in the Great Lakes, in *Physical Behavior of PCBs in the Great Lakes,* D. Mackay, S. Paterson, S. J. Eisenreich, and M. S. Simmons, eds., Ann Arbor Science Publishers, Ann Arbor, Michigan, pp. 213-228.

Eisenreich, S. J., P. D. Capel, and B. B. Looney, 1983, PCB dynamics in Lake Superior water, in *Physical Behavior of PCBs in the Great Lakes,* D. Mackay, S. Paterson, S. J. Eisenreich, and M. S. Simmons, eds., Ann Arbor Science Publishers, Ann Arbor, Michigan, pp. 181-211.

Furukawa, K., and F. Matsumura, 1976, Microbial metabolism of polychlorinated biphenyls: Studies on the relative degradability of polychlorinated biphenyl components by *Alkaligines* sp., *J. Agric. Food. Chem.* **24:**251-256.

Halter, J. T., and H. E. Johnson, 1977, A model system to study desorption and biological availability of PCB in hydrosoils, in *Aquatic Toxicology and Hazard Evaluation,* F. L. Mayer and J. L. Hamelink, eds., American Society for Testing and Materials, Philadelphia, Special Technicial Publication no. **634:**178-195.

Leenheer, M. J., 1981, *Use of Lipids as Indicators of Diagenetic and Source-related Changes in Holocene Sediments,* Ph.D. thesis, University of Michigan, Ann Arbor.

Meyers, P. A., and R. A. Hites, 1982, Extractable organic compounds in Midwest rain and snow, *Atmos. Environ.* **16:**2169-2175.

Meyers, P. A., and O. E. Kawka, 1982, Fractionation of hydrophobic organic materials in surface microlayers, *J. Great Lakes Res.* **8:**288-298.

Meyers, P. A., S. J. Edwards, and B. J. Eadie, 1980, Fatty acid and hydrocarbon content of settling sediments in Lake Michigan, *J. Great Lakes Res.* **6:**331-377.

Meyers, P. A., C. P. Rice, and R. M. Owen, 1983, Input and removal of natural and pollutant materials in the surface microlayer on Lake Michigan, in *Environmental Biogeochemistry,* R. O. Hallberg, ed., *Ecol. Bull.* (Stockholm) **35:**519-532.

Meyers, P. A., J. Leenheer, B. J. Eadie, and S. J. Maule, 1984, Organic geochemistry of suspended and settling particulate matter in Lake Michigan, *Geochim. Cosmochim. Acta.* **48:**443-452.

Murphy, T. J., and C. P. Rzeszutko, 1978, Polychlorinated biphenyls in precipitation in the Lake Michigan basin, EPA-600/3-78-071.

Murphy, T. J., G. Paolucci, A. Schinsky, M. Combs, and J. Pokojowczyk, 1982, Inputs of PCBs from the atmosphere to Lakes Huron and Michigan, Final project report to U.S. EPA, Environmental Research Laboratory, Duluth, Minnesota.

Paerl, H. W., 1975, Microbial attachment to particles in marine and freshwater ecosystems, *Microb. Ecol.* **2:**73-83.

Prahl, F. G., J. T. Bennett, and R. Carpenter, 1980, The early diagenesis of aliphatic hydrocarbons and organic matter in sedimentary particulates from Dabob Bay, Washington, *Geochim. Cosmochim. Acta* **44:**1967-1976.

Rea, D. K., R. M. Owen, and P. Meyers, 1981, Sedimentary processes in the Great Lakes, *Rev. Geophys. Space Phys.* **19:**635-648.

Rice, C. P., and P. J. Samson, 1983, Atmospheric transport of toxaphene to Lake Michigan, Final project report to U.S. EPA, Large Lake Research Station, Gross Ile, Michigan.

Rice, C. P., B. J. Eadie, and K. M. Erstfeld, 1982, Enrichment of PCBs in Lake Michigan surface films, *J. Great Lakes Res.* **8:**265-270.

Rice, C. P., P. A. Meyers, and E. S. Brown, 1983, Role of surface microlayers in the air-water exchange of PCBs, in *Physical Behavior of PCBs in the Great Lakes,* D. Mackay, S. Paterson, S. J. Eisenreich, and M. S. Simmons, eds., Ann Arbor Science Publishers, Ann Arbor, Michigan, pp. 157-179.

12

Nondetrital and Detrital Particulate Metal Transport in Estuarine Systems

Douglas H. Loring and Reijo T. T. Rantala

Bedford Institute of Oceanography

Alan W. Morris

Institute for Marine Environmental Research

Gert Asmund

The Geological Survey of Greenland

INTRODUCTION

Heavy metals and other elements are introduced into estuarine environments in solution and as part of, or in association with, solid particles from natural and anthropogenic sources. This introduction occurs mainly via rivers but can occur through outfalls and dumping. Very little is known about the carriers and transport modes of these particulate metals or their response to estuarine conditions.

Measurement of total particulate metal concentrations is a poor method of determining characteristics because part of the metal load is loosely bound to the particles and part is locked up physically or chemically, or both, in detrital particles and minerals. Selective chemical methods have been developed to partition the particulate metals into their loosely bound (nondetrital) and residual (detrital) phases. Using such techniques, detrital and nondetrital zinc (Zn), copper (Cu), lead (Pb), and cadmium (Cd) concentrations of particulate matter entering and within the St. Lawrence estuary (Canada), the Tamar estuary (United Kingdom), and those artificially introduced into an Arctic fjord system on the west coast of Greenland (Fig. 12-1) are determined. The carriers, transport mode, and potential bio-availability of particulate metals entering and within these different systems are examined.

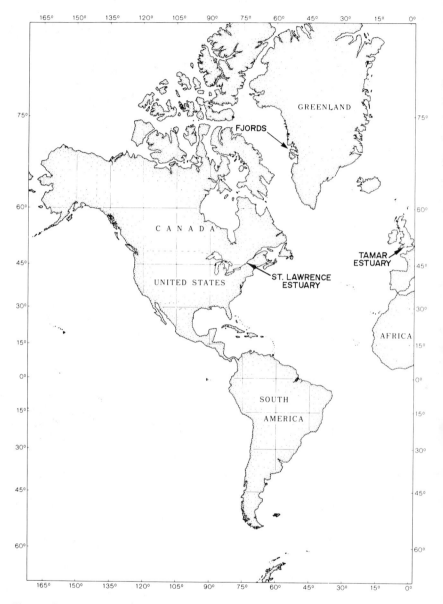

Figure 12–1.   Location of the St. Lawrence estuary, the Tamar estuary, and Arctic fjords.

MATERIALS AND METHODS

The methods of sample collection varied depending on the estuary studied. In the St. Lawrence estuary the samples were collected in modified Niskin bottles and filtered under pressure; in the Tamar estuary the samples were collected using a polyethylene bucket and were filtered under vacuum; in the Arctic fjords the samples were collected by a Hydro-Bios reversible plastic water sampler and were filtered under vacuum. All the suspended particulate matter (SPM) samples were filtered onto preweighed 47-mm, 0.40-μm Nuclepore membranes, after which they were washed with deionized water to remove occluded sea salts. Sample weights ranged from 0.14 to 15 mg.

The chemical technique to partition the particulate metals into their nondetrital and detrital phases used 25% v/v acetic acid and hydrofluoric acid (Rantala and Loring, in press). The acetic acid leaching was carried out in a modified Millipore Sterifil vacuum filtration apparatus. The filter was covered with 5 ml of 25% acetic acid for 24 hr, after which the extractant was drawn off by vacuum into a 10-ml volumetric flask. The filter was then washed twice with 2 ml of deionized water, and the washings were combined with the filtrate and made up to a volume of 10 ml. The residue on the filters was decomposed with a combination of Ultrex grade hydrofluoric acid and *aqua regia* in Lorran decomposition vessels (Rantala and Loring, 1973). Element concentrations were determined by atomic absorption spectrometry either by flame (Zn, Cu) or graphite furnace (Cd, Cu, Pb, Zn). Perkin-Elmer models 303 and 306 atomic absorption spectrophotometers and an HGA-500 graphite furnace were used (Rantala and Loring, 1977, 1980). L'vov platforms were inserted into the graphite tubes for the Cd, Pb, and Zn determinations. The relative accuracy was found to be within 10% of the accepted means for the reference materials: NRC estuarine sediments BCSS-1 and MESS-1 and the USGS marine mud MAG-1.

The acetic acid method was selected for use because this chemical treatment was used to effectively remove the total, weakly-bound metal concentrations in sediments (Loring, 1978) and particulate matter (Loring et al., 1982). This method removed metals held in ion-exchange positions, easily soluble amorphous compounds of iron and manganese, carbonates, and metals weakly bound to organic matter. The concentration of the metal determined in this fraction was operationally defined as the nondetrital (acid soluble) fraction because silicate lattices remained intact and resistant iron and manganese minerals and organic compounds were not attacked. The proportion of a metal remaining in this residual fraction was referred to as the acetic acid–insoluble or detrital fraction of the material.

RESULTS

The St. Lawrence Estuary

The St. Lawrence drainage basin covers an area of 1.32×10^6 km^2 (Jordon, 1973) in eastern Canada and the northeastern United States. Its major features include the Great Lakes and the St. Lawrence River, which discharges an average of 4.4×10^{11} tonnes of water and 3.5×10^6 tonnes of particulate matter (Holeman, 1968) annually at Quebec City. Yeats and Bewers (1982) estimated the total particulate metal inputs from the river to be: 2.3×10^6 kg/yr Zn; 5.7×10^5 kg/yr Cu; 9×10^3 kg/yr Cd. The river is connected by a 350-km-long estuary to the Gulf of St. Lawrence. The upper estuary (Fig. 12-2) extends seaward from the mouth of the river at Quebec City for about 150 km and varies in width from 2 km at Quebec City to about 24 km near the mouth of the Saguenay Fjord. It is a partially mixed estuary with salinities ranging from <.1/mil at its head to over 32/mil; the estuary is dominated by a saltwater wedge beneath the river outflow and a turbidity zone some 45 km below Quebec City (d'Angelejan and Smith, 1973; Kranck, 1979). Bewers and Yeats (1978) found that mixing of fresh and saline water and the distribution and transport of particulate matter were the most important factors controlling the distribution and behavior of dissolved metals. Particulate samples, including those used in this study, were collected in April 1976 as part of a comprehensive study of the factors controlling the distribution and behavior of dissolved and particulate metals in the estuary. Additional parameters measured for these samples included: suspended particulate silicon (Si), aluminum (Al), iron (Fe), manganese (Mn), calcium (Ca), magnesium (Mg) (in $\mu g/g$), salinity, particulate concentrations (mg/l), and sample depth (Table 12-1). Not all of these parameters are reported in this paper.

Analysis of the particulate matter in the river entering the estuary in April reveals the abundance and mode of transport of particulate metals. The data indicate that particulate matter contains 413 ppm Zn, 284 ppm Cu, 111 ppm Pb, 2.52 ppm Cd, 1334 ppm Mn, and 4.76% Fe. Partition of the total metal concentrations into their nondetrital and detrital contributions (Fig. 12-3) indicates that 69% of the total Zn, 79% of the total Cu, 72% of the total Pb, and 87% of the total Cd are present in the nondetrital fraction. After entry to the river the particulate composition changes rapidly for Zn, Cu, and Pb. At Station 67, some 60 km below Quebec City (Fig. 12-3), the particulate matter contains 314 ppm Zn, 121 ppm Cu, 67 ppm Pb, and 2.32 ppm Cd. Partition indicates that 55% of the total Zn, 66% of the total Cu, 66% of the total Pb, and 91% of the total Cd are still held in the nondetrital fraction of the particulate metals. Other changes in particulate composition also occur throughout the estuary in response to changing physico-chemical conditions. These results are reported elsewhere.

The Tamar Estuary

Although the Tamar estuary, located in southwest England, is very small compared to the St. Lawrence estuary, its physical, chemical, and biological characteristics are well known (Butler and Tibbits, 1972; Morris, Bale, and Howland, 1982; Loring et al., 1982). It is 31.7 km long from its seaward boundary with Plymouth Sound to the weir at Weir Head (Fig. 12-4). Normally the limit of salt intrusion is located 5-15 km seaward of the weir.

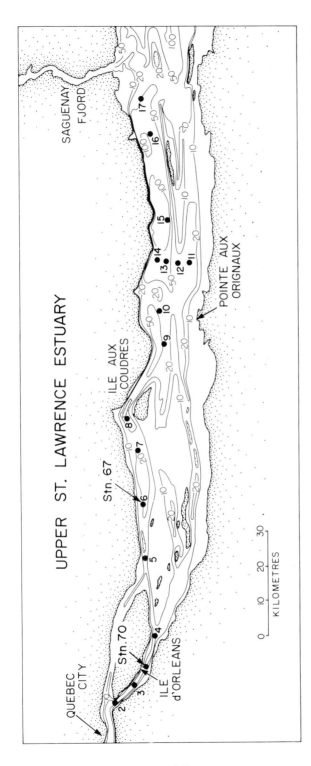

Figure 12-2. Location of Stations 70 and 67 in the upper St. Lawrence estuary. Other dots and numbers refer to locations sampled but not referred to in the text. Bathymetric contours of 10, 20, 50, and 100 m are also shown.

Table 12-1. Geochemical Data for Particulate Matter from the St. Lawrence and Tamar Estuaries and the Arctic Fjords. Concentrations in ppm Except % for Fe.

	St. Lawrence		Tamar Estuary				Arctic Fjords		
	April 1976		December 1979		April 1980			April 1982	
Sample Date:							"A"	"A"	"Q"
Station Location	70	67	20	17	19	15	2A	4	12
Depth	1	1	1	1	1	1	50	50	30
Salinity	.1	.1	.10	.80	.05	.06			
SPM mg/l	9.1	29.2	108	268	2.9	74.5	1.8	2.7	3.0
Zn									
Nondetrital Zn	284	174	368	333	1610	422	2880	2799	272
Detrital ↓Zn	129	140	146	161	153	145	20060	17140	1167
Total Zn	413	314	514	494	1763	567	22940	19939	1439
% nondetrital Zn	68.8	55.4	77.6	67.4	91.3	74.4	12.6	14.0	18.9
Cu									
Nondetrital Cu	224	80	303	276	1854	463			
Detrital Cu	60	41	134	112	91	127			
Total Cu	284	121	437	388	1945	590			
% nondetrital Cu	78.9	66.1	69.3	71.1	95.3	78.5			

Pb									
Nondetrital Pb	80	44					10170	10600	179
Detrital Pb	31	23					184	140	339
Total Pb	111	67					10354	10740	518
% nondetrital Pb	72.1	65.7					98.2	98.7	34.6
Cd									
Nondetrital Cd	2.20	2.10					10.0	10.1	0.30
Detrital Cd	0.32	0.22					110	88.7	3.7
Total Cd	2.52	2.32	2.60	5.30	4.10	1.1	120	98.8	4.0
% nondetrital Cd	87.3	90.5					8.3	10.2	7.5
Fe									
Nondetrital Fe (%)	1.10	0.85	1.07	1.08	3.71	0.94			
Detrital Fe	3.66	4.40	4.31	4.43	4.27	4.36			
% nondetrital Fe	23	16	19.9	19.6	46.5	17.7			
Mn									
Nondetrital Mn	849	563	590	553	1000	943			
Detrital Mn	485	568	305	331	300	332			
% nondetrital Mn	64	50	65.9	62.6	76.9	74.0			

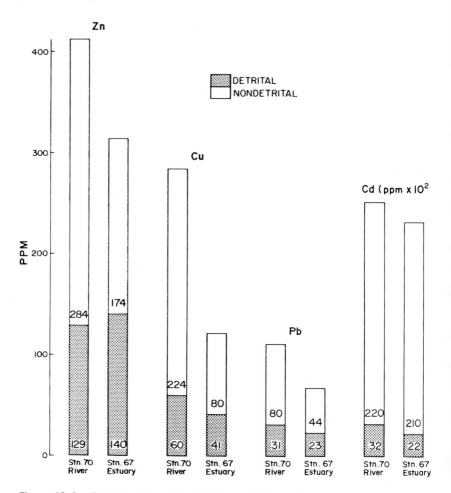

Figure 12-3. Partition of total Zn, Cu, Pb, and Cd into their nondetrital and detrital contributions in the particulate matter from the river flow (Station 70) and estuarine environment (Station 67) in April 1976 in the St. Lawrence SPM.

The middle estuary is generally partially mixed, with a vertical salinity range of 1-2/mil. During the sampling periods, the salinity ranged from 29/mil at the mouth to <0.1/mil at the weir. A pronounced turbidity maximum prevails in the low salinity (0.1-5/mil) region of the estuary irrespective of the geographical location of the saline intrusion. Suspended loads within the turbidity zone range from 50 mg/l to more than 1000 mg/l. These levels contrast with generally low loads (5 mg/l) prevalent in the marine and freshwater inputs to the estuary. Under river spate conditions, however, the influxing particulate loads often exceed 50 mg/l. The estuary receives

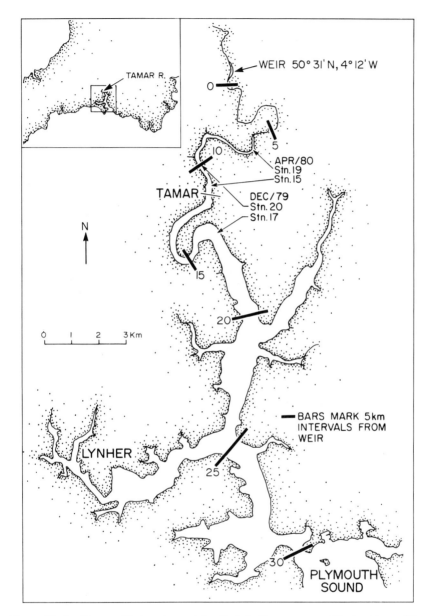

Figure 12-4. Geographic location and detail of the Tamar estuary. The locations of the samples obtained in December 1979 (high river flow) and April 1980 (low river flow) are indicated.

dissolved and particulate metals via the Tamar River, which drains a mineralized (Zn, Cu, Pb) area near its mouth.

When sampled in December 1979 during a period of high river flow (42.3 m^3/sec), the particulate load was about 108 mg/l. The particles entering the estuary (Station 20, Table 12-1) had total concentrations of 514 ppm Zn, 437 ppm Cu, 2.6 ppm Cd, 895 ppm Mn, and an Fe concentration of 5.38%; nondetrital and detrital Pb and Cd concentrations were not measured for these samples. Partition shows (Fig. 12-5a) that 78% of the total Zn and 69% of the total Cu entered the estuary in the nondetrital fraction at this time. Particulate matter samples, collected 7 km below Station 20 at Station 17, contained comparable total metal concentrations of 494 ppm Zn, 388 ppm Cu, 5.3 ppm Cd, 884 pm Mn, and 5.51% Fe (Table 12-1). Partition of this material indicated that 67% of the total Zn and 71% of the total Cu was held

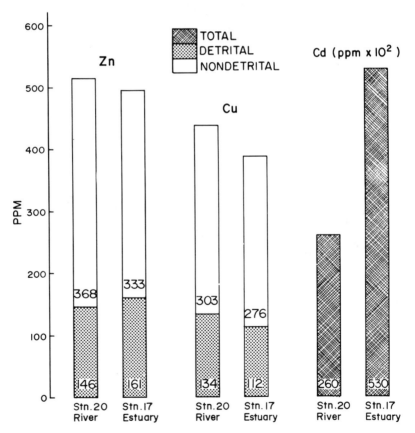

Figure 12-5(a).   Nondetrital and detrital Zn and Cu and total Cd concentrations in particulate matter obtained in the Tamar estuary during a period of high river flow in December 1979.

in the nondetrital fraction of the estuarine material. In fact, a detailed study by Loring et al. (1982) showed that both the detrital and nondetrital metal concentrations remained relatively constant throughout the whole estuary during this period of high river flow. In contrast, the particulate metal levels entering the estuary were different at a period of low river flow (12.8 m^3/sec) in April 1980 (Fig. 12-5b). The particulates were characterized at Station 19

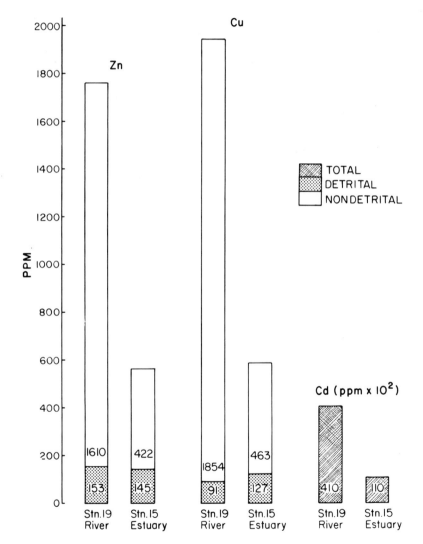

Figure 12-5(b).   Nondetrital and detrital Zn and Cu and total Cd concentrations in particulate matter obtained in the Tamar estuary during a period of low river flow in April 1980.

(Table 12-1) by high total Zn (1763 ppm), Cu (1945 ppm), Cd (4.1 ppm), Mn (1300 ppm), and Fe (7.98%) concentrations. Partition of the total metals (Fig. 12-5*b*) also indicated that a larger proportion of the metals (91% Zn, 95% Cu) entered the estuary in the nondetrital form at this period. Comparison of the riverine SPM (Station 19) with particulate material in the estuary (Station 15) showed marked decreases in total Zn (567 ppm) and Cu (590 ppm) concentrations and a decline in the percentages of total Zn (74%) and Cu (79%) held in the nondetrital fraction. This is mainly the result of internal particle cycling and sediment resuspension in the estuary (Loring et al., 1982). It should be pointed out that detrital Zn and Cu concentrations in the particulates were remarkably consistent during these two sampling periods (mean values of all samples: $n = 20$: detrital Zn 162 ± 26 ppm vs. 132 ± 15 ppm; Cu 114 ± 16 vs. 110 ± 15 ppm) in the river as well as in the estuary; the significance of this is discussed elsewhere in the text.

Arctic Fjords (Greenland)

Since 1973, about 1400 tonnes per day of tailings from a lead-zinc mine at Marmorilik, Greenland (71°06'N, 51°14'W), have been discharged into

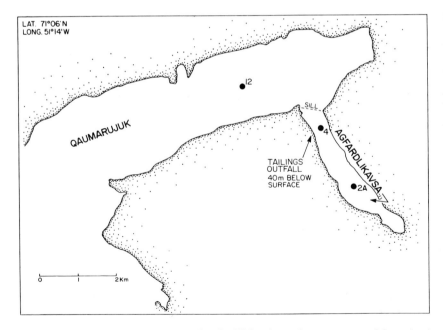

Figure 12-6.   Sample locations in "A" and "Q" fjords on the west coast of Greenland. Numbered stations are referred to in the text. Note location of tailings outfall.

Agfardikavsa ("A") fjord through an outfall at 40 m below the surface (Asmund, 1980). "A" fjord (Fig. 12-6) is 0.5 km wide and about 4 km long with water depths up to 80 m. It is separated from the ocean by Qaumarujuk ("Q") fjord and by a sill with a maximum depth of 27 m. "Q" fjord, which opens to the sea, is about 2 km wide, 10 km long, and has water depths ranging from 100 to 200 m (Bondam, 1978). A series of environmental studies by the Greenland Geological Survey (Bondam and Asmund, 1974; Asmund, 1975; Asmund, Bollingberg, and Bondam, 1976) indicated that the waters of "A" fjord contained extremely high levels of dissolved Zn (300-1000 μg/l), Pb (175-250 μg/l), and Cd (2-5 μg/l) as a result of the dissolution of the Zn-Pb-Cd-rich tailings in the fjord. The dispersal of dissolved metals from "A" fjord to "Q" fjord has been used to trace the water movements in the fjord system (Asmund, 1980) and has resulted in the metal enrichment of macroalgae and invertebrates not only in "A" fjord but also in "Q" fjord.

In April 1982 samples of the particulate matter were obtained from the fjord for analysis. Since the sample weights were very low (0.14-3.02 mg), only five of the samples could be partitioned into their detrital and nondetrital contributions. The total particulate metal concentrations measured throughout the fjord showed that total Pb ranged from 406 to 28,480 ppm, total Zn from 594 to 14,495 ppm, and total Cd from 3.7 to 156 ppm. Approximately 2 km from the tailing outfall (Station 2A), at a depth of 50 m, total concentrations of 22,940 ppm Zn, 10,354 ppm Pb, and 120 ppm Cd were measured in the particulate material (Table 12-1). Partitioning of this material indicated that 98% of the total Pb, 13% of the total Zn, and 8% of the total Cd were present in the nondetrital fraction (Fig. 12-7). This is in sharp contrast to the estuarine data, which showed that most of the Cd and Zn is held in the nondetrital fraction (Fig. 12-8). High particulate metal levels (19,939 ppm Zn, 10,740 ppm Pb, and 98.8 ppm Cd) also recorded in the sample from Station 4, some 0.2 km from the outfall, at a depth of 50 m. These particles are also characterized by a high proportion of nondetrital Pb (99%) and relatively low proportions of nondetrital Zn (14%) and Cd (10%) (Fig. 12-7). Although the total particulate metals were much lower at Station 12 (1,439 ppm Zn, 518 ppm Pb, and 3.7 ppm Cd), the source of the particles can still be identified by their relatively high proportions of nondetrital Pb and low proportions of nondetrital Zn (19% of total) and Cd (8% of total), even across the 27 m sill at a distance of 3.5 km from the outfall.

DISCUSSION

Knowledge of the way in which metals are bound in particles entering estuarine systems is essential to understanding the processes that govern their distribution and behavior, not only during river transport but also in the estuaries themselves. Such information also allows assessment of the potential bio-availability of particulate metal contaminants entering the estuarine environment by allowing more significant mass balance calculations.

Chemical leaching of the particulate matter has revealed, in general, the mode of transport, the relative availability of the metals, and some knowledge of the metal carriers in the three estuarine systems. The data, although of a limited nature, shows that most of the total particulate metal load entering the St. Lawrence and Tamar estuaries is weakly bound in the particles. The proportion of metal bound in the nondetrital fraction was

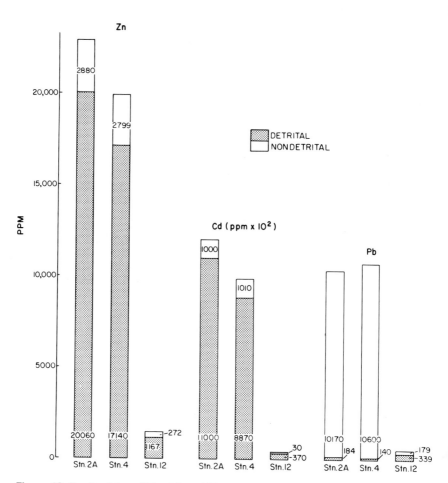

Figure 12–7. Partition of Zn, Cd, and Pb into their nondetrital and detrital contributions in the particulate matter from "A" fjord (Stations 2a and 4) and from "Q" fjord (Station 12) in the Arctic.

found to be in the range of 69%-91% Zn, 69%-95% Cu, 72% Pb, and 87% Cd. Despite the vast differences in size and geographical location between the two estuaries, the proportions of the metals carried in the nondetrital fractions were very comparable. Such high proportions of the total particulate metal contents, carried in the operationally defined nondetrital fraction, have been found in other rivers and estuaries. Duinker and Nolting (1978), Duinker et al. (1980), and Duinker et al. (1982), using a mild chemical leach (0.1 N HCl) comparable to our acetic acid technique, showed that 80%-100% of the total Zn, 75%-95% of the total Pb, 75%-80% of the total Cu, and 80%-100% of the total Cd were held in the acid-soluble (nondetrital) fraction of particulates from the Rhine, Vârde, and Elbe rivers. Such universality of transport mode might imply that most particulate metals entering estuarine systems are weakly bound, but contrary data exist for other systems. Studies by Gibbs (1977) and Presley, Trefry, and Shokes (1980) suggested that the bulk of the particulate metals entering the Amazon, Yukon, and Mississippi rivers were held in the detrital (lattice-bound) fraction. Since the sequential chemical leaching techniques used by these

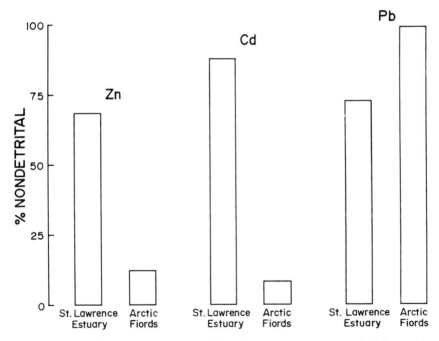

Figure 12-8. Comparison of nondetrital Zn, Cd, and Pb as percentage (%) of the total in particulate matter entering the St. Lawrence estuary with that found in the "A" fjord. Note the low percent of nondetrital Zn and Cd and high percent of nondetrital Pb in particulate matter from "A" fjord.

authors are very different from ours, such results are not directly comparable, but it would be interesting to have comparable data from these systems.

These data imply that most of the natural particulate metal load entering the estuaries is, in a chemical sense, in an exchangeable or reactive position and therefore responsive to changes in the physico-chemical conditions in the estuaries. Although study of the St. Lawrence estuary is not complete, experience from the Tamar area provides some insight into the response to be expected. Under normal flow conditions, the riverine particulate, nondetrital metal composition is modified after entry by internal particle cycling and by resuspension of sediments. Within the turbidity maximum, particulate nondetrital Zn, Cu, and Fe carried into the estuary from the adjacent mineralized area by the river is depleted in response to ambient redox conditions and by changes in the major variables, such as salinity and turbidity, that one might expect to influence particulate metal compositions. In most cases the nondetrital metals are inversely related to particle mass and salinity, not only in the Tamar but also in the St. Lawrence, indicating that other factors such as pH, oxygen-saturation levels, and organic content play a much more important role than was previously thought for controlling particulate metals in estuaries. It is important to note that the detrital metal contributions remained essentially constant irrespective of changes in estuarine conditions that modified the nondetrital composition of the particles. During a period of high river flow, the freshwater particulate inputs predominated to such an extent that they controlled the distribution, abundance, and composition of the estuarine particulate matter. As a result, there was little or no chemical modification of the nondetrital and detrital metal composition throughout the entire estuary except for a seaward increase in calcium and magnesium as the riverine particles mixed increasingly with marine particles of high carbonate content carried by the intruding saline flow. It is doubtful, however, that large estuaries such as the St. Lawrence could be so completely dominated by the influx of riverine particles that chemical modifications of the nondetrital metal concentrations could not, or would not, take place in response to the estuarine conditions. The apparent constancy of the particulate detrital metal fraction is of interest as it might well provide a better estimate of the proportion of riverine particulate metals that might eventually find their way into the deep oceans rather than the transitory total or nondetrital concentrations that are constantly being modified within the estuarine environment.

In contrast, examination of the particulate material artificially introduced into the Arctic fjord system reveals that that system is quite different from the St. Lawrence estuary and is characterized by only a small proportion of Zn and Cd but a very high proportion of Pb in the nondetrital fraction (Fig. 12-8). Such characteristics have been developed as the metals are introduced through the outfall into the fjord. Although some mixing with natural particles undoubtedly occurs in the outer fjord ("Q"), these particles are still identifiable some distance away from the outfall. This allows the spread of metal contamination to be monitored in the fjord system long before such contamination would be noticed in the underlying sediments that are deposited very slowly (0.1 cm/yr).

The nondetrital and detrital metal carriers can be identified by microscopic observation and empirically (Loring et al., 1982). The nondetrital fraction represents the proportion of the total metal concentration that was initially leached from the source rocks or supplied in dissolved form from industrial sources. This nondetrital fraction has been incorporated or become associated with the solid particles at the site of weathering and during river transport, from solution or colloidal form, by precipitation, absorption, or extraction by living and dead organisms. Soluble iron oxide grain coatings have been found (Loring et al., 1982; Loring, 1978) to be the main carriers of nondetrital Zn and Cu in the Tamar particulates and in the estuarine sediments of the St. Lawrence, whereas organic matter appeared to be the main carrier of Cd in the Tamar and the St. Lawrence. The presence of lead carbonate in the "A" and "Q" fjord particulates accounts for the very high nondetrital contribution of Pb in this material. The Pb carbonates (Asmund, 1980) result from the initial oxidation of lead sulfides to sulfates and the subsequent reaction with sea water to form carbonates. The Zn- and Cd-bearing sulfides, however, do not oxidize to the same extent and so are dispersed as detrital particles in the fjord. Elsewhere, nondetrital Pb is most likely carried in iron oxide grain coatings and weakly bound to organic matter.

The detrital carriers are easier to determine. The metals in this fraction are derived from the lattices of silicate, sulfide, and oxide minerals as well as the acid-insoluble secondary compounds and are transported as fine-grained clastic particles. Although small amounts of the metals reside in the silicate lattices, most of the detrital Pb, Zn, and Cu is carried in detrital sulfide minerals such as pyrite (FeS_2), galena (PbS), sphalerite (ZnS), and chalcopyrite ($CuFeS_2$), and sometimes in the ferromagnesium minerals. Such sulfides are found (Asmund, 1980) in the tailings outfall to "A" fjord system and account for the extremely high levels of Pb, Zn, and Cd found in the detrital fraction of the particles throughout the water column.

As a result of this and other studies, it should be clear that there is an ever-increasing awareness that a variety of chemical and biological processes in estuarine and coastal waters contributes to significant exchanges of material between particulate and dissolved states. Differentiation of the essentially unreactive detrital components from the potentially exchangeable nondetrital constituents of particulate materials in these waters is required to facilitate the recognition and quantification of such processes. It also allows an improved understanding and use of local and global mass balances of the elements.

ACKNOWLEDGEMENTS

The authors wish to thank Dr. P. A. Yeats, Atlantic Oceanographic Laboratory, for permission to use some unpublished data from BIO cruise 76-006.

REFERENCES

Asmund, G., 1975, Environmental study in relation to mining operations at Marmorilik, Umanak district, central west Greenland, *Rapp. Gronlands Geol. Unders.* **75**:46-47.

Asmund, G., 1980, Water movements traced by metals dissolved from mine tailings deposited in a fjord in north-west Greenland, in *Fjord Oceanography,* H. J. Freeland, D. M. Framer, and C. D. Levings, eds., Plenum Press, New York and London, pp. 347-353.

Asmund, G., H. J. Bollingberg, and J. Bondam, 1976, Continued environmental studies in the Qaumarujuk and Agfardlikavsa fjords, Marmorilik, Umanak district, central west Greenland, *Rapp. Gronlands Geol. Unders.* **80**:53-61.

Bewers, J. M., and P. A. Yeats, 1978, Trace metals in the waters of a partially mixed estuary, *Estuar. and Coastal Mar. Sci.* **7**:147-162.

Bondam, J., 1978, Recent bottom sediments in Agfardlikavsa and Qaumarujuk fjords near Marmorilik, west Greenland. *Geol. Soc. Denmark. Bull.* **27**:39-45.

Bondam, J., and G. Asmund, 1974, Environmental studies in the Qaumarujuk and Agfardlikavsa fjords, Umanak district, central west Greenland, *Rapp. Gronlands Geol. Unders.* **65**:29-33.

Butler, E. I., and S. Tibbets, 1972, Chemical survey of the Tamar estuary. I. Properties of the water, *Mar. Biol. Assoc. U. K. J.* **52**:681-699.

d'Anglejan, B., and E. C. Smith, 1973, Distribution, transport, and composition of suspended matter in the St. Lawrence estuary, *Can. J. Earth Sci.* **10**:1380-1396.

Duinker, J. C., and R. F. Nolting, 1978, Mixing, removal, and mobilization of trace metals in the Rhine estuary, *Neth. J. Sea Res.* **12**:205-223.

Duinker, J. C., M. T. J. Hillebrand, R. F. Nolting, S. Wellershaus, and N. K. Jacobsen, 1980, The river Vârde: Processes affecting the behaviour of metals and organochlorines during estuarine mixing, *Neth. J. Sea Res.* **14**:237-267.

Duinker, J. C., M. T. J. Hillebrand, R. F. Nolting, and S. Wellershaus, 1982, The river Elbe: Processes affecting the behaviour of metals and organochlorines during estuarine mixing, *Neth. J. Sea Res.* **15**:141-169.

Gibbs, R. J., 1977, Transport phases of transition metals in the Amazon and Yukon rivers, *Geol. Soc. Amer. Bull.* **88**:829-843.

Holeman, J. N., 1968, The sediment yield of major rivers of the world, *Water Resour. Res.* **4**:737-747.

Jordon, F., 1973, The St. Lawrence system run-off estimates, *Bedford Inst. Oceanogr. Rep.* BI-D-73-10.

Kranck, K., 1979, Dynamics and distribution of suspended particulate matter in the St. Lawrence estuary, *Nat. Can.* **106**:163-173.

Loring, D. H., 1978, Geochemistry of zinc, copper and lead in the sediments of the estuary and Gulf of St. Lawrence, *Can. J. Earth Sci.* **16**:757-772.

Loring, D. H., R. T. T. Rantala, A. W. Morris, A. J. Blake, and R. J. M. Howland, 1982, Chemical composition of suspended particles in an estuarine turbidity maximum zone, *Can. J. Fish. Aquat. Sci.* **40:**201-206.

Morris, A. W., A. J. Bale, and R. J. M. Howland, 1982, Chemical variability in the Tamar estuary, southwest England, *Estur. Coastal Shelf Sci.* **14:**649-663.

Presley, B. J., J. H. Trefry, and R. F. Shokes, 1980, Heavy metal inputs to Mississippi delta sediments, *Water Soil Pollut.* **13:**481-494.

Rantala, R. T. T., and D. H. Loring, 1973, New low-cost Teflon decomposition vessel, *At. Absorb. Newl.* **12:**97-99.

Rantala, R. T. T., and D. H. Loring, 1977, A rapid determination of 10 elements in marine suspended particulate matter by atomic absorption spectrophotometry, *At. Absorb. Newl.* **16:**51-52.

Rantala, R. T. T., and D. H. Loring, 1980, Direct determination of cadmium in silicates from a fluoroboric-boric acid matrix by graphite furnace atomic absorption spectrometry, *At. Spectrosc.* **1:**163-165.

Rantala, R. T. T., and D. H. Loring, in press, Partition and determination of cadmium, copper, lead, and zinc in marine suspended particulate matter, in *International Journal of Environmental and Analytical Chemistry.*

Yeats, P. A., and J. M. Bewers, 1982, Discharge of metals from the St. Lawrence River, *Can. J. Earth Sci.* **19:**982-992.

13

The Binding of Metallic Ions to Bacterial Walls and Experiments to Simulate Sediment Diagenesis

Terry J. Beveridge

University of Guelph

INTRODUCTION

The outermost component of the bacterial cell is its wall; it contributes not only cellular shape and form but also an inanimate boundary through which the cell perceives the surrounding environment. Whatever goes into or out of bacteria must eventually percolate through the fabric of the wall. This fabric, to a certain extent, must influence the molecular (Nakae and Nikaido, 1975; Yoshimura and Nikaido, 1982) and ionic (Beveridge, 1981; Beveridge and Murray, 1980; Beveridge and Koval, 1981; Beveridge, Forsberg, and Doyle, 1982; Doyle, Matthews, and Streips, 1980; Marquis, Mayzel, and Carstensen, 1976; Matthews, Doyle, and Streips, 1979) form of the entering or exiting substance. Given the stress and strain of natural microbial habitats, it is conceivable that this boundary layer could act as a microenvironment that completely surrounds the protoplast and that could interact with, buffer, or even modify the stressing influence of the habitat before it comes in contact with the cell.

Like most cell surfaces, the bacterial wall is anionic (Beveridge, 1981). This characteristic is independent of whether or not the bacterium is gram-positive or gram-negative (Beveridge and Murray, 1980; Beveridge and Koval, 1981; Doyle, Matthews, and Streips, 1980; Hoyle and Beveridge, 1983, 1984). *Bacillus* walls of the gram-positive variety consist of these primary components: a peptidoglycan (PG) matrix to which either teichoic (TA) or teichuronic (TUA) acids are covalently bound. Each of these polymers, because of its chemical makeup, is anionic, and the various ratios of each ultimately determine the wall's electronegative charge density (Beveridge and Murray, 1980; Beveridge, Forsberg, and Doyle, 1982; Doyle, Matthews, and Streips, 1980; Matthews, Doyle, and Streips, 1979). Good evidence of the anionic nature of *Bacillus* walls can be seen by their inherent ability to adsorb cationized ferritin (an electron-microscopic marker for electronegative sites) to their surfaces (Fig. 13-1).

The gram-negative envelope is more complex than its gram-positive counterpart. It consists of two membrane bilayers (the outer and plasma membranes) that are chemically distinct from one another and that sandwich a thin PG layer between them (Beveridge, 1981). The outer membrane (OM) and the PG layer constitute the wall of these bacteria and, at least with *Escherichia coli,* these layers are negatively charged (Beveridge and Koval, 1981; Hoyle and Beveridge, 1983, 1984).

Bacteria are ubiquitous in nature and form an often large but variable proportion of the organic load in the sediments. Given that they have an encompassing layer (or layers) that is anionically charged, it is reasonable to assume that they will interact strongly with dilute metal in solution within natural bodies of water. Laboratory experiments have demonstrated that metal accumulation can be substantial within the wall fabric (Bevridge and Murray, 1976; Beveridge and Murray, 1980; Beveridge and Koval, 1981; Beveridge, Forsberg, and Doyle, 1982; Doyle, Matthews, and Streips, 1980;

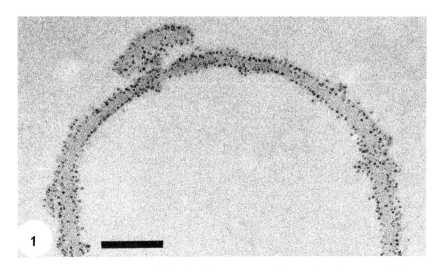

Figure 13-1. Thin section of a *Bacillus thuringiensis* wall stained with cationized ferritin and contrasted with uranyl acetate. The cationized ferritin has bound to the surfaces of the wall.

Marquis, Mayzel, and Carstensen, 1976; Matthews, Doyle, and Streips, 1979). Given a light but constant rain of bacteria and their products throughout the aqueous environment, this could be an important process in the immobilization of metals into the sediments. Here, the accumulated metal could undergo the geochemistry necessary for mineralization.

The described experiments demonstrate the natural metal-binding capacity of walls and the tenacity of the binding during a laboratory simulation of low-temperature (100°C) diagenesis. These biological polymers with their bound metal stimulate and quicken mineral authigenesis.

MATERIALS AND METHODS

The Binding of Metallic Ions to Bacterial Walls

Walls from *Bacillus subtilis* 168, *B. licheniformis* NCTC 6346 *his,* and *Escherichia coli* AB264 were prepared from exponential-growth-phase cultures that had been broken by passage through a French Pressure Cell at 18,000 psi. The suspension of cell fragments was digested with RNAse (100 μg/ml) and DNAse (50 μg/ml) for 60 min at room temperature and then washed with 0.05 M HEPES buffer (pH 6.8) until the supernatant was devoid of 260 nm absorbing material. Details of the purification procedures for *B. subtilis* and *B. licheniformis* walls can be found in Beveridge and Murray (1976), and for *E. coli* envelopes in Beveridge and Koval (1981).

TA was extracted from the gram-positive walls by 0.1 N NaOH treatment at 35°C for 24 hr (Beveridge and Murray, 1980). TUA was removed from *B. licheniformis* walls by treatment with 5% (w/v) trichloroacetic acid for 8 hr at 35°C (Beveridge, Forsberg, and Doyle, 1982).

To partition the *E. coli* envelopes into outer membrane and peptidoglycan, they were processed as follows: For outer membrane, envelopes were treated with 1% Triton X-100 (v/v) for 30 min at 30°C to remove the plasma membrane (Schnaitman, 1971) and with 50 μg lysozyme/ml for 60 min at 23°C to remove the peptidoglycan. This resulted in an outer membrane preparation that was chemically similar to that produced by Osborn's sphaeroplast osmotic lysate procedure (Hoyle and Beveridge, 1983; Osborn et al., 1972). The peptidoglycan layer was isolated from *E. coli* by boiling the cells in 4% (w/v) sodium dodecylsulfate for 45 min. RNAse (100 μg/ml) and DNAse (50 μg/ml) digested associated nucleic acids, whereas trypsin (30 μg/ml) removed bound lipoprotein (Braun and Sieglin, 1970; Braun and Wolff, 1970; Hoyle and Beveridge, 1984).

Each of the distinct wall preparations was incubated under saturating binding conditions in 2 ml of a 5 mM metal chloride solution for 10 min at 23°C (Beveridge and Murray, 1976). Depending on the preparation, 0.1 to

5.0 mg dry weight of walls were used (Beveridge and Murray, 1976; Beveridge and Murray, 1980; Beveridge and Koval, 1981; Beveridge, Forsberg, and Doyle, 1982); the walls were then washed with water by centrifugation until no metal could be detected in the supernatant. Quantitation of the bound metal was obtained by either atomic absorption or X-ray fluorescence spectroscopy. Electron microscopy was performed on both unfixed and fixed (4% glutaraldehyde) samples of whole mounts and thin sections. No stains were used except for the metal that was bound to the walls. Consequently, the electron-scattering images showed the exact positioning of the metal. Energy-dispersive X-ray analyses (EDS) of each sample ensured that the only electron-scattering agent within the walls was the absorbed metal. Elementary deposition products were identified by X-ray diffraction.

For the diagenesis experiments, whole *B. subtilis* cells whose walls had been loaded with metal (as previously described) were used. About 25 mg of dried bacteria were mixed with 250 mg synthetic sediment, and 0.5 ml deionized water was added. Usually 50 mg of either magnetite or elemental sulfur was added as a redox buffer. The synthetic sediments were "spec pure" crystalline quartz, calcite ($CaCO_3$), or a combination of the two. Each mixture was added to a thick Pyrex tube, sealed by flame, and heated at 100°C for periods of 1, 10, 100, and 200 days. The samples were monitored by electron microscopy and EDS throughout the experiment (Beveridge et al., 1983).

RESULTS AND DISCUSSION

Metal Binding by Bacterial Walls

Gram-positive Walls.   Due to growth conditions (phosphate and magnesium were present) these walls consisted of two polymers, teichoic acid (54%) and peptidoglycan (45% dry weight) (Beveridge and Murray, 1980). So much metal was bound to these walls that visible electron-dense aggregates could sometimes be seen by electron microscopy. In general, the more unstable the metal in aqueous solution (e.g., the lanthanides), the more often metal precipitates were encountered within the fabric of the wall (Fig. 13-2 is an extreme example). Precipitates were never seen in the metal solutions without the addition of the walls. Metallic ions that were freely soluble in water (e.g., Na, Mg^{2+}, Cu^{2+}, and so on) were also avidly sequestered from solution by the walls. In these instances, the walls were diffusely stained (Fig. 13-3 is representative).

To determine the chemical sites of metal interaction within the wall, TA was extracted, and metal binding studies were performed on the remaining PG sacculi. Metal uptake decreased in all instances, but it was apparent that most of the binding capacity remained associated with the PG (Table 13-1). Carboxylate groups constituent within the PG are the sites that should be most electronegative in this polymer. These were neutralized by the addition of glycine ethyl ester to the available carboxyl groups, resulting in a substantial decrease in binding capacity (Figs. 13-3 and 13-4 and Table 13-1).

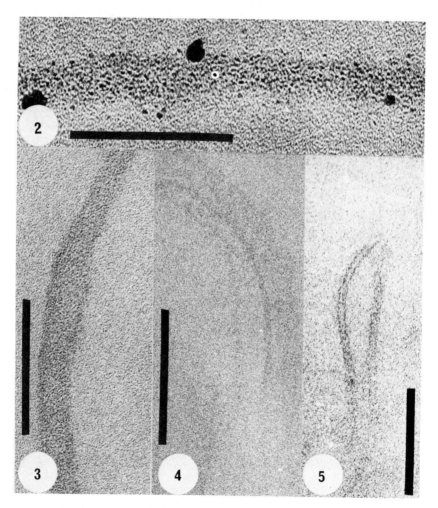

Figure 13-2.   Thin section of a B. subtilis wall with sorbed palladium. No stain other than the palladium bound to the wall has been used (Beveridge and Murray, 1976).

Figure 13-3.   This B. subtilis wall has bound scandium (Beveridge and Murray, 1976).

Figure 13-4.   Same as Figure 13-3, but the carboxyl groups have been neutralized by addition of glycine ethyl ester after carbodiimide activation (Beveridge and Murray, 1980). Very little scandium has been bound.

Figure 13-5.   Thin section of an E. coli outer membrane that has been contrasted with magnesium (Hoyle and Beveridge, 1983b).

Table 13–1. Metal Binding by Bacterial Walls

Metal	B. subtilis			B. licheniformis			E. coli[a]	
	Native[a]	Neutralized COO[a,b]	No Teichoic Acid[c]	Native[a]	No Teichoic or Teichuronic Acids[c]	Envelopes[d]	Murein	Outer Membrane
Na	2.697	0	1.497	0.910	0.080	0.042	0.290	0.081
K	1.944	0	0.782	0.560	0	0.082	0.058	0.025
Mg	8.226	0.520	7.683	0.400	0.024	0.256	0.035	0.019
Ca	0.399	0.380	0.012	0.590	0.096	0.035	0.038	0.020
Mn	0.801	0.732	0.656	0.662	0.004	0.140	0.052	0.012
FeIII	3.581	2.260	1.720	0.760	0.172	0.200	0.100	0.233
Ni	0.107	0.024	0.021	0.520	0	0.002	0.019	0.019
Cu	2.990	0.993	2.488	0.490	0	0.090	n.d.	n.d.
AuIII	0.363	0.214	0.265	0.031	0.012	0.056	n.d.	n.d.

n.d. = not determined
[a] μmol metal per mg dry weight of walls.
[b] Neutralized by the addition of glycine ethyl ester to carbodiimide-activated carboxylate groups.
[c] The dry weight of these walls has been adjusted downwards to reflect the loss of mass due to the extraction. All quantities are expressed in μmol.
[d] The envelope consists of outer membrane, peptidoglycan, and plasma membrane.

161

These types of experiments (Beveridge and Murray, 1976; Beveridge and Murray, 1980; Doyle, Matthews, and Streips, 1980) have demonstrated with the *B. subtilis* 168 wall that most of the metal binding is determined by the PG. Clearly, there is no apparent stoichiometry between the excessive quantity of metal that binds to sites on the walls (see Beveridge and Murray [1980] for details). Yet, these quantities are real (Beveridge and Murray, 1976; Beveridge and Murray, 1980; Doyle, Matthews, and Streips, 1980; Matthews, Doyle, and Streips, 1979). A two-step mechanism is proposed for the deposition process; the first step in time is a stoichiometric interaction between metal ion and active site within the wall. This interaction then acts as a nucleation site for the deposition of more metal ion from solution. The deposition product therefore grows in size within the intermolecular spaces of the wall fabric until it is physically constrained by the polymeric meshwork of the wall. The end result is a bacterial wall that contains copious amounts of metal not easily replaced by water (i.e., hydronium ion).

   B. licheniformis NCTC 6346 *his* walls are not like those of *B. subtilis*. These walls contain TUA (26%) as an additional component to PG (22%) and TA (52%). Since the walls contained less than one-half the PG of *B. subtilis* walls and an additional polymer (TUA), their metal-binding capacity was of interest (Beveridge, Forsberg, and Doyle, 1982).

   B. licheniformis walls were not able to bind as much metal as those of *B. subtilis* (Table 13-1). In fact, when TA and TUA were extracted (Beveridge, Forsberg, and Doyle, 1982), the walls lost most of their binding capacity (Table 13-1). This suggests that, unlike the *B. subtilis* situation, it is the secondary polymers that interact most strongly with metallic ions in solution, and that there is a fundamental charge-distribution difference between the two types of walls.

Gram-negative Walls.   Only one gram-negative wall, that of *E. coli* K-12 strain AB264, has been examined. The OM of this bacterium possesses an LPS that contains a complete core oligosaccharide but lacks O-antigenic side chains. The major OM proteins are the lipo-, OmpF, OmpC, and OmpA proteins. The major phospholipid is phosphatidylethanolamine (Beveridge and Koval, 1981; Hoyle and Beveridge, 1983; White, Lennarz, and Schnaitman, 1972).

   These envelopes did not bind as much metal from solution as their gram-positive counterparts. Typically, the quantities were about one-tenth or more of these amounts (Table 13-1). This also held true when the envelopes were partitioned into their constituent layers. For example, the OM did not react as avidly with metal ions as did gram-positive walls (Table 13-1), but there was still enough metal to give an electron-scattering profile (Fig. 13-5). The bilayer distribution of the metal suggested that the hydrophilic faces of the OM provided the sites of metallic ion interaction. The prime candidates for this interaction are the phosphate groups resident within the polar head groups of the LPS and phospholipids. The validity of this has been proven by ^{31}P-NMR of isolated OM in the presence of stoichiometric amounts of europium (a paramagnetic lanthanide); as the Eu^{3+} concentration was increased, the ^{31}P signal was masked (decreased) and chemically shifted to the right. At a 1:1 ratio of Eu^{3+}:^{31}P, the signal is almost nonexistent (Fig. 13-6).

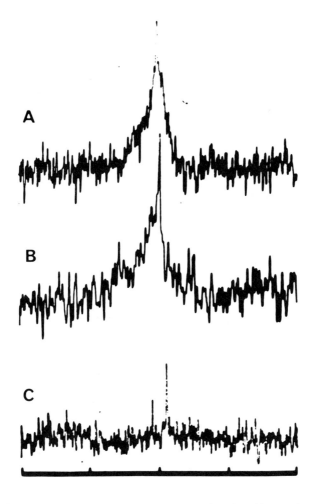

Figure 13-6. ^{31}P-NMR scans of *E. coli* outer membrane after the addition of europium. (*A*) 1 Eu:4 P, (*B*) 1 Eu: 2 P, and (*C*) 1 Eu: 1 P. The europium is binding to the phosphate of the LPS and phospholipid to mask the ^{31}P signal and shift it to the right. Very little signal is left when Eu = P.

The PG of *E. coli* is chemically similar to that of *B. subtilis* and *B. licheniformis* (Schleifer and Kandler, 1972) and most probably exists as a monolayer (Beveridge, 1981). It has a crosslinking efficiency of 30% and has an alanine:glutamic acid:diaminopimelic acid ratio of 2.7:1:1 (Hoyle and Beveridge, 1984). It interracts more strongly with metal ions than does the OM (Table 13-1), presumably because of available carboxylate groups. It is estimated that the total number of carboxyl groups available for interaction with metal ions in this PG is 0.013 ± 0.005 micromol/mg dry weight (Hoyle and Beveridge, 1984). Clearly, there is more metal adsorbed to the PG than simple stoichiometry could explain, but the amounts are reduced from those attributed to gram-positive PG (Table 13-1). Data suggest that a two-step deposition mechanism is at work with *E. coli* PG (hence, greater than stoichiometric quantities of metal) but, since there is only a monolayer of fabric, there are few intermolecular spaces in which metal aggregates can grow (hence, reduced quantities when compared to the 25 layers of *Bacillus* PG).

Laboratory Simulation of Low-Temperature Sediment Diagenesis.   By mixing metal-loaded bacteria with a synthetic sediment of quartz and calcite in the presence or absence of magnetite or sulfur as redox buffers, 100°C sediment diagenesis has been simulated in the laboratory (Beveridge et al., 1983). The metal associated with the bacteria was primarily in the wall fabric (Beveridge and Murray, 1976; Beveridge et al., 1983). The sediment was monitored during diagenesis by electron microscopy to follow the mineralization process. Typically, crystallization was seen as early as day number one since microcrysts developed within the bacteria wall (Fig. 13-7). These grew with time until all of the wall material had been mineralized (Fig. 13-8); next, crystals developed within the cytoplasm (Fig. 13-9). Eventually, the entire cell became crystalline; the composition of the mineral depended on the metal sorbed to the bacterium and mobile ions within the inorganic fraction of the sediment (Beveridge et al., 1983). Energy-dispersive X-ray analysis allowed determination of the composition of mineral, whereas crystal habit provided mineral identity (Beveridge et al., 1983). Table 13-2 outlines the results of these experiments.

Biogeochemical Consequences.   It is now possible to state unequivocally that bacteria interact with metallic ions in solution and bind large amounts to their walls. Gram-positive walls seem more reactive than their gram-negative counterparts, but each in its own right is capable of strong interaction.

Given a light but constant rain of bacteria and their products throughout

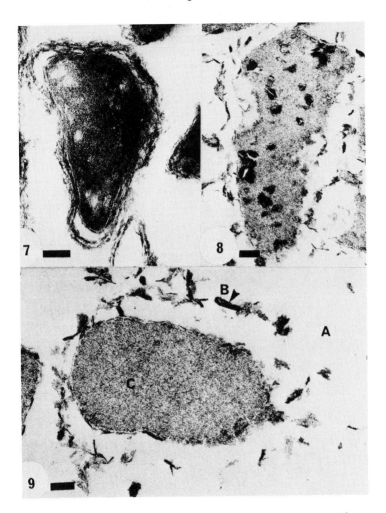

Figure 13-7. Thermal degradation of uranium-loaded cells in quartz with magnetite (as a redox buffer) after one day. The wall has begun crystallization; arrows point to developing microcrysts. *(From Beveridge et al., 1983, p. 1099).*

Figure 13-8. Same as Figure 13-7, but after 10 days. Platy phosphate microcrysts are found in the wall and cytoplasm. *(From Beveridge et al., 1983, p. 1099).*

Figure 13-9. Same as Figure 13-8. The entire wall has been converted to microcrysts, but none (as yet) are found in the cytoplasm. Energy-dispersive X-ray analysis was done at points A, B, and C. B had a high uranium, phosphorus, and potassium level and C had the same elements in reduced quantities, whereas A was completely devoid of them. After 100 days meta-ankoleite ($K_2(UO_2)(PO_4)_2 \cdot 6H_2O$) was the predominant mineral. *(From Beveridge et al., 1983, p. 1099).*

Table 13-2. Mineralization During Experiments [a]

Metal[b]	Sediment[c]	B. subtilis[d]	Redox[e] Buffer	Mineralization
Controls (no	Q	−	−	
sorbed metal)	Q	−	M	
	Q	−	S	
	Q	+	−	Phosphate microcrysts
	Q	+	M	Phosphate microcrysts
	Q	+	S	Phosphate microcrysts
	C	−	−	
	C	−	M	
	C	−	S	
	C	+	−	Phosphate microcrysts[f]
	C	+	M	Phosphate microcrysts[f]
	C	+	S	Phosphate microcrysts[f]
Uranium	−	+	M	Uranium phosphate microcrysts
	−	+	S	Uranium phosphate microcrysts; polymeric uranium-organic sulfide residues
	Q	+	M	Uranium phosphate microcrysts
	Q	+	S	Uranium phosphate microcrysts; polymeric uranium-organic sulfide residues
	Q + C[g]	+	M	Uranium phosphate microcrysts[f]
	Q + C[g]	+	S	Uranium phosphate microcrysts; polymeric uranium-organic sulfide residues
Copper	−	+	M	Phosphate microcrysts[h]
	−	+	S	Dominant copper sulfides + traces of phosphate microcrysts[h]
	Q	+	M	Phosphate microcrysts[h]
	Q	+	S	Dominant copper sulfides + traces of phosphate
	Q + C[g]	+	M	Minimal phosphate microcrysts[f]
	Q + C[g]	+	S	Dominant copper sulfides[f]
Zinc	−	+	M	Phosphate microcrysts[h]
	−	+	S	Dominant zinc sulfides + traces of zinc phosphate microcrysts
	Q	+	M	Phosphate microcrysts
	Q	+	M	Dominant zinc sulfides + traces of phosphate microcrysts
	Q + C[g]	+	M	Minimal phosphate
	Q + C[g]	+	S	Dominant zinc sulfides
Iron	−	+	M	Iron phosphate microcrysts
	−	+	S	Iron phosphate microcrysts

[a]Due to limitations of time and space, not all combinations were run. Those chosen were intuitive and of most interest to the authors.
[b]Metal that was sorbed to bacterial cells.
[c]Q, quartz (SiO_2); C, calcite ($CaCO_3$).
[d]−, Bacteria absent; +, bacteria present.
[e]M, magnetite; S, elemental sulfur.
[f]No calcium was detected associated with these microcrysts.
[g]Calcite was added as a pH buffer, and the extent of phosphate mineralization was reduced as a consequence of the higher pH.
[h]Phosphate mineralization was reduced from that seen in the uranium and iron experiments.

our natural bodies of water, it can be concluded that they could remove from the water dilute metals and partition them into the sediments. Here the tenacity of the binding of metal to biopolymer would so strongly immobilize the metal that mineralization could proceed given the proper temperature, pressure, and redox potential. In time, diagenesis would be complete. The simulation of sediment diagenesis implies that, at least with a low-temperature horizon, bacterial components both stimulate and quicken mineral anthigenesis (Beveridge et al., 1983).

ACKNOWLEDGEMENTS

I am indebted to all of my colleagues who helped with this research and to the Natural Sciences and Engineering Council of Canada, which supported this work.

REFERENCES

Beveridge, T. J., 1981, Ultrastructure, chemistry, and function of the bacterial wall, *Int. Rev. Cytol.* **72:**229-317.

Beveridge, T. J., and S. F. Koval, 1981, Binding of metals to cell envelopes of *Escherichia coli* K-12, *Appl. Environ. Microbiol.* **42:**325-335.

Beveridge, T. J., and R. G. E. Murray, 1976, Uptake and retention of metals by cell walls of *Bacillus subtilis, J. Bacteriol.* **127:**1502-1518.

Beveridge, T. J., and R. G. E. Murray, 1980, Sites of metal deposition in the cell wall of *Bacillus subtilis, J. Bacteriol.* **141:**876-887.

Beveridge, T. J., C. W. Forsberg, and R. J. Doyle, 1982, Major sites of metal binding in *Bacillus licheniformis* walls, *J. Bacteriol.* **150:**1438-1448.

Beveridge, T. J., J. D. Meloche, W. S. Fyfe, and R. G. E. Murray, 1983, Diagenesis of metals chemically complexed to bacteria: Laboratory formation of metal phosphates, sulfides, and organic condensates in artificial sediments, *Appl. Environ. Microbiol.* **45:**1094-1108.

Braun, V., and V. Sieglin, 1970, The covalent murein-lipoprotein structure of the *Escherichia coli* cell wall. The attachment site of the lipoprotein on the murein, *Eur. J. Biochem.* **13:**336-346.

Braun, V., and H. Wolff, 1970, The murein-lipoprotein linkage in the cell wall of *Escherichia coli, Eur. J. Biochem.* **14:**387-391.

Doyle, R. J., T. H. Matthews, and U. N. Streips, 1980, Chemical basis for the selectivity of metal ions by the *Bacillus subtilis* cell wall, *J. Bacteriol.* **143:**471-480.

Hoyle, B. D., and T. J. Beveridge, 1983, The binding of metallic ions to the outer membrane of *Escherichia coli., Appl. Environ. Microbiol.* **46:**749-752.

Hoyle, B. D., and T. J. Beveridge, 1984, Metal binding by the peptidoglycan sacculus of *Escherichia coli* K-12, *Can. J. Microbiol.* **30:**204-211.

Marquis, R. E., K. Mayzel, and E. L. Carstensen, 1976, Cation exchange in cell walls of gram-positive bacteria, *Can. J. Microbiol.* **22:**975-982.

Matthews, T. H., R. J. Doyle, and U. N. Streips, 1979, Contributions of peptidoglycan to the binding of metal ions by the cell wall of *Bacillus subtilis, Curr. Microbiol.* **3:**51-53.

Nakae, T., and H. Nikaido, 1975, Outer membrane as a diffusion barrier in *Salmonella typhimurium*. Penetration of oligo- and polysaccharides into isolated outer membrane vesicles and cells with degraded peptidoglycan layer, *J. Biol. Chem.* **250:**7359-7365.

Osborn, M. J., J. E. Gander, E. Parisi, and J. Carson, 1972, Mechanism of assembly of the outer membrane of *Salmonella typhimurium*. Isolation and characterization of cytoplasmic and outer membrane, *J. Biol. Chem.* **247:**3962-3972.

Schliefer, K. H., and O. Kandler, 1972, Peptidoglycan types of bacterial cell walls and their taxonomic implications, *Bacteriol. Rev.* **36:**407-477.

Schnaitman, C. A., 1971, Solubilization of the cytoplasmic membrane of *Escherichia coli* by Triton X-100, *J. Bacteriol.* **108:**545-552.

White, D. A., W. J. Lennarz, and C. A. Schnaitman, 1972, Distribution of lipids in the wall and cytoplasmic membrane of the cell envelope of *Escherichia coli, J. Bacteriol.* **109:**686-690.

Yoshimura, F., and H. Nikaido, 1982, Permeability of *Pseudomonas aeruginosa* outer membrane to hydrophilic solutes, *J. Bacteriol* **152:**636-642.

14

Microbial Potentiation of Arsenic Transport from Soil and Oil Shale Energy Waste

Robert A. Sanford and Donald A. Klein

Colorado State University

R. R. Meglen

University of Colorado at Denver

INTRODUCTION

Arsenic biotransformation and mobilization by microorganisms has been well documented, with Challenger (1945) and Challenger, Higgenbottom, and Ellis (1933) observing volatile arsenical production from inorganic and organoarsenical substrates by cultures of *Scopulariopsis brevicaulis.* A variety of organism and reaction mechanisms has been documented in prior literature on arsenic transfer in natural systems. Cox and Alexander (1973) isolated cultures of *Candida, Humicola,* and *Penicillium* that methylated and reduced arsenic species, forming trimethylarsine. Transformation by *Methanobacterium* of arsenate to dimethylarsine under anaerobic conditions was observed by McBride and Wolfe (1971). Cheng and Focht (1979) demonstrated that microorganisms in soils, amended with methylated and inorganic forms of arsenic, also resulted in arsenic volatilization by reductive or demethylative pathways; in addition, *Pseudomonas* and *Alcaligenes* isolates reduced arsenate and arsenite to arsine. In comparison, a reductive or methylative pathway was suggested to occur in the production of volatile arsenicals from soils in studies by Woolson (1977). In soils, differing rates of transformation in relation to added arsenic forms have been shown by Holm et al. (1980).

Oil shale development and processing are of particular environmental interest because of arsenic concentrations in the range of 20-60 $\mu g/g$ that are typically found in retorted material (Shendrikar and Faudel, 1978). Recently,

169

Hassler (1982) and Klein and Hassler (1981) suggested that microorganisms may play a role in arsenic volatilization from retorted shales. The present study was initiated to examine in more detail the microbial role in arsenic release from spent leached and unleached shales in comparison with soils. With more detailed information on microbial contributions to arsenic releases from these materials, it should be possible to develop a better understanding of biogeochemical cycling of arsenic in natural systems.

MATERIALS AND METHODS

Retorted Lurgi-6 process shale was acquired from the Rio Blanco Oil Shale Company. Soil was collected at an intensive study site in the Piceance Basin; this site is under the supervision of the Range Science Department, Colorado State University. The retorted shale and soil were sieved through a 32-mesh screen, and leaching was done by passing approximately 2 l of deionized water through each kilogram of shale placed in a pipette washer. After leaching, the material was dried to the same moisture content as the original retorted material.

The arsenic contents, as determined by the Colorado State University Soil Testing Laboratory, of the soil and the leached and nonleached shale were 5, 29, and 31 $\mu g/g$, respectively. Conductivity and pH decreased slightly in leached shale from values obtained with the original materials.

The volatile trapping system employed was a modified version of the system utilized by Woolson (1977) and by Hassler (1982). A 500-ml flask was connected to one 150 mm x 25 mm test tube using polyethylene and glass tubing connections. Premoistened air was passed over shale and soil samples in flasks at a rate of approximately 30 ml/min and then bubbled through 20 ml 0.01 N KI trapping solutions with excess I_2. Neoprene stoppers were used throughout the experiment since rubber reportedly absorbs volatile arsenicals (McBride and Edwards, 1977). Trapping solutions were sampled at three-week intervals over a 15-week period. Analyses were carried out at the analytical laboratory at the University of Colorado at Denver, using graphite-furnace, atomic-absorption spectrometry with nickel matrix modification.

Bacterial numbers and fungal hyphal lengths were estimated microscopically. One-gram samples were taken at three-week intervals concurrently with the volatile (As) trap sampling. Each sample was blended in a bicarbonate buffer for 2 min, using a modification of the procedure described by Babiuk and Paul (1970). From the blended solution subsamples were taken for separate microscopic analyses. Ten μl of the solution were spread onto a 1-cm^2 taped area on a slide, dried, and stained for 3 min with fluorescein isothiocyanate. Additional 1.0-ml subsamples were placed in individual tubes, stained with phenolic aniline blue, and placed as an agar-film onto microscope slides for direct measurement of fungal hyphal lengths (Jones and Mollison, 1948).

Retorted oil shale or soil (100 g) was mixed with soybean meal (6 g) and amendments of sodium arsenate (SA), dimethyl arsinic acid (DMAA), or methylarsonic acid (MAA), then placed in 500-ml flasks. Arsenic stock solutions in 34 ml H_2O containing 2.44 x 10^2 μg SA/ml, 2.29 x 10^2 μg MAA/ml, or 1.09 x 10^2 μg DMAA/ml were added to the leached shale, nonleached shale, and soil samples to give final values of 20 μg/g As. A set of samples without arsenic amendments, but with equivalent moisture contents, was used to monitor innate arsenic volatilization. As a control, similar flasks of shale were amended with nutrients and sterilized prior to the addition of a filtered arsenic solution. These controls were monitored only for volatile arsenical production, and the apparatus remained intact for the duration of the experiment.

RESULTS

Arsenic Volatilization

Cumulative percent evolution of total arsenic over a 15-week period by soil, nonleached and leached shale amended with nutrients, and arsenic compounds is shown in Table 14-1. Sterile nonleached shale with DMAA and MAA amendments showed no cumulative arsenic volatilization over a nine-week period. This result is in contrast to the nonsterile samples in which DMAA-treated samples showed the highest releases of volatile arsenic. DMAA-treated leached shale appeared to release volatile arsenicals before nonleached shale (Fig. 14-1), although the cumulative releases from the two samples were quite similar. Soil treated with MAA showed a fourfold increase in arsenic volatilized over similarly treated shale samples (Fig. 14-2). Sodium arsenate-treated material showed extensive As release from leached shale, while soil and nonleached shale released fairly low amounts of arsenic. The trend of varying arsenic releases with nonleached and leached shale was further supported by results without arsenic amendments; higher amounts of arsenic were trapped from nonleached samples (Table 14-1).

Table 14-1. Percent Total Arsenic Released from Soil, Shale, and Leached Shale Treated with Various Arsenic Compounds

	Material Type		
Treatment	*Soil*	*Leached Shale*	*Shale*
DMAA	5.1%	2.9%	3.0%
MAA	6.3%	0.5%	0.9%
SA	0.8%	1.7%	0.7%
None	0.0%	0.9%	2.8%

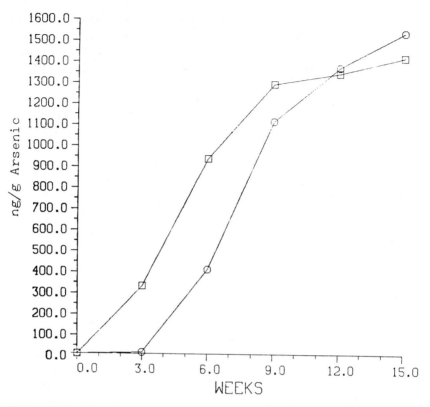

Figure 14-1.   Cumulative arsenic volatilized from DMAA-treated leached (square) and nonleached (circle) shale.

Microbiological Responses

Fluorescent staining of the samples and subsequent microscopic bacterial enumeration revealed that two morphologically distinct types of microorganisms were responding to the added nutrients under these test conditions. One group of organisms had the characteristics of bacteria, while the other consisted of much larger oval organisms similar to yeasts. The final total counts (bacteria plus yeastlike organisms) were different at the end of the experiment, depending on the substrate and treatment. The major difference appeared to be due to the final number of yeastlike organisms, which were consistently higher in leached shale than in nonleached shale (Fig. 14-3).

The fungal hyphal length values varied depending on the treatments. DMAA-treated soil and nonleached shale had twice the fungal hyphal lengths of the leached shale samples. This trend was reversed with SA-treated leached shale, which had four times the hyphal lengths of the other two

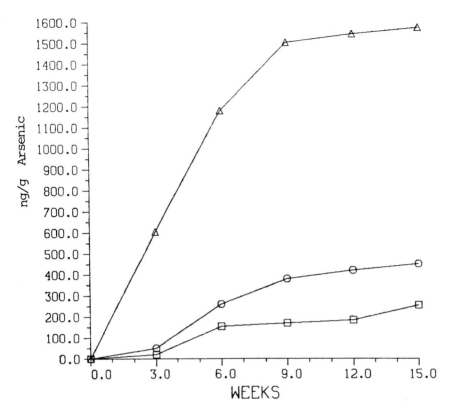

Figure 14-2.   Cumulative arsenic volatilized from MAA-treated leached shale (square), nonleached shale (circle), and soil (triangle).

samples. Hyphal measurements of MAA-amended samples showed soil with twice the quantity of either shale type. This varied fungal growth trend was also observed with the nonamended leached and nonleached shale. Figure 14-4 indicates that the overall microbial responses were higher in nonleached material, but this population developed more slowly.

DISCUSSION

The absence of arsenic volatilization in autoclaved DMAA-treated samples, in contrast to the extensive arsenic release observed with nonsterile materials, indicates a biological role in these mobilization processes. This hypothesis is further supported by the microbiological responses over the course of the experiments in relation to arsenic mobilization. As an example, the SA-treated leached shale, which had four times the fungal development of soil or nonleached shale, showed a much greater arsenic release. In addition, soil

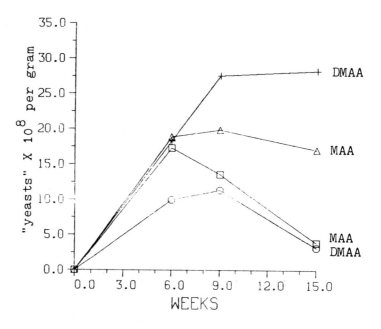

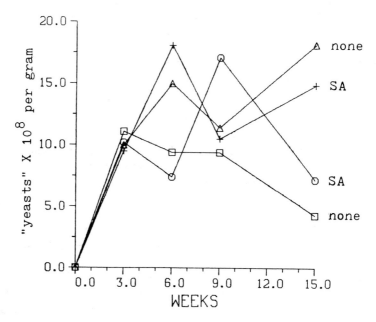

Figure 14–3. Yeastlike organism response to different arsenic treatments in leached (triangle and plus) and nonleached (square and circle) shale.

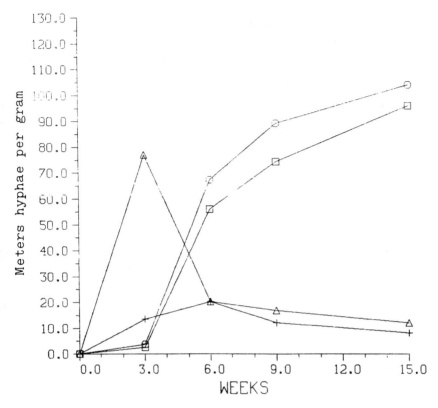

Figure 14-4. Fungal hyphal lengths in untreated leached (triangle and plus) and non-leached (square and circle) shale.

allowed more extensive release of arsenic from MAA treatments than was observed from either shale sample. These results suggest that, should extensive leaching occur, soluble materials, including organoarsenicals, could be mobilized from shale and accumulate in surrounding soils at disposal sites. If the compounds released are similar to MAA, then volatile arsenic may be released from such waste sites.

Leaching of retorted shale prior to treatment with arsenic and nutrients had a definite effect on the amount and rate of arsenic volatilization. Figure 14-1 exemplifies this point; similar total amounts of arsenic were released from DMAA-treated leached and nonleached shale, with the initial rate of release faster in the leached shale material. When SA was applied as an amendment, more arsenic was mobilized from leached than nonleached shale, suggesting that leaching may change the shale's suitability as a substrate for transformation of sodium arsenate. Without arsenic amendments, the trend of higher release from leached material was reversed; more of the innate arsenic was volatilized from nonleached material. This result is expected, as leaching removes from the shale the more soluble arsenic forms that are susceptible to microbial volatilization. A similar effect is observed with arsenic-amended soil. When soil is amended with MAA, the arsenic appears more susceptible to volatilization. As illustrated in Figure 14-2, this observed effect may become important if extensive leaching occurs between deposited spent shale and surrounding soil.

During the period of arsenic release, the bacteria numbers and fungal hyphal lengths initially showed rapid increases; subsequently the values gradually decreased. The values for the yeastlike organisms illustrate this phenomenon (Fig. 14-3). Also illustrated in Figure 14-3 is the change in the microbial community in leached shale. This growth pattern approximately paralleled the arsenic release; the initially rapid rate of release was followed by a declining rate of arsenic volatilization. The differing microbial responses between the various treatments appeared to correspond to differences in arsenical volatilization, especially with the SA-treated and nontreated materials. Leached shale amended with SA demonstrated a fourfold higher hyphal length value and much more extensive volatilization of arsenic than observed with nonleached shale or soil. In contrast (Fig. 14-4), nonleached, unamended shale had a final fungal biomass higher than that of leached shale. This corresponded to the higher volatilization occurring in the nonleached material. The microbial responses appear to be related to the higher volatilization occurring in the nonleached material. The microbial responses appear to be related to the amounts of arsenic volatilized; as biomass increased in the test material, increased volatilization of arsenic was observed.

ACKNOWLEDGEMENTS

This research was supported by the Department of Energy under contract No. DE-AC02-83ER60121 to the University of Colorado, Boulder, Colorado.

REFERENCES

Babiuk, L. A., and E. A. Paul, 1970, The use of fluorescein isothiocyanate in the determination of the bacterial biomass of grassland soil, *Can. J. Microbiol.* **16**:57-62.
Challenger, F., 1945, Biological methylation, *Chem. Rev.* **36**:315-361.
Challenger, F., C. Higgenbottom, and L. Ellis, 1933, The formation of organometalloidal compounds by microorganisms: Part 1. Trimethylarsine and dimethylarsine. *J. Chem. Soc.* **95**:95-101.
Cheng, C. H., and D. D. Focht, 1979, Production of arsine and methylarsines in soil and in culture, *Appl. Environ. Microbiol.* **38**:494-498.
Cox, D. P., and M. Alexander, 1973, Production of trimethylarsine gas from various arsenic compounds by three sewage fungi, *Bull. Environ. Contam. Toxicol.* **9**:84-88.

Hassler, R. A., 1982, *Microbial Contributions to Soluble and Volatile Arsenic Dynamics in Retorted Oil Shale,* M.S. thesis, Colorado State University, Ft. Collins.

Holm, T. R., M. A. Anderson, R. R. Stanforth, and D. G. Iverson, 1980, The influence of adsorption on the rates of microbial degradation of arsenic species in sediments, *Limnol. Oceanogr.* **25:**23-30.

Jones, P. C. T. and J. E. Mollison, 1948, A technique for the quantitative estimation of soil microorganisms, *J. Gen. Microbiol.* **2:**54-69.

Klein, D. A., and R. A. Hassler, 1981, Microbiological mobilization of arsenic from retorted oil shales—speciation and monitoring requirements, in *Environmental Speciation and Monitoring Needs for Trace Metal-Containing Substances from Energy-Related Processes,* F. E. Brinckman and R. H. Fish, eds., Proceedings of the DOE/NBS Workshop, Gaithersburg, Maryland, pp. 286-300.

McBride, B. C. and T. L. Edwards, 1977, Role of the methanogenic bacteria in the alkylation of arsenic and mercury, in *Biological Implications of Metals in the Environment,* H. Drucker and R. E. Wildung, eds., Proceedings of the fifteenth annual Hanford Life Sciences Symposium, Sept. 29-Oct 1, 1975, Richland, Washington, Technical Information Center, Energy Research and Development Administration, Washington, D.C., pp. 1-19.

McBride, B. C., and R. S. Wolfe, 1971, Biosynthesis of dimethylarsine by *Methanobacterium, Biochemistry* **10:**4312-4317.

Shendrikar, A. D., and G. B. Faudel, 1978, Distribution of trace metals during oil shale retorting, *Environ. Sci. Technol.* **12:**332-334.

Woolson, E. A., 1977, Generation of alkylarsines from soil, *Weed Sci.* **25:**412-416.

15

Spark Source Mass Spectrometry of Carapaces of *Cyprideis torosa* (Crustacea, Ostracoda): A Model to Reconstruct Multifarious Saline Environments

Anne-Marie Bodergat

Université Claude Bernard

INTRODUCTION

It is the aim of paleoecology to reconstruct the nature and dynamism of ancient environments. The reconstructions are based on both sedimentological and paleontological data. Key information regarding the fossil content of sediments includes not only the occurrence and numbers of taxa, but also the shape, ornamentation, and size of shells. It is widely believed that the chemical composition and the ornamentation of shells may be related to environmental parameters. This paper presents evidence that this relationship exists (see also Bodergat, 1983). For this study, the author used the ostracod *Cyprideis torosa* (Crustacea), which occurs in a multifarious, saline environment.

MATERIALS AND METHODS

The euryhaline *Cyprideis torosa* (Crustacea, Ostracoda) was collected from ponds with salt concentrations ranging from 0.5 to 120 g/l. Samples were collected from ponds and lagoons located at Isle of Noirmoutier, Camargue, Languedoc, and Le Brusc near Toulon in France. Other samples were obtained from Santa-Pola near Alicante, Spain (Fig. 15-1), and the Dead Sea and Tiberiade Lake in Israel. Salinity was measured when samples were collected.

Ostracod shells have two valves. In this species they may be smooth, reticulated, or have nodular surfaces (Fig. 15-2*a, b,* and *c*). An average valve is 1 mm in length and 0.05 mg in weight. The shells were analyzed using a spark source mass spectrometer allowing detection of trace elements in very small samples. Each analysis was a composite of five valves. For analysis, samples were embedded in silver powder to form small electrodes. Mass-spectrometry data were treated with factorial analysis (Benzecri and Benzecri, 1980; Fenelon, 1981).

RESULTS

In the multifarious environment, sources of pond water are rivers, streams, rain, and the sea. The ostracods may be both fresh and marine species; while some species only exist in fresh water, others are confined to saline environments. However, some species occur in both fresh and saline water. Chemical analysis of the carapaces enables differentiation of these species.

In Figure 15-3 a factorial plane shows ostracod distribution. The F_1 axis represents the relative chemical composition in terms of strontium (Sr) and aluminum (Al) of marine and freshwater environments; Sr is strongly

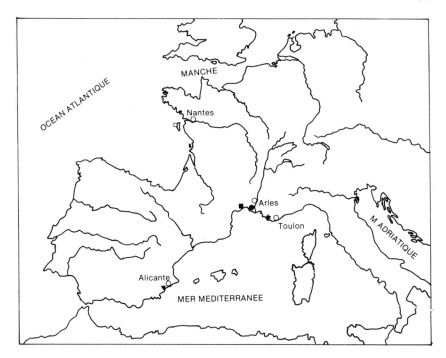

Figure 15-1. Location of samples: ▼Santa-Pola near Alicante, Spain; ■Languedoc, France; *LeBrusc near Toulon, France; ●Camargue, France. Not shown on this map are locations of samples collected from the Dead Sea and Tiberiade Lake in Israel.

associated with marine environments, while Al is representative of waters of continental origin (Mackensie, 1974; Pascal, 1979). Analysis of the shells of *C. torosa* from the lagoon at Santa-Pola were rich in Sr; these data correlated with the marine origin of the lagoon water. Samples from near a small stream at Arnel (Languedoc) were rich in Al, confirming the terrestrial origin of the waters from which these shells were collected.

The F_2 axis in Figure 15-3 defines redox boundary conditions. The presence in samples of chlorine (Cl), sulfur (S), iron (Fe), vanadium (V), chromium (Cr), and zinc (Zn) is indicative of reducing conditions. The latter three elements are associated with plant-derived organic matter (B. Porthault, unpub. data).

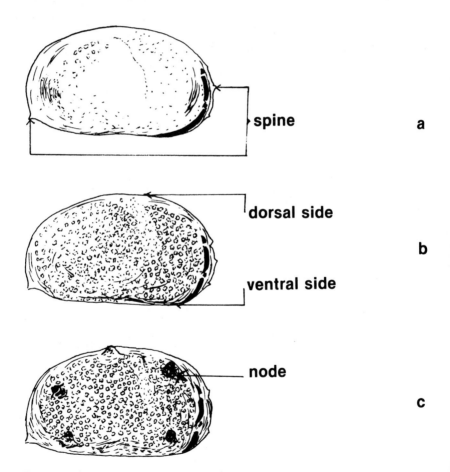

Figure 15-2.   Surface characteristics of the ostracod *Cyprideis torosa: a,* smooth; *b,* reticulated; *c,* nodular.

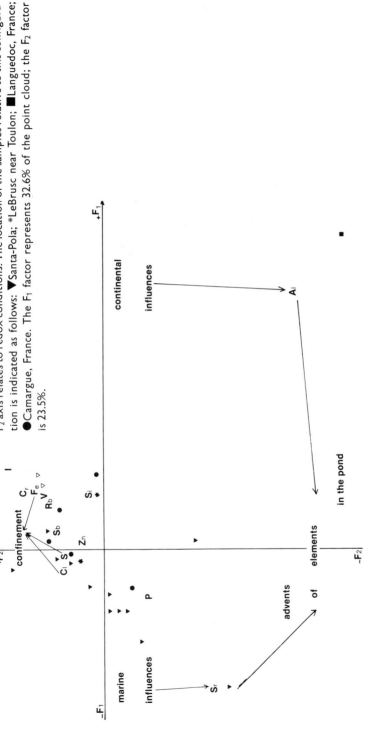

Figure 15-3. The factorial plane F_1–F_2 illustrates the distribution of ostracods relative to their chemical compositon. The F_1 axis relates to marine and freshwater influences; the F_2 axis relates to redox conditions. The location of the samples relative to this configuration is indicated as follows: ▼Santa-Pola; *LeBrusc near Toulon; ■Languedoc, France; ●Camargue, France. The F_1 factor represents 32.6% of the point cloud; the F_2 factor is 23.5%.

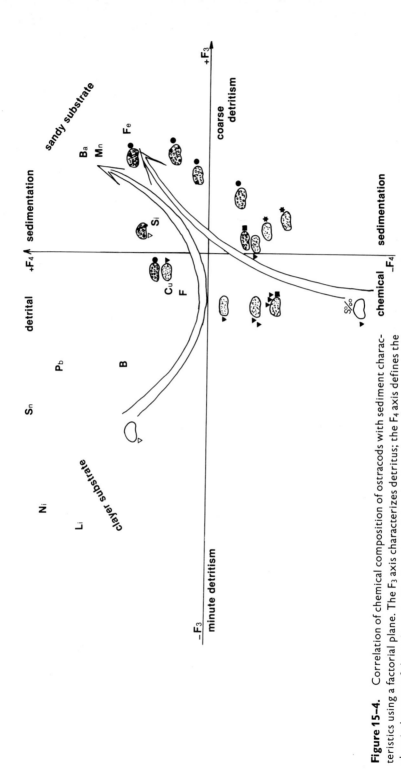

Figure 15–4. Correlation of chemical composition of ostracods with sediment characteristics using a factorial plane. The F_3 axis characterizes detritus; the F_4 axis defines the chemical nature of the sediments. The location of the samples relative to this configuration is indicated as follows: ▼Santa-Pola; ■Languedoc, France; *LeBrusc near Toulon; ●Camargue, France. $F_3 = 13.5\%$; $F_4 = 12\%$.

The F_3 axis in Figure 15-4 is related to the particle size of the substratum. The presence of Fe and manganese (Mn) is associated with oxidizing sediments with a sandy texture (Michard, 1969). The presence of barium (Ba) is characteristic of coarse detritus containing organic matter; Ba is also present in feldspathic minerals (Mosser, 1980). Lithium (Li), nickel (Ni), and boron (B) are associated with argillaceous materials, and Ni and B are present when organic matter occurs (Mosser, 1980). Analyses of samples from two pools at Camargue showed the presence of Fe, Mn, and Ba. These results correlated well with the actual environments, which were oligohaline pools with coarse sediments. The pools were fed with soft water. Three samples from the marine lagoon at Santa-Pola contained Li, Ni, B, and S; this lagoon possessed a fine sediment.

The distinction between chemical and detrital sedimentation is correlated with the F_4 axis shown in Figure 15-4. Chemical sedimentation is associated with S/mil, while detrital sedimentation is represented by Li, B, lead (Pb), Fe, Mn, silicon (Si), Ni, and copper (Cu). Ostracod samples from the Santa-Pola lagoon, representing an evaporitic environment, contain the former element association, while materials from the Camargue pools (oligohaline waters having coarse sediments) have the latter association (Fig. 15-4). This confirms the relationship between sediment character and the chemical composition of ostracod shells.

Ostracod shells with smooth surfaces were collected from highly saline (120 g/l) pools with fine sediments at the Santa-Pola lagoon. Analyses indicated that many of these specimens contained high concentrations of S, Br, and chlorine (Cl). Other shells, collected from ponds with thin, clay sediments, contained elements indicative of this type of environment (Fig. 15-4). Shells collected from oligohaline ponds characterized by coarse sediments were rich in Ba, and many contained Fe and Mn; these shells possessed abundant nodes (see Fig. 15-4 and 15-5). These results indicated a strong correlation among shell design, salt concentrations, and sediment character.

CONCLUSIONS

From the results of this study, it is evident that there is a strong correlation between the chemical composition of ostracod shells and environmental parameters, including freshwater and saline influences, redox potential, character of the sediment, and mode of sediment formation. This study also indicates a relationship between the design of carapaces of ostracods and the chemical composition of these shells; the more ornamental the shell, the higher the barium content. The relationships and correlations described in this paper may be useful in reconstructing paleoenvironments.

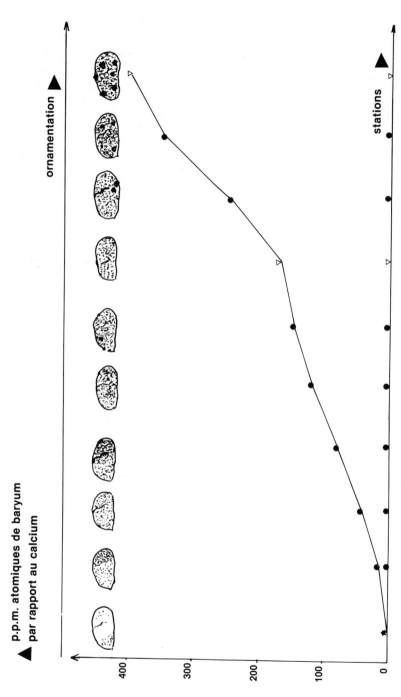

Fig. 15-5. Correlation between the abundance of nodes on *C. torosa* shells and the Ba concentration. Denoted on the chart is the sample location: *LeBrusc near Toulon; ●Camargue, France; ▽Israel.

p.p.m. atomiques de baryum
par rapport au calcium

ornamentation ▲

stations ▲

400
300
200
100
0

ACKNOWLEDGEMENTS

Thanks are due to Anne-Marie Andreani (Centre d'Energie Nuclèaire de Grenoble, Isère, France) for chemical analyses and to Dr. Michel Rio (Centre de Paléontologie Stratigraphique et Paléoécologie de l'Université Claude Bernard-LA 11, CNRS-Villeurbanne, France) for discussions and data reduction.

REFERENCES

Benzecri, J. P., and F. Benzecri, 1980, *Pratique de l'analyse des donnés. 1-Analyse des correspondances, Exposé Élémentaire,* Dunod, Paris.

Bodergat, A. M., 1983, *Les ostracodes, témoins de leur environment: Approache chimique et écologie en milieu lagunaire et océanique,* Ph.D. thesis, Etat, Univ. Lyon I.

Fenelon, J. P., 1981, *Qu'est-ce que l'analyse des données?,* Lefonen, Paris.

Mackensie, F. J., 1964, Strontium content and variable strontium chlorinity relationship of Sargasso sea water, *Science* **146:**517-518.

Michard, G., 1969, *Contribution à l'étude du comportement du manganèse dans la sédimentation chimique,* Ph.D. thesis, Etat Paris.

Mosser, C., 1980, *Etude géochimique de quelques éléments traces dans les argiles des altérations et des sédiments,* Ph.D. thesis, Etat, Univ. L. Pasteur, Strasbourg.

Pascal, A., 1979, Utilisation des éléments traces dans la caractérisation des paléomilieux sédimentaires urgoniens bascocantabriques, in *L'Urgonien des Pays méditerranéens, Géobios,* M.S. **3:**331-345.

16

Mesophilic Manganese-Oxidizing Bacteria from Hydrothermal Discharge Areas at 21° North on the East Pacific Rise

Henry L. Ehrlich

Rensselaer Polytechnic Institute

ABSTRACT: As part of a study of microbial communities around hydrothermal vents at midocean rift zones, water samples, filter paper from an *in situ* pumping system, the mantle edge of a clam (Mollusca), and pieces of exposed glass slides, all collected at 21° North latitude on the East Pacific Rise, were examined for the presence of Mn (II)-oxidizing bacteria. Culturing portions of the samples in tryptone-dextrose-yeast extract medium prepared in sea water and supplemented with manganous sulfate revealed the presence of such bacteria in all samples. Most isolates resembled earlier ones obtained from samples from the Galapagos Rift Zone. However, at least one isolate was different. It resembled isolates from ferromanganese nodules and associated sediments from the central and eastern Pacific and the Atlantic oceans in that it could not oxidize Mn(II) unless it was bound to Mn(IV) oxide. Yet, unlike the manganese-oxidizing isolates from ferromanganese nodules and associated sediments, but like all other isolates from the hydrothermal vent areas, its manganese oxidase was inducible by Mn(II). The characteristics of some of the isolates from the hydrothermal vent area at 21° North on the East Pacific Rise are described.

INTRODUCTION

Hydrothermal solution, originating from sea water having penetrated porous basaltic rock to depths of more than 1 km at midocean rift zones, and having reacted with the rock at 300°C or more and well in excess of 250 bars of hydrostatic pressure (Bischoff and Rosenbauer, 1983), has been shown to carry dissolved hydrogen sulfide, ferrous iron, manganous manganese, cupric copper, zinc, hydrogen, methane, and other substances (see Baross, Lilley, and Gordon, 1982). Such solution has been found to discharge from hydrothermal vents at 21° North on the East Pacific Rise (Jannasch, 1984). This hydrothermal solution has been found to support communities of organisms in a temperature regime in the range of <4° to 21°C that are specially adapted to live around these vents (Rise Project Group, 1980; Jannasch, 1984). The reduced forms of sulfur appear to be a prime energy source for a food chain in these communities, the base of which consists of chemolithotrophic bacteria that can use hydrogen sulfide as their prime energy source. Invertebrates such as vestimentiferan tube worms and molluscs represent the second trophic level, feeding at the expense of the primary producers mainly by way of secretions in intimate or loose symbiotic association (Cavanaugh et al., 1981; Rau and Hedges, 1979). A third trophic level consists of predators and scavengers (crustaceans, fish, and so forth) that appear to live at the expense of the primary feeders. At the Galapagos Rift Zone the bacterial community around the vents includes not only hydrogen sulfide–oxidizing bacteria but also Mn(II)-oxidizing bacteria (Ehrlich, 1983).

Recently Baross, Lilley, and Gordon (1982) and Baross and Deming (1983) have reported the presence of bacteria in hydrothermal fluid in vent chimneys before the fluid mixed with ocean-bottom sea water at 21° North on the East Pacific Rise. The temperature of the fluid in the chimneys at the moment of sampling exceeded 300°C. They found in it chemolithotrophic bacteria that could grow at 250°C and 205 atmospheres of hydrostatic pressure. Some apparently used Mn(II) as an energy source, while others used reduced sulfur or ferrous iron. Some of the manganese oxidizers may be able to use nitrate as terminal electron acceptor in place of oxygen, which seems reasonable in view of the probable lack of available oxygen in the hot hydrothermal fluid. These organisms were extreme thermophiles that at 1 atmosphere of hydrostatic pressure could not grow below 75°C.

In the present work, water samples, water filters, clam-mantle tissue, and experimentally exposed glass slides from bottom water very close to hydrothermal discharge points were examined for the presence of mesophilic, manganese-oxidizing bacteria.

MATERIALS AND METHODS

Samples

The samples from which manganese-oxidizing bacteria were isolated for this study came from 21° North on the East Pacific Rise off the Mexican coast and were collected by H. W. Jannasch and C. O. Wirsen of the Woods Hole Oceanographic Institution in 1982.

Samples consisted of water from hot vents collected during dives 1212, 1214, and 1220, pieces of filter membranes from an *in situ* pumping system at these diving sites, glass chips in sea water from slides from an array left on the bottom for five months next to a black smoker, and a piece of clam-mantle edge in sea water. All samples were maintained at around 25°C.

Although enrichments in TDYM-SW broth were made for manganese-oxidizing bacteria from each of the samples, it was also possible to streak directly from each sample to TDYM-SW agar to obtain colonies of manganese-oxidizing bacteria. Such colonies were recognizable by their browning or by the browning of the surrounding medium. Pure cultures were prepared from enrichments and from directly streaked plates. All original samples yielded manganese-oxidizing bacteria. For this study three strains were examined in detail. They were strain E_{13} from a water sample collected on dive 1212 and streaked directly onto TDYM-SW agar; strain HCM-41 from clam-mantle edge streaked directly onto TDYM-SW; and strain CFP-11 from filter paper of a pumping system used during dive 1220 streaked directly onto TDYM-SW.

Culture Media

The culture media used in this study were the same as used by Ehrlich (1983). TDYM-SW medium contained in grams per liter: Bacto-tryptone, 5; Bacto-dextrose, 1; Bacto-yeast extract, 2.5; $MnSO_4 \cdot H_2O$, 1.0; and for solid medium, Bacto-agar, 15. TDY-SW medium had the same composition as TDYM-SW medium, except that the manganous sulfate was omitted. For resting-cell and cell-extract preparations, the cultures were grown on Roux slants of TDYM-SW or TDY-SW agar. Incubation was at 25° C except where specified. Roux slants were incubated for 24 hr.

Resting Cell and Cell Extracts

These preparations were made as described by Ehrlich (1983).

Assay for Mn(II)-Oxidizing Activity and Inhibition Tests

The procedures for these tests were the same as used by Ehrlich (1983). The assays measured the Mn(II) removed from solution by intact cells or cell extract after 4 hr of incubation at 25°C or other temperature, as specified. Correction was made for Mn(II) removal in the absence of cells or cell extract. Mn(II) sorption by cells, cell extract, and manganese oxide was distinguished by measuring Mn(II) removal after pretreatment of samples with equal volumes of 10 mM copper sulfate solution according to the method of Bromfield and David (1976). Mn(II) removal measured after copper sulfate treatment was taken to represent Mn(II) oxidation, whereas Mn(II) removal before copper sulfate treatment measured both adsorption and oxidation. Effects of pH on Mn(II) oxidation were determined by adjusting the bicarbonate buffer of the reaction mixture to appropriate values. The adjusted pH in any reaction vessel did not change measurably during the course of an experiment. The oxygen requirement for Mn(II) oxidation was determined by incubation of reaction mixtures in a nitrogen atmosphere.

Cell and Protein Measurements

Cell concentrations were determined turbidimetrically at 450 nm on a Coleman Junior 6A spectrophotometer and expressed in optical density units. Cell protein was measured with Folin reagent by the method of Henry (1965) using bovine albumin as standard.

RESULTS

Culture Traits

Some basic culture characteristics of strains E_{13}, HCM-41, and CFP-11 are summarized in Table 16-1. Of particular note are the observations that all three strains are gram-negative with a temperature growth range from 5°C to 45°C and an inability to grow in freshwater nutrient medium. Strain CFP-11 grew distinctly more slowly than the other two strains. As will be shown, it also oxidized Mn(II) differently from the other two strains.

Mn(II)-Oxidizing Activity

Intact cells and cell extracts of all strains oxidized Mn(II) (Table 16-2). Strains E_{13} and HCM-41 oxidized Mn(II) in the absence of initially added manganic oxide. Strain CFP-11, on the other hand, oxidized Mn(II) only in the initial presence of manganic oxide. The Mn(II)-oxidizing activity of cell extract but not intact cells of strain E_{13} showed some stimulation when manganic oxide was present initially. Strain HCM-41 showed no such stimulation. The copper sulfate treatment showed that intact cells and cell extract of strains E_{13} and HCM-41 bound significant amounts of Mn(II) without oxidizing it. Intact cells of strain CFP-11 did not bind measurable amounts of Mn(II), nor did cell extract.

Oxygen Requirement for Mn(II) Oxidation

Oxygen was required by cell extracts of all three strains for Mn(II) oxidation (Table 16-3). Interestingly, no Mn(II) was bound by two of the cell extracts

Table 16-1. Traits of Three Isolates of Mesophilic Mn^{2+}-Oxidizing Bacteria from 21° North Latitude on the East Pacific Rise

Trait	Culture		
	E_{13}	HMC-41	CFP-11
Morphology	short rod	spirillum	short rod
Motility	+	+	+
Gram reaction	−	−	−
Colony type	brown, round 3 mm	off-white 1 mm	off-white 1 mm
Temperature growth range[a]	5°C-45°C	5°C-45°C	5°C-45°C
Growth in:			
SWNB[b]	+	+	+
FWNB[c]	−	−	−
3% NaCl NB[d]	+[e]	+[e]	+[e]
Doubling time[f]	1.85 hr	1.30 hr	3.72 hr

[a]Weak growth by all three strains at 5°C and 45°C.
[b]Seawater nutrient broth (Difco).
[c]Freshwater nutrient broth (Difco).
[d]Nutrient broth (Difco) prepared in 3% NaCl.
[e]Pellicle formation.
[f]In TDYM broth in stationary culture at 25°C.

Table 16-2. Mn^{2+}-Oxidizing Activity in the Presence and Absence of MnO$_2$ of Three Bacterial Isolates from 21° North Latitude on the East Pacific Rise

Culture	Test Material	Cell or Enzyme Concentration[a]	Mn^{2+} Oxidizing Activity (nmol/4 hr)			
			Without MnO$_2$		With MnO$_2$	
			−Cu[b]	+Cu[c]	−Cu[b]	+Cu[c]
E_{13}	intact cells	0.395	500	300	500	300
	cell extract	6.9 mg	1000	600	1300	800
HMC-41	intact cells	1.000	200	100	200	100
	cell extract	10.2 mg	1000	800	1000	800
CFP-11	intact cells	0.990	0	0	300	300
	cell extract	5.1 mg	0	0	300	300

[a]Cell concentration expressed in absorbance units at 450 nm of a 10-fold dilution of cell suspension.
[b]Assayed without CuSO$_4$ treatment.
[c]Assayed with CuSO$_4$ treatment.

Table 16-3. Oxygen Requirement of the Mn^{2+} Oxidase of Three Bacterial Isolates from 21° North Latitude on the East Pacific Rise

		Mn^{2+}-Oxidizing Activity (nmol/4 hr)			
		In Air		Under N_2	
Culture	Protein Concentration (mg)	−Cu	+Cu	−Cu	+Cu
E_{13}	6.2	900	400	0[a]	0
	6.4	700	400	0	0
HCM-41	8.1	700	200	0	0
	7.2	800	400	0	0
CFP-11	7.2	N.D.[b]	400	ND	100
	7.7	N.D.	400	ND	100

[a]Manganese oxidation not measurable.
[b]Not done.

under nitrogen, suggesting that Mn(II) binding by any cell extract in air is somehow related to Mn(II) oxidation. Residual Mn(II)-oxidizing activity by strain CFP-11 extract under nitrogen was attributable to the difference in method of testing this strain. While Thunberg tubes were used with strains E_{13} and HCM-41, 50-ml Erlenmeyer flasks in a vacuum desiccator had to be used with strain CFP-11 to allow for optimal surface exposure of the manganic oxide required in the reaction mixture in this case. Evacuation and repeated flushing with nitrogen of the flasks in the desiccator took about 10 min as compared to 3 to 4 min with Thunberg tubes. It has previously been found that equilibrium of the dissolved gas in the flasks with the nitrogen atmosphere in the desiccator is established slowly. The slow evacuation and equilibration thus accounted for the residual oxidizing activity with the extract from strain CFP-11.

pH Optimum

Mn(II)-removal activity of intact cells and cell extract of all three strains was optimal between pH 7.0 and 7.5. This has also been found with Galapagos strains SSW_{22} and S_{13} (Ehrlich, unpublished results).

Inducibility of Mn(II) Oxidase

The manganese oxidase was inducible in all three strains because cell extracts of all three cultures exhibited Mn(II) oxidase activity when grown on manganese-supplemented TDYM-SW medium but not on unsupplemented TDY-SW medium (Table 16-4).

Table 16–4. Inducibility of Mn^{2+} Oxidase in Three Bacteria Isolates from 21° North Latitude on the East Pacific Rise

Culture	Medium	State of Induction	Protein (mg)	Mn^{2+} Oxidized (nmol/4hr)
E$_{13}$	TDY	−	10.5	0[a]
	TDYM	+	10.5	100
HCM-41	TDY	−	10.6	0
	TDYM	+	10.7	900
CFP-11	TDY	−	7.6	0
	TDYM	+	7.5	800

[a]No manganese-oxidizing activity measurable.

Table 16–5. Effect of Electron Transport Inhibitors on Mn^{2+} Removal by Cell Extracts from Three Bacterial Isolates from 21° North Latitude on the East Pacific Rise

Culture	Inhibitor	Concentration	Percent Inhibition of Mn^{2+} Removal
E$_{13}$	Rotenone	0.2 mM	53
	Antimycin A	14.6 μM	43, 50
	NOQNO[a]	70.0 μM	47, 43
	NaN$_3$	1.0 mM	37
	KCN[b]	0.1 M	60
HCM-41	Rotenone	0.2 mM	45
	Antimycin A	14.6 μM	50, 35
	NOQNO	70.0 μM	44, 58
	NaN$_3$	1.0 mM	47
	KCN	0.1 M	67
CFP-11	Rotenone	0.2 mM	38
	Antimycin A	14.6 μM	93
	NOQNO	70.0 μM	60
	NaN$_3$	1.0 mM	45
	KCN	0.1 M	64

[a]2n-nonyl-4-hydroxyquinoline-N-oxide.
[b]Intact cells pretreated with 0.1 M KCN (see Ehrlich, 1983).

Involvement of Electron-Transport System

Mn(II)-removal activity of cell extracts of all three strains was sensitive to rotenone, antimycin A, 2n-nonyl-4-hydroxyquinoline-N-oxide (NOQNO), and sodium azide (Table 16-5). Mn(II)-removal activity of intact cells of all three strains was sensitive to potassium cyanide (Table 16-5).

Effect of Temperature

Mn(II)-removal activity of all three strains was exhibited only at 15°C and 25°C, not at 5°, 30°, or 37°C.

DISCUSSION

Strains E_{13} and HCM-41 in this work resemble earlier isolates from a hydrothermal discharge area on the Galapagos Rift (Ehrlich, 1983) in morphology and physiological activity in relation to Mn(II) oxidation. Strain CFP-11, however, differs from strains E_{13} and HCM-41 and from the Galapagos isolates in being able to oxidize Mn(II) only when bound to manganic oxide. In that respect it behaves like bacterial isolates from ferromanganese nodules and associated sediments (Ehrlich, 1968, 1975, 1984) with one exception. Whereas the manganese oxidase in all isolates from nodules and associated sediments has been found to be constitutive, it was clearly inducible in strain CFP-11.

Although not yet demonstrated directly, strains E_{13}, HCM-41, and CFP-11, appear to be able to couple adenosine 5'-triphosphate (ATP) synthesis to Mn(II) oxidation. This is inferred from the results with electron-transport inhibitors that were similar to those used with strain SSW_{22} and strain S_{13} from the Galapagos Rift, for which ATP synthesis coupled to Mn(II) oxidation was shown directly as well as with the electron-transport inhibitors (Ehrlich, 1983).

The temperature range of activity of the manganese oxidase of strains E_{13}, HCM-41, and CFP-11 was narrower (15°C to 25°C) than for growth (5°C to 45°C). The narrow temperature dependence is supporting evidence that manganese removal by these strains is enzymatic.

It will be important to determine how numerous the organisms are at the hydrothermal vent site and how intense their manganese-oxidizing activity is *in situ* to assess their importance to the vent community of organisms.

ACKNOWLEDGEMENTS

I wish to thank H. W. Jannasch and C. O. Wirsen for supplying the samples for this study and A. R. Ellett for her expert technical assistance.

REFERENCES

Baross, J. A., and J. W. Deming, 1983, Growth of "black smoker" bacteria at temperatures of at least 250°C, *Nature* **303:**423-426.

Baross, J. A., M. D. Lilley, and L. I. Gordon, 1982, Is the CH_4, H_2, and CO venting from submarine hydrothermal systems produced by thermophilic bacteria? *Nature* **298:**366-368.

Bischoff, J. L., and R. J. Rosenbauer, 1983, A note on the chemistry of seawater in the range 350-500°C, *Geochim. Cosmochim. Acta* **47:**139-144.

Bromfield, S. M., and D. J. David, 1976, Sorption and oxidation of manganous oxide by cell suspensions of a manganese-oxidizing bacterium, *Soil Biol. Biochem.* **8:**37-43.

Cavanaugh, C. M., S. L. Gardiner, M. L. Jones, H. W. Jannasch, and J. B. Waterbury, 1981, Prokaryotic cells in the hydrothermal vent tube worm *Riftia pachyptila* Jones: Possible chemoautotrophic symbionts, *Science* **213:**340-342.

Ehrlich, H. L., 1968, Bacteriology of manganese nodules. II. Manganese oxidation by cell-free extract from a manganese nodule bacterium, *Appl. Microbiol.* **16:**197-202.

Ehrlich, H. L., 1975, The formation of ores in the sedimentary environment of the deep sea with microbial participation: The case for ferromanganese concretions, *Soil Sci.* **119:**36-41.

Ehrlich, H. L., 1983, Manganese-oxidizing bacteria from a hydrothermally active area on the Galapagos Rift, in Environmental Biogeochemistry, R. Hallberg, ed., *Ecol. Bull.* **35:**357-366.

Ehrlich, H. L., 1984, Different forms of bacterial manganese oxidation, in *Microbial chemoautotrophy,* W. R. Strohl and O. H. Tuovinen, eds., Ohio State University Biosciences Colloquia, Ohio State University Press, Columbus, pp. 47-56.

Henry, R. J., 1965, *Clinical Chemistry, Principles and Techniques,* Harper & Row, New York.

Jannasch, H. W., 1984, Microbial processes at deep-sea hydrothermal vents, in *Hydrothermal Processes at Seafloor Spreading Centers,* P. A. Rona, K. Bostrom, L. Lauvier, and K. L. Smith, eds., Plenum Press, New York, pp. 677-709.

Rau, G. H., and J. I. Hedges, 1979, Carbon-13 depletion in the hydrothermal vent mussel: Suggestion of a chemosynthetic food source, *Science* **203:**648-649.

Rise Project Group: Spiess, F. N., K. C. Macdonald, T. Atwater, R. Ballard, A. Carranza, D. Cordoba, C. Cox, V. M. Diaz Garcia, J. Francheteau, J. Guerrero, J. Hawkins, R. Haymon, R. Hessler, T. Juteau, M. Kastner, R. Larson, B. Lykendyk, J. D. Macdougall, S. Miller, W. Normark, J. Orcutt, and C. Rangin, 1980, East Pacific Rise: Hot springs and geothermal experiments, *Science* **207:**1421-1433.

17

Microbial Manganese Oxidation and Trace Metal Binding in Sediments: Results from an *in situ* Dialysis Technique

Paul E. Kepkay

Bedford Institute of Oceanography

ABSTRACT: Sediments associated with freshwater ferromanganese concretions in Lake Charlotte, Nova Scotia, contained microscopic precipitates of manganese and iron. These precipitates were dispersed within the sediments and were found to be as rich in nickel, cobalt, and copper as deep-sea concretions. In addition, the development of the precipitates was associated with the microbial oxidation of manganese. Results from the deployment of poisoned and unpoisoned dialysis probes, or "peepers," demonstrated that microbial manganese oxidation and nickel binding were closely associated, causing a fivefold enhancement of abiotic processes such as adsorption. The microbial enhancement of copper binding was less pronounced due to organic-metal interactions in competition with manganese oxidation. Organic-metal interactions and oxidation both may have been in effect during iron binding, but the relative importance of the two processes could not be determined from the kinetic data.

INTRODUCTION

Ferromanganese concretions are economically valuable, not only for their manganese content, but for the nickel, cobalt, and copper that coprecipitate with or bind to the manganese oxides. Many laboratory studies (e.g. Beijerinck, 1913; Ali and Stokes, 1971; Ehrlich, 1976; Douka, 1980; and Kepkay, Burdige, and Nealson 1984) have demonstrated that manganese-oxidizing bacteria exist, but only recently have field studies (Emerson, Cranston, and Liss, 1979; Emerson et al., 1982) taken their activities into account. Manganese-oxidation rates have been estimated in sediments from steady-state models of pore water and solid-phase manganese profiles (see Burdige and Gieskes, 1983). However, the assumptions used to obtain these estimates remain untested, and models alone provide no information on the relative importance of biological as opposed to abiotic or purely chemical processes. As an alternative to modeling, direct measurements of manganese oxidation have been carried out by Emerson et al. (1982) in the water column of Saanich Inlet, British Columbia, and with appropriate poison controls these measurements have distinguished between abiotic and microbially mediated oxidation. While a similar approach has not been applied to sediments, it would probably not be useful because concentration gradients of dissolved metals must be taken into account. Natural gradients are not always maintained over the times necessary to determine oxidation rates from cores in the laboratory, and samples incubated separately are completely isolated from gradients.

To overcome these difficulties, Burdige and Kepkay (1983) developed a technique for introducing horizontal concentration gradients to a sediment through a dialysis membrane. The equipment, a dialysis probe, or "peeper" (Hesslein, 1976), has previously been used to investigate the kinetics of abiotic and microbial manganese oxidation in artificial laboratory sediments. This paper reports its use in the sediments of Lake Charlotte, Nova Scotia, which contain extensive freshwater ferromanganese concretions (Honeyman, 1881; Kindle, 1933, 1935; Beals, 1966; and Harriss and Troup, 1970). In addition to manganese (Mn) oxidation, the binding characteristics of iron, (Fe), nickel (Ni), and copper (Cu) have been examined to determine the extent to which these metals were affected by oxidative processes.

FIELD AREA

All data were collected from sediments in the region of Lake Charlotte referred to as Concretion Cove by Kindle (1935). The cove (Fig. 17-1) is situated at the southern end of the lake and is protected from wave and current action in the deeper, northern region by a constriction of the shorelines. Kindle (1935) has suggested that the added protection of the cove on three sides by land (Fig. 17-1) prevents the extensive transport of sediment, even though the water depth is less than 2 m. He has also pointed out that the protected locale, the absence of nearby outlet streams, and runoff through soils rich in metals and organics all contribute to relatively high concentrations of dissolved metals and dissolved organics (as humic and fulvic acids) in the bottom water. Two sites were chosen (Fig. 17-1), one of which (site 5) was characterized by a hard, sandy bottom covered with disclike ferromanganese concretions that were 5 cm to 15 cm in diameter. The distribution, composition, and structure of the concretions have been described in detail by Kindle (1933, 1935), Beals (1966), and Harriss and

Troup (1970). A second area (site M) was located where large concretions were not present and where the macrophyte *Eriocaulon septangulare* (known locally as pipe wort) covered a soft mud bottom.

MATERIALS AND METHODS

Sampling

All sampling and *in situ* experiments were carried out between September 10 and October 30, 1982, when bottom-water temperature was between 12°C and 15°C and the pH was between 6.9 and 7.1. All temperatures were measured *in situ* with a standard laboratory thermometer, and pH was measured with a Metrohm 605 pH meter and combination electrode. At both sites cores were collected for dissolved metal, dissolved organic carbon, and solid-phase metal analyses. Cores were subsampled as 1-cm slices within 1 hr

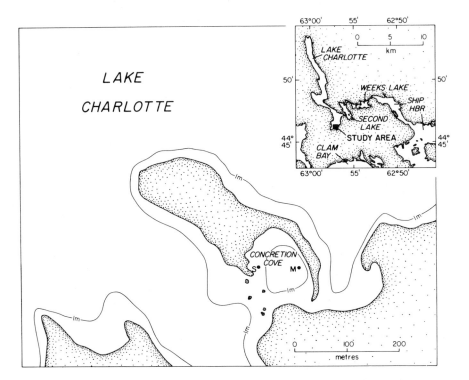

Figure 17-1. Location of sites M and S in the Concretion Cove area of Lake Charlotte, Nova Scotia. Site S is a hard, sandy bottom covered by ferromanganese concretions. Site M is a mud bottom covered by the macrophyte *Eriocaulon septangulare* and no concretions. Most of the cove is less than 1 m deep.

of collection, and pore waters were separated from the solid phase using a six-manifold Reeburgh (1967) squeezer described by Helder, deVries, and Van der Loeff (1982). Bottom-water and pore-water samples were then filtered through 0.22 μm membrane filters (Millipore) and stored at 5°C for analysis. During sampling of cores from site M, rootlets of the dominant macrophyte, *Eriocaulon septangulare,* had to be cut within the upper 2 cm of sediment to be included in the appropriate 1-cm slices.

Dissolved Metals and Dissolved Organic Carbon

Filtered pore-water and bottom-water samples were analyzed for dissolved organic carbon (DOC) the same day as collection using the U.V. oxidation method of Gershey et al. (1979). Triplicate 20-μl water samples were analyzed by direct injection on a Varian 975 flameless atomic absorption unit for dissolved manganese, iron, cobalt (Co), and copper. In the case of dissolved nickel, 40-μl samples were injected to bring the analyses within the sensitivity range of the instrument.

Solid-Phase Metals

Subsamples of squeezed sediments were dried at 60°C for 24 hr and leached using the methods of Willey and Fitzgerald (1980). Metal oxides were solubilized by the acid reduction of 0.5 g sediment in 5 ml of 0.2% hydroxylamine hydrochloride in 20% acetic acid (pH 4.0). After 24 hr at 20°C the supernatant leachates were filtered through 0.22-μm membrane filters (Millipore) and analyzed in triplicate for dissolved manganese, iron, nickel, cobalt, and copper on a Perkin-Elmer 503 flameless atomic absorption unit.

Peeper Experiments

An earlier version of the peeper and a laboratory test of its design and operation have been described by Burdige and Kepkay (1983). The design is based on the experimental constraint that dissolved material must diffuse laterally and in one dimension out of a dialysis cell (as a constant source) and into sediment enclosed by an associated core box (Fig. 17-2). The dialysis cell is connected to a large (500-ml) reservoir to ensure that it remains a constant source, and the close match of the cell edges to the inner walls of the core box allows diffusion to progress as a one-dimensional rather than a radial front.

All experiments were carried out at site M because of the ease of peeper emplacement in the muds and the ease of pore-water extraction from small samples of the sediment. The restricted cross section of the core box

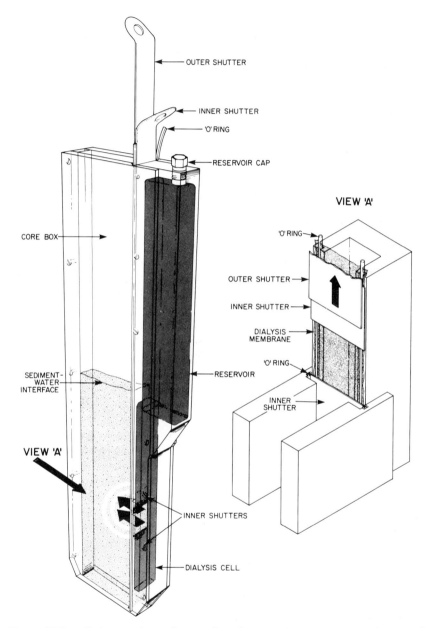

Figure 17-2. Dialysis probe, or "peeper," used to examine manganese oxidation and metal binding in sediments. The core box and dialysis cell are clear acrylic held together by stainless steel screws. The enlargement of the front view of the dialysis cell shows how the dialysis membrane is held in place by an O ring and the inner shutters. The core box is 70 cm long by 10 cm wide by 10 cm deep.

(Fig. 17-2) allowed the recovery of only a few centimeters of sand, so that site S was not suitable for deployment of the peeper. The same peeper was emplaced twice at site M to carry out two experiments. The reservoir contained dissolved metals during the first run and contained both metals and sodium azide to act as a poison control during the second run. Dissolved metal concentrations in the reservoir (20 μM manganese, 60 μM iron, 0.2 μM nickel, and 0.3 μM copper) corresponded to bottom-water concentrations as the highest values measured. Cobalt was not included because pore-water concentrations of the metal were low (Figs. 17-3 and 17-4), often less than the detection limit of flameless atomic absorption. Reservoir solutions were made up from atomic absorption standards (Fisher) diluted in deionized, distilled water at pH 7.0 to match the average bottom-water pH. In the case of the poison-control run, sodium azide was chosen as one of the few poisons that inhibits microbial manganese oxidation but does not interfere with the redox chemistry of manganese (Emerson et al., 1982). Even then, azide could be used at a maximum concentration of only 20 μM because higher concentrations caused the iron in solution to precipitate. Each time the peeper was emplaced, an outer shutter was removed so that metals could diffuse from the cell and into a section of sediment predetermined by the position of two inner shutters (Fig. 17-2). The inner shutters were set to expose sediment that was 1 to 4 cm subbottom because pore-water and solid-phase profiles (Fig. 17-4) indicated that manganese oxidation was at a

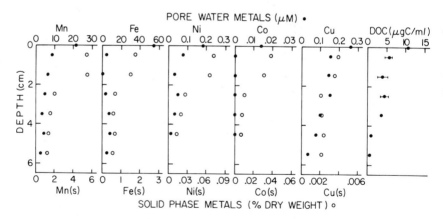

Figure 17-3. Dissolved metal profiles (closed circles) and solid-phase metal oxide profiles (open circles) of the sands at site S. The dissolved organic carbon (DOC) profile of the sediment is included for comparison with the dissolved copper and copper oxide profiles. The solid-phase analyses do not include the large ferromanganese concretions at the sediment surface, and the values plotted at the sediment-water interface are from bottom-water analyses.

maximum within this horizon. During each experiment the peeper was left in place for five days and capped at the bottom before removal from the sediment. The core box was then dismantled in the laboratory and the sediment horizon from 2 cm to 3 cm subbottom sampled as 0.25-cm slices away from the dialysis cell. Following the recommendations of Burdige and Kepkay (1983), only the horizon from 2 cm to 3 cm was sampled, not the complete 1 cm to 4 cm. This allowed sediment affected by vertical rather than lateral diffusion and sediment containing macrophyte roots (from 0 cm to 2 cm) to be discarded. Each 1-ml sediment slice (0.25-cm long by 1-cm wide by 4-cm deep) was centrifuged for 2 min at 20°C in a Beckman Microfuge 11, and the supernatant pore water was removed with an acid-washed Pasteur pipette. The pore water was then stored at 5°C for dissolved metal analysis.

A detailed justification of the mathematics applied to the diffusion of dissolved metals and azide between dialysis cell and sediment has been presented by Burdige and Kepkay (1983). They have also proven experimentally that azide diffuses 2.8 to 3.1 times faster than manganese in laboratory clays and natural sediments. As a result, the abiotic environment developed in the poisoned peeper allows dissolved manganese diffusion to

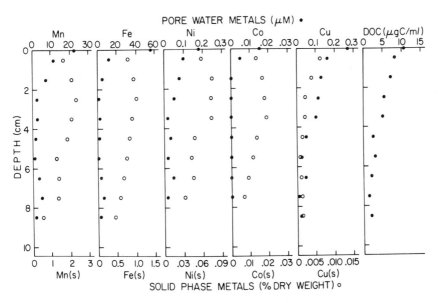

Figure 17-4. Dissolved metal profiles (closed cirlcles) and solid-phase metal oxide profiles (open circles) of the muds at site M. The dissolved organic carbon (DOC) profile of the sediments is included for comparison with the dissolved copper profile. Values plotted at the sediment-water interface are from bottom-water analyses.

follow a modified form of Fick's second law (Crank, 1975; Berner, 1980),

$$\frac{\partial c}{\partial t} = \frac{D_s}{(1+K)}\left(\frac{\partial^2 c}{\partial x^2}\right) \qquad \text{(Eq. 17-1)}$$

where c is the concentration of dissolved metals in the pore waters, D_s is the bulk sediment diffusion coefficient, t is time and x, the distance between dialysis membrane and sample. Adsorption has been accounted for in Eq. 17-1 by assuming that the process is rapid and reversible and can be explained in terms of a dimensionless adsorption coefficient (K) defined from a linear isotherm (Duursma and Hoede 1967; Berner 1980). In the unpoisoned peeper, where biological activity is not inhibited, it is assumed that manganese oxidation or metal binding follows Michaelis-Menten kinetics, so that Eq. 17-1 is modified to

$$\frac{\partial c}{\partial t} = \frac{D_s}{(1+K)}\left(\frac{\partial^2 c}{\partial x^2}\right) - \frac{1}{(1+K)}\left(\frac{V_{max}c}{K_m+c}\right) \qquad \text{(Eq. 17-2)}$$

where V_{max} is the maximum binding or oxidation rate and K_m the half saturation constant. The solutions to Eq. 17-1 and Eq. 17-2, as well as the procedures for fitting curves to the pore-water data produced by the diffusion of dissolved metals in a peeper, have already been presented by Burdige and Kepkay (1983). In laboratory clays they have demonstrated that manganese adsorption coefficients from this curve-fitting procedure agree with values determined independently from linear adsorption isotherms. They also found from curve fits that Michaelis-Menten parameters for manganese oxidation by clays inoculated with bacterial cultures are in good agreement with the oxidation kinetics of the same cultures in chemostats (Kepkay, Burdige, and Nealson, 1984).

RESULTS

Dissolved Mn, Fe, Ni, Co, and Cu decreased rapidly with depth at both sites (Figs. 17-3 and 17-4). The oxides of Mn, Fe, Ni, and Co attained maximum values between 1 cm and 4 cm subbottom and decreased with depth below these horizons, indicating that dissolved Mn, Fe, Ni, and Co were oxidized as they diffused downward from the bottom water. The behavior of copper was distinctly different, where dissolved Cu profiles followed dissolved organic carbon (Figs. 17-3 and 17-4) rather than dissolved Mn, Fe, Ni, and Co.

Kinetically distinct microbial and abiotic processes were observed in the poison and live peepers, and while the terms "poison" and "live" may not be microbiologically correct, they have been adopted for the sake of brevity. The term "metal binding" is used here to describe the processes restricting the migration of dissolved metals from the dialysis cell. In the case of manganese, the data of Burdige and Kepkay (1983) indicate that adsorption was the abiotic process in the poison peeper and microbial oxidation was predominant in the live peeper. These simplistic definitions of abiotic and microbial metal binding may not be entirely correct in the case of manganese (Burdige and Kepkay, 1983) and certainly cannot be applied to the binding of Fe, Ni, and Cu. Until data are available to determine the oxidation state(s) of these metals in organic-rich waters, "abiotic binding" will be used to describe processes in the poison peeper and "microbial binding" to describe the biological processes apparent in the live peeper. The term "apparent adsorption coefficient" has also been used to define the coefficient produced from curve fits to live peeper data (Table 17-1). It is essentially a fictitious coefficient because it includes biological processes that do not necessarily follow simple, linear adsorption. Despite its fictitious nature, the apparent adsorption coefficient is useful because it allows the direct comparison of microbial and abiotic binding.

At the end of the peeper experiments, dissolved Mn, Fe, Ni, and Cu in the dialysis cells matched the dissolved metal concentrations in the reservoirs and original laboratory solutions. Pore-water profiles of these metals did not change appreciably over the two five-day emplacement periods. Curves fitted to the pore-water data (Fig. 17-5) indicate that dissolved Mn, Fe, Ni, and Cu all diffused farther in the poison peeper than in the live peeper. The data resulting from the curve fits (Table 17-1) indicate that dissolved Mn and Ni diffused at almost the same rate in the poison peeper, with adsorption coefficients of 6.6 and 6.9, respectively. These values are in good agreement with a coefficient of 6.9 determined for maganese adsorption by a laboratory clay (Burdige and Kepkay, 1983). In the live peeper, dissolved Mn and Ni also appeared to be subject to the same microbial oxidation or binding regime, with apparent adsorption coefficients of 32.1 and 34.3, respectively. However, dissolved Fe did not follow Mn or Ni (Table 17-1), and the enhancement of iron binding by microbial activity (from an adsorption coefficient of 4.0 to an apparent adsorption coefficient of 12.3) was only 3.1 times the abiotic binding. In contrast, this microbial enhancement was 5.2 times for Mn and 4.9 times for Ni. The behavior of dissolved Cu was even more divergent from the behavior of Mn and Ni, with a microbial enhancement of Cu binding that was only 1.9 times the abiotic effect (increasing an adsorption coefficient of 3.9 to an apparent adsorption coefficient of 7.5). When Michaelis-Menten curves were generated from best fit K_m and V_{max} values, first-order or linear oxidation and binding kinetics were predicted for dissolved metal concentrations (Table 17-1) that were greater than pore-water values (Fig. 17-4). This means that any quantitative description of microbial metal oxidation or metal binding at site M can be based on the assumption that first-order kinetics are in operation.

Table 17-1. Parameters Used to Generate the Pore-Water Concentration Curves Fitted to the Data from in situ Dialysis Experiments

	Whole Sediment Diffusion Coefficient[a] $D_s (cm^2 s^{-1})$	Sediment Porosity at 2-3 cm Subbottom ϕ	Formation Factor[b] F	Dialysis Cell Concentration[n] $C_0 hr (\mu M)$	Background Pore-Water Concentration[n] $C_i hr (\mu M)^c$	Adsorption Coefficient K	Apparent Adsorption Coefficient[e] K_a	Curve Fit Variables[d]		
								K_m (μM)	V_{max} $(\mu M.h^{-1})$	Limit of First-Order Oxidation[f] or Binding (μM)
Mn	3.48×10^{-6}	0.78	1.8	20.361	3.110	6.62 ± 0.06	32.11 ± 1.30	723.2 ± 53.1	75.7 ± 5.6	53.1
Ni	3.48×10^{-6}	0.78	1.8	0.198	0.031	6.93 ± 0.11	34.32 ± 1.93	6.83 ± 0.52	0.75 ± 0.06	0.52
Fe	3.72×10^{-6}	0.78	1.8	60.482	4.163	4.00 ± 0.16	12.33 ± 0.94	1792.4 ± 206.1	134.5 ± 15.5	206.1
Cu	3.77×10^{-6}	0.78	1.8	0.308	0.050	3.91 ± 0.08	7.54 ± 0.23	3.60 ± 0.39	0.21 ± 0.06	0.39

[a]Calculated from $D_s = D/\phi F$, where D is the free-iron diffusion coefficient at infinite dilution (Li and Gregory, 1974) and 14°C (the average sediment temperature at site M).
[b]Calculated as an average value from $F = \phi^{-1.8} = 1.6$ (Mannheim and Waterman, 1974; Berner, 1980) and $F = \phi^{-2.8} = 2.0$ (Andrews and Bennett, 1981).
[c]Determined from the 2-3-cm horizon in cores and in core boxes at 9-10 cm away from the dialysis cells.
[d]All best-fit values are presented with the 1 σ values for the fits.
[e]The apparent adsorption coefficient (K_a) includes abiotic adsorption and the enhanced or apparent adsorption brought about by microbial activity.
[f]When the dissolved metal concentrations defining the limits of first-order or linear binding are exceeded, then the full Michaelis-Menten equation is required to describe the kinetics of metal oxidation or metal binding. The estimation of this limit is based on the minimum uncertainty of each K_m determination.

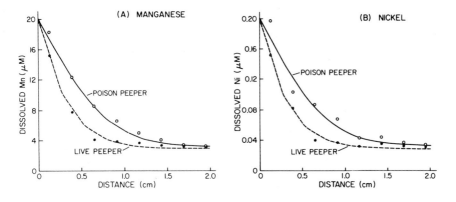

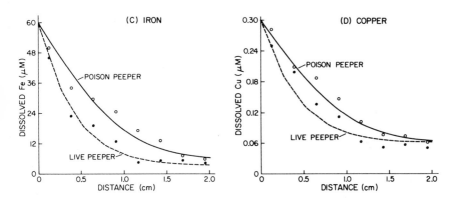

Figure 17-5. Best-fit curves for pore water Mn, Ni, Fe, and Cu resulting from the diffusion of the dissolved metals in live peepers (open circles) and poison peepers (closed circles). All horizontal axes define distance from a dialysis cell as the source of dissolved metals. Best-fit adsorption coefficients and Michaelis-Menten parameters for manganese oxidation or metal binding are summarized in Table 17-1.

DISCUSSION

At both sites all of the dissolved metals appeared to be diffusing downward from the bottom waters (Figs. 17-3 and 17-4) and were subject to oxidation within site M sediments (Fig. 17-4). At site S oxidative processes were apparent at the sediment surface (indicated by large ferromanganese concretions) as well as within the sediment itself (Fig. 17-3). These data agree with the concept of Kindle (1933, 1935) and Harriss and Troup (1970) that concretions are formed as dissolved metals from ground water or more reducing parts of the lake come in contact with shallow-water, oxidizing environments such as Concretion Cove. However, data from the peeper experiments represent metal oxidation and metal binding in the dark. This does not agree with their contention that the photosynthetic activities of benthic diatoms regulate metal oxidation.

It is not clear why manganese oxidation is at a maximum at 2-3 cm subbottom at site M (Fig. 17-4) when higher dissolved manganese and oxygen in the bottom water would favor oxidation at the sediment surface. Macrophyte roots may have introduced oxygen to greater depths in the sediment and may have been involved in the uptake or oxidation of manganese (Wium-Andersen and Andersen, 1972). As a result, microbial oxidation appears to have been restricted to regions unaffected by roots—that is, to deeper than 2 cm subbottom. This contention is supported by the presence of large ferromanganese concretions at site S, indicating that if macrophytes are not present, metal oxidation will take place preferentially at the sediment-water interface.

Results from the peeper experiments suggest that the kinetics of manganese oxidation and nickel binding were closely associated (Table 17-1) and that microbial activity enhanced the two processes fivefold. The microbial enhancement of iron and copper binding was less pronounced, only 3.1 times in the case of iron and 1.8 times in the case of copper. This result may have been due to the complexation of the metals with organic material that would effectively remove them from the influence of manganese oxidation. The interaction of copper with organic material is relatively well defined, with dissolved Cu following dissolved organic carbon (Figs. 17-3 and 17-4). The case for iron-organic interaction is less clear because oxidative processes play a significant role in the speciation of the metal (Figs. 17-3 and 17-4). High concentrations of dissolved Fe in the bottom water may have allowed it to be affected by manganese oxidation, metal-organic complexation, or some other process such as microbial iron oxidation (Nealson, 1983). Data must be collected to define the speciation of iron and copper in organic-rich waters before the relative contributions of these processes can be defined.

With the upper limits of first-order manganese oxidation and metal binding in live peepers (Table 17-1) greater than the pore-water concentrations of the metals (Fig. 17-4), first-order kinetics can be applied at site M. This observation is in agreement with the assumption built into the models of Burdige and Gieskes (1983) and others (see Berner, 1980) that manganese oxidation follows first-order kinetics in sediments. The data in Table 17-1 also indicate that the microbial mediation of manganese oxidation and trace-metal binding is a very important component of metal cycling in Lake Charlotte, outweighing abiotic processes such as adsorption by about two to five times. These processes may well be episodic and subject to seasonal

Table 17-2. Solid-Phase Chemistry of Ferromanganese Precipitates from Nova Scotian lakes and the Pacific Ocean

	Fe:Mn	*Ni:Mn*	*Co:Mn*	*Cu:Mn*
Lake Charlotte muds (average from 10 sediment horizons at site M)	0.63	4.3×10^{-2}	1.2×10^{-2}	2.7×10^{-3}
Lake Charlotte sands (average from 6 sediment horizons at site S)	0.59	2.0×10^{-2}	1.1×10^{-2}	1.5×10^{-3}
Lake Charlotte concretion (Harriss and Troup, 1970)	0.63	4.2×10^{-4}	9.8×10^{-4}	2.6×10^{-5}
Grand Lake concretion (Harriss and Troup, 1970)	0.51	5.9×10^{-4}	8.9×10^{-4}	4.2×10^{-5}
"Average" Nova Scotian lake concretion (Kindle, 1933, 1935)	0.60	—	—	—
"Average" Pacific concretion (Mero, 1965; Cronan, 1977)	0.58	4.1×10^{-2}	1.5×10^{-2}	2.1×10^{-2}

variation, but their importance can be further demonstrated by comparing the metal oxides in the sediments to Lake Charlotte and Pacific concretions (Table 17-2). Lake Charlotte concretions have ratios of Ni, Co, and Cu to Mn that are typical of the low ratios found in lacustrine ferromanganese concretions (Harriss and Troup, 1970; Calvert and Price, 1977). In contrast, the oxides within the sediments have Ni-, Co-, and Cu-to-Mn ratios that are two orders of magnitude higher, very near an average ratio for deep-sea concretions (Mero, 1965; Cronan, 1977). The negligible environmental and economic value normally attributed to lacustrine metal precipitates may therefore be due to the concentration of research efforts on large concretions rather than on the sediments associated with the concretions.

REFERENCES

Ali, S. H., and J. L. Stokes, 1971, Stimulation of heterotrophic and autotrophic growth of *Sphaerotilus discophorus* by manganese ions, *Antonie van Leeuwenhoek* **37:**519-528.

Andrews, D., and A. Bennett, 1981, Measurements of diffusivity near the sediment-water interface with a fine-scale resistivity probe, *Geochim. Cosmochim. Acta* **45:**2169-2175.

Beals, H. L., 1966, Manganese-iron concretions in Nova Scotia lakes, *Marit. Sediments* **2:**70-72.

Beijerinck, M. W., 1913, Oxydation des manganbikarbonates durch bakterien and schimmelpilze, *Folia Microbiol.*(Delft) **2:**123-134.

Berner, R. A., 1980, *Early Diagenesis: A Theoretical Approach,* Princeton University Press, Princeton, New Jersey.

Burdige, D. J., and J. M. Gieskes, 1983, A pore water/solid phase diagenetic model for manganese in marine sediments, *Am. J. Sci.* **283:**29-47.

Burdige, D. J., and P. E. Kepkay, 1983, Determination of bacterial manganese oxidation rates in sediments using an *in situ* dialysis technique. I. Laboratory studies, *Geochim. Cosmochim. Acta* **47:**1907-1916.

Calvert, S. E., and N. B. Price, 1977, Shallow water, continental margin and lacustrine nodules: Distribution and geochemistry, in *Marine Manganese Deposits,* G.P. Glasby, ed., Elsevier, New York, pp. 45-86.

Crank, J., 1975, *The Mathematics of Diffusion,* 2nd ed., Oxford University Press, London.

Cronan, D. S., 1977, Deep-sea nodules: Distribution and geochemistry, in *Marine Manganese Deposits,* G. P. Glasby, ed., Elsevier, New York, pp. 11-44.

Douka, C. E., 1980, Kinetics of manganese oxidation by cell-free extracts of bacteria isolated from manganese concretions from soil, *Appl. Environ. Microbiol.* **39:**74-80.

Duursma, E. K., and C. Hoede, 1967, Theoretical, experimental and field studies concerning molecular diffusion in sediments and suspended sediment particles, *Neth. J. Sea Res.* **3:**423-457.

Ehrlich, H. L., 1976, Manganese as an energy source for bacteria, in *Environmental Biogeochemistry,* Vol. 2, *Metals, Transfer and Ecological Mass Balance,* O. Nriagu, ed., Ann Arbor Science Publishers, Ann Arbor, Michigan, pp. 633-644.

Emerson, S., R. E. Cranston, and P. S. Liss, 1979, Redox species in a reducing fjord: Equilibrium and kinetic considerations, *Deep-Sea Res.* **26A:**859-878.

Emerson, S. A., S. Kalhorn, L. Jacobs, B. M. Tebo, K. H. Nealson, and R. A. Rosson, 1982, Environmental oxidation rate of manganese (II): Bacterial catalysis, *Geochim. Cosmochim. Acta* **46:**1073-1079.

Gershey, R. M., M. D. McKinnon, P. J. Le B. Williams, and R. M. Moore, 1979, Comparison of three oxidation methods for the analysis of dissolved organic carbon in seawater, *Mar. Chem.* **7:**289-306.

Harriss, R. C., and A. G. Troup, 1970, Chemistry and origin of freshwater ferromanganese concretions, *Limnol. Oceanogr.* **15:**702-712.

Helder, W., R. T. P. deVries, and M. M. R. Van Der Loeff, 1982, Behaviour of nitrogen nutrients and silica in the Ems-Dollard estuary, *Can J. Fish. Aquat. Sci.* **40**(Suppl. 1):188-200.

Hesslein, R. H., 1976, An *in situ* sampler for close interval pore water studies, *Limnol. Oceanogr.* **21:**912-915.

Honeyman, D., 1881, Nova Scotia geology (superficial), *Trans, N.S. Inst. Natural Sci. Proc.* **5:**328.

Kepkay, P. E., D. J. Burdige, and K. H. Nealson, 1984, Kinetics of bacterial manganese binding and oxidation in the chemostat, *Geomicrobiol. J.* **3:**245-261.

Kindle, E. M., 1933, Lacustrine concretions of manganese, *Am. J. Sci.* **24:**496-504.

Kindle, E. M., 1935, Manganese concretions in Nova Scotia lakes, *Roy. Soc. Can. Trans.* Sect. 4, **29:**163-180.

Li, Y. H., and S. Gregory, 1974, Diffusion of ions in seawater and deep sea sediments, *Limnol. Oceanogr.* **33:**703-714.

Mannheim, F. T., and L. S. Waterman, 1974, Diffusimitry (diffusion constant estimation) on sediment cores by resistivity probe, *Initial Reports of DSDP 22:* 663-670.

Mero, J. L., 1965, *The Mineral Resources of the Sea,* Elsevier, Amsterdam.

Nealson, K. H., 1983, Microbial oxidation and reduction of manganese and iron, in *Biomineralization and Biological Metal Accumulation,* P. Westbroek and E. W. deJong, eds., D. Reidel, Boston, pp. 459-580.

Reeburgh, W. S., 1967, An improved interstitial water sampler, *Limnol. Oceanogr.* **12:**163-165.

Willey, J. D., and R. A. Fitzgerald, 1980, Trace metal geochemistry of sediments from the Miramichi estuary, New Brunswick, *Can. J. Earth Sci.* **17:**254-265.

Wium-Andersen, S., and J. M. Andersen, 1972, The influence of vegetation of the redox profile of the sediment of Grane Langso, a Danish *Lobelia* lake, *Limnol. Oceanogr.* **17:**948-952.

18

In situ Manganese (II) Binding Rates at the Oxic/Anoxic Interface in Saanich Inlet, B.C., Canada

Bradley M. Tebo

Scripps Institution of Oceanography

Craig D. Taylor

Woods Hole Oceanographic Institution

Kenneth H. Nealson

Scripps Institution of Oceanography

Steven Emerson

University of Washington

ABSTRACT: The rate of manganese [Mn(II)] binding at the oxic/anoxic interface in Saanich Inlet, B.C., Canada, is enhanced by bacterial activities (Emerson et al., 1982; Tebo, 1983). *In situ* rates of manganese binding were measured in short-term experiments with a sampler incubation device (SID) that collects a water sample at depth, simultaneously mixes it with a radioactive tracer [^{54}Mn(II)], and in a time-course fashion removes subsamples into syringes containing poisons to inhibit further biological activity. Such measurements confirmed the rapid removal of Mn(II) in the region just above the oxic/anoxic interface. The rate constants for dissolved Mn(II), computed from these direct rate measurements, were between 35.4/yr and 350.4/yr in agreement with those calculated from chemical models (Emerson, Cranston, and Liss, 1979; Emerson et al., 1982). The rates of Mn(II) binding in these SID deployments were significantly faster than the rates measured on ship in closed-bottle incubations in which no exogenous air was introduced (oxygen-limiting conditions), but were slower than the initial rates observed when the incubations were saturated with air. Thus, in such suboxic zones, where chemical parameters may change rapidly on collection of a water sample, the SID provides a method of accurately estimating reaction rates *in situ* under conditions that closely approximate the real environment.

INTRODUCTION

The geochemical processes controlling trace-metal distributions in the marine environment have been a subject of increasing interest in the past few years. Whereas the distribution and speciation of many trace metals can be explained by inorganic chemical processes, redox processes, or both, the importance of biological activity in the behavior of some of these metals [e.g., Mn(II) and Co(II)] has been demonstrated (Emerson et al., 1982; Wollast, Billen, and Duinker, 1979; Chapnick, Moore, and Nealson, 1982; Tebo, 1983; Tebo et al., in press); however, such studies have not directly demonstrated these biological activities *in situ.*

Previous studies using poisoned controls in shipboard incubations with water collected from the oxic/anoxic interface in Saanich Inlet, B.C., Canada, have shown that the rate of manganese binding is enhanced by bacterial activities (Emerson, et al., 1982; Tebo, 1983; Tebo et al., in press). This report describes Mn(II) binding and oxidation studies performed in Saanich Inlet with a sampler incubation device (SID) (Taylor, Molongoski, and Lohrenz, 1983) that takes a water sample, simultaneously mixes it with a radioactive tracer [in this case ^{54}Mn(II)], and removes subsamples with time to syringes containing poisons. Thus, the SID provides a way of measuring the time course of Mn(II) binding *in situ* in a water sample that closely approximates the natural environment. The rates measured in these SID incubations have been compared to the rates measured on ship in incubations under oxygen-limiting and air-saturating conditions.

MATERIALS AND METHODS

Study Site

All work described in this report was performed in Saanich Inlet, British Columbia, Canada, during a cruise on the R/V *Alpha Helix* in May 1981. Saanich Inlet is a fjord with a maximum depth of 220 m and a sill depth of 70m; this fjord is located on the southeast side of Vancouver Island. Water is trapped behind the sill during late winter and summer. The basin stratifies, becoming anoxic below about 130 m for approximately six months, until dense oxygenated water overflows the sill in the fall due to strong coastal upwelling. This results in mixing (turnover) and destratification (Anderson and Devol, 1973; Richards, 1965). Recently, bacteria have been shown to remove Mn(II) rapidly from solution above the oxygen/hydrogen sulfide interface that occurs below the euphotic zone in the water column (Emerson et al., 1982; Tebo, 1983; Tebo et al., in press).

Analytical Methods

Oxygen concentrations were measured by the method of Broenkow and Cline (1969), and hydrogen sulfide was determined by the spectrophotometric method of Cline (1969). Dissolved Mn(II) was determined by the formaldoxime method of Brewer and Spencer (1971) on water samples that had been pressure filtered (nitrogen) through 0.4-μm Nuclepore membrane filters.

Particulate manganese (retained on 0.4-μm Nuclepore filters) concentrations were measured by neutron activation and gamma counting or by dissolution and atomic absorption spectroscopy. Ammonia, nitrate, nitrite, phosphate, and silica were measured colorimetrically in an autoanalyzer (Strickland and Parsons, 1972).

^{54}Mn(II) Binding Experiments

The potential of bacteria to catalyze Mn(II) precipitation was determined by incubating water samples with a radioactive tracer in the presence and absence of poisons, filtering subsamples with time through 0.2-μm membrane filters, and measuring the percentage of radioactivity trapped by the filters. The poison used in the experiments (azide) was selected on the basis of laboratory control experiments. These experiments demonstrated that azide was effective in inhibiting Mn(II) binding and oxidation by pure culture isolates of bacteria in the laboratory and did not interfere with the chemistry of manganese. The criteria that established this latter point were that azide (1) did not interfere in the adsorption of Mn(II) onto synthetic manganates, (2) caused no desorption of Mn(II) that had been preadsorbed to synthetic manganates, and (3) did not reduce synthetic manganates (Tebo, 1983). Other control experiments on Mn-rich particulates from Saanich Inlet showed that the poison did not inhibit adsorption onto concentrated and sonicated particulate material from Saanich Inlet (Emerson et al., 1982).

Water samples collected by Go-flo bottles were incubated with gamma-emitting ^{54}Mn(II) (New England Nuclear). For aerobic incubations, samples were contained in polystyrene or polypropylene tubes or bottles. For oxygen-limiting incubations, samples were placed in 60-ml glass serum bottles (Wheaton) with aluminum crimp-seal caps and Teflon-silicon liners (Pierce). The serum bottles were pretreated with Surfasil (Pierce) to minimize adsorption on the vessel walls, and Surfasil-coated glass beads (3-mm diameter) were placed in the bottles to allow proper mixing of the incubation mixtures. The serum bottles were flushed with three volumes of water from the Go-flo bottle and capped without an airspace. Azide was then added to some of the tubes or bottles to a final concentration of 0.1% from a 10% stock solution. In the case of oxygen-limiting conditions, the azide was first sparged with nitrogen gas to remove as much of the oxygen as possible. Syringes with needles were used to inject the azide into the bottles, with a second needle inserted to allow water to escape. The experiments were initiated by adding the radioactive substrate (carrier free) to a specific activity that would result in about 10^5 cpm per subsample volume (the exact activity was different from experiment to experiment). The experiments were incubated at 8°C-10°C (the temperature of the water at the oxic/anoxic interface is about 9°C). Subsamples were removed at various times and filtered in duplicate or triplicate through 0.2-μm membrane filters (Gelman) and washed by filtering 5 to 25 times the subsample volume with prefiltered (0.2-μm) sea water. In the oxygen-limiting incubations, subsamples were removed by syringe, and the volume was replaced with nitrogen gas. Filters and filtrates were counted using a Beckman Biogamma II gamma counter, and the percentages of the radioactivity trapped on the filters as a function of time were determined. In addition, subsamples from all incubations (without filtering) were taken for direct measurement of total radioactivity to ensure complete recovery of the added label.-

In situ Incubation Experiments with the Sampler Incubation Device (SID)

The design and use of the SID has been previously described (Taylor, Molongoski, and Lohrenz, 1983). The check valve of the SID was filled with 0.1 ml of ^{54}Mn(II) (0.2 mCi/ml), and 0.5 ml of 10% sodium azide was included in the 50-ml subsampling syringes. The SID was deployed at 115 m or 120 m on a piece of stainless steel hydrowire attached to a set of floats and tethered to the ship with nylon rope. The incubation time for different experiments varied from 3.3 hr to 6.4 hr with subsampling syringes triggering every 15 or 31.5 min, respectively. At the end of the incubation time, the instrument was brought on board ship, and the water in the chamber and subsampling syringes was analyzed for total radioactivity and filtered in triplicate for determination of the percentage of bound radioactivity. This cruise represented the first deployment of the SID at depths greater than 100 m, and several problems were encountered. The device was deployed eight times; leakage occurred in the main chamber on three of those occasions, and the results of these experiments were discarded. On the other deployments leakage occurred in some of the subsampling syringes. When counts of total radioactivity indicated that significant leakage (dilution) had occurred, these data points were deleted from the final analyses. Although some leakage commonly occurred, it is felt that, since it was confined to the subsampling syringes that were poisoned, leakage of even oxygen-saturated sea water would not have significantly affected the amount of ^{54}Mn(II) bound (which is calculated as the percentage of total counts trapped on the filter). Nevertheless, only the data from experiments in which the total counts of ^{54}Mn(II) in the subsampling syringes were greater than 70% of the total counts in the main chamber were used in the final analyses. Samples from the chamber were analyzed for Mn(II) and nutrients after each deployment as described above.

RESULTS AND DISCUSSION

Profiles of oxygen, hydrogen sulfide, nitrate, ammonia, Mn(II), and particulate manganese are presented in Figure 18-1. The chemistry of the water column during this cruise in May 1981 was characterized by a broad oxygen minimum zone (100 m to 175 m), with the oxygen/hydrogen sulfide interface occurring at around 175 m. The region of high particulate manganese occurs between about 100 m and 140 m, well below the euphotic zone. The profile of soluble Mn(II) is a biphasic curve with steep curvature above 160 m, indicative of removal in the region of high particulate manganese. Below 160 m the curve is much more gradual and increases to the bottom, indicating possible diffusion of Mn(II) from the sediments.

Microbial experiments using [54]Mn(II) as a radiotracer in the presence and absence of poisons that inhibit bacterial activities without interfering with the chemistry of manganese have demonstrated that a peak in the biological Mn(II)-binding occurs in the region of high particulate manganese (Emerson, et al., 1982; Tebo, 1983; Tebo, et al., in press). The effect of oxygen tension on the [54]Mn(II) binding was tested in incubations of water from 115 m (Fig. 18-2). In these incubations, air-saturation conditions were compared to conditions of limiting oxygen (no additional oxygen was added). The results (Fig. 18-2) indicate that, when oxygen is limiting, significantly less Mn(II) is

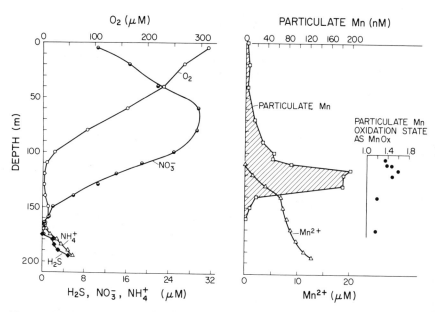

Figure 18-1. Profiles of oxygen, sulfide, nitrate, ammonia, mananese(II), particulate manganese, and the oxidation state of manganese in particulates trapped on 0.4 μm membrane filters during a cruise to Saanich Inlet in May 1981. The chemistry of the water column during this cruise was characterized by a broad oxygen minimum zone (100 m to 175 m) with the oxygen/hydrogen sulfide interface occurring around 175 m. The region of high particulate manganese occurs below the euphotic zone between about 100 m and 140 m. A peak in the oxidation state of the particulate manganese occurs at 115 m.

bound. The initial rates (the initial 1.5-hr time interval) of ^{54}Mn(II) binding in these experiments are given in Table 18-1.

The results for five deployments of the SID are given in Table 18-2, and the kinetics of Mn(II) removal for these deployments are plotted in Figure 18-3. Included in Table 18-2 are the concentrations of nutrients and Mn(II) at the end of the experiment, the Mn(II) binding rates, the residence time of Mn(II), and the first-order rate constants for Mn(II) binding. The manganese binding rate was calculated from the measured manganese concentration and the slope of the line based on the percentage of radioactivity trapped on 0.2-μm membrane filters as a function of time. The residence time $(t_{1/2})$ was calculated as the time in days it would take to trap half of the measured Mn(II) concentration on filters. The first-order rate constant for Mn(II) binding (k_{Mn}) was calculated as the rate/[Mn(II)]. Also given in Table 18-2 is the correlation coefficient determined from the data points making up the slope of the binding rate.

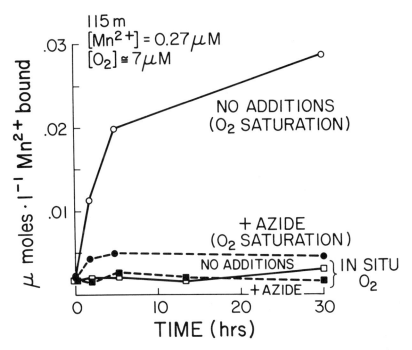

Figure 18-2. A comparison made during a cruise to Saanich Inlet in May 1981 of ^{54}Mn(II) binding under conditions of oxygen saturation (in equilibrium with the atmosphere) and *in situ* oxygen (at the start of the incubation, this experiment was performed in a closed bottle to which no oxygen was introduced, and thus oxygen is considered to be limiting). The results show that when oxygen is limiting, significantly less Mn(II) is bound.

Table 18-2 reveals that even though the SID was positioned at the same depth in several experiments, the chemical parameters were quite variable. This is undoubtedly the result of vertical oscillation of the interface in Saanich Inlet. The interface moves up and down by as much as 5 m or more as a result of either tidal fluctuations or internal waves. This movement

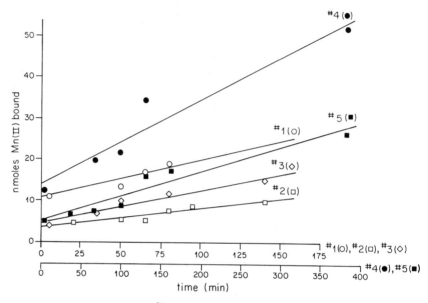

Figure 18-3. Time course of ^{54}Mn(II) binding in the SID incubations. The data points plotted correspond to the results presented in Table 18-2. Separate regression lines are drawn for the results of each SID deployment. Experiments numbers 1–3 and numbers 4–5 represent data in which subsampling syringes triggered every 15 min and 31.5 min, respectively. Note the different time scale for the longer time course experiments.

Table 18-1. Shipboard Incubation Experiments

	Binding Rate (nmol/l/h)
Air saturation	
No additions	6.3
+ azide	2.3
Oxygen limitation	
No additions	0.9
+ azide	0.1

Note: depth = 115 m.
[Mn(II)] = 0.27 μM.
[O$_2$] $\simeq$ 7 μM.

Table 18–2. Sampler Incubation Device (SID) in situ Experiments

Experiment Number	Depth (m)	Elapsed Time* (hr)	NH_4 (µM)	NO_2 (µM)	NO_3 (µM)	Si (µM)	PO_4 (µM)	Mn(II) (µM)	Binding Rate (nmol/l/n)	Number Data Points	Correlation Coefficient (r)	$t_{1/2}$ (d)	k_{Mn} yr^{-1}
2	120	4.75	0.10	0.15	12.07	114.94	4.52	0.66	2.7	7	.938	5.1	35.8
5	115	6.4	0.14	0.13	13.22	124.98	4.66	0.89	3.6	7	.96	5.2	35.4
3	115	3.3	0.15	2.11	16.34	104.01	4.25	0.11	4.9	5	.967	0.5	350.4
1	120	3.3	0.22	0.03	16.71	110.16	4.63	1.21	5.5	4	.917	4.6	39.8
4	115	6.4	0.09	0.29	14.19	117.27	4.52	1.35	6.1	5	.962	4.6	39.6

*The time at which the chamber and the syringes were sampled and filtered.

217

could cause variation in the results of experiments in which water is taken from Niskin bottles on different casts; consequently these experiments are difficult to compare with one another.

In spite of the variability due to the nonuniformity of the particulate layer, the rates of [54]Mn(II) binding in these SID incubations were much faster than the rates observed on ship in closed-bottle incubations in which oxygen became limiting. These rates were only slightly slower than the rates measured in incubations exposed to the atmosphere.

Table 18-2 also demonstrates that the manganese-binding rate is not clearly correlated with the concentrations of ammonia, nitrite, nitrate, silica, phosphate, or Mn(II). The correlation of binding with the concentration of nitrate would be good with the exception of the experiment where the fastest rate was calculated (Experiment number 4). Since the chemical species were measured at the end of the experiment, some utilization or production of nitrate (or any of the other nutrients, for that matter) may have occurred.

There is also a tendency for binding rate to increase with the concentration of Mn(II); this is in agreement with the proposed reactions and kinetics of manganese oxidation (Stumm and Morgan, 1981). The calculated first-order rate constants for Mn(II) binding are well within the range of values published for a variety of sedimentary and aquatic environments (Table 18-3; modified from Kepkay, Burdige, and Nealson, 1984) and in good agreement with values calculated from chemical models for the particulate layer in Saanich Inlet (Emerson, Cranston, and Liss, 1979; Emerson et al., 1982). In comparing the results from the different SID deployments, it is difficult to explain the increasing manganese concentration with increasing nitrate concentration unless nitrate was produced during the incubation. Normally just the opposite relationship is expected; that is, as Mn(II) concentration increases, the nitrate should decrease. This has been clearly shown for chemical profiles of these species across the oxic/anoxic interface in both Saanich Inlet (Emerson, Cranston, and Liss, 1979) and in sediments (Emerson et al., 1980; Froelich et al., 1979; Jahnke et al., 1982). Since nitrate is expected to more closely parallel oxygen, one could argue that the binding rate is proportional to oxygen tension, as has been suggested by previous experiments (see Fig. 18-2; Tebo, 1983; Tebo et al., in press). However, the proportionality between nitrate and oxygen in Saanich Inlet during this cruise was not perfect, so it is not possible to extrapolate the nitrate data to the oxygen tension.

Although some difficulties with the SID incubations were encountered—for example, leakage of some of the subsampling syringes (a problem that subsequently was corrected)—this method appears promising for future studies of *in situ* rates of Mn(II) oxidation in aquatic environments.

Table 18-3. A Comparison of First-Order Rate Constants for Manganese Oxidation Published for a Variety of Natural Environments with the Rate Constants Determined from *in situ* Incubations Presented in this Work

Location	$k(yr^{-1})$
Lower limit for marine sediments based on laboratory data (Boudreau and Scott, 1978)	3.2
E. Equatorial Atlantic sediments (Burdige and Gieskes, 1982)	10.7-12.0
Long Island Sound sediments (Aller, 1980)	50
Narragansett Bay sediments (Elderfield *et al.*, 1981)	1000
Particulate layer of Saanich Inlet water column (Emerson, Cranston, and Liss, 1979)	180
Particulate layer of Saanich Inlet water column estimated by SID incubations (this work)	35.4-350.4

One problem in correlating the binding rate with the various chemical parameters is that the specific adsorption and auto-oxidation rate was not measured in these experiments. However, since azide does not interfere with the adsorption of Mn(II) onto particulates from the interface in Saanich Inlet (Emerson et al., 1982), the total binding due to adsorption at the end of the deployment may be inferred by extrapolating the rate of binding to the zero time point. Nevertheless, variation in the total particulate concentration could certainly have imposed some of the variability in the data in Table 18-2. One way to circumvent this problem would be to utilize a dual-chambered SID, with one chamber poisoned from the onset of the incubation and the other unpoisoned. This device would then have a built-in mechanism for establishing the variable contribution of adsorption by particulate surfaces. Such a device is currently being manufactured and should prove useful in further studies of Mn(II) precipitation at oxic/anoxic interfaces occurring in the water column.

ACKNOWLEDGEMENTS

We acknowledge the expertise of J. Molongoski and D. Wiebe for operating the SID. Lucinda Jacobs provided the particulate manganese data. Nutrient analyses were performed by Kathy Krogslund. This work was supported by National Science Foundation grants, #OCE-80-18189 to SE, IDOE (MANOP) #OCE-81-00641 to KHN.

REFERENCES

Aller, R. C., 1980, Diagenetic processes near the sediment water interface of Long Island Sound sediments II. Fe and Mn, *Adv. Geophys.* **22:**351-415.

Anderson, J. J., and A. H. Devol, 1973, Deep water renewal in Saanich Inlet, an intermittently anoxic basin, *Estuar. and Coastal Mar. Sci.* **1:**1.

Boudreau, B. P., and M. R. Scott, 1978, A model for the diffusion-controlled growth of deep sea manganese nodules, *Am. J. Sci.* **278:**903-929.

Brewer, P. G., and D. W. Spencer, 1971, Colorimetric determination of manganese in anoxic waters, *Limnol. Oceanogr.* **16:**107-112.

Broenkow, W. W., and J. D. Cline, 1969, Colorimetric determination of dissolved oxygen at low concentrations, *Limnol. Oceanogr.* **14:**450.

Burdige, D. J., and J. M. Gieskes, 1982, A pore water/solid phase diagenetic model for manganese in marine sediments, *Am. J. Sci.* **283:**29-47.

Chapnick, S. D., W. S. Moore, and K. H. Nealson, 1982, Microbially mediated manganese oxidation in a freshwater lake, *Limnol. Oceanogr.* **27:**1004-1014.

Cline, J. D., 1969, Spectrophotometric determination of hydrogen sulfide in natural waters, *Limnol. Oceanogr.* **14:**454-458.

Elderfield, H., N. Luedtke, R. J. McCaffrey, and M. Bender, 1981, Benthic flux studies in Narragansett Bay, *Am. J. Sci.* **281:**768-787.

Emerson, S., R. E. Cranston, and P. S. Liss, 1979, Redox species in a reducing fjord; equilibrium and kinetic considerations, *Deep-Sea Res.* **26A:**859-873.

Emerson, S., R. Jahnke, M. Bender, P. Froelich, G. Klinkhammer, C. Bowser, and G. Setlock, 1980, Early diagenesis in sediments from the eastern equatorial Pacific. I. Pore water nutrient and carbonate results, *Earth Planet. Sci. Lett.* **49:**57-80.

Emerson, S., S. Kalhorn, L. Jacobs, B. M. Tebo, K. H. Nealson, and R. A. Rosson, 1982, Environmental oxidation rate of manganese (II): Bacterial catalysis, *Geochim. Cosmochim. Acta* **46:**1073-1079.

Froehlich, P. N., G. P. Klinkhammer, M. L. Bender, N. A. Luedtke, G. R. Heath, D. Cullen, P. Dauphin, D. Hammond, B. Hartman, and V. Maynard, 1979, Early oxidation of organic matter in pelagic sediments of the eastern equatorial Atlantic: Suboxic diagenesis, *Geochim. Cosmochim. Acta* **43:**1075-1090.

Jahnke, R., D. Heggie, S. Emerson, and V. Grundmanis, 1982, Pore waters of the central Pacific Ocean: Nutrient results, *Earth Planet. Sci. Lett.* **61:**233-256.

Kepkay, P. E., D. J. Burdige, and K. H. Nealson, 1984, Kinetics of bacterial manganese binding and oxidation, *Geomicrobiol. J.* **3:**245-262.

Richards, F. A., 1965, Anoxic basins and fjords, in *Chemical Oceanography,* Vol. I, J. P. Riley and G. Skirrow, eds., Academic Press, New York, pp. 611-641.

Strickland, J. D. H., and T. R. Parsons, 1972, A Practical Handbook of Seawater Analysis, *Fish. Research Board of Can. Bull.* **125:**1-203.

Stumm, W., and J. J. Morgan, 1981, *Aquatic Chemistry,* John Wiley and Sons, New York.

Taylor, C. D., J. J. Molongoski, and S. E. Lohrenz, 1983, Instrumentation for measurements of phytoplankton production, *Limnol. Oceanogr.* **28:**781-787.

Tebo, B. M., 1983, *The Ecology and Ultrastructure of Marine Manganese Oxidizing Bacteria,* Ph.D. thesis, Scripps Institution of Oceanography, University of California, San Diego.

Tebo, B. M., K. H. Nealson, S. Emerson, and L. Jacobs, in press, Bacterial mediation of Mn(II) and Co(II) precipitation at oxic/anoxic interfaces in the marine environment, *Limnol. Oceanogr.*

Wollast, R., G. Billen, and J. C. Duinker, 1979, Behavior of manganese in the Rhine and Scheldt estuaries, *Estuar. Coastal Mar. Sci.* **9:**161-169.

19

Seasonal Sulfate Reduction and Iron Mobilization in Estuarine Sediments

Mark E. Hines, Mary Jo Spencer, Joyce B. Tugel, W. Berry Lyons, and Galen E. Jones

University of New Hampshire

INTRODUCTION

Rapid deposition of organic material in estuaries produces highly anoxic sediments that support active populations of bacteria (Penchel and Blackburn, 1979). Sulfate reduction is the predominant bacterial respiratory process in these sediments (Jorgensen, 1982). Reduced sulfur generated during sulfate reduction is influential in regulating the sedimentary chemistry of iron and other metals (Berner, 1980; Goldhaber and Kaplan, 1974; Elderfield et al., 1981). Temperate estuarine sediments can exhibit a pronounced seasonality in the rate of sulfate reduction and the extent to which oxygen penetrates into the sediments (Jorgensen, 1977 *b;* Nedwell and Abram, 1978; Sansone and Martens, 1982; Hines et al., 1982). Therefore, the seasonal variations in the production and oxidation of iron and sulfur compounds increase the potential for rapid temporal changes in the precipitation, dissolution, and, hence, remobilization of iron (Hines et al, in press).

Particle reworking and irrigation by bioturbating infauna strongly influence the sedimentary chemistry of iron and sulfur by enhancing redox reactions in sediments (Aller, 1977). Seasonal variations in the presence, composition, abundance, and activity of these infauna may produce large temporal and spatial changes in the mobilization of metals in sediments (Hines et al., 1982).

Iron, abundant in estuarine sediments, is highly reactive and exerts a strong influence on the biogeochemistry of other elements (Berner, 1980).

Therefore, a thorough knowledge of the temporal variation in iron mobilization and the factors regulating its mobilization is essential if one is to understand the role of estuaries and estuarine sediments in regulating the cycling of bioactive elements.

The study described here represents the compilation of selected data collected from a single site during 1978, 1979, 1980, and 1982. These data are from three separate but related studies and are presented here to demonstrate the occurrence of large seasonal and annual variations in the remobilization of iron and the reduction of sulfate in an estuarine sediment. In addition, the influence of bioturbation on iron dissolution is discussed.

MATERIALS AND METHODS

Study Site

Sediment samples were collected from a shallow subtidal area near the Jackson Estuarine Laboratory in Great Bay, New Hampshire, U.S.A. The sediments consisted of silt, clay, and fine sand-sized particles. Bioturbation is usually active in these sediments from June through November (Armstrong et al., 1979; Hines et al., 1982); the dominant macroorganisms are the capetellid polychaete *Heteromastus filiformis* and the tellinid bivalve *Macoma balthica* (Black, 1980). Sediment temperatures vary from about $-0.5°C$ in February to $25°C$ in July. Seasonal temperature variations were similar from year to year throughout the present study.

Sampling and Sample Handling

Sediments were collected using hand-held Plexiglas box corers as described by Armstrong, Lyons, and Gaudette (1979). Upon return to the laboratory and removal of the overlying water, cores were extruded under nitrogen and sliced into 2- or 3-cm horizontal sections. The sediment was centrifuged and the pore water was filtered through $0.4\text{-}\mu m$ Nuclepore filters. Anoxic (N_2) conditions were maintained throughout processing. Extreme precautions were taken to minimize metal contamination during sample handling and analysis. Details of cleaning procedures are outlined in Hines et al. (in press).

Analysis

The rate of sulfate reduction was determined, in triplicate, by measuring the production of dissolved and acid-volatile ^{35}S generated during the reduction of $^{35}SO_4^{2-}$ (Hines and Lyons, 1982). Dissolved iron was determined colorimetrically using Ferrozine (Stookey, 1970). Pore-water samples for iron analyses were pre-acidified with Ultrex nitric acid (1% final concentration).

Dissolved sulfide was determined colorimetrically (Cline, 1969) on samples prefixed with zinc acetate. Using pore water freshly collected from sediment squeezers (Robbins and Gustinis, 1976), dissolved sulfide was also determined from the difference in absorbance at 230 nm between acidified and unacidified samples.

RESULTS

Dissolved Iron

Dissolved iron concentrations varied greatly throughout the entire sampling period (Figs. 19-1 and 19-2). Low iron concentrations generally were observed during winter. This was due to sediment oxidation. A spring maximum in dissolved iron concentration was consistently noted and on some occasions was higher than 150 μM. A rapid decrease in iron levels followed these maxima. High iron concentrations occurred during the summer of 1979, 1980, and 1982 (Figs. 19-1 and 19-2) with values as high as 150-180 μM. Conversely, dissolved iron was nearly absent from these sediments during the summer of 1978 (Fig. 19-1). An unusually high concentration of iron was recorded in June 1980 (Fig. 19-1); this maximum was accompanied by a large increase in bioturbation and anaerobic microbial activity (Hines et al.,

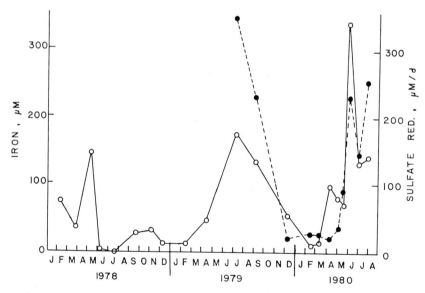

Figure 19-1. Variations in dissolved iron (open circles) from 1978 to 1980 and sulfate reduction rates (closed circles) from 1979 to 1980 in Great Bay sediments. Values represent averages of upper 6 cm of sediment.

1982). The summer increase in dissolved iron occurred gradually during 1982 (Fig. 19-2) as opposed to the dramatic increase of 1980.

Rates of sulfate reduction in these sediments were measured between June 1979 and August 1980 (Fig. 19-1) and again during the 1982 study period (Fig. 19-2).Rates were slow during winter (approximately 30 nmol/ml/d), but increased significantly during spring and summer. The maximum rates recorded occurred in July of each year. These July rates decreased each successive year in which they were measured: 350 nmol/ml/d in 1979, 230 in 1980, and 150 in 1982. In general, these sulfate-reduction rates were similar to those rates reported for near-shore marine sediments by others (Jorgensen, 1977*a*, 1977*b*; Nedwell and Abram, 1978; Winfrey et al., 1982; Hines and Lyons, 1982). However, these rates are slower than those reported for salt-marsh sediments (Howarth and Teal, 1979) and microbial mats (Jorgensen and Cohen, 1977, Lyons et al., 1984). Interestingly, the rapid summer rates reported here occurred simultaneously with high concentrations of dissolved iron.

Dissolved Sulfide

The concentrations of dissolved sulfide were low in these Great Bay sediments (Fig. 19-2) and often were lower than the detection limit of the methods employed (approximately 2-3 μM). Dissolved sulfide minima were encountered when dissolved iron concentrations were highest. Most important, dissolved sulfide was not detected during July, when sulfate reduction was most rapid. Although dissolved sulfide was not measured during 1979 and 1980, the smell of sulfide was not present in summer samples from those years. However, samples collected during late fall and winter occasionally smelled of sulfide.

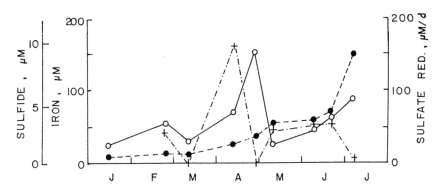

Figure 19-2. Variations in dissolved iron (open circles), sulfate reduction rates (closed circles), and dissolved sulfides (crosses) in Great Bay sediments from January to July, 1982. Values represent averages of the upper 6 cm.

DISCUSSION

Few studies have measured the annual variations in anaerobic microbial activity in estuarine depositional environments. Jorgensen (1977a) reported that sulfate reduction rates in Danish sediments varied approximately two-fold from one year to the next. The greater-than-twofold decrease in sulfate reduction in Great Bay sediments from 1979 to 1982 is not easily explained. There were insufficient data to determine the influence of varying overlying water productivity, river discharge, and winter ice scouring on rates of microbial metabolism. However, these results demonstrated that researchers should exercise caution when using data from short-term studies to explain long-term processes.

The dissolution of iron in these sediments occurred first during early spring (April), when sedimentary temperatures began to rise. Increasing microbial activity during this period probably caused a decrease in redox potential by the increased consumption of oxygen and production of reduced end products. This process caused the reduction and dissolution of iron. Another possibility is that iron dissolution occurred as a result of direct iron reduction by bacteria (Sorensen, 1982). It was not determined whether the iron remobilized in April diffuses into the overlying water. Others have demonstrated that iron tends to remain in sediments because of rapid iron oxidation and precipitation upon contact with oxygenated waters (Aller, 1977; Luther et al., 1982). Even in the absence of any iron efflux, this rapid remobilization results in the redistribution of iron in the sediments. Concentration of other metals, including manganese, molybdenum, and copper, also exhibited April maxima in these sediments (Hines et al., in press). The spring iron maximum was short-lived because of the removal of iron as a sulfide precipitate once sulfate reduction activity increased.

The second event responsible for iron dissolution occurred during the summer. One would expect that dissolved iron concentrations would remain low throughout the summer due to rapid sulfate reduction and the precipitation of iron sulfide minerals. This occurred only during 1978. The continued availability of iron in the presence of sulfide formation during the summers of 1979, 1980, and 1982 was probably a result of the increased cycling of iron and sulfur during infaunal reworking and irrigation activities (Hines et al., 1982). Bioturbation increased rapidly in these sediments during early summer. This was particularly evident during June 1980, when the visual and textural characteristics and the porosity of the sediments changed dramatically (Hines et al., 1982), and the dissolved iron concentration increased fivefold (Fig. 19-1). Apparently, bioturbation increased more slowly during 1982, since dissolved iron values increased slowly from May into July (Fig. 19-2).

Since the previous winter was extremely severe, the anomalously low iron values recorded during June and July 1978 (Fig. 19-1) probably were due to the absence of an active bioturbating community at that time. Yeo and Risk

(1979) described the long-term (about one-year) infaunal recolonization of sediments after severe storms. X-radiographs of cores collected at the Great Bay site during 1978 revealed that rapid infaunal activity did not occur until the fall (Armstrong et al., 1979). Therefore, the iron data appeared to reflect the following bioturbation differences occurring annually in the sediments: (1) bioturbation inactive (1978); (2) bioturbation initiating rapidly (June 1980); (3) bioturbation initiating gradually (1982).

The absence of dissolved sulfide and the occurrence of high concentrations of dissolved iron in July 1982 indicated that, even though sulfate reduction exceeded 150 nmol/ml/d, sulfide was quantitatively removed from solution as an iron sulfide precipitate. Jorgensen (1977*a*) reported that sulfide accumulation in Danish fjord sediments displayed a one- to two-week lag behind the onset of rapid sulfate reduction in early summer. He attributed the lag to the fact that sulfide was removed continually as FeS as long as iron was available. Once iron was consumed, dissolved sulfide concentrations increased. In the present sediments, iron was supplied continually throughout the summer by bioturbation (Fig. 19-1). FeS precipitation prevented sulfide accumulation.

In conclusion, sulfate reduction rates and dissolved iron concentrations in these estuarine sediments varied dramatically both seasonally and annually. Frequent sampling was necessary to delineate relatively short-lived iron remobilization events. During periods of active bioturbation, iron concentrations were high even though sulfide production was rapid. High iron concentrations in summer resulted in rapid precipitation of sulfide, thus maintaining extremely low dissolved sulfide concentrations during periods of the year when bacteria were most active.

ACKNOWLEDGEMENTS

We appreciate the analytical expertise of P. B. Armstrong and the sampling help of S. Knollmeyer and R. N. Foley. This research was supported by National Science Foundation grants DES 75-04790, OCE 77-20484, and OCE 80-18460.

REFERENCES

Aller, R. C., 1977, *The Influence of Macrobenthos on Chemical Diagenesis of Marine Sediments,* Ph.D. thesis, Yale University, New Haven, Connecticut.

Armstrong, P. B., C. Fischer, W. B. Lyons, and H. E. Gaudette, 1979, Seasonal variation in bioturbation activities in a northern temperate estuary as determined by x-radigraphic techniques, Paper presented at Estuarine Research Federation meeting, Jekyll Island, Georgia.

Armstrong, P. B., W. B. Lyons, and H. E. Gaudette, 1979, Application of formaldoxime colorimetric method for the determination of manganese in the pore water of anoxic estuarine sediments, *Estuaries* **2:**198-201.

Berner, R. A., 1980, *Early Diagenesis: A Theoretical Approach,* Princeton University Press, Princeton, New Jersey.

Black, L. F., 1980, The biodeposition cycle of a surface deposit-feeding bivalve, *Macoma balthica* (L.), in *Estuarine Perspectives,* V. S. Kennedy, ed., Academic Press, New York, pp. 389-402.

Cline, J. D., 1969, Spectrophotometric determination of hydrogen sulfide in natural waters, *Limnol. Oceanogr.* **14:**454-458.

Elderfield, H., R. J. McCaffrey, N. Luedtke, M. Bender, and V. W. Truesdale, 1981, Chemical diagenesis in Narragansett Bay sediments, *Amer. J. Sci.* **281:**1021-1055.

Fenchel, T., and T. H. Blackburn, 1979, *Bacteria and Mineral Cycling,* Academic Press, London.

Goldhaber, M. B., and I. R. Kaplan, 1974, The sulfur cycle, in *The Sea,* Vol. 5, E. D. Goldburg, ed., John Wiley and Sons, New York, pp. 569-655.

Hines, M. E., and W. B. Lyons, 1982, Biogeochemistry of nearshore Bermuda sediments. I. Sulfate reduction rates and nutrient generation, *Mar. Ecol. Progr. Ser.* **8:**87-94.

Hines, M. E., W. H. Orem, W. B. Lyons, and G. E. Jones, 1982, Microbial activity and bioturbation-induced oscillations in pore water chemistry of estuarine sediments in spring, *Nature* **299:**433-435.

Hines, M. E., W. B. Lyons, P. B. Armstrong, W. H. Orem, M. J. Spencer, H. E. Gaudette, and G. E. Jones, in press. Seasonal metal remobilization in the sediments of Great Bay, New Hampshire, *Marine Chemistry.*

Howarth, R. W., and J. M. Teal, 1979, Sulfate reduction in a New England salt marsh, *Limnol. Oceanogr.* **24:**999-1013.

Jorgensen, B. B., 1977*a,* The sulfur cycle of a coastal marine sediment (Limfjordan, Denmark), *Limnol. Oceanogr.* **22:**814-823.

Jorgensen, B. B., 1977*b,* Bacterial sulfate reduction within reduced microniches of oxidized marine sediments, *Mar. Biol.* **41:** 7-17.

Jorgensen, B. B., 1982, Mineralization of organic matter in the sea bed—the role of sulphate reduction, *Nature* **296:**643-645.

Jorgensen, B. B., and Y. Cohen, 1977, Solar Lake 5. The sulfur cycle of the benthic cyanobacterial mats, *Limnol. Oceanogr.* **22:**657-666.

Luther, G. W. III, A. Giblin, R. W. Howarth, and R. A. Ryans, 1982, Pyrite and oxidized iron mineral phases formed from pyrite oxidation in salt marsh and estuarine sediments, *Geochim. Cosmochim. Acta* **46:** 2665-2669.

Lyons, W. B., M. E. Hines, and H. E. Gaudette, 1984, Major and minor element pore water geochemistry of modern marine sabkhas: The influence of cyanobacterial mats, in *Microbial Mats: Stromatolites,* Y. Cohen, R. W. Castenholz, and H. O. Halvorson, eds., A. R. Liss, Inc., New York, pp. 411-423.

Nedwell, D. B., and J. W. Abram, 1978, Bacterial sulphate reduction in relation to sulphur geochemistry in two contrasting areas of saltmarsh sediment, *Estuar. Coastal Mar. Sci.* **6:** 341-351.

Robbins, J. A., and J. Gustinis, 1976, A squeezer for efficient extraction of pore water from small volumes of anoxic sediment, *Limnol. Oceanogr.* **21:**905-909.

Sansone, F. J., and C. S. Martens, 1982, Volatile fatty acid cycling in organic-rich marine sediments, *Geochim. Cosmochim. Acta* **46:**1576-1589.

Sorensen, J., 1982, Reduction of ferric iron in anaerobic, marine sediment and interactions with reduction of nitrate and sulfate, *Appl. Environ. Microbiol.* **43:**319-324.

Stookey, L. L., 1970, Ferrozine—A new spectrophotometric reagent for iron, *Anal. Chem.* **42:**779-781.

Winfrey, M. R., D. G. Marty, A. J. M. Bianchi, and D. M. Ward, 1982, Vertical distribution of sulfate reduction, methane production, and bacteria in marine sediments, *Geomicrobiol. J.* **2:**341-362.

Yeo, R. R., and M. J. Risk, 1979, Intertidal catastrophies: Effect of storms and hurricanes on intertidal benthos of the Minas Basin, Bay of Fundy, *Fish. Res. Bd. Can. J.* **36:**667-669.

20

Effect of Dissimilatory Sulfate Reduction on Speciation of Zinc in Freshwater Sediments

Sally G. Hornor and Sung Ok Yoon

Virginia Polytechnic Institute and State University

INTRODUCTION

The distribution and fate of heavy metals in freshwater ecosystems are controlled by complex interactions of physical, chemical, and biological processes. Key reactions occurring at the sediment-water interface include metal adsorption, chelation, and precipitation. Redox potential and pH strongly influence microbial community activity, which in turn influences the rates of metal transformations and mobility within aquatic ecosystems. The occurrence of microbial dissimilatory sulfate reduction may profoundly affect metal speciation due to the strong scavenging ability of released sulfide to precipitate metals as insoluble metallic sulfides.

The objective of this study was to determine the effect of dissimilatory sulfate (SO_4^-) reduction on the speciation of zinc (Zn) in freshwater sediment. Dissimilatory sulfate reduction and metallic sulfide precipitation have been well documented in marine and estuarine systems, but such activity in fresh water has not been thoroughly studied (e.g., Khalid, Patrick, Jr., and Gambrell, 1978; Ramamoorthy and Rust, 1978; Howarth and Giblin, 1983). Although naturally occurring sulfate levels in fresh water are often insufficient to permit extensive sulfate reduction, lake sediments as well as ground water and streams receiving industrial input from coal-related industries may contain sulfate levels great enough to permit such activity (Cappenberg and Jongejan, 1977; Dokins et al., 1980; Nriagu and Hem, 1978; Lovley and Klug, 1983).

One of the most critical problems to be addressed in heavy-metal contamination of fresh water is the bio-availability and potential toxicity of metals to aquatic biota. Numerous investigators have attempted to chemically fractionate metals in sediments and to determine which of these fractions are available (Pickering, 1980; Agemian and Chau, 1976; Gadd and Griffiths, 1978). A scheme has been developed for the chemical fractionation of Zn and iron (Fe) in sediments in this study, and the study attempts to determine the stability of these fractions under varying redox conditions. It was of interest to determine the extent to which inorganic Zn, when supplied as zinc sulfate ($ZnSO_4^-$), would be removed from the water column and precipitated to the sediment surface as zinc sulfide (ZnS). Removal of toxic heavy metals by metallic sulfide precipitation has been demonstrated in effluents from mining operations (Tuttle, Dugan, and Randles, 1969; Ilyaletdinov, Enker, and Loginova, 1977), but has not been investigated in surface waters receiving allochthonous input of sulfate and soluble heavy metals.

MATERIALS AND METHODS

Sample Collection and Microcosm Incubation

Sediment and surface-water samples were collected from Adair Run, a second-order mountain stream in Glen Lyn, southwestern Virginia. This stream had received the effluent from a coal fly ash-settling basin up until one year prior to this study. Water and sediment were brought to the laboratory and stored at 5°C until they were used for microcosm construction; care was taken to minimize exposure to atmospheric O_2. Contamination by metals during collection and handling was minimized by acid washing of all glassware according to standard methods (U.S. Environmental Protection Agency, 1979). Organic content of the sediments ranged from 2% to 3% of dry weight, which was 75%-80% of the wet weight.

Microcosms consisted of four 9-1 gas-tight Plexiglas chambers 40 x 15 x 15 cm in length, width, and height. Each chamber was divided into four identical compartments and fitted with a gas-tight lid with a gas-sampling port closed with a rubber septum. Each compartment received approximately 400 g of wet sediment and 500 ml of stream water. Each of the four microcosms received a different treatment: (1) 200 mg/l Na_2SO_4; (2) 200 mg/l Na_2SO_4 plus 1% (wt/dry wt sediment) native fibrous cellulose (Baker TLC); (3) 200 mg/l $SO_4^=$ plus 100 mg/l $ZnSO_4$ $-Zn$; and (4) 200 mg/l $SO_4^=$, 100 mg/l $ZnSO_4$ $-Zn$, plus 1% cellulose. All microcosms were incubated at 22°C for 24 hr, subsampled, sealed, and incubated for three weeks at 22°C.

Chemical Analyses

The pH and redox potential (Eh) in the water column were measured with an Orion 601A Ionanalyzer. Eh measurements were performed with a platinum electrode calibrated with Zobell standard solution (Zobell, 1946). Sulfate concentration was measured using the BaCl turbidometric method (APHA, 1971); dissolved oxygen in water was detected using the Winkler method (APHA, 1971); and chemical oxygen demand (COD) in water was measured by standard methods (U.S. Environmental Protection Agency, 1979). Zinc and iron in the water column were measured with a Perkin-Elmer Model 460 atomic absorption spectrophotometer.

Percent dry weight of sediments was determined by drying at 85°C–95°C for 24 hr, and percent organic content was determined by loss of weight on combustion of dried sediment at 500°C for 3 hr. Acid-volatile sulfide in sediments was released as H_2S by addition of 6 N HCl at a solid-to-acid ratio of 1:1. The evolved H_2S was quantified with a Varian 3700 gas chromatograph equipped with a flame photometric detector (FPD) sensitive to ng quantities of S compounds, as previously described (Hornor and Mitchell, 1981).

Sediment Metal Analyses

Chemical fractionation of Zn and Fe in sediments was performed by sequential leaching techniques according to Figure 20-1. The metal content of each fraction was determined by atomic absorption spectrophotometry (AA). Details of metal speciation are given below.

Water-Soluble Fraction. Fresh sediment was treated with deionized water at a solid-to-water ratio of 1:8 and centrifuged at 3170 × gravity for 25 min. The supernatant was filtered through a 0.45-μm Gelman membrane filter and preserved with 3 ml concentrated HNO_3 per liter of sample.

Loosely Bound Exchangeable Fraction. 1 N sodium acetate was added at a solid-to-water ratio of 1:8 to the residual solid from the water-soluble fraction (Khalid, Gambrell, and Patrick, Jr., 1976). The mixture was centrifuged and filtered as described for the water-soluble fraction.

Tightly Bound Organic Fraction. The residual solid material from the exchangeable fraction was extracted for 24 hr at 22°C with 5 N NH_4OH at a solid-to-water ratio of 1:8. The supernatant was centrifuged and filtered as described for the water-soluble fraction.

Sulfide Fraction. The residue from the organic fraction was dried and finely ground. One gram of sediment was then digested with 25 ml of concentrated HNO_3 and extracted in 25 ml of 6 N HCl (Holmes, Slade, and McLerran,

1974). The solution was centrifuged and filtered as described for the water-soluble fraction.

The percent recovery of the loosely bound exchangeable Zn was more than 90% when tested with $ZnSO_4$, $ZnCl_2$, and $ZnCO_3$. The percent recovery of tightly bound organic Zn with Zn-acetate was more than 95%, and the percent recovery for reagent-grade ZnS was more than 80%.

RESULTS AND DISCUSSION

Indices of Microbial Sulfate Reduction

Chemical parameters indicating the extent of microbial sulfate reduction are shown in Table 20-1. After 24 hr of microcosm incubation, the water column

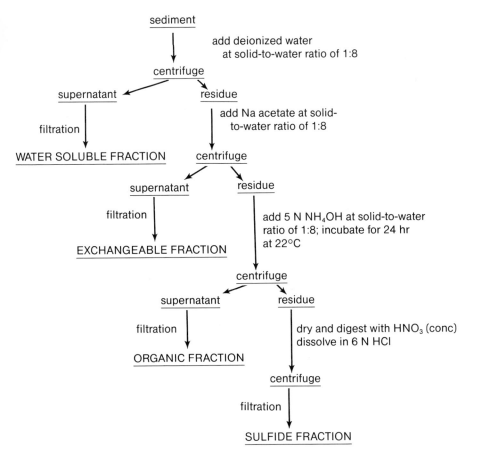

Figure 20-1. Flow diagram of metal-extraction method.

Table 20–1. Characteristics of Microcosms

	Amended with 200 ppm $SO_4^=$			Amended with 100 ppm Zn and 200 ppm $SO_4^=$		
	Oxidized[a]	Moderately Reduced[b]	Reduced[c]	Oxidized[a]	Moderately Reduced[b]	Reduced[c]
$SO_4^=$ (mg/l)	210.0	86.1	<5	219.0	7.4	<5
COD (mg/l)	4.8	310.8	1089.0	4.0	650.0	1338.1
Eh (mV)	530	349	119	550	431	170
pH	7.3	7.6	7.2	6.7	7.6	6.9
Dissolved O_2 (mg/l)	12.0	1.6	0.2	6.5	1.4	0.2
Zn (mg/l)	<0.03	<0.03	<0.03	76.0	<0.03	<0.03
Fe (mg/l)	<0.1	<0.1	<0.1	<0.1	0.5	9.9
$S^=$ (μg/g)	1.8	73.1	150.5	1.2	142.5	169.8
Zn (μg/g)	73.1	57.5	50.5	60.9	436.4	442.1
Fe (mg/g)	16.6	16.6	16.8	17.5	18.2	18.0

[a]Sampled 24 hr after treatment.
[b]Sampled after 3 weeks incubation, no organic carbon added.
[c]Sampled after 3 weeks incubation, 1% (wt/dry wt) cellulose added.

remained well oxidized. The extent of oxidation is readily seen in the high Eh and dissolved oxygen levels and in the low measurements of chemical oxygen demand (COD). Sulfate concentrations indicate that no measurable sulfate reduction had occurred, and sulfide concentrations in the sediments were negligible. This state of oxidation was found in all microcosms, regardless of amendment with Zn or cellulose. The only difference between the two oxidized microcosms was the high level of Zn in the water column of the Zn-amended system. The loss of 24 mg/l Zn in the first 24 hr of incubation may be accounted for by adsorption by sediments. Earlier studies in which sediments were omitted from the microcosms showed no significant loss of Zn from the water column after three weeks of incubation.

Those microcosms that did not receive 1% cellulose were moderately reduced following the incubation period. The best indices of partial reduction are COD and dissolved oxygen in the water column, as well as sulfate in the water and sulfide in the sediments. Although the Eh of the water column was lower than that seen initially, Eh values in the moderately reduced systems were greater than would be anticipated. Eh was measured approximately 1 cm below the headspace-water interface, and this value provides an estimate of the oxidation state of the least reduced portion of the systems. Due to the microcosm design, Eh could not be measured at the redoxocline (Hallberg, 1977), but estimates of Eh at the sediment-water interface would be approximately 200 mV.

The Zn-amended microcosms without cellulose addition were shown to be much more active in sulfate reduction than the non-Zn-amended microcosms without cellulose. This greater microbial activity in the presence of Zn was evident from the loss of sulfate in the water column and the accumulation of twice as much sulfide, as well as high levels of Zn, in the sediments. COD levels twice as great in the Zn-amended microcosms may have resulted from an increased production of organic acids and reduced sulfur by the microbial community. Sulfate-reducing bacteria are known to be inhibited by sulfide accumulation and to be enhanced by removal of sulfide as metallic sulfide (Ilyaletdinov, Enker, and Loginova, 1977). The accumulation of sulfide as ZnS in the sediments of the Zn-amended systems may account for the greater extent of sulfate reduction in these microcosms.

Complete reduction of sulfate in the water column occurred in both Zn-amended and non-Zn-amended microcosms incubated with cellulose for three weeks. The high degree of reduction was indicated by COD values greater than 1000 mg/l and by high sulfide concentrations in the sediments. Eh values indicate the extent of reduction in the least-reduced portion of the water column. Zn amendment was again seen to be stimulatory to rates of sulfate reduction. Zn-amended microcosms exhibited a greater COD, lower dissolved oxygen, and greater sediment sulfide concentrations than non-Zn-amended systems. Release of Fe to the water column of cellulose and Zn-enriched systems provided further evidence for the extensive reduction in these microcosms.

Effect of Sulfate Reduction on Metal Speciation

The percent distribution of Fe and Zn in the sediments of oxidized, moderately reduced, and reduced microcosms is presented in Table 20-2. Iron behavior was similar in both the presence and absence of added Zn. Water-soluble fractions of Fe were negligible throughout the study. The fraction of total Fe that was released from sediments by sodium acetate extraction (exchangeable Fe) increased with increasing degree of reduction, while the iron sulfide fraction decreased with increasing degree of reduction. Thus it appears that iron sulfide was converted to exchangeable Fe during the course of incubation. It is unlikely that acidophilic iron-oxidizing thiobacilli would account for this conversion due to the consistently circumneutral pH; however, sulfide-oxidizing thiobacilli capable of growth and sulfide oxidation at a pH of 6.8 have been shown to occur within these sediments (Hornor, 1983). Abiotic transformations of FeS to exchangeable Fe are unlikely in this reduced, carbonate-buffered, sedimentary system. The low fraction of Fe associated with organic compounds in this study indicates that chelation of Fe with organic acids produced by the microbial community was not an important mechanism for enhancing Fe mobility.

The effect of a reducing environment on the speciation of native Zn with-

Table 20-2. Speciation of Iron and Zinc in Microcosm Sediments

	Amended with 200 ppm $SO_4^=$			Amended with 100 ppm Zn and 200 ppm $SO_4^=$		
	Oxidized[a]	Moderately Reduced[b]	Reduced[c]	Oxidized[a]	Moderately Reduced[b]	Reduced[c]
Fe						
% water soluble	0.01	0.05	0.01	0.04	0.01	<0.01
% exchangeable	0.87	4.01	5.96	0.78	6.24	8.00
% organic	<0.01	0.13	0.07	0.04	0.04	<0.01
% sulfide	99.12	95.18	93.96	99.14	93.71	92.00
Zn						
% water soluble	0.27	<0.01	1.61	0.29	0.07	0.05
% exchangeable	13.88	8.87	8.72	12.15	31.02	29.40
% organic	0.12	1.29	3.11	1.03	2.24	2.00
% sulfide	85.76	89.84	86.57	86.53	66.68	68.50

[a]Sampled 24 hr after treatment.
[b]Sampled after 3 weeks incubation, no organic carbon added.
[c]Sampled after 3 weeks incubation, 1% (wt/dry wt) cellulose added.

in stream sediments can be seen in the results from the non-Zn-amended microcosms. The primary trend noted in these results was the transformation of exchangeable Zn to either water-soluble or organic Zn complexes. The ZnS fraction remained consistently high, while the exchangeable fraction decreased with increasing reduction. The organic Zn fraction increased with increasing reduction, while the water-soluble fraction was increased in the fully reduced state but was not detectable in the moderately reduced state.

The behavior of added Zn can be seen in the Zn-amended microcosms. In contrast to the native Zn speciation, added Zn as $ZnSO_4$ became incorporated primarily into the exchangeable and sulfide fractions. By subtracting the native Zn pool from the added Zn, it can be seen that approximately 32% of the added $ZnSO_4$ was transformed to the exchangeable pool, 66% was converted to ZnS, and approximately 2% became bound to organic matter. This behavior was similar in both the presence and absence of added cellulose. No added Zn was found in the water-soluble fraction; this fraction decreased as Eh was decreasing due to dilution of native Zn with added Zn. The fact that not all of the added Zn became transformed to ZnS may have been due to depletion of sulfate in the water column and subsequently insufficient supply of sulfide for ZnS formation.

The large fraction of $ZnSO_4$ that was transformed to ZnS may be predicted to remain in the sedimentary environment as an unavailable metal. The exchangeable and organic fractions, in contrast, represent labile and biologically available forms of the metal. In sulfate-limited freshwater systems, allochthonous water-soluble Zn may be incorporated into labile pools within surface sediments and may then be available to both microbial and invertebrate communities. In carbonate-buffered sediments, the labile Zn pool is primarily adsorbed within the sediment-pore water matrix by cation-exchange mechanisms. When sulfate is not limiting, however, as in the case of coal-related allochthonous input, water-soluble Zn would be predicted to be rapidly removed from the water column as an insoluble and unavailable metallic sulfide.

The addition of organic carbon to stream sediments did not significantly alter the fate and distribution of Zn or Fe in this study, although it did enhance the rate of sulfate reduction and subsequent metallic sulfide precipitation. The high percentage of ZnS-Zn in the fresh stream sediment provides evidence for extensive *in situ* sulfate reduction, despite a normally well oxygenated water column. The primary factor controlling precipitation of metallic sulfides in such an environment is the sulfate concentration available for diffusion into the subsurface sediments.

REFERENCES

Agemian, H., and A. S. Y. Chau, 1976, Evaluation of extraction techniques for the determination of metals in aquatic sediments, *Analyst* **101:**761-767.

American Public Health Association, 1971, *Standard Methods for the Examination of Water and Wastewater,* 13th ed., APHA, Washington, D.C.

Cappenberg, T. E., and E. Jongejan, 1977, Microenvironments for sulfate reduction and methane production in freshwater sediments, in *Environmental Biogeochemistry and Geomicrobiology,* W. E. Krumbein, ed., Ann Arbor Science, Ann Arbor, Michigan, pp. 129-138.

Dokins, W. S., G. J. Olson, G. A. McPeters, and S. C. Turbak, 1980, Dissimilatory bacterial sulfate reduction in Montana groundwaters, *Geomicrobiol. J.* **2:**83-98.

Gadd, G. M., and A. J. Griffiths, 1978, Microorganisms and heavy metal toxicity, *Microb. Ecol.* **4:**303-317.

Hallberg, R., 1977, Metal-organic interaction at the redoxocline, in *Environmental Biogeochemistry and Geomicrobiology,* W. E. Krumbein, ed., Ann Arbor Science, Ann Arbor, Michigan, pp. 947-953.

Holmes, C. W., E. A. Slade, and C. J. McLerran, 1974, Migration and redistribution of zinc and cadmium in marine estuarine system, *Environ. Sci. Tech.* **8:**255-259.

Hornor, S. G., 1983, Toxicity of zinc concentrate to stream bacteria, Abstract, First International Symposium of Toxicological Testing Using Bacteria, May 17-19, 1983, Burlington, Ontario, Canada.

Hornor, S. G., and M. J. Mitchell, 1981, Effect of the earthworm, *Eisenia foetida* (Oligochaeta), on fluxes of volatile carbon and sulfur compounds from sewage sludge, *Soil Biol. Biochem.* **13:**367-372.

Howarth, R. W., and A. Giblin, 1983, Sulfate reduction in the salt marshes at Sapelo Island, Georgia, *Limnol. Oceanogr.* **28:**70-82.

Ilyaletdinov, A. N., P. B. Enker, and L. V. Loginova, 1977, Role of sulfate-reducing bacteria in the precipitation of copper, (Engl. trans.), *Microbiology* **46:**92-95.

Khalid, R. A., R. P. Gambrell, and W. H. Patrick, Jr., 1976, Chemical transformations of cadmium and zinc in Mississippi River sediments as influenced by pH and redox potential, in *Environmental Chemistry and Cycling Processes,* Proceedings of 2nd Mineral Cycling Symposium, CONF-760429, D. C. Adriano and I. L. Brisbin, eds., U. S. Department of Commerce, Springfield, Va., pp. 417-433.

Khalid, R. A., W. H. Patrick, Jr., and R. P. Gambrell, 1978, Effect of dissolved oxygen on chemical transformation of heavy metals, phosphorus, and nitrogen in an estuarine sediment, *Estuar. Coastal Mar. Sci.* **6:**21-35.

Lovley, D. R., and M. J. Klug, 1983, Sulfate reducers can outcompete methanogens at freshwater sulfate concentrations, *Appl. Environ. Microbiol.* **45:**187-192.

Nriagu, J. O., and J. D. Hem, 1978, Chemistry of pollutant sulfur in natural waters, in *Sulfur in the Environment, Part II,* J. O. Nriagu, ed., John Wiley and Sons, New York, pp. 211-270.

Pickering, W. F., 1980, Zinc interaction with soil and sediment components, in *Zinc in the Environment, Part I,* J. O. Nriagu, ed., John Wiley and Sons, New York, pp. 71-112.

Ramamoorthy, S., and B. R. Rust, 1978, Heavy metal exchange processes in sediment-water systems, *Environ. Geol.* **2:**165-172.

Tuttle, J. H., P. R. Dugan, and C. I. Randles, 1969, Microbial sulfate reduction and its potential utility as an acid mine water pollution abatement procedure, *Appl. Microbiol.* **17:**297-302.

United States Environmental Protection Agency, 1979, Methods for chemical analysis of water and wastes, EPA-600/4-79-020, Environmental Monitoring and Support Laboratory, Cincinnati, Ohio, pp. 410.1-410.2.

Zobell, C. E., 1946, Studies in redox potential of marine sediments, *Am. Assoc. Petrol. Geologists Bull.* **30:**477-513.

21

Mineralogical, Morphological, and Microbiological Characteristics of Tubercles in Cast Iron Water Mains as Related to Their Chemical Activity

Jerry M. Bigham and Olli H. Tuovinen

Ohio State University

INTRODUCTION

Drinking-water pipelines constructed of bitumen-coated cast iron are susceptible to interior corrosion, which results in the leaching of iron from the pipe walls and the development of graphitized areas that are too weak to withstand hydraulic stress. Coincident precipitation of solubilized iron and minor elements leads to the formation of tubercles on the pipe interior. These tubercles cause a reduction in hydraulic capacity, discolor the water supply, and may result in a loss of residual chlorine.

Pipeline tuberculation in potable-water distribution systems was recognized in this century. While the problem has largely been alleviated in new construction by the introduction of cement-lined, ductile iron pipe, it persists in old distribution sections. Some water utilities have resorted to the use of corrosion inhibitors (e.g., zinc phosphates) to decelerate corrosion, but no universal solution other than pipe replacement is available to end advanced corrosion and tuberculation. Practical guidelines for addressing water-quality problems in cast iron distribution systems have recently been published by the Water Research Centre in England (Ainsworth, 1981).

Despite the economic significance of pitting corrosion and pipeline tuberculation, relatively few studies have been directed at evaluating the agents and processes responsible for these phenomena. Likewise, pipeline tubeculation is known to result in both the chemical and microbiological degradation of water quality, but such events have rarely been characterized by actual analysis of tubercle material in failed distribution sections. The primary objective of this study was to examine the morphological, mineralogical, and microbiological characteristics of natural corrosion tubercles in order to better understand the geochemical and biotic interactions responsible for their formation. Evidence is also presented to indicate that tubercle materials show oxidant demand and support trihalomethane formation upon suspension in water.

MATERIALS AND METHODS

Sampling

Tuberculated sections of cast iron pipeline, constructed 40 to 50 years ago and exhibiting signs of extensive internal corrosion, were excavated and removed from the Columbus, Ohio, water distribution system. Estimated corrosion rates in these sections averaged about 0.4 mils/yr. Tubercles were mechanically removed from the pipe interior surface and examined either within 6 hr after collection or after air drying for several months.

Micromorphology

Whole, air-dried tubercles were vacuum-impregnated with Scotchcast #3 epoxy resin using a method similar to that described by Innes and Pluth (1970). In the modified procedure, tubercles held in 250-ml polyethylene cups were placed in a vacuum-impregnating unit and evacuated to 5 mm Hg. After 0.5 hr under vacuum, preheated (95°C), two-part epoxy resin was mixed and slowly added to the samples through a separatory funnel. A vacuum was maintained until all air in the tubercles had been displaced as determined by cessation of bubbling. The samples were then placed in an oven and cured for 12 hr at 95°C. After curing, the impregnated tubercles were sectioned, polished, and mounted on 2.5 x 7.5 cm glass slides. A Hillquist thin-section machine was used to trim and polish the sections to a final thickness of 50 μm. All samples were examined under reflected light using a binocular microscope.

Mineralogy

Air-dried tubercles were dissected with a micro-spatula and needle under a low-power binocular microscope. Various morphological features were isolated and ground using an agate mortar and pestle to pass a 45-μm sieve. Powdered sample material was then gently pressed into a standard cavity holder to yield a randomly oriented specimen for X-ray diffraction (XRD) analysis. All analyses were conducted using a Philips PW 1316/90 wide-range diffractometer employing Cu Kα X-radiation under ambient conditions. The diffraction assembly included a theta-compensating-beam divergence slit, a medium-resolution Soller slit, and a diffracted-beam monochromator. All diffractograms were recorded from 5° to 65°2θ.

Chlorine and Trihalomethane Determinations

Fresh tubercle material removed from the interior of pipe specimens was manually ground to a slurry using a mortar and pestle. The chlorine demand was determined at ambient temperature using approximately 0.5 g/l tubercle material with either 2 mM acetate (pH 4), 4 mM phosphate (pH 7), or 2 mM carbonate-bicarbonate (pH 10) buffer as a suspending medium. Both the free and total chlorine were determined by amperometric titration using phenylarsine oxide as a titrant (American Public Health Association, 1980). The free and total chlorine measurements were in close agreement ($< 5\%$), indicating that the tubercles contained negligible amounts of ammonia or amine-nitrogen. In parallel holding tests, trihalomethanes were analyzed in chlorinated samples in accordance with the U.S. EPA (1979) liquid-liquid extraction procedure followed by a gas chromatographic analysis of trihalomethanes. Standard mixtures of all four trihalomethanes—chloroform, bromodichloromethane, chlorodibromomethane, and bromoform—were analyzed for reference and instrument calibration.

Scanning Electron Microscopy

Subsamples were chipped from tubercle material and fixed in the dark with 1% glutaraldehyde followed by fixation with osmium vapors for 12 hr. Specimens were then either air dried or dehydrated with ethanol, critical point-dried with liquid carbon dioxide, and coated with gold before examination with a Model S-500 Hitachi scanning electron microscope operated at 20 kV.

Microbiological Examinations

Media and procedures for bacterial enumeration of tubercle samples have been previously described (Tuovinen et al., 1980; Tuovinen and Hsu, 1982).

RESULTS AND DISCUSSION

Morphology

Pitting corrosion was the predominant form of metal decay observed in the pipeline samples acquired for this study. Surface encrustations ranged from small, hollow blisters to convoluted tubercles with internal structure. Most of the tubercles examined in detail were 10-15 cm in circumference, stood 1-2 cm above the inner pipe surface, and overlaid graphitized wall cavities (Fig. 21-1). Fresh tubercles were brittle, reddish black in color, and contained black, fluid interiors (Eh = -400 to -600 mV). With oxidation and aging, these tubercles became ochreous and developed a hard, crusty texture.

Thin sections of dried, oxidized specimens (Fig. 21-1) usually revealed a complex internal structure. Almost invariably, a porous precipitate forms at the tubercle-water interface and overlies a thin (0.25-0.5-mm-thick) but continuous magnetic membrane that provides a protective shell for the corrosion tubercle. In fully developed tubercles, the black magnetic membrane often possesses a number of internal branches that create a chambered structure. However, the chamber walls differ from the external membrane in that they are reddish brown in color, porous in appearance, and weakly magnetic to nonmagnetic. These characteristics suggest that the chamber walls are former tubercle surfaces that have been developed and degraded with tubercle growth. They also suggest that tubercle formation proceeds at a nonuniform rate that may be influenced by periodic fluctuations in water quality or hydraulic conditions and, perhaps, by rupture of exterior membranes.

Mineralogy

Mineralogical analyses of the corrosion tubercles revealed an array of iron oxides and oxyhydroxides, most of which were associated with specific

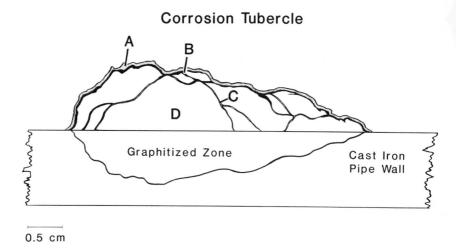

Figure 21-1. Schematic presentation of a tubercle cross section: A, surface crust; B, magnetic membrane; C, internal chamber wall; D, fluid interior.

morphological features. For example, the exterior magnetic membranes (Fig. 21-1, B) are composed almost entirely of the mineral magnetite. When completely isolated these membranes frequently yield XRD spectra (Fig. 21-2, B) characteristic of synthetic or museum-grade samples.

Above the membrane, a thin crust of Fe(III)-oxyhydroxides forms at the water-tubercle interface (Fig. 21-1, A). This crust is intimately associated with the underlying magnetic shell, and physical separation of the two is normally difficult to achieve. An XRD spectrum from a composite sample

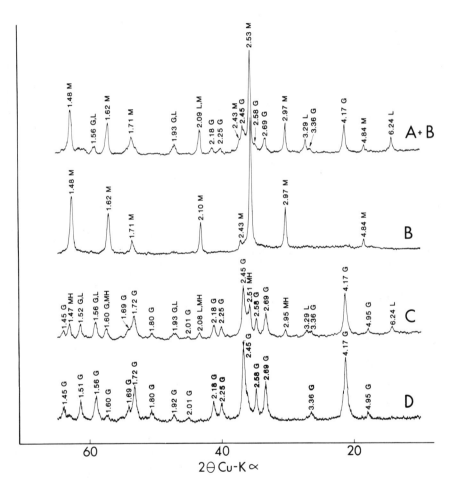

Figure 21-2. X-ray diffraction specra from air-dried tubercle materials. Letters A, B, C, and D refer to morphological components described in Figure 21–1. Letters G, L, M and MH refer to goethite, lepidocrocite, magnetite, and maghemite, respectively. Numerical values are in Angstrom units (Å).

(Fig. 21-2, A + B) shows characteristic diffraction lines for goethite (αFeOOH) and lepidocrocite (γFeOOH) in addition to magnetite. The surface crust is porous, and its constituent particles are granular in appearance (Fig. 21-3A). Both goethite and lepidocrocite frequently occur as fibrous crystals; however, their morphology and crystallinity are readily altered by precipitation from solutions containing foreign organic and inorganic compounds (Schwertmann and Taylor, 1977). Thus, the anhedral crystals shown in Figure 21-3A probably result from the sorption of metals or certain anions (e.g., silicate, sulfate, and phosphate) during formation of the iron oxide particles. Concentrations of such species at tubercle surfaces have been previously reported (Tuovinen et al., 1980).

Intense chemical gradients occur across the membranes of intact corrosion tubercles (Tuovinen et al., 1980), and mineral speciation reflects this rapid change in microenvironments. While goethite and lepidocrocite are stable in the oxidizing conditions associated with the tubercle-water interface, they would not be expected to occur in the reduced environment common to tubercle interiors. Previous work (Tuovinen et al., 1980) has demonstrated that the major crystalline iron form in the fluid interiors of fresh, cast iron corrosion tubercles is green rust, an Fe (II, III) hydroxy-

Figure 21-3. Scanning electron micrographs of A, surface crust containing granular particles of goethite and lepidocrocite, scale bar = 10 μm; B, hexagonal crystals of green rust from tubercle interior, scale bar = 5 μm.

compound with a pyroaurite structure (Brindley and Bish, 1976) and an Fe (II)/Fe (III) chemical ratio that varies between 0.8 and 4.0 (Bernal, Dasgupta, and MacKay, 1959). Green rust commonly crystallizes in the form of hexagonal plates or prisms that preferentially cleave perpendicular to the prism axis (Figs. 21-3*B*, 21-4*A*). The partial oxidation and hydrolysis of ferrous iron to form green rust may account for the slightly acidic character (pH 6.0-6.5) of the interstitial fluids as compared to the distribution waters possessing a pH of 7.5-8.5.

Green rust has been described as a metastable precursor to several other iron oxides and oxyhydroxides, including magnetite, lepidocrocite, and maghemite (γFe_2O_3) (Kassim, Baird, and Fryer, 1982). The final reaction product is determined by environmental conditions and kinetic parameters (e.g., oxidation rate). Direct transformation (with partial oxidation and dehydration) of green rust to magnetite is apparently favored by slow oxidation (Kassim, Baird, and Fryer, 1982). On the other hand, fast oxidation leads to complete Fe (III) formation before partial dehydration can occur, thereby enhancing the formation of lepidocrocite (Misawa, Hashimoto, and Shimodaira, 1973). If the dehydration of green rust precedes its oxidation, then maghemite is apparently favored over lepidocrocite (Schwertmann and Taylor, 1977).

 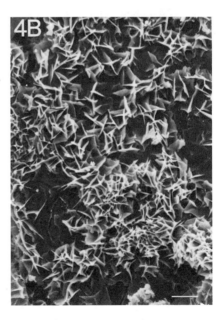

Figure 21-4. Scanning electron micrographs of *A*, lamellar crystals and hexagonal prisms of green rust from tubercle interior, scale bar = 5 μm; *B*, goethite crystals precipitated from air-dried internal slurry, scale bar = 10 μm.

Maghemite is ferromagnetic and possesses a spinel-like structure similar to that of magnetite. However, the cubic lattice parameter is smaller ($a = 0.834$ nm) so that XRD lines are displaced toward slightly higher angles than are obtained for magnetite with $a = 0.839$ nm. In this study, maghemite was commonly found in the degraded chamber walls of air-dried corrosion tubercles (Fig. 21-1, C) in association with both lepidocrocite and goethite (Fig. 21-2, C).

Goethite has not generally been recognized as a direct oxidation product of green rust; however, it is the dominant iron oxide formed when the internal slurry of a corrosion tubercle is exposed to air and allowed to oxidize (Fig. 21-2, D). Under these conditions, goethite is most likely produced by direct oxidation of Fe(II) followed by slow hydrolysis of soluble entities such as $[Fe(OH)_2]^+$, which feed the growing FeOOH particles (Knight and Sylva, 1974). Well-developed crystals of goethite precipitated on the internal surface of an air-dried section of magnetic membrane are shown in Figure 21-4B.

Previous investigators (e.g., Feigenbaum, Gal-Or, and Yahalom, 1978; Sontheimer, Kölle, and Snoeyink, 1981) have commonly reported the occurrence of iron sulfides, calcite, and siderite in scale deposits from cold-water distribution systems. Sontheimer, Kölle, and Snoeyink (1981) have placed special emphasis on the role of siderite in the formation of uniform scale deposits. The sulfur level in the tubercles examined in this study was approximately 3%, suggesting that iron sulfides were not major corrosion products when compared to the iron oxides. Likewise, neither calcite nor siderite were detected in the tubercle samples examined. It should be noted, however, that the distribution system was not extensively sampled, and uniform scale deposits adjacent to the pipeline walls were not characterized in the present work. In these areas, calcareous scale deposits were frequently observed.

Chlorine Demand and Trihalomethane Formation

The rates of chlorine demand were characterized by a relatively short but rapid first phase (100-1000 mg chlorine/l/hr) and a slow second phase (<0.20 mg/l/hr) that sometimes ensued for several days. The rates varied depending on the sample fraction examined. The major part of chlorine demand must be attributed to the presence of Fe(II) in the tubercle material, where it accounts for about 50% of the elemental composition. The chlorine demand is increased at pH 4 compared with that measured at pH 7 and pH 10 (Fig. 21-5). The inverse relationship between chlorine demand and pH is in agreement with the rate of chemical oxidation (by oxygen) of Fe(II); this oxidation rate decreases as the pH decreases, and thus the reaction of ferrous iron with chlorine is enhanced at acidic pH values. Conversely, an increase in the pH enhances the chemical oxidation of Fe(II) by oxygen and decreases the chlorine demand correspondingly. The inverse relationship between the chlorine demand and pH may also indicate that undissociated hypochlorous acid, which is the predominant form of chlorine at pH 4 (White, 1972), penetrates the particulate material better than does dissociated hypochlorite ion at pH 7 and pH 10.

Other compounds in tubercles responsible for the chlorine demand include sulfide, Mn(II), and organic carbon. For the sulfide-chlorine reaction, a low pH favors the complete oxidation of sulfide to sulfate contrasted with incomplete oxidation at high pH values. Chlorine-organic carbon reaction is promoted by high pH (U.S. Environmental Protection Agency, 1981) and results in the formation of chlorinated by-products such as trihalomethanes that were analyzed in the present work. Figure 21-5 shows that the formation of trihalomethanes was enhanced at pH 7 and pH 10, compared with that observed at pH 4. Chloroform was the exclusive trihalomethane compound

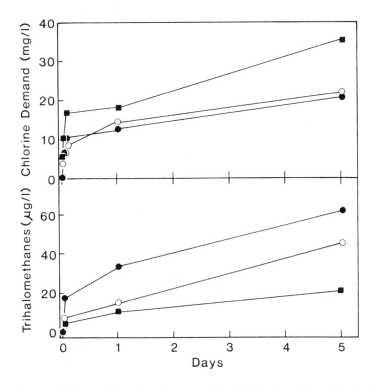

Figure 21-5. Chlorine demand and associated trihalomethane formation by fresh, ground tubercle material (approx. 0.5 g/l) at pH 4 (■), pH 7(○), and pH 10 (●).

identified in the chlorinated sample materials. As with chlorine demand, the rates of chloroform formation were biphasic.

Because chloroform was detected as a chlorination product, it can be concluded that tubercle material contains precursor compounds, such as humic residues, that are known to be responsible for the formation of trihalomethanes in chlorinated drinking-water distribution systems. The chlorine demand and the formation of chloroform also suggest that tubercle material will deteriorate water quality should it become dislodged and suspended in treated, disinfected drinking water. Normally, the tubercle's magnetic membrane is an effective diffusion barrier against residual chlorine and oxygen, but its rupture leads to exposure and mixing of fluids in the reduced tubercle interior with oxidants in the water.

Microbiological Investigations

Microorganisms were enumerated in tubercle samples using five different media (two plate-count media for aerobic heterotrophs, nitrate broth incubated aerobically and anaerobically, and a selective medium for sulfate-reducing bacteria). The enumeration results have been presented elsewhere (Tuovinen and Hsu, 1982). Correlation coefficients and the respective P values were calculated for pairwise comparison of the five media tested. Statistically significant associations were found between sulfate-reducing and heterotrophic denitrifying bacteria, suggesting that these organisms cohabit similar niches in tubercle material. Statistical analyses also indicated a moderate to strong association between denitrifying and aerobic heterotrophs. The available data indicate that the denitrifying population constituted about 3.5% of the heterotrophic microbial population that could be enumerated. The viable counts varied depending on the sample material, ranging from very low concentrations to about 7×10^5 heterotrophic bacteria per gram of tubercle material. In one extreme case, 3×10^8 bacteria per gram were detected. Because fresh tubercles could not be fractionated for microbial enumeration without some contamination from the bulk matrix, it was not possible to evaluate the possible enrichment of different physiological groups of bacteria in specific tubercle fractions with the data available.

The characterization of 43 heterotrophic isolates indicated the absence of opportunistic pathogens and bacteria of enteric origin. Previous work has also qualitatively demonstrated the presence of sulfur-oxidizing and nitrifying chemolithotrophic microbes in fresh tubercle material (Tuovinen et al., 1980). Filamentous iron bacteria, such as *Gallionella* sp., which are sometimes associated with distribution systems receiving ground water (e.g., Ridgway, Means, and Olson, 1981), were not detected in any of the samples examined.

The distribution system studied in the present work uses treated surface water and maintains a chlorine residual (0.1-1.0 mg/l) in the finished water.

The presence of bacteria, sometimes in high numbers, in tubercles collected from the distribution pipeline highlights the protective effect of tubercle mass against disinfection. The presence of heterotrophic microbes also indicates that organic carbon flux occurs in tubercle material to support the microbial population. The primary source of organic carbon is the distribution water because virgin cast iron contains carbon only as graphite and carbide. Microbiological degradation of bitumen pipe coatings may also occur, but, more important, the cycling of metabolites and decomposition of dead cells are presumed to contribute to the carbon flow in the tubercle microcosm. In addition, the cycling of at least sulfur and nitrogen involves biotic reactions in this system because sulfur-oxidizing, sulfate-reducing, nitrifying, and denitrifying bacteria were detected in tubercles. Whether bacteria such as the sulfate-reducers can accelerate the corrosion of cast iron in water pipelines is not known. Previous laboratory studies (Miller and Tiller, 1970; Iverson, 1972) have implicated microorganisms in pipeline tuberculation, but no direct evidence is available from field samples or from the examination of cases of pipe failure. Unlike laboratory experiments that have provided ample evidence for the role of microbes as catalysts of localized corrosion, the causative relationships in water distribution systems are difficult to determine because of the numerous factors, including chemical and electrochemical reactions, surface-associated phenomena, interfacial fluxes, and hydraulic conditions, that influence the corrosion rates of cast iron pipe.

ACKNOWLEDGMENTS

We thank Jarmila Banovic for technical assistance in trihalomethane measurements. We are also grateful to David M. Mair for providing sample materials for this investigation.

REFERENCES

Ainsworth, R. G., ed., 1981, *A Guide to Solving Water Quality Problems in Distribution Systems,* Technical Report 167, Water Research Centre, Medhenham-Stevenage-London.

American Public Health Association, 1980, *Standard Methods for the Examination of Waste and Wastewater,* 15th ed., APHA-AWWA-WPCF, Washington, D.C.

Bernal, J. L., D. R. Dasgupta, and A. L. Mackay, 1959, The oxides and hydroxides of iron and their structural interrelationships, *Clay Miner. Bull.* **4:**15-30.

Brindley, G. W., and D. L. Bish, 1976, Green rust: A pyroaurite type structure, *Nature* **263:**353.

Feigenbaum, C., L. Gal-Or, and J. Yahalom, 1978, Microstructure and chemical composition of natural scale layers, *Corrosion* **34:**65-70.

Innes, R. P., and D. J. Pluth, 1970, Thin section preparation using an epoxy impregnation for petrographic and electron microprobe analysis, *Soil Sci. Soc. Amer. Proc.* **34**:483-485.

Iverson, W. P., 1972, Biological corrosion, *Adv. Corros. Sci. Technol.* **2**:1-42.

Kassim, J., T. Baird, and J. R. Fryer, 1982, Electron microscope studies of iron corrosion products in water at room temperature, *Corros. Sci.* **22**:147-158.

Knight, R. J., and R. N. Sylva, 1974, Precipitation in hydrolysed iron (III) solutions, *J. Inorg. Nucl. Chem.* **36**:591-597.

Miller, J. D. A., and A. K. Tiller, 1970, Microbial corrosion of buried and immersed metals, in *Microbial Aspects of Metallurgy,* J. D. A. Miller, ed., American Elsevier, New York, pp. 61-105.

Misawa, T., K. Hashimoto, and S. Shimodaira, 1973, Formation of $Fe(II)_1$-$Fe(III)_1$ intermediate green complex on oxidation of ferrous iron in neutral and slightly alkaline sulphate solutions, *J. Inorg. Nucl. Chem.* **35**:4167-4174.

Ridgway, H. F., E. G. Means, and B. H. Olson, 1981, Iron bacteria in drinking-water distribution systems: Elemental analysis of *Gallionella* stalks using x-ray energy-dispersive microanalysis, *Appl. Environ. Microbiol.* **41**:288-297.

Schwertmann, U., and R. M. Taylor, 1977, Iron oxides, in *Minerals in Soil Environments,* J. B. Dixon and S. B. Weed, eds., Soil Science Society of America, Madison, Wisconsin, pp. 145-180.

Sontheimer, H., W. Kölle, and V. L. Snoeyink, 1981, The siderite model of the formation of corrosion-resistant scales, *Amer. Water Works Assoc. J.* **73**:572-579.

Tuovinen, O. H., and J. C. Hsu, 1982, Aerobic and anaerobic microorganisms in tubercles of the Columbus, Ohio, water distribution system, *Appl. Environ. Microbiol.* **44**:761-764.

Tuovinen, O. H., K. S. Button, A. Vuorinen, L. Carlson, D. M. Mair, and L. A. Yut, 1980, Bacterial, chemical, and mineralogical characteristics of tubercles in distribution pipelines, *Amer. Water Works Assoc. J.* **72**:626-635.

U.S. Environmental Protection Agency, 1979, *Federal Register* **44**:68683-68690.

U. S. Environmental Protection Agency, 1981, *Treatment Techniques for Controlling Trihalomethanes in Drinking Water,* EPA-600/2-81-156, Cincinnati, Ohio.

White, G. C., 1972, *Handbook of Chlorination,* Van Nostrand Reinhold Co., New York.

22

Autotrophic and Mixotrophic Growth and Metabolism of Some Moderately Thermoacidophilic Iron-Oxidizing Bacteria

Ann P. Wood and Don P. Kelly

University of Warwick

INTRODUCTION

In recent years a number of moderately thermophilic, acidophilic bacteria capable of iron oxidation have been isolated from diverse environments in which oxidizable metal sulfides such as pyrite were present (Brierley, 1978; Brierley and Lockwood, 1977; Brierley et al., 1978; Golovacheva and Karavaiko, 1978). These organisms oxidize metal sulfides, iron, or elementary sulfur. The taxonomic status of most of them is still unclear. An early isolate (Le Roux, Wakerley, and Hunt, 1977; Brierley et al., 1978) was described as a *Thiobacillus,* and a Russian isolate was named *Sulfobacillus* (Golovacheva and Karavaiko, 1978), even though the ability of the various isolates to grow autotrophically on soluble sulfur compounds has not been well established and in some cases may not be possible. The best development of most isolates described so far seems to occur when the organisms are provided with organic nutrients as well as pyrite, iron, or other sulfide minerals. The TH1 and related organisms (Brierley and Lockwood, 1977; Brierley et al., 1978; Norris, Brierley, and Kelly, 1980) exhibited a need for both organic carbon and a source of reduced sulfur. Autotrophic growth is, however, possible if atmospheres enhanced in CO_2 are provided together with an inorganic sulfur source such as thiosulfate or tetrathionate (Marsh and Norris, 1983; Wood and Kelly, 1983).

Virtually nothing has been reported concerning the relationship of iron oxidation and the assimilation of organic substrates, the nature of the carbon dioxide-fixing mechanism during autotrophic growth, or of the intermediary metabolic processes by which organic compounds may be dissimilated. This paper provides a summary report of experiments designed to clarify these metabolic problems using three recently isolated thermophiles (Marsh and Norris, 1983).

MATERIALS AND METHODS

Organisms, Culture Media, and Measurement of Growth

Three moderate thermophiles isolated by Marsh and Norris (1983) were used. They are referred to as strain ALV, strain BC, and strain K (Wood and Kelly, 1983). They were cultured at 50°C in liquid medium containing (g/l) K_2HPO_4, 0.1; $(NH_4)_2SO_4$, 0.2; $MgSO_4 \cdot 7H_2O$, 0.4; KCl, 0.1 supplemented with 50 mM $FeSO_4$ and adjusted to pH 1.7. For mixotrophic growth, organic additions were made as described in the Results. When required, reduced sulfur was provided as $Na_2S_2O_3$ or $K_2S_4O_6$ at 1.0 or 0.5 mM. Autotrophic and mixotrophic cultures were provided with additional CO_2 normally by injecting 1 ml 0.5 M $NaHCO_3$ into sealed flasks (total volume 300 ml containing 100 ml medium). Incorporation of ^{14}C-labeled CO_2 or organic substrates was measured by the membrane-filter technique (Wood and Kelly, 1983). Iron oxidation during growth was determined by ceric sulfate titration.

Assay of ^{14}C-Substrate Uptake, ^{14}C-Substrate Radiorespirometry, and Enzymes of Carbohydrate Degradation

Strain ALV, adapted to grow without reduced sulfur, was grown on 50 mM $FeSO_4$ with 1 mM glucose, glutamate, or citrate. Four to six liters were harvested by centrifugation, washed with dilute H_2SO_4, pH 2.0, and resuspended in K_2HPO_4 at pH 2.0 for transport (3-5.5 mg protein/ml) and radiorespirometry (10-13 mg protein/ml) experiments or in 0.1 M PIPES buffer, pH 7.0, for cell-free extract preparation. Sugar and organic acid uptake were measured at 47°C by the membrane-filter method (Wood and Kelly, 1982); radiorespirometry at 47°C used the techniques described previously (Wood, Kelly, and Thurston, 1977) except that carbonic anhydrase was unnecessary. Cell-free extracts were prepared, and enzymes assayed at 50°C by the methods described by Wood, Kelly, and Thurston (1977), and Wood and Kelly (1980). Ribulose bisphosphate (RuBP) carboxylase was assayed essentially as described by Smith, Kelly, and Wood (1980): Organisms were filtered onto membrane filters (25 mm, 0.45-μm pore size) to give 0.25-0.55 mg dry wt/membrane, and the membranes were immersed in 0.4 ml 5% (v/v) Triton X-100 in scintillation vials at 20°C for 20 min. The standard reaction mixture (0.9 ml) containing ^{14}C-bicarbonate was added, and the mixture was incubated at 47°C for 10 min to activate the enzyme before initiating reaction with RuBP (0.3 ml, 10 mM). Controls without RuBP and zero-time blanks were also set up. Reaction was terminated after 15 min with 3 ml 5% (v/v) acetic acid in methanol, evaporated to dryness at 60°C, residues taken up in 1 ml water, and fixed ^{14}C counted using 10 ml Aqualuma Plus scintillant (LKB). Specific activity of $^{14}CO_2$ was measured by counting aliquots of standard reaction mixture in 2 ml ethanolamine + methoxyethanol (3 + 7 v/v) and 10 ml scintillant.

RESULTS

Autotrophic Growth on Ferrous Sulfate

All three strains required small amounts (0.05 - 1.0 mM) of thiosulfate or tetrathionate in order to grow as chemolithotrophic autotrophs using the oxidation of 50 mM $FeSO_4$ as the sole source of energy. Under these conditions, the growth rates, expressed as specific rate of iron oxidation, were 0.075 ± 0.015/hr, 0.072 ± 0.017/hr, and 0.108 ± 0.041/hr for strains ALV, BC, and K, respectively, equivalent to doubling times of 9.5, 9.6, and 6.4 hr.

Growth yields were estimated from the relationship between iron oxidation and fixation of ^{14}C-labeled CO_2 in growing cultures (Fig. 22-1). Differential incorporation rates (mg CO_2-carbon fixed per mole $FeSO_4$ oxidized) were 137 ± 27, 74 ± 11, and 173 ± 26 for ALV, BC, and K, respectively. Assuming a carbon content of 47% of the dry weight (Kelly, 1982), these are equivalent to growth yields (mg dry wt/mole $FeSO_4$) of 291, 157, and 368. The values for ALV and K are very similar to those obtained in the same media (but at 30°C) using *Thiobacillus ferrooxidans* (Fig. 22-1) and *Leptospirillum ferrooxidans*—namely 240 and 276. Earlier work demonstrated the autotrophic yield for *T. ferrooxidans* to be typically around 350 (Kelly and Jones, 1978). The yield obtained with strain BC was consistently and repeatedly only about half of that of the other iron bacteria.

Carbon dioxide fixation in all three strains was indicated to be by means of the Benson Calvin cycle, as ribulose bisphosphate carboxylase activity was present at similar specific activity (Table 22-1). Mean values from a number of estimations (nmoles CO_2 fixed/mg dry wt organism) were 13.0 ± 7.7, 29.9 ± 1.1, and 15.3 ± 5.5 for strains ALV, BC, and K, respectively, indicating that the poor yield of strain BC was probably not due to deficiency of this enzyme. The activities observed were of the right order to support CO_2-dependent growth at the rates observed.

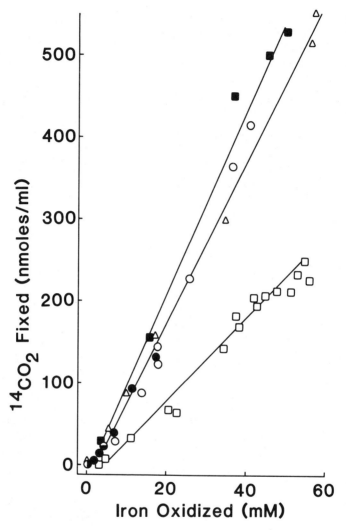

Figure 22-1. Correlation of iron oxidation and fixation of $^{14}CO_2$ during growth under autotrophic conditions by organisms ALV (O, $S_2O_3^{2-}$; ●, $S_4O_6^{2-}$), BC (□, $S_2O_3^{2-}$), and K (■, $S_4O_6^{2-}$) grown at 50 C and *T. ferrooxidans* (△) grown at 30 C, initial pH 1.7 on 50 mM $FeSO_4$.

Mixotrophic Growth on Iron and Glucose

All three strains were previously shown to grow rapidly in $FeSO_4$ medium with yeast extract (Marsh and Norris, 1983; Wood and Kelly, unpublished data). Glucose produced much the same effect, giving doubling times of 5.5 ± 0.8, 7.9 ± 2.3, and 7.0 ± 1.7 hr for strains ALV, BC, and K, respectively.

Table 22-1. Ribulose Bisphosphate Carboxylase Activity in Triton X-100 Permeabilized Cells of Strains ALV, BC, and K

Organism	*Culture Conditions[a]*	*RuBPC Activity[b]* (nmol CO_2 fixed/min/mg dry wt)
ALV	Autotrophic $(+CO_2)$	13.0 ± 7.7
BC	Autotrophic $(+CO_2)$	29.9 ± 1.1
K	Autotrophic $(+CO_2)$	15.3 ± 5.5
T. ferrooxidans[c]	Autotrophic (air)	39.5 ± 18.5
ALV	1 mM citrate $(+CO_2)$	9.8 ± 0.4
ALV	1 mM citrate (air)	14.6 ± 8.8
ALV	1 mM glucose (air)	3.3 ± 1.9
ALV	4 mM glucose (air)	13.3 ± 10.5

[a] All cultures were grown with 50 mM $FeSO_4$.
[b] Mean ($\pm$S.D.) values from 2-9 determinations.
[c] Grown in the same medium but at 30°C in air.

Growth was generally somewhat faster in cultures supplemented with CO_2 rather than incubated only in air. This was particularly true of strain K, which exhibited a doubling time of 17 hr in air and 5.5 hr in 3.5% CO_2 in one experiment. Relative incorporation of CO_2 and glucose was determined by providing parallel cultures with both substrates, with one or the other labeled with ^{14}C. This showed that between 19% and 36% of the total carbon assimilated arose from CO_2 fixation. Yields (mg carbon from CO_2 + glucose/mole $FeSO_4$ oxidized) were maximally 265, 240, and 154 for strains ALV, BC, and K, indicating yields (mg dry wt/mole $FeSO_4$) of 564, 511, and 328. The presence of glucose did not stimulate yield greatly over that seen autotrophically, with strain K showing least effect. The high CO_2 incorporation suggests that autotrophic carbon assimilation was proceeding even in the presence of glucose.

Mixotrophic Growth in the Absence of Reduced Sulfur by Strain ALV

Transfer of strain ALV from $FeSO_4$-yeast extract culture into a medium containing 50 mM $FeSO_4$ + 1 mM citrate allowed growth without added thiosulfate or tetrathionate. The culture could then be subcultured indefinitely without reduced sulfur. The strain was also able to grow in $FeSO_4$ media supplemented with a number of other organic acids and sugars (Table 22-2). It is apparent that the growth rate was determined by the organic supplement, with fructose, sucrose, and acetamide giving the fastest growth. Differential incorporation rates from ^{14}C-labeled glucose and CO_2 gave the same result as seen previously, with a total incorporation of 264 mg carbon/mole $FeSO_4$, of which 18.5% came from CO_2. Similarly, up to 200 mg carbon/mole $FeSO_4$

Table 22-2. Growth Rates of Strain ALV on 50 mM FeSO$_4$ Media Supplemented with Various Organic Substrates

Substrate (mM)	Doubling Time (h)	Specific Iron Oxidation Rate (μ_{Fe}, h^{-1})
Glucose (1)	4.8 ± 0.4	0.144
Fructose (1)	3.6 ± 0.9	0.193
Sucrose (0.5)	4.6	0.151
Sucrose (1)	3.6 ± 0.2	0.193
Ribose (1)	11.5 ± 0.6	0.060
Acetate (1)	10.6 ± 0.3	0.065
Acetamide (2)	4.5 ± 0.1	0.154
Succinamide (1)	4.9 ± 1.6, 15.1 ± 1.1 (biphasic)	0.141, 0.046

were assimilated during growth on glutamate (1 mM) + CO$_2$, of which about 70 mg came from CO$_2$ fixation; similar yields and CO$_2$ fixation occurred with citrate. CO$_2$ fixation in cultures growing mixotrophically on glucose or citrate with or without added CO$_2$ was due to the Calvin cycle, as they all exhibited ribulose bisphosphate carboxylase activities comparable to those found in autotrophic cultures (Table 22-1).

Mixotrophic growth in air appeared to be obligatory since growth on glucose in the absence of FeSO$_4$ (with or without added thiosulfate) was extremely slow. This indicated a doubling time of at least 20 hr with the production of about 40 μg dry wt/ml after 100 hr growth with 1 mM glucose, representing the assimilation of about 55% of the added glucose-C.

Comparison of the incorporation of ^{14}C from glucose labeled in either the first (C-1) or sixth (C-6) carbon atom showed that C-6 was preferentially incorporated at least during the early part of growth, indicating assimilation of breakdown products from glucose rather than solely of CO$_2$ released by its oxidation.

Uptake of Sugars and Organic Acids by Suspensions of Strain ALV

Glucose Transport and the Effect of Metabolic Inhibitors.   Using U-^{14}C-glucose concentrations of 3-100 μM, amounts taken up by glucose-grown cells after 15 sec were dependent on concentration, and Lineweaver-Burk plots indicated a high affinity, low V_{max} transport system (K_s, 5.4 μM and V_{max} 13 nmoles/min/mg dry wt). At concentrations of 150-264 μM, glucose-grown ALV appeared to take up glucose by passive diffusion, as the data plotted according to the Lineweaver-Burk reciprocal relationship extrapolated to zero.

Uptake of U-^{14}C-glucose (132 μM) was inhibited by fluoride (55% and 94% inhibition at 30 and 50 mM) and by uncoupling agents [25% inhibition by 0.1 mM pentachlorophenol or 1 mM 2,4-dinitrophenol (DNP) and about 50% by 4 μM carboxyl-cyanide p-trifluoromethoxyphenylhydrazone (FCCP) and 20 μM 4,5,6,7-tetrachloro-2'-trifluoromethylbenzimidazole (TTFB)]. Cyanide and the analog 0-methylglucose had little effect.

Uptake of ^{14}C-Labeled Organic Acids. Strain ALV grown on glutamate could transport acetate, succinate, and citrate at greater rates than glutamate. Lineweaver-Burk or Cornish-Bowden direct linear plots demonstrated transport of glutamate and citrate by glutamate-grown organisms to be effected by active transport systems with K_s values around 5 and 20 μM and V_{max} values around 4 and 56 nmoles/min/mg dry wt, respectively. Uptake of glutamate and citrate by citrate-grown organisms indicated K_s values of 115 and 39 μM and V_{max} values of 40 and 34 nmoles/min/mg dry wt. Uptake of glutamate and citrate was inhibited (>50%) by the uncouplers FCCP, TTFB, and DNP, indicating an energy-dependent active transport mechanism.

^{14}C-labeled succinate was accumulated by strain ALV previously grown on glutamate or citrate, but Lineweaver-Burk plots of succinate (66-333 μM) indicated uptake to be largely by passive diffusion, as the apparent K_s was very high (sometimes extrapolating to zero on the reciprocal plot), with the apparent V_{max} also being at least 10 times higher than that for citrate.

Radiorespirometric Evaluation of Glucose Oxidation by Strain ALV Grown on Glucose

Release of ^{14}CO$_2$ from glucose variously labeled in its constituent carbon atoms indicated that C-1 of glucose was released as CO$_2$ most rapidly, followed by C-2, then C-3. Plotting interval ^{14}CO$_2$ release (Fig. 22-2a) showed C-1, C-2, and C-3 output as CO$_2$ to peak sequentially within the first 3.5 min of oxidation. Release of C-4 was somewhat delayed, only commencing at about 4 min and peaking in rate at 5 min. The patterns of cumulative ^{14}CO$_2$ release confirmed these trends (Fig. 22-2b), with the total amount of ^{14}CO$_2$ released (from an initial 66.6 nmoles) in 30 min to be 49, 37, 34, and 40 for C-1, C-2, C-3, and C-4, respectively. Release of C-6 was relatively very slow, with only 16 nmoles CO$_2$ produced in 30 min. By contrast, about 23 nmoles of C-5 were released. Glucose oxidation, measured as C-1 release, was completely inhibited by 1 mM fluoride. Ribose (1 mM) had little effect on C-1 release but stimulated cumulative C-2 release by about one third.

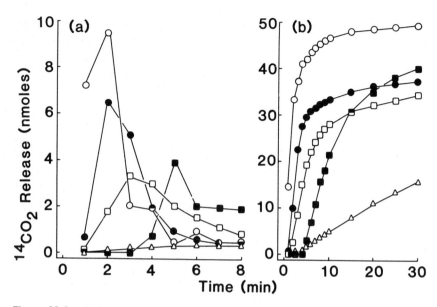

Figure 22-2. (a) Interval and (b) cumulative release of C-1 (○), C-2 (●), C-3 (□), C-4 (■) and C-6 (△) from specifically labeled glucose by glucose-grown ALV.

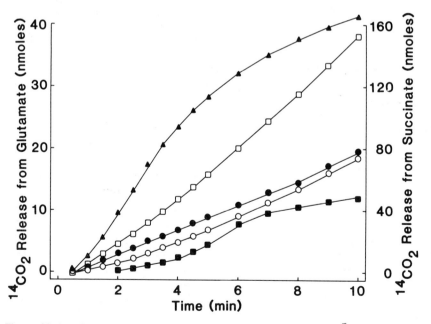

Figure 22-3. Cumulative release of C-1 (○), C-5 (●), C-1,5 (□), and C̄-2,3,4 (■) from specifically labeled glutamate, and C-1,4 (▲) from ^{14}C-1,4-succinate, by glutamate-grown ALV.

Enzymes of Carbohydrate Metabolism in Strain ALV Grown on 50 mM FeSO₄ and 1 mM Glucose

Crude cell-free extracts contained high levels of activity of glucose 6-phosphate dehydrogenase and 6-phosphogluconate (6-PG) dehydrogenase. These enzymes appeared to function equally well with NAD or NADP and gave specific activites around 70 and 40 nmoles NAD(P) reduced/min/mg protein, respectively. No activity could be detected for KDPG aldolase or 6-phosphogluconate dehydratase using direct or coupled assays with 6-PG or KDPG as substrates.

Oxidation of Glutamate and Succinate by Strain ALV Grown on 50 mM FeSO₄ and 1 mM Glutamate

Using the standard radiorespirometric technique, cumulative release of $^{14}CO_2$ from the C-1 and C-5 of glutamic acid was equivalent (Fig. 22-3). Release of C-2,3,4 was also similar during the first 10 min of oxidation, but after 45 min CO_2 released was 51, 53, and 109 nmoles from C-1, C-5, and C-2,3,4, respectively. Release of ^{14}C-glutamate was completely inhibited by 1 mM succinate or 50 μM fluoracetate, and acetate (1 mM) was slightly (10%) inhibitory.

Glutamate-grown cells oxidized 1,4-^{14}C-succinate very rapidly. In one experiment, the relative rates of CO_2 release from succinate and glutamate were about 6 and 0.7 nmoles/min/mg protein, respectively.

DISCUSSION

These studies have demonstrated unequivocally that the iron-oxidizing thermoacidophiles isolated from three different environments in which pyrite was available as a substrate are all capable of purely autotrophic growth using iron oxidation as the sole source of chemolithotrophic energy. The Benson Calvin cycle appears to be the main mechanism for CO_2 fixation. Possibly the most striking feature of the intermediary metabolism of these organisms is that CO_2 is also fixed in significant quantities by ribulose bisphosphate carboxylation in organisms growing mixotrophically with FeSO₄ and organic substrates. This apparently simultaneous functioning of autotrophic, chemolithotrophic, and heterotrophic processes for energy generation and carbon assimilation is unusual in autotrophs growing in nutrient-sufficient batch culture and has previously been demonstrated principally for thiobacilli growing in nutrient-limited chemostats (Matin, 1978; Gottschal and Kuenen, 1980; Smith, Kelly, and Wood, 1980; Wood and Kelly, 1981).

Mixotrophic growth yields on iron with glucose or organic acids were greater than those seen autotrophically, but rapid growth and carbon assimilation were linked to the concurrent oxidation of $FeSO_4$. Incorporation of ^{14}C from organic substrates virtually ceased when added iron was all oxidized even though excess substrate was present in solution. Uptake and oxidation of glucose, glutamate, and citrate by washed cell suspensions of strain ALV were, however, not dependent on the presence of $FeSO_4$. This, together with the significant role of CO_2 as a metabolite and the relatively slight stimulation of yield by organic compounds, seem to indicate that organic-compound oxidation may yield relatively less energy than would be expected in a typical heterotroph. It is proposed that these organisms are representatives of a physiological type whose metabolism is best described as preferentially mixotrophic, in that growth is less rapid in wholly autotrophic or heterotrophic conditions. Optimal autotrophic growth occurred only when the organisms were provided with additional CO_2. Mixotrophic use of organic carbon thus enabled growth in air by alleviating the deficiency in ability to fix CO_2 from atmospheric concentrations. The rapid and extensive oxidation of substrates like glucose and glutamate by strain ALV could be interpreted to indicate that organic substrates were oxidized primarily to generate CO_2 for refixation. While this was in part true, the greater assimilation of C-6 than of C-1 from glucose indicated that multi-carbon fragments from glucose oxidation (e.g., triose sugars or acetate) were clearly used for biosynthesis.

Glucose uptake by strain ALV was shown to be catalyzed by a high-affinity system whose inhibition by fluoride and uncouplers could indicate requirements for both an energy-dependent proton–motive-force transport process and a phosphoenolpyruvate-requiring reaction, such as the PEP-phosphotransferase system known in other bacteria. Fluoride principally inhibits the latter by blocking PEP production from glycolysis. The mechanism of glucose oxidation was therefore evaluated radiorespirometrically to assess the importance of glycolysis.

The kinetics of $^{14}CO_2$ release from specifically labeled glucoses were not consistent with a significant role for the Embden-Meyerhog (glycolytic) pathway because there was no correlation in rates of release of the C-3 and C-4 atoms. Similarly, lack of correlation of release of C-1 and C-4 suggested little contribution from the Entner-Doudoroff pathway. The data (Fig. 22-2) and the presence of the two essential dehydrogenase enzymes are wholly consistent with the virtually exclusive oxidation of glucose by the oxidative pentose phosphate cycle (Krebs and Kornberg, 1957). By the operation of this cycle, CO_2 is released exclusively from the C-1, C-2, and C-3 of glucose during the first, second, and third turns of the cycle, respectively. The fourth turn of the cycle releases C-2 and C-3 alone in the ratio 1:3. Four turns of the cycle reduce the initial concentration of any given amount of glucose by about 80%, the glucose entering the fifth cycle being the result of resynthesis. Commencing with 10 moles of glucose, therefore, four turns of the cycle will release about 24 moles CO_2, regenerate 2 moles of glucose, and produce about 8 moles of C_3-units (as glyceraldehyde 3-phosphate). One can calculate that the predicted release of C-1, C-2, and C-3 will have been in the ratio 100:81:59, compared to experimentally observed proportions of 100:79:69, indicating this pathway to have predominated. If the C_3-units also generated by the cycle were further oxidized by conversion (via enolase and PEP) to pyruvate, the labeling pattern within it would result in release of $^{14}CO_2$ (from the original C-4 of glucose) by pyruvate decarboxylation and the tricarboxylic acid cycle in an experimentally observable ratio of release of C-1:C-4 of 100:80. The actual ratio observed is 100:81, confirming this hypothesis. Since all the C_3-units generated (equivalent to 40% of the initial glucose provided) will thus pass through PEP during conversion to pyruvate, a significant role for PEP-phosphotransferase in glucose uptake and an explanation of the effect of fluoride is obvious.

At present, the mechanism of glutamate oxidation cannot be easily explained as there is substantial and equivalent release of both carboxyl groups as CO_2. Rapid oxidation of glutamate and succinate as well as ready growth on acetate, succinate, and citrate all indicate the operation of a tricarboxylic acid cycle in strain ALV.

ACKNOWLEDGEMENTS

We are indebted to Biogen S.A. for financial support for this work.

REFERENCES

Brierley, J. A., 1978, Thermophilic iron-oxidizing bacteria found in copper leaching dumps, *Appl. Env. Microbiol.* **36**:523-525.

Brierley, J. A., and S. J. Lockwood, 1977, The occurrence of thermophilic iron-oxidizing bacteria in a copper leaching system, *FEMS Microbiol. Lett.* **2**:163-165.

Brierley, J. A., P. R. Norris, D. P. Kelly, and N. W. Le Roux, 1978, Characteristics of a moderately thermophilic and acidophilic iron-oxidizing *Thiobacillus*, *Eur. J. Appl. Microbiol. Biotechnol.* **5**:291-299.

Golovacheva, R. S., and G. I. Karavaiko, 1978, *Sulfobacillus*, a new genus of thermophilic spore-forming bacteria *Mikrobiologiya* (Moscow) **47**:815-822.

Gottschal, J. C., and J. G. Kuenen, 1980, Mixotrophic growth of *Thiobacillus* A2 on acetate and thiosulfate as growth limiting substrates in the chemostat, *Arch. Microbiol.* **126**:33-42.

Kelly, D. P., 1982, Biochemistry of the chemolithotrophic oxidation of inorganic sulphur, in *Sulphur Bacteria,* J. R. Postgate and D. P. Kelly, eds., The Royal Society, London, pp. 69-98.

Kelly, D. P., and C. A. Jones, 1978, Factors affecting metabolism and ferrous iron oxidation in suspensions and batch cultures of *Thiobacillus ferrooxidans:* Relevance to ferric iron leach solution regeneration, in *Microbiological Applications of Bacterial Leaching and Related Microbiological Phenomena,* L. E. Murr, A. E. Torma, and J. A. Brierley, eds., Academic Press, New York, pp. 19-44.

Krebs, H. A., and H. L. Kornberg, 1957, *Energy Transformations in Living Matter,* Springer-Verlag OHG, Berlin.

Le Roux, N. W., D. S. Wakerley, and S. D. Hunt, 1977, Thermophilic thiobacillus-type bacteria from Icelandic thermal areas, *J. Gen. Microbiol.* **100:**197-201.

Marsh, R. M., and P. R. Norris, 1983, The isolation of some thermophilic, autotrophic iron- and sulphur-oxidizing bacteria, *FEMS Microbiol. Lett.* **17:**311-315.

Matin, A., 1978, Organic nutrition of chemolithotrophic bacteria, *Ann. Rev. Microbiol.* **32:**433-468.

Norris, P. R., J. A. Brierley, and D. P. Kelly, 1980, Physiological characteristics of two facultatively thermophilic mineral-oxidizing bacteria, *FEMS Microbiol. Lett.* **7:**119-122.

Smith, A. L., D. P. Kelly, and A. P. Wood, 1980, Metabolism of *Thiobacillus* A2 grown under autotrophic, mixotrophic and heterotrophic conditions in chemostat culture, *J. Gen. Microbiol.* **121:**127-138.

Wood, A. P., and D. P. Kelly, 1980, Carbohydrate degradation pathways in *Thiobacillus* A2 grown on various sugars, *J. Gen. Microbiol.* **120:** 333-345.

Wood, A. P., and D. P. Kelly, 1981, Mixotrophic growth of *Thiobacillus* A2 in chemostat culture on formate and glucose, *J. Gen. Microbiol.* **125:**55-62.

Wood, A. P., and D. P. Kelly, 1982, Kinetics of sugar transport by *Thiobacillus* A2, *Arch. Microbiol.* **131:**156-159.

Wood, A. P, and D. P. Kelly, 1983, Autotrophic and mixotrophic growth of three thermoacidophilic iron-oxidizing bacteria, *FEMS Microbiol. Lett.* **20:**107-112.

Wood, A. P., D. P. Kelly, and C. F. Thurston, 1977, Simultaneous operation of three catabolic pathways in the metabolism of glucose by *Thiobacillus* A2, *Arch. Microbiol.* **113:**265-274.

23

Autotrophic Growth and Iron Oxidation and Inhibition Kinetics of *Leptospirillum ferrooxidans*

Martin Eccleston, Don P. Kelly, and Ann P. Wood
University of Warwick

INTRODUCTION

Leptospirillum ferrooxidans was first isolated from a sulfide ore deposit in Armenia (Markosyan, 1972) and has been shown to be an organism significant in the leaching of pyrite as well as being capable of autotrophic growth dependent on iron oxidation (Balashova et al., 1974; Norris and Kelly, 1978, 1982; Norris, 1983). Little work has been reported on its autotrophic growth of ferrous iron, potential for heterotrophy or mixotrophy, or kinetics of iron oxidation. This paper demonstrates it to be an obligately chemolithotrophic autotroph, showing many physiological similarities to *Thiobacillus ferrooxidans* and similar to the latter in its iron-oxidation kinetics and sensitivity to inhibition by ferric iron.

MATERIALS AND METHODS

Organisms and Culture Conditions

Leptospirillum ferrooxidans (Markosyan, 1972; Balashova et al., 1974) was obtained from Dr. G. A. Zavarzin (Moscow). *Thiobacillus ferrooxidans* was the strain previously used (Tuovinen and Kelly, 1973). The moderately thermophilic, iron-oxidizing acidophile, strain ALV, is described elsewhere (Wood and Kelly, Paper 22). All three were grown in liquid medium containing (g/l): $(NH_4)_2SO_4$, 0.2; $MgSO_4.7H_2O$, 0.4; KCl, 0.1; K_2HPO_4, 0.1, adjusted to pH 1.7 using 2 NH_2SO_4. Ferrous sulfate was routinely supplied at 50 mM. Cultures (100 ml in 250-ml Erlenmeyer flasks) were incubated on orbital shakers at 150 rmp and 30°C for *T.* and *L. ferrooxidans* and 50°C for strain ALV. Inocula used were routinely 10% by volume. *L. ferrooxidans* was also cultured in an LH Engineering CC1500 fermenter in continuous-flow culture (2 liter), pH 1.8, on 100 mM ferrous sulfate, aerated with 3% (v/v) carbon dioxide in air at 500 ml/min at dilution rates of 0.021-0.095/hr. For growth on ^{14}C-labeled substrates, cultures were supplemented with $NaH^{14}CO_3$ (1 ml of 0.5 M per 100 ml culture at a specific activity of 20,000 cpm/μmol) in flasks sealed with Suba-seals, or with U-^{14}C-glucose at a 1 mM final concentration and specific activity of 90,000 cpm/μmol. When the interactive effects of glucose, yeast extract, and carbon dioxide were examined, cultures were grown in sealed flasks (see Results). Incorporation of ^{14}C was determined by removing samples from flasks with syringes, filtering duplicate 2-ml amounts through Sartorius or Whatman membrane filters (0.45-μm pore size), washing with dilute H_2SO_4, pH 1.7 (5 ml), and water (5 ml), before drying and estimating fixed ^{14}C by scintillation counting in 5 ml 0.5% (w/v) butyl PBD in toluene + methanol (3 + 1). Total ^{14}C in cultures was estimated by adding 0.1 ml aliquots to 10 ml scintillant. Iron oxidation during growth was monitored by titration of 1-ml samples in 2-ml 2 N H_2SO_4 with ceric sulfate to the "ferroin" end point, or by titration with $KMnO_4$.

Measurement of Iron Oxidation Using Suspensions of *Leptospirillum ferrooxidans*

Organism suspensions were prepared by centrifuging batch cultures (2.5 l) grown with 100 mM ferrous sulfate, pH 1.6 or pH 1.74 (initially), as described previously for *T. ferrooxidans* (Eccleston and Kelly, 1978). Oxidation of ferrous sulfate was measured using a Clark oxygen electrode cell. Experiments were begun immediately after harvesting and completed as soon as possible as iron-oxidizing ability was rapidly lost both at room temperature and at 4°C (see Results). Oxygen electrode experiments were as described previously (Eccleston and Kelly, 1978; Kelly and Jones, 1978). The cell (stirred at 30°C) contained 2.7 ml substrate with or without inhibitor at pH 1.6, and the reaction was started by injecting 0.3 ml *L. ferrooxidans* (1 mg dry wt) suspension, pH 1.6. Apparent K_m, V_{max}, and K_i values were also estimated from double-reciprocal plots of ferrous sulfate concentration against oxidation rates. K_i values were also estimated from plots of reciprocals of oxidation rate against inhibitor concentration.

Other Analyses.   Elemental analysis (C,H,N) of *L. ferrooxidans* was carried out using a Perkin-Elmer analyzer on organisms previously dried at 108°C.

The guanine + cytosine (% GC) content of the DNA from *L. ferrooxidans* was kindly determined using buoyant density by Dr. C. Dow (Warwick).

RESULTS

Autotrophic Growth of *Leptospirillum ferrooxidans* and Comparison with *Thiobacillus ferrooxidans* and Strain ALV

L. ferrooxidans grown autotrophically in chemostat culture on 100 mM ferrous sulfate (D, 0.05/hr) contained carbon, hydrogen, and nitrogen at 47.7%, 7.1%, and 10.3% of the dry weight, respectively. The guanine + cytosine content of its DNA was 54% GC.

When grown in batch culture on 50 mM ferrous sulfate with $^{14}CO_2$, *L. ferrooxidans* gave a doubling time (as ^{14}C-fixation and specific rate of iron oxidation) of 10.5 hr, compared with 5.5 hr for *T. ferrooxidans*. Plotting $^{14}CO_2$-incorporation against iron oxidized for both organisms (Figs. 23-1*a*, 23-1*b*) enabled calculation of yield for each organism as 270 and 236 mg dry wt/mole ferrous sulfate for *Leptospirillum* and *Thiobacillus ferrooxidans*, respectively, on the basis of a 47% carbon content for each. Mean yield of *L. ferrooxidans* from eight steady states in the chemostat (Table 23-1) was 407 ± 57 mg/mole ferrous sulfate, with the production of 2.27 (±0.32) × 10^{12} bacteria/mole ferrous sulfate. This compares with about 350 mg and 2.2 × 10^{12} organisms/mole for *T. ferrooxidans* (Tuovinen and Kelly, 1973; Kelly and Jones, 1978). Autotrophic growth of strain ALV was comparable (Fig. 23-1*b*).

Influence of Organic Substrates on *Leptospirillum* and *Thiobacillus ferrooxidans* and Strain ALV and of Ferrous Sulfate Concentration on *L. ferrooxidans*

The growth rate of neither *L. ferrooxidans* nor *T. ferrooxidans* on 50 or 100 mM ferrous sulfate was affected at all by yeast extract (0.005, 0.01, or 0.02% w/v) or glucose (1 mM); ^{14}C-glucose was not significantly assimilated (<70 μmol/mole ferrous sulfate oxidized, equivalent to no more than 0.2%-0.4% of the total cell carbon), and neither yeast extract nor glucose had any effect on $^{14}CO_2$ fixation by either organism (Fig. 23-1). In contrast, mixotrophic strain ALV showed considerable depression of $^{14}CO_2$-fixation by yeast extract, while glucose and carbon dioxide were assimilated simultaneously when both were available (Fig. 23-1*b*).

Table 23-1. Growth of Leptospirillum Ferrooxidans in Chemostat Culture

Dilution Rate (1/hr)	% Oxidation Rate of Input FeSO₄	Organism Concentrations		Yield	
		mg dry wt/l	Total numbers/ml	g dry wt/ mole FeSO₄	Organisms/ mole FeSO₄
0.021	100.0	36.3	—	333.3	—
0.037	99.5	53.7	—	518.8	—
0.043	99.9	43.0	2.8×10^8	419.5	2.73×10^{12}
0.046	99.3	39.3	2.6×10^8	386.1	2.55×10^{12}
0.049	99.6	41.1	1.9×10^8	395.2	1.83×10^{12}
0.052	99.9	44.7	2.2×10^8	433.6	2.13×10^{12}
0.068	99.2	36.9	2.3×10^8	354.1	2.21×10^{12}
0.095	99.0	44.7	2.2×10^8	433.6	2.13×10^{12}
0.124	washout of culture	—	—	—	—

Note: Cultures (2 l) in a fermenter stirred at 1000 rpm, aerated with 500 ml 3% (v/v) CO_2 in air/min, controlled to 30°C and pH 1.78 (±0.04) with 4N H_2SO_4, was fed with medium containing 104 (±3) mM FeSO₄ at the dilution rates shown. Values given are steady states established by passage of at least five culture volumes of medium. Biomass (dry wt/l) was determined as weight of centrifuged culture samples after drying at 108°C. D_{crit} was approximately 0.1/hr.

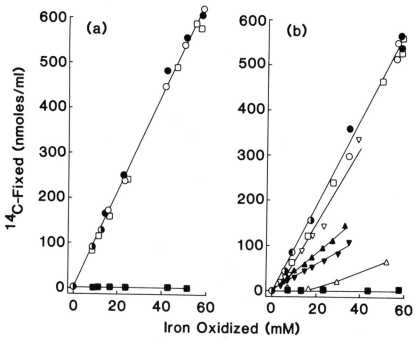

Figure 23-1. Correlation of iron oxidation and incorporation of ^{14}C during growth of (a) *Leptospirillum ferrooxidans* and (b) *Thiobacillus ferrooxidans* on 50 mM FeSO₄ with $NaH^{14}CO_3$ (○), $NaH^{14}CO_3$ + 1 mM glucose (●), $NaH^{14}CO_3$ + 0.02% yeast extract (□) and U-^{14}C-glucose (mM) + NaHCO₃ (■); and strain ALV on 50 mM FeSO₄ with $NaH^{14}CO_3$ (+ 1 mM Na₂S₂O₃) (▽), $NaH^{14}CO_3$ + 1 mM glucose (▲), $NaH^{14}CO_3$ + 0.02% yeast extract (△) and U-^{14}C-glucose (1 mM) + NaHCO₃ (▼).

Effect of Ferrous Iron Concentration on Growth of L. ferrooxidans. Ferrous sulfate was an inhibitory substrate to *Leptospirillum* growing in air; both length of lag and length of doubling time increased with increased concentration. With 25, 50, 75, and 100 mM ferrous sulfate (initial concentrations), respectively, lags following a 5% (v/v) inoculum were 22, 33, 51, and 60 hr and doubling times were 14.3, 15.7, (±0.5), 19.8, and 32 (±0.6) hr, respectively. Yeast extract (0.01% or 0.02%) had no effect on these rates, but the presence of additional (3% v/v) carbon dioxide allowed doubling times of 10.5 hr and 11.3 hr on 50 and 100 mM ferrous sulfate, respectively.

Kinetics of Oxidation of Ferrous Sulfate by Suspensions of *Leptospirillum ferrooxidans*

Suspensions of *L. ferrooxidans* oxidized ferrous sulfate at pH 1.6 or pH 1.74 over the range 0.1-600 mM (Fig. 23-2). Specific rate of oxidation increased linearly over the range 0.1-0.5 mM, but progressive inhibition was produced over the range 50-600 mM, with oxidation of 800 mM ferrous sulfate not being enhanced by the bacteria (Fig. 23-2). Ferrous sulfate was thus proved to be an autoinhibitory substrate, as found for *T. ferrooxidans* (Kelly and Jones, 1978). Oxidation rate (25 mM ferrous sulfate) was not significantly affected over the pH range 1.0-2.0.

Organisms rapidly lost the ability to oxidize ferrous sulfate after harvesting from culture. Loss of oxidative capacity was exponential over several days (Fig. 23-3a). Loss of activity was associated with cell clumping, lysis, and release of nucleic acid into solution. Consequently, control rates of ferrous sulfate oxidation were included at frequent intervals during a series of tests in order to correct for any decay in activity.

Double-reciprocal plots of specific oxidation rate against ferrous sulfate concentration were used to estimate K_s and V_{max} values for oxidation (Figs. 23-3b, 23-4). Loss of oxidative activity was found to decrease V_{max} only, with K_s being unaffected (Fig. 23-3b). In this experiment, apparent K_m for ferrous sulfate was found to be 0.2 mM. From six separate determinations, apparent K_s was found to be 0.5 ± 0.3 mM ferrous sulfate. The greatest V_{max} found was 125 nmol O_2/min/mg protein, but this was inevitably an underestimate.

Ferric iron was a competitive inhibitor of ferrous iron oxidation by *L. ferrooxidans* in that it altered K_s for ferrous sulfate without affecting V_{max} (Fig. 23-4). K_i for ferric iron was calculated from a number of experiments from K_p (inhibited K_s) values obtained from $1/v$ x $1/FeSO_4$ plots using different ferric iron concentrations from the expression $K_i = i/[(K_p/K_s)-1]$. It was also determined by the direct graphical procedure given in Figure 23-4b, where the negative crossover point of the data extrapolations indicates K_i. In the one experiment shown in Figure 23-4, calculation and direct plot methods gave values of 27 and 46 mM, respectively, as K_i for ferric iron. K_s and K_i were not significantly different at pH 1.6 and pH 1.74, and the mean value for K_i at these pH values from seven separate determinations was 33 ± 13 mM ferric iron.

Effect of Other Metal Ions on Iron Oxidation by *Leptospirillum ferrooxidans* and *Thiobacillus ferrooxidans*

Using the oxygen electrode method, inhibition of ferrous sulfate (25 mM) oxidation at pH 1.74 was compared with the two organisms (Table 23-2). *L. ferrooxidans* was generally more sensitive to the metals tested, with the possible exception of molybdate ion, where both were comparably inhibited.

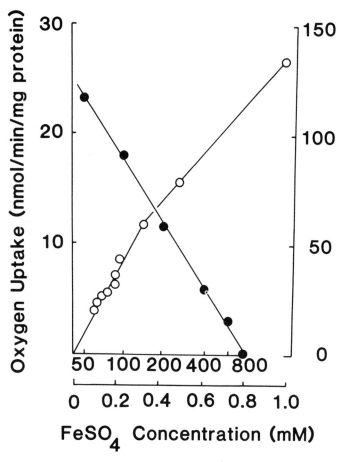

Figure 23-2.   Effect of FeSO₄ concentration on its oxidtion by *Leptospirillum ferrooxidans*, using the oxygen electrode technique. Oxygen uptake rate is given for a low concentration range (0–1 mM, ○; right-hand scale) and an inhibitory concentration range (50–800 mM, log scale, ●; left-hand rate scale).

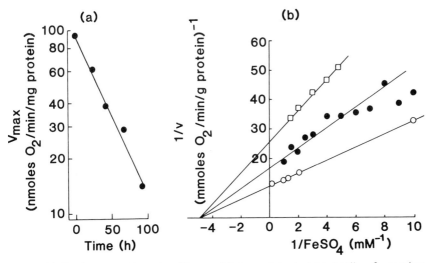

Figure 23-3. Decay and kinetics of iron-oxidizing activity in *Leptospirillum ferrooxidans* following harvesting from culture and storing at 20°C, pH 1.7. (a) Logarithmic decline in activity during storage expressed as decrease in V_{max}, calculated as in Figure 23-3b. (b) Double-reciprocal plots of iron oxidation rate against $FeSO_4$ concentration for a suspension assayed immediately after harvesting (○), after 24 hours (●), and after 48 hours (□).

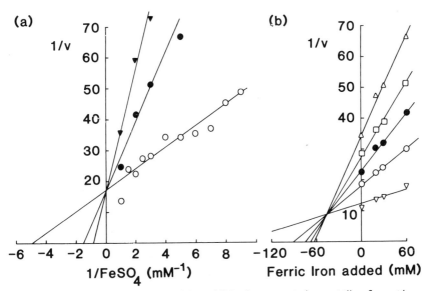

Figure 23-4. Estimation of K_m for $FeSO_4$ and K_i for ferric iron in *Leptospirillum ferrooxidans*. (a) Double reciprocal plot of iron-oxidation rates against $FeSO_4$ concentration in the absence (○) and presence of 60 mM Fe^{3+} (●) or 120 mM Fe^{3+} (▼). (b) Reciprocal of $FeSO_4$ oxidation rate plotted against concentration of added ferric iron for $FeSO_4$ concentrations (mM) of 0.2 (△), 0.33 (□), 0.5 (●), 1.0 (○), and 10.0 (▽).

Table 23-2. Effect of Metals and Some Anions on Iron Oxidation by Suspensions of Leptospirillum and Thiobacillus Ferrooxidans, Measured with the Oxygen Electrode Cell

Additions	Concentration (mM)	Inhibition of $FeSO_4$ L. ferrooxidans[a]	Oxidation Rate (%) T. ferrooxidans[b]
Cu^{2+}	65	30 ± 16	2
	326	85 ± 6	14
	652	96 ± 0	20
Ni^{2+}	100	36 ± 14	4
	500	92 ± 5	43[b]
	1000	99 ± 0	74[b]
Co^{2+}	100	40 ± 20	5
	500	90 ± 6	19
	1000	99 ± 0	69[b]
Zn^{2+}	100	32 ± 17	3
	500	87 ± 5	17
	1000	99 ± 0	61[b]
UO_2^{2+}	50	19 ± 14	—
	100	63 ± 32	—
Cd^{2+}	100	24 ± 10	—
	500	95 ± 0	14
Ag^+	0.1	30	8
	1.0	—	60
MoO_4^{2-}	10	4 ± 5	—
	100	66 ± 19	58
	500	—	94
NO_3^-	10	10 ± 8	15 ± 4[a]
	20	1	84 ± 10
	50	55 ± 14	100
	100	95 ± 2	100
	100	29 ± 19	35 ± 20[a]
Cl^-	1000	—	100

[a]Data are mean $\pm$ standard deviation (σ_{n-1}) of three separate experiments.
[b]Oxidation preceded by a lag.

DISCUSSION

These experiments serve to establish unequivocally that *Leptospirillum ferrooxidans* is an obligately autotrophic, obligately chemolithotrophic, iron-oxidizing bacterium producing growth yields comparable to those of other iron bacteria grown under autotrophic conditions (Kelly and Jones, 1978; Hanert, 1975; Wood and Kelly, 1983). They also establish that it shows iron-oxidation kinetics similar to those exhibited by *T. ferrooxidans* (Kelly and Jones, 1978) in that inhibition by excess substrate and by ferric iron occur. The apparent K_s for ferrous sulfate is very similar to that of *T. ferrooxidans* (0.5 ± 0.3 mM compared to 0.7 ± 0.14). In the chemostat it also exhibits growth rates (when supplied with ferrous sulfate at an input concentration of 100 mM) comparable to those obtainable with *T. ferrooxidans* under identical conditions. With S_R, 100 mM ferrous sulfate, both organisms indicate critical dilution rates around 0.1/hr, above which washout of the cultures occur. These facts all demonstrate that *L. ferrooxidans* is well adapted to survive in an acidic leaching environment, where *T. ferrooxidans* is also likely to occur. Competition between the two could be very close, and coexistence of the two organisms would be likely, given the heterogeneity of their mineral-leaching habitats. Survival and competition could be influenced by temperature, as *L. ferrooxidans* strains probably generally tolerate higher temperatures than do *T. ferrooxidans* (Norris, 1983). The effect of metal ions could be crucial to *in situ* competition, as *L. ferrooxidans* seems generally more sensitive to metal inhibition of iron oxidation, but resistance would presumably be selective in organisms exposed to particular metals in a natural situation. These results provide a physiological basis supporting the contention that one would expect an organism like *Leptospirillum ferrooxidans* to be widely distributed among leaching environments containing oxidizable iron.

ACKNOWLEDGEMENTS

This work was primarily supported by the Natural Environment Research Council under Research Grant GR3/2693. Support from Biogen S. A. (Geneva) is also gratefully acknowledged.

REFERENCES

Balashova, V. V., I. Ya. Verdinina, G. E. Markosyan, and G. A. Zavarzin, 1974, The autotrophic growth of *Leptospirillum ferrooxidans, Mikrobiologiya* **43:**381-385.

Eccleston, M., and D. P. Kelly, 1978, Oxidation kinetics and chemostat growth kinetics of *Thiobacillus ferrooxidans, J. Bacteriol.* **134:**718-727.

Hanert, H., 1975, *Entwicklung, Physiologie and Oekologie des Eisenbakteriums Gallionella ferruginea Ehrenberg,* Beitrage zu einer Monographie, Technical University of Braunschweig.

Kelly, D. P., and C. A. Jones, 1978, Factors affecting metabolism and ferrous iron oxidation in suspensions and batch cultures of *Thiobacillus ferrooxidans* — relevance to ferric iron leach solution regeneration, in *Metallurgical Applications of Bacterial Leaching and Related Microbiological Phenomena,* L. E. Murr, A. E. Torma, and J. A. Brierley, eds., Academic Press, New York, pp. 19-44.

Markosyan, G. E., 1972, A new acidophilic iron bacterium *Leptospirillum ferrooxidans, Biol. Zh. Armenii* **25:**26.

Norris, P. R., 1983, Iron and mineral oxidation with *Leptospirillum*-like bacteria, in *Recent Progress in Biohydrometallurgy,* G. Rossi and A. E. Torma, eds., Associazione Mineraria Sarda, Inglesias, Italy, pp. 83-96.

Norris, P. R., and D. P. Kelly, 1978, Dissolution of pyrite by pure and mixed cultures of some acidophilic bacteria, *FEMS Microbiol. Lett.* **4:**143-146.

Norris, P. R., and D. P. Kelly, 1982, The use of mixed microbial cultures in metal recovery, in *Microbial Interactions and Communities,* A. T. Bull and J. H. Slater, eds., Academic Press, London, pp. 443-474.

Tuovinen, O. H., and D. P. Kelly, 1973, Studies on the growth of *Thiobacillus ferrooxidans, Arch. Mikrobiol.* **88:**285-298.

Wood, A. P., and D. P. Kelly, 1983, Autotrophic and mixotrophic growth of three thermoacidophilic iron-oxidizing bacteria, *FEMS Microbiol. Lett.* **20:**107-112.

Wood, A. P., and D. P. Kelly, 1984, Autotrophic and mixotrophic growth and metabolism of some moderately thermoacidophilic iron-oxidizing bacteria, in *Planetary Ecology: Selected Papers from the Sixth International Symposium on Environmental Biogeochemistry,* D. E. Caldwell, J. A. Brierley, and C. L. Brierley, eds., Van Nostrand Reinhold Co., Inc., New York (Paper 22).

24

Influence of Microorganisms in the Weathering Stability of (Ba, Ra)SO₄ in Uranium Mining and Milling Wastes

Jorge Pérez and Arpad E. Torma

New Mexico Institute of Mining and Technology

ABSTRACT: It is well known that seasonal changes affect the solubility of barium-radium sulfate [(Ba, Ra)SO₄]. In early spring higher Ra-226 concentrations are in solution than by the end of the fall season. The present study has developed an intensified leach technique for the assessment of the influence of microorganisms on the stability of (Ba, Ra)SO₄ contained in synthetic and actual uranium mill-tailings. In this process, a period of 10.5 simulated weathering years is equivalent to three weeks of wet and two weeks of dry treatment of tailings. The results obtained indicate that in the presence of microorganisms, *Thiobacillus ferrooxidans,* the Ra-226 concentration was decreased as a function of time compared to that concentration obtained in the sterile experiments. The decrease in the Ra-226 concentration was found to be the consequence of sulfuric acid and ferric sulfate production from pyrite by bacterial action. The excess of sulfate in solution resulted in a decrease in the solubility of (Ba, Ra)SO₄. In the sterile experiments, the pyrite oxidation was considerably slower, as indicated by pH measurements, and fewer sulfates were produced. Therefore, the solubility of (Ba, Ra)SO₄ could proceed towards the dissociation resulting in higher Ra-226 concentrations in solutions.

INTRODUCTION

Uranium is usually extracted from ores by alkaline or acid leaching. During these operations approximately 95% of the other radionuclides, such as Th-230, Ra-226, and Pb-210, originally present in the ores, remain in the leach residue and are discarded in the tailings (Itzkovitch and Ritcey, 1979;

Raicevic, 1980; Torma et al., 1982). A typical uranium mill produces about 1800 metric tons of tailings per day and about 2×10^5 m^3 per year liquid effluent over a period of 23 years (Rahn and Mabes, 1978).

The presence of radionuclides in tailings and liquid effluents is creating serious problems for the uranium industry. Ra-226 has been identified as the major isotope of concern for environmental control of liquid discharges from uranium mills. The neutralization of tailings with lime or limestone (Merritt, 1971) has the short-term objective of reducing the concentrations of heavy metals, Th, and other radionuclides. The neutralized tailings are transported to a settling pond where the solids are deposited and the decant liquid is recycled in the process or treated with $BaCl_2$ for further Ra-226 removal in the form of $(Ba, Ra)SO_4$. Then the treated effluent can be discarded to the natural water courses (Moffett and Vivyurka, 1982), if necessary.

It has been questioned whether neutralization practice will be effective over the long term. Tailings ponds are influenced by seasonal weathering that will affect the chemical composition of the tailings over the long term. For instance, after a heavy rain the effluents may become acidic due to chemical and microbial oxidation of metal sulfides present in the tailings (Riding et al., 1979; Ritcey and Silver, 1981a, 1981b). There have been new approaches described in the literature (Averrill et al., 1980; Anonymous, 1980) for the treatment of radionuclides in the effluents of uranium mill operations. Some investigators have used microorganisms to remove pollutants from uranium mill waste water. Living organisms may absorb heavy metals and radionuclides from mill effluents (Brierley and Brierley, 1980; Brierley, Brierley, and Dreher, 1980; Tsezos and Keller, 1983; Tsezos and Volesky, 1981). Other workers reported that sulfate-reducing bacteria release radium from stabilized tailings (McCready, Bland, and Gonzalez, 1980). The present study investigates the chemical and biological leachability of radionuclides from uranium leach residues and mill tailings.

MATERIALS AND METHODS

Microorganisms

A pure strain of *Thiobacillus ferrooxidans* (Torma, 1971) used in this study was routinely maintained in the laboratories on an iron-free nutrient medium (Torma, 1977) in which a pyrite concentrate replaced ferrous sulfate as the source of energy. When bacterial growth reached the stationary phase, a portion of the leach suspension was transferred into fresh medium to maintain the stock culture or was used as an experimental inoculum.

Industrial Uranium Mill Tailings

Actual uranium mill tailings were provided by the United Nuclear Corporation from their Church Rock, New Mexico, operation in a slurry containing 70% weight/volume of solids. A sample of the tailings was filtered, washed with water, and dried by solar evaporation. The particle size was found to be less than 200 μm. The solid tailings were composed of sandstone and minor amounts of clay-type minerals and contained 105 pCi/g Ra-226.

Synthetic Tailings

The synthetic tailings contained hydrochloric acid-washed Ottawa sand with 15% weight/weight of pyrite.

Percolator Experiments

In this series of experiments, four identical air-lift percolators (40 x 7.5 cm) were used (Torma and Bosecker, 1982). The columns were charged with 600 g dry uranium mill tailings and 750 cm^3 liquid medium, as follows: column 1 — 750 cm^3 tap water; column 2 — 700 cm^3 tap water, 50 cm^3 active culture of *T. ferrooxidans,* and 7.5 mg BaCl$_2$; column 3 — 750 cm^3 original mill tailings effluent; and column 4 — 700^3 cm original mill tailings effluent, 50 cm^3 of active culture of *T. ferrooxidans,* and 7.5 mg BaCl$_2$. To simulate wet cycles, the liquid was recycled by application of pressurized air in the side arm of each column for the wet cycle; the liquid was removed from each column through the bottom tube to imitate dry cycles. The duration of the wet cycle was three weeks and that of the dry cycle, two weeks. Distilled water was periodically supplied to compensate the water loss due to evaporation. At the end of each wet cycle, liquid samples were removed from the columns and analyzed for dissolved Ra-226 content and pH.

Shake Flask Experiments

All experiments were carried out in 250-cm^3 Erlenmeyer flasks containing 21 g of wet uranium mill tailings or 13 g of synthetic tailings, 61 to 68 cm^3 of iron-free basal salts nutrient medium (Torma and Bosecker, 1982), 1 to 8 cm^3 of standard Ra-226 (nitrate solution containing 1416 pCi dm^{-3} Ra-226), one cm^3 of a 10 mg dm^{-3} BaCl$_2$ solution, and 5 cm^3 of pyrite-grown, active culture of *T. ferrooxidans.* The sterile controls contained 5 cm^3 of 2% thymol in methanol instead of the bacterial suspension. The pH of the solutions was adjusted to 2.3 with H$_2$SO$_4$, and the flasks were incubated on a gyratory incubator shaker apparatus (model G-26 of the New Brunswick Science Co.,

New Jersey) at room temperature and 250 rpm. Periodically the water lost due to evaporation was compensated with distilled water. The liquid samples were analyzed for the dissolved Ra-226 content after 1, 3, 6, and 9 weeks of treatment.

Determination of Ra-226 Concentration

The Ra-226 content of samples was determined on an alpha particle spectrometer (model 576 from EG&G Ortec) connected to a multichannel analyzer (model ND 66 from Nuclear Data, Inc.) (Smithson, Dalton, and Mason, 1978; Zimmerman and Armstrong, 1975).

RESULTS AND DISCUSSION

Typical results obtained in the shake-flask experiments are shown in Figure 24-1. The pH and the dissolved Ra-226 concentration were higher in the sterile controls than in the inoculated runs for both series of experiments, which were carried out with synthetic and actual tailings, respectively. This tendency, as can be seen from Figures 24-2 and 24-3, is characteristic for these experiments. The initial amounts of Ra-226 added to the experimental solutions were precipitated by $BaCl_2$:

$$BaCl_2 + H_2SO_4 \longrightarrow (BaSO_4) + 2\,HCl \qquad \text{(Eq. 24-1)}$$

The insoluble $BaSO_4$ coprecipitated radium in the form of a double sulfate, $(Ba, Ra)SO_4$. It can be seen from Figures 24-1 through 24-3 that the final dissolved Ra-226 concentration decreased more rapidly in the presence of *Thiobacillus ferrooxidans* than in their absence. These findings can be explained by the fact that the pyrite associated with the tailings was oxidized faster in the presence of bacteria than in their absence (Silverman, 1967). The pyrite oxidation can be given by the following equations:

$$FeS_2 + 7Fe_2(SO_4)_3 + 8H_2O \longrightarrow 15\,FeSO_4 + 8H_2SO_4 \qquad \text{(Eq. 24-2)}$$

$$2FeSO_4 + 1/2O_2 + H_2SO_4 \xrightarrow{\text{bacteria}} Fe_2(SO_4)_3 + H_2O \qquad \text{(Eq. 24-3)}$$

The formation of sulfuric acid has the effect of reducing the pH; furthermore, the excess of sulfate increased the stability of the $(Ba, Ra)SO_4$ precipitate by shifting its dissociation towards the left side of the equation:

$$(Ba, Ra)SO_4 \; \underset{\longrightarrow}{\longleftarrow} \; Ba^{2+} + Ra^{2+} + SO_4^{2-} \qquad \text{(Eq. 24-4)}$$

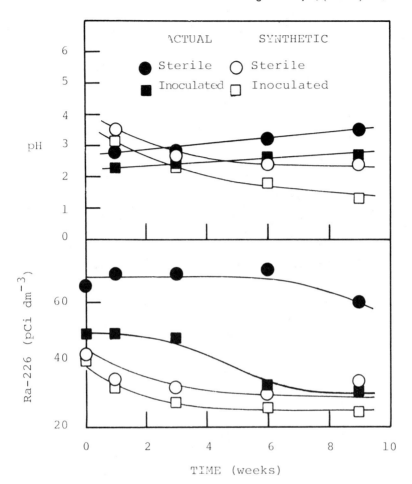

Figure 24-1.   Variation of pH and dissolved radium-226 concentration in presence and absence of *Thiobacillus ferrooxidans* during shake flask experiments with synthetic and actual uranium tailings (initial standard Ra-226 = 8 cm^3).

The chemical oxidation of pyrite was reported to be considerably slower than the biological one (Silver and Torma, 1974). Consequently, there was less sulfuric acid and ferric sulfate formed in the sterile controls than in the inoculated experiments. As a result, reaction 4 was shifted to the right side, and more radium was released into the leach media.

The data represented in Figures 24-4 and 24-5 are the results obtained in the percolator column experiments. The pH and Ra-226 concentrations in

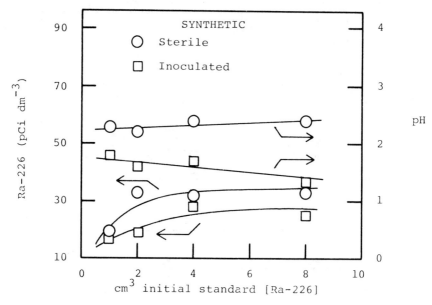

Figure 24-2. Final pH and Ra-226 concentrations in presence and absence of *Thiobacillus ferrooxidans* at different initial Ra-226 concentrations, using synthetic tailings.

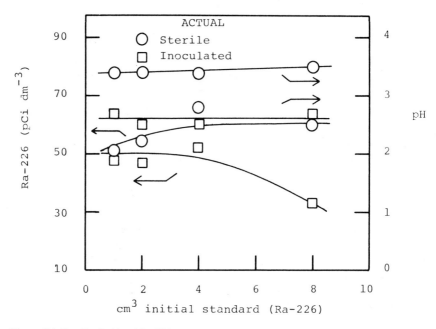

Figure 24-3. Final pH and Ra-226 concentrations in presence and absence of *Thiobacillus ferrooxidans* at different initial Ra-226 concentrations using actual tailings.

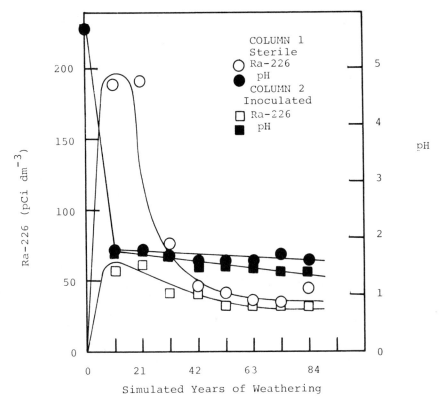

Figure 24-4. Percolator experiments simulating intensified seasonal changes for the weathering stability of (Ba,Ra)SO₄ contained in uranium tailings (leaching with water).

the inoculated experiments were relatively lower than they were in the sterile controls. The conditions used in columns 1 and 2 (Fig. 24-4) simulated a situation in which the tailings ponds dried during the summer season, and this event was followed by heavy rain. After the first wet period, the effluent became very acidic, decreasing considerably the pH of the solution. The concentrations of sulfate for columns 1 and 2 were 42 and 50 ppm at the end of the treatment. Columns 3 and 4 (Fig. 24-5) experiments simulated the situation in which tailings were contacted with actual acidic effluent containing high concentrations of heavy metals, 816 pCi dm^{-3} Ra-226 and 59,000 ppm of SO$_4^{2-}$. The sulfate concentrations in columns 1 and 2 were 141 and 185 ppm at the end of the treatment. During the first wet period, the solid tailings adsorbed about 80% and 97% of the total Ra-226 content of the effluent, corresponding to the inoculated and sterile experiments, respectively; this confirms the ion-exchange capability of the fine clays associated with the

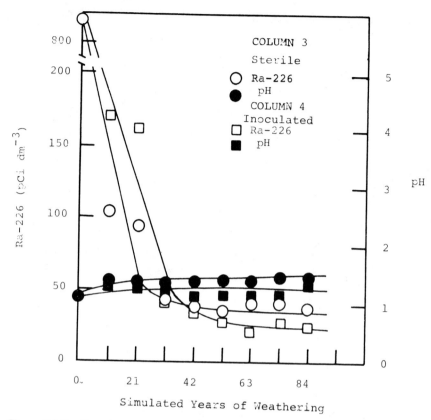

Figure 24–5. Percolator experiments simulating intensified seasonal changes for the weathering stability of (Ba,Ra)SO₄ contained in uranium tailings (leaching with actual effluents).

solid tailings. Data obtained from shake-flask experiments confirm the findings reported for percolator column studies.

This study demonstrated the impact that microorganisms have on the weathering process of uranium mill tailings. However, further studies are needed to better the understanding of the movement of chemical constituents in abandoned uranium mine tailings.

ACKNOWLEDGMENTS

This study was supported in part by U.S. Department of the Interior Grants G1115353 and G1125132, and by New Mexico Energy Research and Development Institute Grant No. EMD 2-69-1308. One of the authors, J Pérez, would like to thank CONACyT and I.P.N.-ESIQIE of Mexico for their financial support. Furthermore, the authors wish to express their appreciation to Mr. Thomas M. Hill, Director of Environmental Affairs, United Nuclear Corporation at Church Rock, New Mexico, for his interest and contributions.

REFERENCES

Anonymous, 1980, *The Current Status of Radium Removal Technology*, Bulletin 607, Trace Metal Data Institute, El Paso, Texas, pp. 1-71.

Averrill, D. W., G. W. Kassakhian, D. Moffet, and R. T. Weber, 1980, Development of radium-226 removal processes for uranium effluents, in *First International Conference on Uranium Mine Waste Disposal*, C. O. Brawner, ed., SME-AIME, New York, pp. 333-350.

Brierley, J. A., and C. L. Brierley, 1980, Biological methods to remove selected inorganic pollutants from uranium waste water, in *Biogeochemistry of Ancient and Modern Environments*, P. A. Trudinger, M. R. Walter, and B. J. Ralph, eds., Australian Academy of Science, Canberra, pp. 661-667.

Brierley, J. A., C. L. Brierley, and K. T. Dreher, 1980, Removal of selected inorganic pollutants from uranium mine waste water by biological methods, in *First International Conference on Uranium Mine Waste Disposal*, C. O. Brawner, ed., SME-AIME, New York, pp. 69-76.

Itzkovitch, I. J., and G. M. Ritcey, 1979, *Removal of Radionuclides from Process Streams—A Review*, CANMET Report 79-21, Energy, Mines and Resources Canada, Ottawa.

McCready, R. G. L., C. J. Bland, and D. E. Gonzalez, 1980, Preliminary studies on the chemical, physical and biological stability of (Ba, Ra)SO₄ precipitates, *Hydromet.* **5:**109-116.

Merritt, R. C., 1971, *The Extractive Metallurgy of Uranium*, Colorado School of Mines Research Institute, Golden.

Moffett, D., and A. J. Vivyurka, 1982, Control of radium-226 releases in liquid effluents, in *Fifth Annual Uranium Seminar*, Albuquerque, Sept. 20-23, 1981, AIME, New York, pp. 259-269.

Rahn, H. P., and D. L. Mabes, 1978, Seepage from uranium tailings ponds and its impact on ground water, in *Proceedings of NEA Seminar on Management, Stabilization and Environmental Impact of Uranium Mill Tailings*, Albuquerque, July 1978, Nuclear Energy Agency, Paris, pp. 127-140.

Raicevic, D., 1980, Removal of radionuclides from uranium ores and tailings to yield environmentally acceptable waste, in *First International Conference on Uranium Mine Waste Disposal*, C. O. Brawner, ed., SME-AIME, New York, pp. 351-360.

Riding, J. R., F. J. Rosswog, G. Buma, and D. R. Tweeton, 1979, Ground water restoration for *in situ* solution mining of uranium, in *In situ Uranium Mining and Ground Water Restoration*, W. J. Schlit and D. A. Dhock, eds., SME-AIME, New York, pp. 67-79.

Ritcey, G. M., and M. Silver, 1981*a*, Lysimeter investigations on uranium tailings at CANMET, Divisional Report MRP/MSL 81-36, Energy, Mines and Resources Canada, Ottawa.

Ritcey, G. M., and M. Silver, 1981*b*, Lysimeter investigations on uranium tailings at CANMET, *Can. Min. Met. Bull. C.I.M.* **75:**134-143.

Silverman, M. P., 1967, Mechanism of bacterial pyrite oxidation, *J. Bacteriol.* **94**:1046-1051.

Silver, M., and A. E. Torma, 1974, Oxidation of metal sulfides by *Thiobacillus ferrooxidans,* grown on different substrates, *Can. J. Microbiol.* **20**:141-147.

Smithson, G. L., J. L. Dalton, and G. L. Mason, 1978, *Radiochemical Procedures for Determination of Selected Members of the Uranium and Thorium Series,* CANMET Report 76-11, Energy, Mines and Resources Canada, Ottawa.

Torma, A. E., 1971, Microbiological oxidation of synthetic cobalt, nickel and zinc sulfides by *Thiobacillus ferrooxidans, Rev. Can. Biol.* **30**:209-216.

Torma, A. E., 1977, The role of *Thiobacillus ferrooxidans* in hydrometallurgical processes, *Adv. Biochem. Eng.* **6**:1-37.

Torma, A. E., and K. Bosecker, 1982, Bacterial leaching, *Progr. Ind. Microbiol.* **16**:77-118.

Torma, A. E., J. J. Santana, A. Singh, and W. M. Fleming, 1982, A novel approach to tailings disposal, in *Fifth Annual Uranium Seminar,* Albuquerque, Sept. 20-23, 1981, AIME, New York, pp. 59-67.

Tsezos, M., and D. M. Keller, 1983, Adsorption of radium-226 by biological origin absorbents, *Biotechnol. Bioeng.* **25**:201-215.

Tsezos, M., and B. Volesky, 1981, Biosorption of uranium and thorium, *Biotechnol. Bioeng.* **23**:583-604.

Zimmerman, J. B., and V. C. Armstrong, 1975, *The Determination of Radium-226 in Uranium Ores and Mill Products by Alpha Energy Spectrometry,* CANMET Report 76-11, Energy, Mines and Resources Canada, Ottawa.

Part IV

SULFUR TRANSFORMATIONS

25

Fractionation of Stable Sulfur Isotopes in a Closed Sulfuretum

Rolf O. Hallberg and L. E. Bågander

University of Stockholm

ABSTRACT: Fractionation of stable isotopes at the sediment-water interface was studied, *in situ,* in closed systems over a period of 300 days. Oxygen was consumed and a negative redox turnover took place after about 50 days. The reducing conditions were established at an Eh near -200 mv and a pH close to 7. The initial rate of sulfate reduction was 5.7×10^{-3} moles/l/yr and 8.5×10^{-3} moles/l/yr for experiments I and II, respectively. The bacteria generated sulfide into the supernatant at a rate corresponding to the depletion of sulfate. However, the very first release of sulfide was trapped in the sediment as metal sulfide. Photosynthetic sulfur bacteria developed and oxidized the sulfide at about twice the rate of sulfide formation. A balance between oxidizing and reducing processes of sulfur was established after about 200 days. Stable isotopes of sulfate and sulfide in the supernatant were analyzed at about weekly intervals. The average k_1/k_2 for the sulfate reduction during the initial state was 1.023. This result is in good agreement with theoretical evaluations of the fractionation process, but a lower value for the rate of sulfate reduction would be expected in comparison with other investigations.

INTRODUCTION

The bacterial fractionation of sulfur isotopes during sulfate reduction has been studied by many scientists since the early 1950s. The observed fractionation effect has been compared with pure chemical reactions and theoretical evaluations. All reported data are from laboratory experiments, mainly with pure cultures of sulfate-reducing microorganisms, either batch cultures or continuous cultivations. The main conclusions from these studies is that the fractionation varies with reduction rate, favoring the lighter isotope, ^{32}S, over the heavier, ^{34}S, at an increasing ratio with decreasing rate. The rate-limiting factors can be temperature, quantity, and quality of metabolizable compounds, concentration of sulfate, or a combination of these and other environmental factors. Variations of fractionation between 10/mil and 25/mil are common and agree well with what could be expected from a theoretical point of view. Occasionally, high fractionation values (up to about 60/mil) have been reported (Deevey, Nakai, and Stuiver, 1963). Such values can be explained with a very low reduction rate. They could also be the result of a multiple reduction reaction—that is, if the sulfide first reduced is oxidized to sulfate and then is reduced a second time, and so forth. The intention of the present study was to form a sulfuretum where the sulfur isotopes were cycled between oxidized and reduced states in order to bring about an apparent fractionation comparable to the high values reported by other scientists. In this paper, however, only the first part of the experiments are examined in order to stipulate the isotopic rate-constant ratio of the bacterial sulfate reduction for the natural habitat of the *in situ* experiments. The constant ratio will be used in future interpretations of the more complex continuation of the *in situ* sulfuretum.

FIELD METHODS

Duplicate cubic transparent boxes (edge length 40 cm) made of Plexiglas were gently pressed down into a sediment depth of 20 cm. The box was open toward the sediment. Each box formed a closed experimental system (Hallberg et al., 1972; Hallberg, Bågander, and Engvall, 1976) that confined 32 liters of water and a sediment area of 0.16 m^2. The box system makes it possible to simulate different natural processes with minimal disturbances of the sedimentary ecosystem.

Samples from the enclosed water were taken with 50-ml plastic syringes through self-sealing rubber membranes placed at the top of the box. The water loss at sampling was compensated for by a simultaneous addition of sea water through a gas lock and was considered when the results were calculated.

LABORATORY METHODS

Sulfate was analyzed turbidimetrically in accordance with Vogel (1961). Sulfate was precipitated as BaSO$_4$, which was stabilized in the solution by gum arabic and detected with a Bausch and Lomb Spectronic 100 spectrophotometer at 450 nm.

Sulfide was analyzed by the methylene blue method (Cline, 1969). To avoid oxidation by air, the sample and reagents were transferred directly from syringes into an incubation flask. The flask had been flushed with nitrogen gas and then sealed with the same type of rubber membrane used in the experimental box. The blue complex was measured on a Bausch and Lomb Spectronic 100 spectrophotometer at 670 nm.

Samples for isotope measurements were prepared by first bubbling nitrogen gas for 1 hr through 150 ml of a water sample and subsequently through a solution of 20 mM $CdCl_2$. Water from the first bottle was mixed 1:1 with 50 mM $BaCl_2$ to form $BaSO_4$. In the second bottle H_2S was trapped as CdS. $BaSO_4$ and CdS were later converted into Ag_2S by using Sn(II) in a solution of HI, H_3PO_2, and HCl at a temperature of 280°C to produce H_2S (Sasaki et al., 1979).

Ag_2S was combusted with an excess of Cu_2O to produce SO_2 gas that was measured with a 602 C Micromass. All data are given with reference to CDT.

Reporting the results as $\delta^{34}S$ indicates the per-mil deviation of the heavier isotope of a sample as compared with troilite sulfur in the Canon Diablo meteorite. The standard sample, Canon Diablo troilite (CDT), is assumed to have an absolute value for $^{32}S/^{34}S$ of 22.220 ($^{34}S/^{32}S$ 0.0450045) that is zero per mil (Nakai and Jensen, 1963).

EXPERIMENTAL CONDITIONS

Two experiments were performed *in situ* on a soft sediment at a water depth of 10 m in the Baltic Sea. The experimental area is located in the archipelago 80 km south of Stockholm at the Askö laboratory. Mean salinity at the experimental site is about 0.7%. The uppermost sediment layer (0.5 cm) has an average water content of 80% (wet weight). The median grain size is about 0.2 mm and total organic content is about 7% (dry weight). Experiment I started in May 1976 and was terminated by an accident that destroyed the equipment in April 1977. Experiment II started in May 1978 and was terminated in October 1981. At the beginning of the experiments, the top part of the sediment had a brown color that changed to grayish black in the lower, reduced-sediment layers. As the systems were closed, the conditions changed from oxic to anoxic in a maximum of two weeks. The top cm of the sediment then contained about 10^6 sulfate reducers/cm^3.

RESULTS AND DISCUSSION

During the course of the experiment, different types of microorganisms dominated the biogeochemistry of the closed system at different periods. Because of these chemical events, when one system is taking over from another, the experiment can be divided into separate eco-units. These units are indicated in Figures 25-1 and 25-2 showing the sulfate and sulfide variations during the course of the experiments.

Deoxygenation Unit (DO)

Once the system was closed, the access to free oxygen was limited and the aerobic organisms, dominating the chemistry of the uppermost sediment layer and the supernatant, became extinct. As the boxes were transparent, oxygen was still produced by photosynthetic organisms during the day but decreased rapidly during the night. The oxygen concentration was thus oscillating on a diurnal basis but eventually decreased to undetectable concentrations (Hallberg, Bågander, and Engvall, 1976). This period usually lasted for less than two weeks and could be easily observed by the change in color of the sediment from brown to black.

Redox Turnover Unit (RT)

When oxygen reached a concentration of less than 1 mg/l, an extreme redox turnover took place. The Eh of the system dropped from about +400 mV to about −200 mV. It took only a few days, and this eco-unit was a transition phase to a period when sulfate-reducing microorganisms dominated the biogeochemistry. The drop in Eh was probably due to a shift in redox reactions from the Fe^{2+}/Fe^{3+} to the HS^-/SO_4^{2-} redox couple (Hallberg, Bubela, and Ferguson, 1980; Bågander, 1980) where the latter was bacteriologically mediated.

Hydrogen Sulfide Unit (HS)

After the redox turnover, sulfate reduction was easily observed from the analytical data. The rate of the initial sulfate reduction in experiments I and II was 5.7×10^{-3} moles/l/yr and 8.5×10^{-3} moles/l/yr, respectively. In equivalent but opaque systems the redox turnover takes place sooner but at a somewhat lower rate of about 4×10^{-3} moles/l/yr (Bågander, 1980). Bågander concludes that this phenomenon is not the result of the extinction of light but is mainly due to a lower temperature in May than June.

Hydrogen sulfide was observed in the water phase about two weeks after the redox turnover. The hydrogen sulfide appeared in the water phase with some lag compared to the decrease of sulfate concentration. This was probably because the first hydrogen sulfide formed was fixed in the sediment, mainly as iron sulfide. When all iron in the brown top layer of the sediment was converted into iron sulfide, hydrogen sulfide began to appear in the supernatant of the box. Six weeks later the sulfide concentration of experiment II had reached a level of around 100 mg/l. That was the peak of sulfide

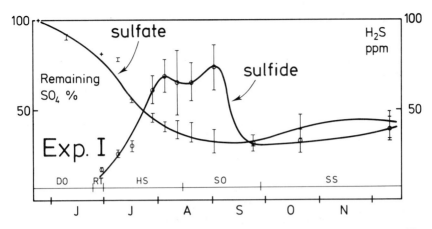

Figure 25–1. Sulfate and sulfide concentrations during the course of experiment I. The eco-units (see text) are indicted at the bottom of the figure.

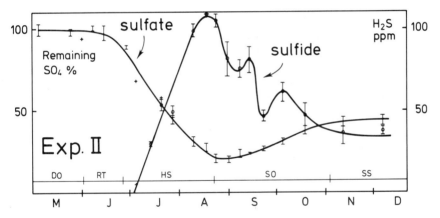

Figure 25-2. Sulfate and sulfide concentrations during the course of experiment II. The eco-units are indicated at the bottom of the figure.

concentration, and a sudden decrease took place. Sulfide dropped rapidly to concentrations around 50 mg/l. The DO and RT units of experiment I were extremely short, and the sulfate reduction started almost immediately. The sulfate graph, though, has the common features of experiment II (Figs. 25-1 and 25-2). They level out at the same concentration, somewhat less than 50 mg sulfate/l. The bimodal shape of the sulfide graph is due to very good weather conditions during August in combination with low turbidity of the water.

Sulfide Oxidation Unit (SO)

From the peak day of sulfide, purple sulfur bacteria of the genus *Chromatium* were identified. They consumed sulfide at a rate about twice that of its formation. The transparent boxes were situated in the photic zone, where such anaerobic photo-autotrophic microorganisms as *Chlorobium* and *Chromatium* proliferated on the hydrogen sulfide formed. Hydrogen sulfide was oxidized to intra- and extracellular elemental sulfur in the first step. In the second step, when the hydrogen sulfide concentration had dropped and become a limiting component in metabolism, the elemental sulfur was oxidized to sulfate. The sulfur cycle was thereby completed and a sulfuretum created. The inside of the box was colored purple from the *Chromatium* sp. As they covered the walls of the box, the insulation diminished with time. A balance between oxidizing and reducing processes was established, and six months from the start the biogeochemistry of the sulfuretum was in a steady state (SS unit).

REACTION ORDER

The reaction order of the sulfate reduction in the same type of experimental system has been studied by Bågander (1977). The data agreed very well with a first-order reaction, although the first part could also be interpreted as a zero-order reaction. The change from a zero-order to a first-order reaction took place at a sulfate concentration of 2 mM. This is in agreement with the theoretical evaluation by Rees (1973), who states that the process of bacterial sulfate reduction is of first order only when sulfate concentration is insufficient to saturate enzyme-activation sites, which takes place at a concentration of about 2 mM.

In a first-order reaction, the relation of reactant and product is indicated by the following equation:

$$\ln(\text{sulfate}_o) - \ln(\text{sulfate}_o - \text{sulfide}_t) = kt \qquad \text{(Eq. 25-1)}$$

sulfate_o = initial concentration of sulfate

sulfide_t = amount of sulfide produced at time t

t = time (days)

k = rate constant

If the remaining sulfate concentration at time t is equal to $\text{sulfate}_o - \text{sulfide}_t = \text{sulfate}_t$ then,

$$\ln(\text{sulfate}_o) - \ln(\text{sulfate}_t) = kt \qquad \text{(Eq. 25-2)}$$

If the sulfate reduction is assumed to be a first-order reaction, the rate constant, k, can be calculated from the observed ratios of sulfate$_o$/sulfate$_t$. The average k values of the duplicates for experiments I and II are 1.4×10^{-2}/day and 2.5×10^{-2}/day, respectively. Although the observation data in Figure 25-3 of the HS-unit scatter a little around the least-square fit equation, they are well within experimental error and indicative of a first-order reaction. The reduction of sulfate to sulfide for the two isotopic species ^{32}S and ^{34}S can be applied separately in accordance with Equation 25-2.

$$\ln(^{32}\text{sulfate}_o) - \ln(^{32}\text{sulfate}_t) = k_1 t \qquad \text{(Eq. 25-3)}$$

$$\ln(^{34}\text{sulfate}_o) - \ln(^{34}\text{sulfate}_t) = k_2 t \qquad \text{(Eq. 25-4)}$$

k_1 and k_2 are rate constants of the respective isotopes.

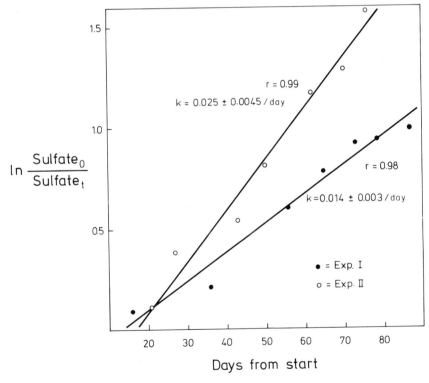

Figure 25-3. Rate constants during reduction.

THE RATIO OF ISOTOPIC
RATE CONSTANTS

The ratio k_1/k_2, usually denoted a, is expected to be constant for the bacterial sulfate reduction. It is suggested that the reduction of sulfate to sulfide is unidirectional and that the lighter sulfur isotopes react 2.0% faster than the heavier isotopes (Nakai and Jensen, 1964).

The constant can be calculated according to the following equations (Nakai and Jensen, 1964; Chambers and Trudinger, 1979):

$$a^{-1}\mathrm{xlog}\, F = \log FR_1 \qquad\qquad \text{(Eq. 25-5)}$$

$$a^{-1}\mathrm{xlog}\, F = \log\,(1-(1-F)R_2) \qquad\qquad \text{(Eq. 25-6)}$$

F is the ratio of remaining sulfate at time t to sulfate at time O; R_1 and R_2 are the $\delta^{34}S$ of sulfate and sulfide at time t, respectively, vis-à-vis $\delta^{34}S$ of initial sulfate.

Because of analytical errors, only a few sulfate isotopic data could be used to deduce a-values according to Eq. 25-5 from experiment I. Data from the two experiments are given in Table 25-1. The a-values are 1.020 for the sulfate data but are significantly higher for the sulfide data. One explanation for this result may be the fact that the sulfide samples were extremely small (usually less than 10 mg) because of the low sulfide concentrations of the supernatant. Larger water samples could not be obtained as collecting them would have upset the chemical equilibria inside the box. At the beginning of the HS-unit in experiment II, in particular, the isotopic data are uncertain due to sample size. The standard deviations are also larger for the sulfide than for respective sulfate data.

The fractionation of sulfur isotopes during dissimilatory sulfate reduction has been attributed to isotopic effects associated with different boundary conditions of the sulfur pathway, which, according to Rees (1973), is:

$$\text{external sulfate} \xleftrightarrow{\;1\;} \text{internal sulfate} \xleftrightarrow{\;2\;} \text{APS} \xleftrightarrow{\;3\;} SO_3^{2-} \xrightarrow{\;4\;} S^{2-}$$

Rees (1973) assumed that only the forward steps 1, 3, and 4 involved fractionation. He also considered the sum of steps 1 and 3 to give rise to a fractionation of -2.2%. However, at the present state the fractionation pathway has not been entirely validated, and, therefore, we will not relate our total average a-value of -2.3% to any specific step of the sulfur pathway. Also, our data were not obtained during steady-state conditions, as we increased the facilities for the anaerobic organisms in the sediment when we

Table 25-1. Isotopic Data

	Experiment I						Experiment II				
Days from Start	F	δ^{34} Sulfate	$\alpha(5)$	δ^{34} Sulfide	$\alpha(6)$	Days from Start	F	δ^{34} Sulfate	$\alpha(5)$	δ^{34} Sulfide	$\alpha(6)$
0	1.00	20.0				0	1.00	19.52			
16	0.89	22.0	1.0171			21	0.90	21.75	1.0212		
	0.92	21.7	1.0204				0.88	21.30	1.0138		
36	0.81			+0.3	1.0219	27	0.68	26.58	1.0182		
	0.81						0.68	25.02	1.0141		
45	0.79			+0.6	1.0219	36	0.62	29.59	1.0210	−9.26	1.0372
	0.77			−2.2	1.0254		0.62	29.99	1.0215	−5.15	1.0318
53	0.56	33.5	1.0218	+1.5	1.0250	43	0.58	34.82	1.0281	−9.43	1.0388
	0.54			−1.2	1.0293		0.58	31.38	1.0217	+0.60	1.0251
65	0.43			+3.4	1.0259	50	0.42	35.18	1.0179	−1.78	1.0339
	0.48			+2.1	1.0263		0.47	36.31	1.0221	+3.74	1.0234
72	0.38			+6.2	1.0231	62	0.28	39.65	1.0156	+6.24	1.0266
	0.43			+3.6	1.0256		0.34	43.33	1.0219	+7.86	1.0208
79	0.34			+9.4	1.0189	70	0.25	47.34	1.0198	+6.73	1.0274
	0.44						0.30	46.20	1.0219	+7.60	1.0229
87	0.32			+6.2	1.0255						
	0.43			+4.5	1.0242						
100	0.26			+10.0	1.0208						
	0.39			+4.1	1.0262						
$\bar{\alpha}$			1.020		1.024				1.020		1.029
s			1.002		1.003				1.004		1.006

293

closed the system. Oxygen disappeared, the aerobic organisms died, their organic material was made available for the sulfate reducers, and so on. Despite the fact that the environmental conditions could not be controlled as in a pure culture in a laboratory, the encountered data agree well with the theoretical. The scattering of the data can be explained by several factors. There was no single energy source for the bacteria as in a synthetic medium. The natural habitat has a mixed culture of sulfate-reducing bacteria, and, therefore, the H_2S production per unit cell varies.

Although the bacteria are acclimatized to the prevailing temperature, their activity must have changed with the increase in temperature from about 5°C to about 18°C during the summer period. However, Harrison and Thode (1957) could not verify any variation of isotopic fractionation with temperature.

It has been demonstrated by several researchers that a varies inversely with the rate of reduction (e.g., Jones et al., 1956; Harrison and Thode, 1958; Kaplan and Rittenberg, 1964). Krause et al. (1968) could, however, in their study find no consistent pattern between a and the sulfite reduction rate per unit cell.

The reduction rates of our experiments differ from each other, although the difference is only a factor of about 2. A significant difference between the a-values of our two experiments is not revealed in Table 25-1. According to the data presented by Goldhaber and Kaplan (1975), an a-value around 1.010 would be more likely to find for the Baltic sediments, but this is obviously not the case.

FRACTIONATION FACTOR

Many who have studied microbiological isotope fractionation have used the term "fractionation factor" for the ratio between $\delta^{34}S$ of the total product and the remaining reactant. For first-order reactions, Nakai and Jensen (1964) have shown that:

$$r = \frac{\delta^{34}\text{sulfide}_t}{\delta^{34}\text{sulfate}_t} = \frac{F^{(\frac{1}{a} - 1)} - F}{1 - F} \qquad \text{(Eq. 25-7)}$$

This equation can be deduced from Eq. 25-5 and Eq. 25-6.

At values of F very close to 1, r is equal to a. This is also the case in a so-called open system where the isotopic composition of the reactant remains practically constant (e.g., in the ocean during short geological periods). In closed systems, however, the isotopic composition of the reactant will change with time. In the case of sulfate reduction it will be enriched in ^{34}S. This situation is illustrated in Figure 25-4. Thus, during the course of the

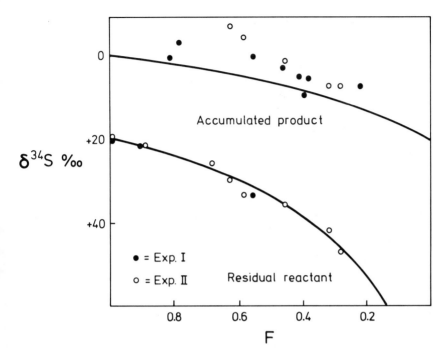

Figure 25–4.   Variation of isotope fractionation between sulfate and sulfide. The graphs are calculated with an α-value of 1.020.

experiment the sulfate in the boxes was getting heavier, and the fractionation during the sulfate reduction resulted in sulfur isotopic ratios of the sulfide that changed with time. The result is an apparent fractionation factor that can be calculated for any observation by using Eq. 25-7. Our observations are illustrated in Figure 25-4. Because the α-value of the graphs has been chosen equal to 1.020, the sulfate data resembles the graph better than the sulfide data. The fractionation between sulfate and sulfide of experiment II is almost constant, probably because of erratic sulfide data, especially at high F-values.

ACKNOWLEDGEMENTS

We thank Lyn Plumb, Agneta Pettersson, and Peter Torssander for doing the isotope measurements.

REFERENCES

Bågander, L. E., 1977, *In situ* studies of bacterial sulfate reduction at the sediment-water interface, *Ambio* special report **5:**147-155.

Bågander, L. E., 1980, Bacterial cycling of sulfur in a Baltic sediment: An *in situ* study in closed systems, *Geomicrobiol. J.* **2:**141-159.

Chambers, L. A., and P. A. Trudinger, 1979, Microbiological fractionation of stable sulfur isotopes: A review and critique, *Geomicrobiol. J.* **1:**249-293.

Cline, J. D., 1969, Spectrophotometric determination of hydrogen sulfides in natural waters, *Limnol. Oceanog.* **14:**454-458.

Deevey, E. S., N. Nakai, and M. Stuiver, 1963, Fractionation of sulfur and carbon isotopes in Meromictic Lake, *Science* **139:**407-408.

Goldhaber, M. B., and I. R. Kaplan, 1975, Control and consequences of sulfate reduction rates in recent marine sediments, *Soil Sci.* **119:**42-45.

Hallberg, R. O., L. E. Bågander, A. -G. Engvall, and F. A. Schippel, 1972, Method for studying geochemistry of sediment-water interface, *Ambio* **1:**71-72.

Hallberg, R. O., L. E. Bågander, and A. -G. Engvall, 1976, Dynamics of phosphorous, sulfur and nitrogen at the sediment-water interface, in *Environmental Biogeochemistry*, Vol. 1, J. O. Nriagu, ed., Ann Arbor Science, Ann Arbor, Michigan, pp. 295-308.

Hallberg, R. O., B. Bubela, and J. Ferguson, 1980, Metal chelation in sedimentary systems, *Geomicrobiol. J.* **2:**99-113.

Harrison, A. G., and H. G. Thode, 1958, Mechanism of the bacterial reduction of sulphate from isotope fractionation studies, *Faraday Soc. Trans.* **54:**84-92.

Harrison, A. G., and H. G. Thode, 1975, The kinetic isotope effect in the chemical reduction of sulphate, *Faraday Soc. Trans.* **53:**1648-1651.

Jones, G. E., R. L. Starkey, H. W. Feely, and J. L. Kulp, 1956, Biological origin of native sulfur in salt domes of Texas and Louisiana, *Science* **123:**1124-1125.

Kaplan, I. R., and S. C. Rittenberg, 1964, Microbiological fractionation of sulfur isotopes, *J. Gen. Microbiol.* **34:**195-212.

Krause, H. R., R. G. L. McCready, S. A. Husain, and J. N. Campbell, 1968, Sulfur isotope fractionation and kinetic studies of sulfate reduction in growing cells of *Salmonella Heidelberg, Biophys. J.* **8:**109-124.

Nakai, N., and M. L. Jensen, 1964, The kinetic isotope effect in the bacterial reduction and oxidation of sulfur. *Geochim. Cosmochim. Acta* **28:**1893-1912.

Rees, C. E., 1973, A steady state model for sulfur isotope fractionation in bacterial reduction processes, *Geochim. Cosmochim. Acta* **37:**1141-1162.

Sasaki, H., E. Gunnlaugsson, J. Tomasson, and J. E. Rouse, 1980, Sulfur isotope systematics in Icelandic geothermal systems and influence of seawater circulation at Reykjanes, *Geochim. Cosmochim. Acta* **44:**1223-1231.

Vogel, A., 1961, *A Textbook of Quantitative Inorganic Analyses.* Longmans Ltd., London.

26

Fate of Sulfate in a Soft-Water, Acidic Lake

Lawrence A. Baker and Patrick L. Brezonik

University of Minnesota

ABSTRACT: Laboratory experiments and field studies show that sulfate is not a conservative ion in McCloud Lake, a small acidic lake in north-central Florida. From 50% to 72% of the input SO_4^{2-} was removed by sediment columns designed to simulate ground-water flow to the lake. Although column-inflow water ranged in pH from 3.2 to 6.8, column effluent was buffered by internal mechanisms to pH 5-6. Batch experiments indicate that SO_4^{2-} adsorption by the lake sediments is unimportant. Pore-water analyses of dissected profundal sediment cores showed SO_4^{2-} decreased from 150 to 15μeq/l in the top 10 cm. Calculations based on Fick's law suggest a diffusive flux of 72 meq/m^2-yr of SO_4^{2-} in the sediments. Mass-balance calculations show that 37% to 72% of the input SO_4^{2-} to the lake is lost to internal sinks. These results suggest that biological reduction is an important sink for SO_4^{2-} and an important mechanism for acid neutralization in soft-water lakes.

INTRODUCTION

Models that predict the pH response of surface waters to acid precipitation generally assume that cation exchange and mineral dissolution are the principal mechanisms that neutralize acidity in watersheds (e.g., Henriksen, 1980; Thompson, 1982). Although reactions involving sulfate have been ignored or considered negligible in these models, sulfate transformations in watersheds and lake sediments are potentially important in neutralizing acid inputs.

Two mechanisms of sulfate adsorption may result in accumulation of sulfate in watersheds: Reversible, or nonspecific, adsorption occurs when SO_4^{2-} is adsorbed as a counterion to protonated surfaces. Nonreversible, or specific, adsorption occurs when SO_4^{2-} displaces OH^- groups on metal hydroxide surfaces (Rajan, 1978). Johnson et al. (1980) discussed sulfate adsorption mechanisms and the types of soils with which they are associated. They summarized results of sulfate mass balances for several watersheds in the eastern United States and reported that sulfate was accumulating in all except Hubbard Brook, New Hampshire. Accumulation rates ranged from 6.2 to 9.3 kg/ha-yr. Insoluble sulfate (equivalent to specifically adsorbed sulfate) comprised 25% to 77% of the total sulfate in these soils. Chen, Gherini, and Goldstein (1979) postulated reversible sulfate adsorption as the mechanism for short-term pH buffering in Panther Lake watershed, New York. During summer, relatively acidic precipitation is neutralized by sulfate adsorption in the subsoil. During snowmelt, relatively neutral water passes through the soil, desorbing sulfate. Christopherson and Wright (1981) reported that reversible sulfate adsorption also results in seasonally stable sulfate concentrations in stream drainage from the Birkeness watershed in Norway.

The role of biological sulfate reduction as an acid-consuming (or alkaline-generating) process is well known. However, the significance of this process in lakes sensitive to acidification by atmospheric deposition has received little attention. Kelley et al. (1982) reported that sulfate reduction was the major mechanism for acid neutralization in Lake 223 of the Experimental Lakes Area, Ontario. Sulfate reduction in littoral sediments and in the lake hypolimnion neutralized nearly half of the H_2SO_4 added to the lake in an artificial acidification experiment. To date, this is the only study in which sulfate loss (and consequent acid neutralization) has been quantified in a soft-water (acid-sensitive) lake.

The objective of the present study was to evaluate the extent and mechanisms of SO_4^{2-} loss in soft-water acidic lakes of northern Florida. The site of the investigation was McCloud Lake, a small (approximately 5 ha) seepage lake in a sandy watershed in the Trail Ridge region of north-central Florida (Brezonik et al., 1983). This lake was chosen because its watershed has no human habitation, its water is very poorly buffered, and it has become increasingly acidic during the past 15 years. Water chemistry data collected in 1968-1969, 1978-1979, and 1981 (Table 26-1) show that the pH has declined from 5.0 to 4.6 during that time period. A water budget conducted during the present study shows that the lake receives about 90% of its water directly from precipitation; the remainder comes from subsurface seepage through porous (sandy) soils. The littoral sediments are sandy and interspersed with pockets of peat deposits, while the profundal sediments are an organic ooze (73% volatile solids) with high (91.4%) water content.

Table 26-1. Physical and Chemical Characteristics of McCloud Lake, Florida

Morphometry	
Surface area	$5.0 \times 10^4 \text{ m}^2$
Mean depth	2.4 m
Volume	$1.2 \times 10^5 \text{ m}^3$
Water residence time: V/Qin	1.3 yr
V/Qout	~8 yr*

Chemical Composition ($\mu eq/l$)**			
	1968-1969	*1978-1979*	*1981-1982*
H^+	14	19	32
Ca^{2+}	30	47	56
Mg^{2+}	47	51	65
Na^+	122	121	182
K^+	6	15	5
Cl^-	167	145	175
SO_4^{2-}	104	142	190

*Based on outseepage only; does not include evaporation.
**Source: Data from 1968-1969, Brezonik and Shannon (1971); 1978-1979, Brezonik et al. (1983); 1981-1982, this study.

MATERIALS AND METHODS

Sulfate adsorption by McCloud Lake sediments was determined in sediment-water "titrations" by adding 20 g wet sediment, 50 ml distilled, deionized water, 1 ml chloroform (to inhibit microbial activity), and varying amounts of H_2SO_4 to 125-ml polyethylene bottles. Duplicate bottles each received from 0.2 to 1.0 ml of 0.1 N H_2SO_4 and were placed on a shaker table for one week. Water from each was then filtered through a 0.4-μm Millipore filter and analyzed for SO_4^{2-} by the automated methylthymol blue method (APHA, 1981). The sediment was dried at 105°C and weighed to determine water content and dry weight. The added SO_4^{2-} was compared with the measured SO_4^{2-} to compute SO_4^{2-} adsorption (on a dry-weight basis).

An upflow column study was conducted to simulate the flow of ground water through McCloud Lake littoral sediments. Synthetic ground water used in the experiment simulated the composition of soil water from a depth of 200 cm at a nearby site that had been irrigated repeatedly with simulated pH 3.0 rain (J. Byers, University of Florida, personal communication, 1983). This water was circumneutral (pH 6.8) and contained 270 μeq/l of alkalinity (estimated from ion balance). Synthetic ground water was passed through three 30-cm columns of littoral sediment at a rate of approximately 40 ml/day, corresponding to observed seepage rates in the lake. Three additional columns received ground water acidified with H_2SO_4 to pH 3.2. A third set of columns received ground water initially at pH 6.0 and later (end of week 6) acidified to pH 4.0 to observe effects of rapid acidification. Following a month of pH stabilization, eluate samples were collected weekly from each column and analyzed for major anions by ion chromatography (IC), alkalinity (titration to pH 4.5), and major cations (flame atomic absorption).

Littoral and profundal sediment cores were obtained from the lake by a Livingston-type corer in February 1983. Three cores (3.8-cm diameter) from each site were dissected into 2.5-cm segments in a glove box purged with oxygen-free nitrogen gas. Individual segments were placed into centrifuge tubes, treated with 0.1 ml 2 N zinc acetate and 0.06 ml 6 N NaOH to precipitate ZnS (APHA, 1981), and centrifuged for 20 min at 15,000 rpm. The supernatant was filtered through 0.45-μm Millipore filters and analyzed by IC for major anions. The flux of sulfate to the sediment was computed from the sulfate gradient by Fick's law:

$$F = D_c (C_i - C_{i+1})/z \qquad \text{(Eq. 26-1)}$$

where D_c = diffusion coefficient, C_i, C_{i+1} = sulfate concentration at depth $i+1$, and z = distance between depths i and $i+1$. Lerman (1978) reported that D_c approximately equals $D_0 (\phi)$, where ϕ is the sediment porosity (estimated from water content) and D_0 is the bulk water diffusion coefficient. For SO_4^{2-} flux calculations, $D_0 = 8.9 \times 10^{-6}$ cm^2/sec was used (Li and Gregory, 1974).

During 1981-1982, measurements were made to compute mass balances and thus identify sources and sinks of all major ions associated with pH-neutralizing mechanisms in McCloud Lake. Wet-only precipitation was collected at the site with an Aerochemetrics Model 101 wet/dry collector. Concentrations of SO_2 and aerosol SO_4^{2-} were measured during twenty 24-hr intervals in August and September 1982. Dry deposition rates were estimated from ambient air concentrations (C_i) of SO_4^{2-} and SO_2 using the equation:

$$F = C_i \times v_d \qquad \text{(Eq. 26-2)}$$

where v_d is the deposition velocity. For SO_2 we used $v_d = 1.0 \pm 0.5$ cm/sec (Sehmel, 1980), and for aerosol SO_4^{2-} we used $v_d = 0.4 \pm 0.2$ cm/sec. Dry deposition of chloride was estimated by subtracting measured wet-only Cl$^-$ deposition at the site from bulk deposition for 1978-1979 at three other northern Florida sites (Jasper, Gainesville, and Hastings) (Brezonik et al., 1983). Dry deposition of SO_4^{2-} estimated this way agreed well with calculations based on ambient air concentrations and the above equation.

Seepage to and from McCloud Lake was measured using 22 seepage meters (Fellows and Brezonik, 1980) placed along six transects perpendicular to the shoreline (Baker, 1983). Flow measurements were made monthly. When flow was into the lake, samples were collected from the meters and analyzed for major ions. Since seepage rates decreased with distance from shore, the lake was divided into five concentric regions whose boundaries were formed by points equidistant from adjacent meters along each transect. Flow in each region was computed as the product of the mean seepage rate for the meters within the region and the area of the region. Inflow chemical composition did not exhibit systematic trends with distance from shoreline; therefore, mean concentrations $[C_i]$ from all 22 meters were used to calculate major ion fluxes.

Lake-water samples collected at 1-m intervals monthly during 1981-1982 were analyzed for major cations and anions by atomic absorption and AutoAnalyzer techniques (APHA, 1981). Since the lake was unstratified, storage of each constituent was determined from the mean concentration $[C_i]$ and lake volume for each month.

RESULTS

A plot of predicted vs. measured sulfate (Fig. 26-1) shows that sulfate adsorption was negligible for the first four acid additions. At the highest dose of acid (final pH = 3.7), the observed $[SO_4^{2-}]$ was 0.73 meq/l less than the predicted value. The amount adsorbed was only 0.9 meq/100 g (dry wt) and is unimportant as an acid-neutralizing mechanism (Baker, 1983).

Buffering mechanisms in the sediment columns resulted in eluate pH values of 5.0 to 6.0 in all three sets of columns (Table 26-2). Two mechanisms are believed to be responsible for the observed results: cation exchange (Baker, 1983) and sulfate reduction. During the month following pH stabilization, sulfate levels in the eluates were reduced by 50%-75% (Table 26-2). For the high pH columns, the change in $[SO_4^{2-}]$ through the columns had little effect on pH since inflow $[SO_4^{2-}]$ was low (80 μeq/l); cation exchange of sediment protons by divalent cations from the inflow water was more important in determining eluate pH and alkalinity. In the low pH columns, a decrease in $[SO_4^{2-}]$ of 0.410 μeq/l in the eluate paralleled a decrease of 0.402 μeq/l in $[H^+]$. Although either sulfate adsorption or sulfate reduction could account for the observed results, the sulfate adsorption experiment indicates that this process is not important as a pH-buffering mechanism in these sediments; we hypothesize that sulfate reduction was responsible for the results. In the intermediate pH columns, a greater decrease in eluate $[SO_4^{2-}]$ occurred immediately upon acidification of the inflow to pH 4.0. Both cation exchange and sulfate reduction were significant mechanisms of pH neutralization in these columns.

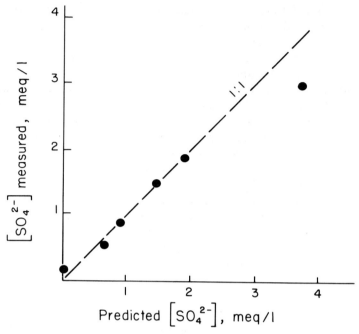

Figure 26-1. Measured vs. predicted sulfate concentrations for addition of H_2SO_4 to McCloud Lake littoral sediments. 1:1 line implies no loss of sulfate by adsorption.

Table 26-2. Composition of Synthetic Groundwater Passing Through McCloud Lake Sediment Columns *

		Inflow pH			
Parameter	*Sample***	*3.2*	*4.0*	*6.2*	*6.6*
H^+	I	0.403	0.11	0.001	0.002
	O	0.001	0.001	0.004	0.006
Alk	I	0.00	0.00	—	0.27
	O	0.07	0.07	—	0.09
SO_4^{2-}	I	0.74	0.47	0.28	0.08
	O	0.33	0.16	0.14	0.02
ΣM^+	I	0.533	0.52	0.52	0.52
	O	0.584	0.40	0.39	0.32

*Values in $\mu eq/l$.
**I = inflow to columns; O = outflow from columns.

Figure 26-2 shows that pore water $[SO_4^{2-}]$ decreased to approximately 14 $\mu eq/l$ within the top 10 cm of all three profundal cores and then remained constant. From the observed gradient the diffusive flux of SO_4^{2-} to the sediment was estimated. For $z = 0$ cm, $C_i = 150$ $\mu eq/l$ and for $z = 6.25$ cm, $c_{i+1} = 30$ $\mu eq/l$; the computed flux thus is 72 meq/m^2-yr.

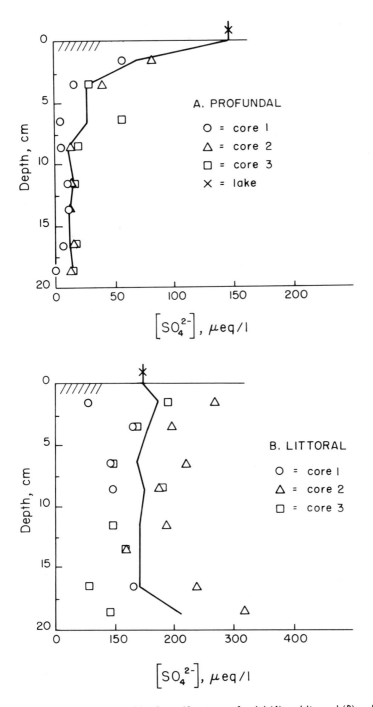

Figure 26-2. Pore-water profiles for sulfate in profundal (A) and littoral (B) sediment cores from McCloud Lake. Solid lines represent means of three cores.

Table 26-3. Mass Loadings and Storage of Sulfate and Chloride in McCloud Lake*

Component	SO_4^{2-}	Cl^-	SO_4^{2-}/Cl^-
Lake storage, eq	22,500	20,600	1.1
Wet precipitation, eq/yr	2080	914	2.3
Bulk precipitation, eq/yr:			
Sulfate dry dep. = 180 eq/ha-yr	2990	1120-1520	2.0-2.7
Sulfate dry dep. = 690 eq/ha-yr	5580	1120-1520	3.7-5.0
Seepage, eq/yr	325	371	0.88
Total inflow, eq/yr:			
Sulfate dry dep. = 180 eq/ha-yr	3310	1490-1890	1.7-2.2
Sulfate dry dep. = 690 eq/ha-yr	5910	1490-1890	3.1-4.0

*Data for September 1981-August 1982.

For the littoral sediments, pore water $[SO_4^{2-}]$ was irregular (Fig. 26-2), suggesting that pockets of sulfate reduction and sulfide oxidation may occur. This hypothesis is supported by the irregular pattern of mottling observed on four silver-coated rods inserted into the sediment for 24 hr at the sampling site. Mean $[SO_4^{2-}]$ for the three littoral cores remained approximately constant with depth, implying no net reduction. It is not known whether sulfate reduction occurs at other littoral sites or at this site during other times of the year. When the cores were taken, water was probably moving out of the lake (Baker, 1983), and it is possible that a gradient of $[SO_4^{2-}]$ caused by sulfate reduction was obliterated by water movement.

Initially an attempt was made to determine whether McCloud Lake was a sulfate sink by the mass balance equation: $\Delta S = P + F - S$, where ΔS = change in storage of sulfate, P = precipitation inputs (wet + dry), F = net seepage flux of sulfate, and S = sulfate sink. Preliminary calculations indicated that small errors in ΔS for sulfate would produce large errors in the calculated sink term. To illustrate, the mean volume of McCloud Lake during the study was $1.18 \times 10^5 m^3$, the mean $[SO_4^{2-}]$ was 190 μeq/l, and the total storage was 2.25×10^4 eq. An error of only 5% in the storage of sulfate at the beginning or end of the annual budget period thus would produce an error in ΔS of 1130 eq, or 19%-34% of the annual input. Although mass-balance calculations indicated a net sink for SO_4^{2-} in the lake, the above-mentioned uncertainties in storage diminish the reliability of the estimates.

An alternative method to evaluate the fate of sulfate is to compare SO_4^{2-}/Cl^- ratios (eq/eq basis) for the total inflow and in the lake water. The input SO_4^{2-}/Cl^- ratio was computed based on total sulfate and chloride loads:

$$[SO_4^{2-}]_{in}/[Cl^-]_{in} = (P_{SO_4^{2-}} + S_{SO_4^{2-}})/(P_{Cl^-} + S_{Cl^-}) \qquad \text{(Eq. 25-3)}$$

where P_i = total precipitation loading (wet + dry) and S_i = seepage input of species i. Precipitation loadings vary depending upon the estimated dry deposition rates used. The estimated total dry sulfate deposition, which includes SO_2 and aerosol SO_4^{2-}, was 180-690 eq/ha-yr, while the dry deposition of Cl^- was 40-120 eq/ha-yr. As seen in Table 26-3, seepage inputs of SO_4^{2-} and Cl^- were relatively minor. Although the range of SO_4^{2-}/Cl^- in total inflow is broad (1.8-3.0) and depends on the dry deposition values used, no reasonable estimates could produce a ratio as low as that of the lake water (1.1). Calculations based on these ratios show that in-lake sinks removed 1225-4250 eq/yr of sulfate, or 37%-72% of the total input.

DISCUSSION

Calculations of sulfate flux based on pore-water profiles and the sulfate mass balance both indicate a large in-lake sink for sulfate. On an areal basis, the mass balance shows a sink of 25-85 meq/m^2-yr. This range compares well with the estimate derived from profundal pore-water profiles (72 meq/m^2-yr), suggesting that the lake sediment is actively involved as a sulfate sink.

Data do not conclusively prove that biological sulfate reduction is the mechanism responsible for the observed sulfate sink since the end products of sulfate reduction (FeS or H_2S) have not been identified. Sulfate adsorption, however, does not account for the observed loss of sulfate. It is unlikely that other mechanisms, such as assimilation by the biota, could account for the loss. Furthermore, conditions in McCloud Lake, including warm tempera· ture (annual mean = 20°C), an organic-rich sediment, and a long hydraulic residence time (12-14 years for Cl^-) are conducive to sulfate reduction. Although there is some evidence that sulfate reduction is inhibited at pH levels <5 (Zinder and Brock, 1978), the pH of sediments in McCloud Lake usually is above 5, perhaps as the result of sulfate reducers.

Sulfate reduction consumes protons and thus is a potentially important mechanism of acid neutralization. In-lake processes such as sulfate reduction (discussed above) and sediment cation exchange (Baker, 1983) appear to be important neutralizing processes in "Type I" lakes (Schnoor, Eilers, and Glass, unpub. data) that receive most of their water directly from precipitation. Such lakes, which are considered highly susceptible to acidification (Schnoor, Eilers, and Glass, unpub. data), are common in Florida and northern Wisconsin.

ACKNOWLEDGEMENTS

This work was supported by the University of Florida, University of Minnesota Graduate School, and a grant from the U.S. Environmental Protection Agency-North Carolina State University program in acid precipitation research. Assistance of E. Edgerton and R. W. Ogburn in sampling and analyses is gratefully appreciated.

REFERENCES

American Public Health Association, 1981, *Standard Methods for the Examination of Water and Wastewater,* APHA Publication Office, Washington, D.C.

Baker, L. A., 1983, *Sediment-Water Interactions in an Acidic Lake,* Ph.D. thesis, University of Florida, Gainesville, Florida.

Brezonik, P. L., and E. E. Shannon, 1971, *Tropic State of Lakes in North-Central Florida,* Water Resources Research Center, Publ. 13, University of Florida, Gainesville, Florida.

Brezonik, P. L., C. D. Hendry, E. S. Edgerton, R. L. Schultze, and T. L. Crisman, 1983, *Acidity, Nutrients and Minerals in Atmospheric Precipitation over Florida: Deposition Patterns, Mechanisms, and Ecological Effects,* EPA 600/3-83-004, U.S. Environmental Protection Agency, Washington, D.C.

Chen, C. W., S. A. Gherini, and R. A. Goldstein, 1979, in *Ecological Effects of Acid Precipitation,* M. J. Wood, ed., prepared for the Electric Power Research Institute, Palo Alto, California.

Christopherson, N., and R. F. Wright, 1981, Sulfate budget and a model for sulfate concentration in stream water at Birkeness, a small, forested catchment in southernmost Norway, *Water Resour. Res.* **17:**377-389.

Fellows, C. R., and P. L. Brezonik, 1980, Seepage flows into Florida lakes, *Water Resour. Bull.* **16:**635-641.

Henriksen, A., 1980, Acidification of freshwater—a large-scale titration, in *Ecological Impact of Acid Precipitation,* D. Drablos and A. Tollan, eds., SNSF, Oslo, Norway.

Johnson, D. W., J. W. Hornbeck, J. M. Kelley, W. T. Swank, and D. E. Todd, 1980, in *Atmospheric Sulfur Deposition: Environmental Impact and Health Effects,* D. S. Shriner, C. R. Richmond, and S. E. Lindberg, eds., Ann Arbor Press, Ann Arbor, Michigan.

Kelley, C. A., J. W. Rudd, R. B. Cook, and D. W. Schindler, 1982, The potential importance of bacterial processes in regulating rate of lake acidification, *Limnol. Oceanogr.* **27:**868-882.

Lerman, A., 1978, *Geochemical Processes: Water and Sediment Environments,* John Wiley and Sons, New York.

Li, Y. H., and S. Gregory, 1974, Diffusion of ions in sea water and in deep-sea sediments, *Geochim. Cosmochim. Acta* **38:**703-714.

Rajan, S. S. S., 1978, Sulfate adsorbed on hydrous alumina, ligands displaced, and changes in surface charge, *Soil Sci. Soc. Am.* **42:**39-44.

Sehmel, G. A., 1980, Particle and gas dry deposition: A review, *Atmos. Environ.* **14:**983-1011.

Thompson, M. E., 1982, The cation denudation rate as a quantitative index of sensitivity of eastern Canada rivers to acidic atmospheric precipitation, *Water, Air, Soil, Pollut.* **18:**215-226.

Zinder, S. H., and T. D. Brock, 1978, in *Sulfur in the Environment, Part II: Ecological Impacts,* J. Nriagu, ed., John Wiley and Sons, New York.

27

The Geomicrobiology of Sulfur Occurrences in West Texas

Christine L. Miller
Colorado School of Mines

Leo J. Miller
Miller Exploration Co.

David M. Updegraff
Colorado School of Mines

INTRODUCTION

The large-scale formation of elemental sulfur is one of the most outstanding examples of the influence of bacteria on a subsurface, solid-rock environment. Species of sulfate reducers, such as *Desulfovibrio desulfuricans,* utilize sedimentary sulfate (e.g., gypsum or anhydrite) as an electron acceptor during the course of their metabolism of organic matter. The end products are sulfide and carbon dioxide (Postgate, 1979; and others). In the presence of oxygenated ground waters, the sulfide will react to form colloidal sulfur that, over a period of time, converts to yellow crystalline sulfur. Carbon dioxide forms carbonate anions in water that react with certain cations to form solid carbonates, most commonly calcium carbonate (secondary calcite). It is this calcite that forms the matrix in which sedimentary sulfur is usually found (Davis and Kirkland, 1970).

Within the sediments, the organic substrate associated with the process of large-scale sulfur formation is thought to be petroleum or its associated gases (Davis, 1967; Ivanov, 1968). A petroliferous residue is commonly found in sulfur-bearing rock (Davis and Kirkland, 1979). Isotopic analysis of the secondary calcite found with sulfur shows it to be enriched in ^{13}C, which links the calcite carbon to the isotopically light carbon (^{12}C) characteristic of petroleum and particularly petroleum gases (Kirkland and Evans, 1976). Although no sulfate-reducing bacteria have been shown to consume long-chain hydrocarbons directly, they can use methane (Davis, 1967; Hanson, as cited by Whittenbury and Kelly, 1977) or oxidative breakdown products of petroleum, such as organic acids that could result from the activities of other microorganisms.

The geologic history that has created the flat-lying sulfur occurrences in west Texas includes an unusual combination of events. Two of these events, faulting and karsting, increased the porosity of the sediments and hence have provided a possible avenue for dispersion of sulfur compounds from the subsurface sulfur to the overlying soils. For this reason, it seemed reasonable to expect that microbial populations in the soils might be affected by the existence of sulfur at depth (up to 700 ft deep). This study investigates microbiological indications of sulfur occurrences in the soils of a region underlaid by numerous sulfur occurrences and at least one massive sulfur deposit (the Duval Sulfur Mine, approximately 80 million tons). The field area is shown in Figure 27-1.

GEOLOGIC CONTROLS

The field area is located in a region of predominantly Permian sediments, composed of sands, shales, limestones, dolomites, gypsum, and anhydrites deposited within a large inland sea. The formations of interest are illustrated by the stratigraphic section in Table 27-1.

Figure 27-1.   Location of the field area.

Table 27-1. Stratigraphic Section of the Regional Sediments

Age	Formation			Thickness (ft)
Cretaceous	Cox Sandstone			up to 200
Permian	Dewey Lake	silty sand (red bed)		up to 500
	Rustler Formation 300-560 ft	Forty-niner	= gypsum + anhydrite	50
		Magenta	= dolomite	20
		Tamarisk	= gypsum + anhydrite	60-100
		Culebra	= dolomite	25-50
		Impersistent	= gypsum + anhydrite	up to 50
		Unnamed	= dolomitic sandstone	30-125
	Salado	—	gypsum + anhydrite	500-700
	Castile	—	gypsum + anhydrite	1100-1300
	Bell Canyon	—	sandstone	700-1200

Three principal developments determined the regional geology:

1. The formation of an inland evaporite basin (the Delaware Basin) during the Permian, with subsequent Permian deposition of chemical and clastic sediments (Snider, 1966).
2. A period of deep weathering with no recorded deposition for 90 million years between the deposition of the Dewey Lake Formation (Upper Permian) and the Cox sandstone of the Cretaceous (field observation). This long Mesozoic period of weathering produced extensive solution of anhydrite and halite, resulting in prolific development of underground caverns and the associated karst topography. The solution activity occurred down to the level of the Castile and Salado contact (field observation).
3. Cenozoic tectonic activity resulting in regional eastward dipping of the sediments and some extensive faulting, striking northeast (Davis and Kirkland, 1970). The eastward dip of the sediments provided the necessary hydrostatic pressure for flow of liquids (e.g., petroleum and water) through porous sandstone aquifers. Upward movement of liquids was possible along the faults, which extended through the petroliferous Bell Canyon sandstone up to its contact with the more flexible Castile anhydrite. The Castile formation is monoclinally folded over the faults and exhibits no actual displacement (field observation).

The Bell Canyon formation is an important aquifer and is also the source of hydrocarbons for numerous oil- and gasfields within the Delaware Basin (Davis and Kirkland, 1970). Movement of the hydrocarbons through gypsiferous sediments can be traced by the presence of isotopically light, secondary calcite. Where faulting has occurred in the Bell Canyon formation, petroleum and ground water have migrated upward to the Castile contact. Solution of the Castile anhydrite and subsequent sulfate-reducing activity have in some cases resulted in a secondary calcite vein extending 20 ft or so down into the Bell Canyon formation, as well as in a sheet of secondary calcite along the Castile/Bell Canyon contact, extending laterally from the fault for up to 1,000 ft (field observation). Frequently, there has been up to 500 ft or so of further penetration of the petroleum into the overlying folded beds of the Castile, resulting in remarkable secondary calcite ridges (where exposed by weathering) that extend northeast/southwest along the faults (field observation). Less frequently, pipe zones (dilatancy) must have developed along the faults, allowing greater vertical penetrations of hydrocarbons and resulting in randomly spaced secondary calcite buttes that occur along the ridges. The major portion of the sulfur once associated with these aboveground secondary calcite ridges and buttes has long since weathered away (only traces of sulfur may be found).

The aforementioned Mesozoic cavern formation in the Salado and Rustler formations may have been the final factor necessary for the formation of large sulfur deposits. A vertically and laterally extensive network of caverns would have provided the perfect environment for upward flow of petroleum into and through the Salado and Rustler beds. These caverns are often marked by Cox sandstone, which was deposited as the Cretaceous sea transgressed upon the karsted land. It is for this reason that Cox sandstone can be found lying beneath the Salado and Rustler formations of the Permian (field and drilling observations).

Where deep caverns intersected the petroleum-bearing pipes extending up from faults in the Bell Canyon formation, extensive formation of sulfur and secondary calcite was highly probable. Unlike the relatively anaerobic system of the Bell Canyon faults and Castile pipes that precludes extensive breakdown of petroleum, the network of caverns (connected to the surface by sink holes) would have provided a better milieu for aerobic degradation of petroleum. More substrate would then have been available for sulfate reduction that would have occurred at the walls of the caverns (presumably in anaerobic microenvironments). Oxygenated surface waters, percolating through the sink holes and caverns, would have oxidized the sulfide (generated by sulfate reduction) to sulfur. This model accommodates the fact that massive accumulations of sulfur are found in the Salado and Rustler formations, but not in the Castile formation.

For a more comprehensive discussion of the regional geology, the reader is referred to Snider (1966), Davis and Kirkland (1970 and 1979), and Kirkland and Evans (1976).

GEOMICROBIOLOGICAL SOIL-SAMPLING PROGRAM

The existence of geochemical anomalies over sulfur occurrences in west Texas has been established by Texasgulf Minerals Exploration Co. (1981). Carbonyl sulfide (COS), carbon disulfide (CS_2), and, more rarely, hydrogen sulfide (H_2S) can be detected at anomalous, though low, concentrations (ppb levels). In a few areas, the flux of sulfur gas is, or has been in the past, high enough to generate an acid soil [via production of sulfuric acid (H_2SO_4) by thiooxidizing bacteria].

Since certain thiooxidizing bacteria are known to utilize sulfur gases for energy (see Miller, 1983, for a review of this subject), a flux of COS, CS_2, and H_2S through the soil would be expected to prompt their growth. The concomitant consumption of the sulfur gases would theoretically lower the net flux of the gas reaching the atmosphere. For this and other reasons (Miller, 1983), an assay for the numbers of thiooxidizing bacteria in a particular soil could be orders of magnitude more sensitive than assays for the sulfur gas itself.

In order to find and enumerate bacteria that might be flourishing as a result of consuming dispersion products from a sulfur occurrence, soil samples were cultured in media containing sulfur compounds as the energy source. Production of colonies (agar media) and oxidation of the sulfur-containing energy source (liquid media) were considered indicative of growth.

An assay for sulfate-reducing bacteria was incorporated into the survey to serve as an index of the local generation of reduced sulfur in the soil. Through the production of H_2S, high numbers of sulfate-reducing bacteria might be expected to prompt the growth of high numbers of thiooxidizing bacteria. The resulting anomaly would be unrelated to the dispersion of sulfur compounds from depth and would, in effect, be a spurious anomaly.

One of the pitfalls of geomicrobiological assays is the number of variables that influence a bacterial ecosystem. These include Eh, pH, and mineral, moisture, organic, and oxygen content. Sampling at constant depth certainly does not ensure constancy in these factors. Rather than attempting to assay all the possible variables, it was thought that a general assay for aerobic heterotrophic microorganisms would reveal the overall status of the soil. Nutrient agar plate counts were included in the survey to meet this requirement.

Soil samples were collected over four different areas: three known sulfur occurrences and one area thought to be barren of sulfur. The important characteristics of each are summarized in Table 27-2. With two exceptions

Table 27-2. Characteristics of the Sampling Areas

Sampling Area	Proximity of S to Surface	Surface Geology	Faulting	Vegetation
Maverick Draw sulfur occurrence	within 200 ft	Dewey Lake and upper Rustler Formation outcrops	not mapped	sparse
U-K sulfur occurrence	within 700 ft	post-Cretaceous gypsum; Cox sandstone	projected from southeast	sparse
Kyle sulfur occurrence	within 250 ft	Dewey Lake and upper Rustler Formation outcrops	not mapped	sparse
High Windmill area	none discovered	Dewey Lake and upper and lower units of the Rustler Formation	projected from southeast	sparse

(as noted in Figures 27-2 and 27-3), the soils were alkaline, ranging in pH from 7.5 to 8.8.

MATERIALS AND METHODS

Aseptic technique was adhered to throughout sample collection and sample processing.

Sample Collection

The sample-collection period extended from June through September 1981. Samples were collected below the grass-root zone, which was generally at a depth of 15-18 in (lower B or upper C horizon). Shallower samples were collected where hardpan prevented deeper digging. All samples were immediately removed from direct sunlight and placed in an ice chest within 2 hr, to be kept refrigerated until prepared for culturing (18 months later).

Sample Preparation

Manual grinding with a mortar and pestle was carried out for 1.5 min on 1 g of the soil sample to give a soil water suspension. Serial dilutions were made from this suspension and were inoculated into liquid media for the three-tube most-probable-number (MPN) technique and onto agar media for colony counts.

Assay Media

Except for the nutrient agar, all of the following media contained 1 ml/l of Tuovinen's trace element solution (Tuovinen and Kelly, 1973):

1. Two liquid thiosulfate-containing media (as per Hutchinson and White, 1965), one supplemented with formate (0.68 g/l) and the other supplemented with biotin (5×10^{-5} M), both brought to pH 7. Positive growth was defined by the consumption of at least 10% of the thiosulfate in the media. The resulting MPN counts for thiosulfate-oxidizing bacteria in each media were then averaged to yield the combined results presented in Figures 27-2 through 27-5. The percent error of the combined values was determined to be ±50%.
2. A solid thiosulfate agar medium at pH 7 (Tuovinen and Kelly, 1973) supplemented with 10^{-6}M biotin, for the purpose of identifying thiosulfate-oxidizing bacteria.
3. A solid sulfide/sulfur agar medium, adapted from Wieringa (1966). This medium was acidified to pH 6 prior to the addition of Oxoid purified agar and the autoclaving process. Growth was identified and bacteria were

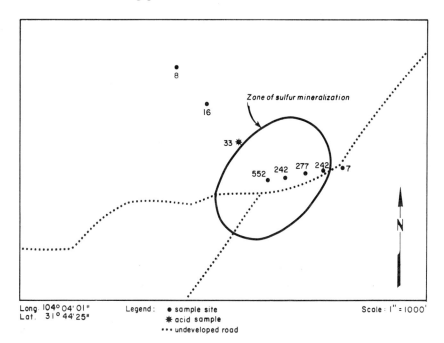

Figure 27-2. Average count of thiosulfate-oxidizing *Thiobacillus* (per gm soil), The Maverick Draw Sulfur Occurrence.

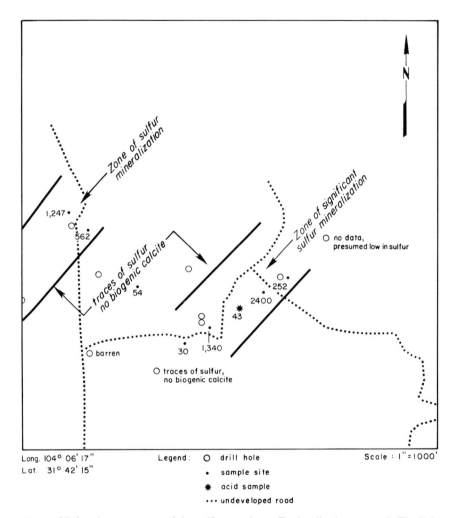

Figure 27-3. Average count of thiosulfate-oxidizing *Thiobacillus* (per gm soil), The Kyle Sulfur Occurrence.

enumerated by the counting of healthy, *Thiobacillus*-like colonies (well-rounded; some coloration; relatively large, at least 1 mm in size). The percent error of the colony enumeration was determined to be ±25% of the log value.

4. Liquid lactate-containing medium, for the enumeration of sulfate-reducing bacteria. This medium was adapted from Postgate (1979) medium E (minus the agar). Growth was considered positive when the medium

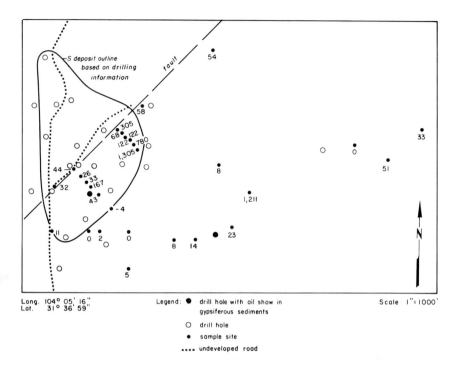

Figure 27-4. Average count of thiosulfate-oxidizing *Thiobacillus* (per gm soil), The U-K Sulfur Occurrence.

turned black. The percent error of the MPN enumeration was determined to be ±30% of the log value.

5. Nutrient agar for the enumeration of aerobic, heterotrophic microorganisms (Pasco Laboratory, Wheatridge, Colorado). All colonies were counted, and the percent error of the count was determined to be ±7% of the log value.

Percent Error

The percent error for each assay was calculated by plotting duplicate versus original values (10) and measuring the percent spread of the resulting points.

RESULTS

The nutrient agar assay revealed that all except the acid soils (noted in Figures 27-2 and 27-3) were similar in their ability to support microbial

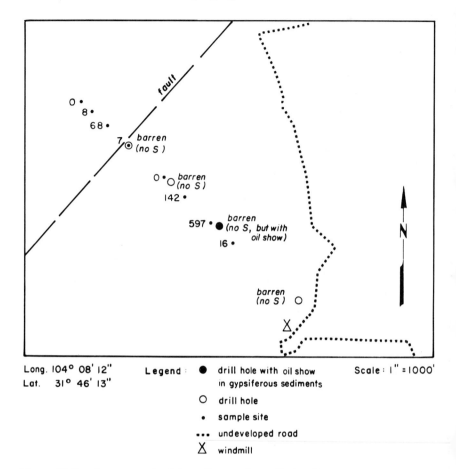

Figure 27-5.   Average count of thiosulfate-oxidizing *Thiobacillus* (per gm soil), The High Windmill Area.

growth. Among the alkaline soils, there were no gross variations in total bacterial counts (approximately 10^5 to 10^6 per g range) or in the basic microbial diversity. *Bacillus* and *Streptomyces* were the two predominant genera present, and no definite pattern was apparent with respect to the predominance of one or the other over sulfur occurrences. Furthermore, there was no correlation (inverse or direct) between the total numbers of microorganisms and numbers of thiosulfate oxidizers or sulfide/sulfur oxidizers.

Sample depth is an important parameter that was varied at three sample sites. Aerobic heterotrophic bacteria increased by up to a factor of 10 between a 15-in sample (lower B, upper C horizon) and a 5-in sample (lower A horizon). Perhaps as a reflection of the obligately aerobic nature of

Streptomyces and the facultatively anaerobic nature of many *Bacillus,* the *Streptomyces/Bacillus* ratio was markedly higher in the 5-in samples than in the 15-in samples. Where sulfate-reducing bacteria were still present at measurable levels in the shallow sample, the count was approximately 1/10 of that present in the deeper sample. Sulfate-reducing bacteria were apparently able to inhabit these predominantly aerobic soils, possibly in anaerobic microenvironments.

The number of sulfate-reducing bacteria exhibited a pronounced low over the sulfur occurrences (average = 2/g) and much higher counts in background areas (average = 20/g, but as high as 1,000/g). A correlation of 0.84 was discovered between the numbers of sulfate-reducing bacteria in background areas and the numbers of *Thiobacillus* adapted to oxidize sulfur, sulfide, or both in background areas. Presumably this correlation is the result of *Thiobacillus* growth on the H_2S produced by sulfate-reducing bacteria. As expected, there was no such correlation over sulfur occurrences, where the supply of sulfur gases coming from depth could prompt the growth of *Thiobacillus* irrespective of the H_2S produced by sulfate-reducing bacteria within the soil. Unexpectedly, there was no correlation between the numbers of sulfate-reducing bacteria and the numbers of thiosulfate-oxidizing bacteria present in the soil, both over the sulfur occurrences and in the background areas.

All of the bacteria isolated from the enrichment media containing thiosulfate or sulfide and sulfur were mixotrophic *Thiobacillus.* Two of the species were tentatively identified as *Thiobacillus A2* (see description of Taylor and Hoare, 1969) and the biotin-requiring *Thiobacillus novellus* (see Matin, Kahan, and Leefeldt, 1980, and the description in Buchanan and Gibbons, 1974).

Thiosulfate-oxidizing *Thiobacillus* appear to correlate positively with the sulfur occurrences (Figs. 27-2 through 27-5). In contrast, *Thiobacillus* growing on sulfide/sulfur agar were high both over the sulfur occurrences and in background areas, probably because of the effect of sulfate-reducing bacteria within the soil. There were some high counts of thiosulfate-oxidizing *Thiobacillus* in background areas (1,211 in Figures 27-4, and 597 in Figure 27-5), but in both these instances the samples were close to drill holes that had intersected oil in gypsiferous sediments. Although no sulfur was intersected, the oil may well have been prompting sulfur-generating activity in adjacent beds along some northeast trending fault.

Low counts of thiosulfate-oxidizing *Thiobacillus* are presented in Figures 27-2 and 27-3 for two acid soils, which most certainly have been subject to a high degree of sulfur gas flux and should contain high numbers of *Thiobacillus.* However, *Thiobacillus* adapted to such acid conditions would not be expected to yield high counts in the neutral media of this survey.

DISCUSSION

It is curious that *Thiobacillus* adapted to oxidize sulfide, sulfur, or both, correlate with counts of sulfate-reducing bacteria, while *Thiobacillus* adapted to oxidize thiosulfate do not. Perhaps the H_2S generated within the soil lacks the opportunity to speciate into more oxidized compounds (e.g., thiosulfate) before it is consumed by sulfide-oxidizing *Thiobacillus* in the soil. In this scenario the ability to oxidize thiosulfate would not develop as frequently as over a sulfur occurrence where, before encountering the first soil *Thiobacillus*, the sulfur gas from the sulfur occurrence has a long distance to travel through the sedimentary column, becoming speciated into more oxidized compounds (eventually thiosulfate) in the process.

The markedly low numbers of sulfate-reducing bacteria over sulfur occurrences was an unexpected finding. One possible explanation relates to the model of sulfur formation. If sulfate-reducing bacteria have consumed, or are currently consuming, suitable fractions of the petroleum in the matrix of the sulfur occurrence, then these fractions are no longer available for dispersion to the surface, to be consumed by (and prompt the growth of) sulfate-reducing bacteria in the overlying soils.

An important lesson of this survey is the obvious need for a detailed understanding of microbial ecology. The ability to characterize a soil comprehensively on the basis of its microbiology would be a tremendous asset to geomicrobiological sampling. Of what significance is the *Streptomyces:Bacillus* ratio in an alkaline, arid soil? It could be a clue to the prevailing oxygen content, the moisture content, and the organic content. Although these factors can be tested independently, they do not provide as much information about the day-in, day-out microenvironment status of the soil as do microorganisms. Furthermore, the effects that microorganisms have on each other are quite important. The antibiotics excreted by *Streptomyces* may adversely affect *Thiobacillus*, or the acid often excreted by *Thiobacillus* may adversely affect *Streptomyces*.

In principle, at least, the overriding determinant of microbial ecology is probably the available energy source. Rarely does an energy source go wasted, as is demonstrated in this survey by the comparatively high counts of *Thiobacillus* found over three sulfur occurrences in west Texas.

ACKNOWLEDGEMENTS

We extend our sincere appreciation to Texasgulf Minerals Exploration Co. for providing us with funding, data, and guidance during the course of this project.

REFERENCES

Buchanan, R. E., and N. E. Gibbons, 1974, *Bergey's Manual of Determinative Microbiology,* Van Nostrand Reinhold Co., Baltimore.

Davis, J. B., 1967, *Petroleum Microbiology,* Elsevier Publishing Co., Amsterdam.

Davis, J. B., and D. W. Kirkland, 1970, Native sulfur deposition in the Castile Formation, Culberson County, Texas, *Econ. Geol.* **65:**107-121.

Davis, J. B., and D. W. Kirkland, 1979, Bioepigenetic sulfur deposits, *Econ. Geol.* **74:**462-468.

Hutchinson, M., K. L. Johnstone, and D. White, 1965, Taxonomy of certain thiobacilli, *J. Gen. Microbiol.* **41:**358.

Ivanov, M. V., 1968, *Microbiological Processes in the Formation of Sulfur Deposits,* CFSTI, Springfield, Virginia, trans. from Russian.

Kirkland, W., and R. Evans, 1976, Origin of limestone buttes, gypsum plain, Culberson County, Texas, *Am. Assoc. Petr. Geol. Bull.* **60:**2005-2018.

Matin, A., F. J. Kahan, and R. H. Leefeldt, 1980, Growth factor requirement of *Thiobacillus novellus, Arch. Microbiol.* **124:**91-95.

Miller, C. L., 1983, *The Geomicrobiology of Sulfur Occurrences in West Texas,* M. S. thesis #2774, Colorado School of Mines, Golden.

Postgate, J. R., 1979, *The Sulfate-Reducing bacteria,* Cambridge University Press, Cambridge.

Snider, H. W., 1966, *Stratigraphy and Associated Tectonics of the Upper Permian Castile-Salado-Rustler Evaporite Complex, Delaware Basin, West Texas and Southeast New Mexico,* Ph.D. thesis, University of New Mexico, Albuquerque.

Taylor, B. F., and D. S. Hoare, 1969, New facultative *Thiobacillus* and a reevaluation of the heterotrophic potential of *Thiobacillus novellus, J. Bacteriol.* **100:**487-497.

Tuovinen, O. H., and D. P. Kelly, 1973, Studies on the growth of *Thiobacillus ferrooxidans, Arch. Microbiol.* **88:**285-298.

Whittenbury, R., and D. P. Kelly, 1977, Autotrophy: A conceptual phoenix, in *Microbial Energetics,* 27th Symposium of the Society for General Microbiology, B. A. Haddock and W. A. Hamilton, eds., Cambridge University Press, Cambridge, England, pp. 121-149.

Wieringa, K. T., 1966, Solid media with elemental sulphur for the detection of sulphur oxidizing microbes, *Antonie van Leeuwenhoek J. Microbiol. Serol.* **32:**183-186.

28

Thermophilic Archaebacteria Occurring in Submarine Hydrothermal Areas

Karl O. Stetter

Universität Regensburg

ABSTRACT: Extremely thermophilic anaerobic archaebacteria have been isolated from submarine solfatara fields in Italy. They are sulfur-hydrogen autotrophs, sulfur respirers, methanogens, and fermentative organisms. The most extremely thermophilic isolates grow between 80° and 110°C, with an optimum around 105°C.

INTRODUCTION

During the last few years, microorganisms growing optimally between 80°C and 88°C have been isolated from continental solfataric hot springs and mud holes. These extremely thermophilic organisms all belong to the archaebacteria, the third kingdom of life (Fox et al., 1980). They are both aerobes and strict anaerobes, representing the new genera *Sulfolobus* (Brock et al., 1972), *Thermoproteus* (Zillig et al., 1981), *Thermofilum* (Zillig et al., 1983a), *Desulfurococcus* (Zillig et al., 1982), and *Methanothermus* (Stetter et al., 1981), and they obtain their energy by sulfur oxidation, hydrogen sulfur autotrophy (Fischer et al., 1983), sulfur respiration, and methanogenesis, respectively. These organisms show an upper temperature limit for growth between 90°C and 97°C, depending on the isolates. Above 100°C and under normal pressure, water exists in the gaseous phase and is therefore not accessible for life. Under elevated pressure, however, water remains liquid even at temperatures above 100°C. Such conditions exist naturally in hydrothermal systems on the sea floor, where the boiling point is raised due to the hydrostatic pressure. At a depth of only 10 m, the boiling point of water is elevated to 120°C. In order to check the possibility of life existing above 100°C, samples were taken from submarine geothermally-heated areas close to Vulcano, Ischia, and Naples, all situated in Italy.

Sampling

Samples from the hot sea sediments, sulfur crusts, and waters were sucked into syringes equipped with an enlarged inlet. Due to the elevated temperatures and the presence of H_2S, practically no oxygen was present in the original sites, as indicated by an immediate spontaneous reduction of the redox dye resazurin added to the samples drawn. Therefore, they were stored anaerobically in stoppered serum bottles (Balch et al., 1979). Oxygen, sometimes invading during the collection procedure, was reduced by addition of sodium dithionite until resazurin became colorless again. The samples were taken to the laboratory without temperature control and were then stored at 4°C.

A New Thermophilic Methanogen. *Methanococcus thermolithotrophicus,* a new thermophilic member of the Methanococcales (Huber, et al., 1982), could be isolated from a marine sediment with an original temperature of 50°C in the bay of Stufe di Nerone, close to Baia (Naples, Italy). This organism is motile due to unusual, short (3 μm) flagella inserted in a ribbonlike area on the cell surface (Fig. 28-1). Growth occurs within a broad range between about 37°C and 70°C, indicating a less extreme thermophily than in *Methanothermus fervidus* (Stetter et al., 1981). The fastest cell multiplication is at 65°C, with a doubling time of only 36 min in mineral medium at optimal hydrogen supply. *Methanococcus thermolithotrophicus* also grows on formate, either in the presence of yeast extract (2 g/l) or of tungsten (1 μmol/l), suggesting that the stimulation by yeast extract may be due to its tungsten content. A stimulatory effect of tungsten and selenium, most likely due to activation of formate dehydrogenase, is already known from studies of *Methanococcus vanniellii* (Jones and Stadtman, 1977). However, selenium did not stimulate growth of *Methanococcus thermolithotrophicus.*

In the presence of molecular sulfur, H_2 and CO_2, *Methanococcus thermolithotrophicus* formed large amounts of H_2S, while methane production decreased drastically (Fig. 28-2). As was the case with molecular sulfur, sulfite (8 mmol/l) was also reduced to H_2S but to a smaller extent, while sulfate and thiosulfate were scarcely reduced, if at all (data not shown). Since the switch to H_2S formation is incomplete (Fig. 28-2), it remains unclear whether cells are able to obtain energy by sulfur reduction. Geothermally heated marine sediments contain up to 20% molecular sulfur (Wauschkuhn and Gröpper, 1975), indicating that this reaction also occurs in the natural habitat. Control experiments revealed that sulfur reduction was a common property in methanogens (Stetter and Gaag, 1983), which, by virtue of this feature, are similar to the anaerobic sulfur-reducing thermophilic archaebacteria (e.g., *Thermoproteus*). Sulfur reduction may possibly be a primitive form of energy conservation in methanogens, a forerunner of energetically more efficient methanogenesis.

Pyrodictium, A Novel Genus of Disc-Shaped Organisms.   Disc-shaped anaerobic bacteria, isolated from hot waters and sediments with original temperatures of up to 103°C, from the sea floor close to Porto di Levante, Vulcano, Italy, could be grown at 100°C in the presence of H_2, CO_2, and molecular sulfur (Stetter, 1982). Simultaneously, large amounts of H_2S were formed, indicating a hydrogen sulfur autotrophy similar to that of *Thermoproteus neutrophilus* (Fischer et al., 1983). Six isolates from enrichment cultures of different samples could be obtained after serial dilution. They all grew between 85°C and 110°C with a temperature optimum around 105°C (110 min doubling time). Even at 110°C, the doubling time was only about 120 min if the cultures remained unshaken. At 120°C no growth could be

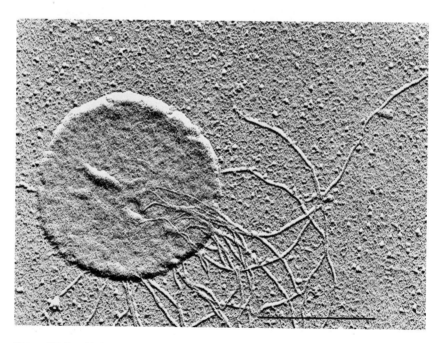

Figure 28–1. *Methanococcus thermolithotrophicus.* Electron micrograph, platinum shadowed. Bar, 1 μm.

observed, possibly because sulfur was less accessible to the organisms. No growth occurred at 80°C or below. However, these bacteria survive at 4°C in the anaerobic state for years. In the presence of oxygen at 4°C, a varying proportion of the population (Stetter, König, and Stackebrandt, 1983) is

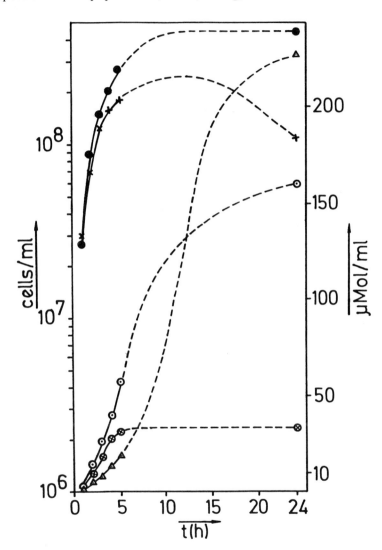

Figure 28-2. Growth of *Methanococcus thermolithotrophicus* on H₂/CO₂ with and without molecular sulfur. ●, growth without sulfur; x, growth in the presence of sulfur; ⊙, methane formation without sulfur; ⊗, methane formation in the presence of sulfur; △, H₂S formation with sulfur. Growth occurred under optimal conditions (Huber et al., 1982; Stetter and Gaag, 1983).

oxygen resistant, enabling the bacteria to survive oxygen-rich cold waters during dispersal on the sea floor. The reason for this oxygen resistance is unclear at the moment. The organisms show an unusually broad salt tolerance between 0.2% and 12% NaCl, with optimal growth around 1%. During growth of mass cultures in a fermenter, pyrite was formed and covered the plastic-coated parts with a rigid layer. It is as yet unclear whether pyrite is formed from the metabolic end product H_2S by reaction with ferrous ions and sulfur without biocatalysis, or whether the cells catalyze this reaction in order to gain additional energy. Deposits of pyrite were recently seen to be formed from molecular sulfur in the hydrothermal system at Porto di Levante (Wauschkuhn and Gröpper, 1975), where these bacteria were isolated. In unshaken cultures, the organisms grew like cobwebs on the layer of sulfur and could be removed by gentle shaking.

Under the phase-contrast microscope, disc-shaped cells varying in diameter between 0.3 and 2.5 μm and about 0.2-μm thick were visible, some having an indented, dishlike appearance. In the electron microscope, the discs appeared to be fixed within a huge network of ultrathin fibers that formed the cobwebs and that was invisible in the light microscope due to the small diameter of the fibers (only 0.04 to 0.08 μm). The single fibers showed a streaky surface structure (Fig. 28-3). They were often associated in bundles.

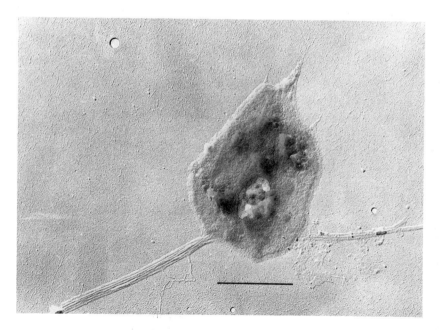

Figure 28-3. *Pyrodictium occultum.* Electron micrograph; platinum shadowed. Bar, 1 μm.

As is the case with flagella, the fibers are hollow and are composed (U. Sleytr, pers. comm.) of protein subunits 5 nm in diameter, arranged in a helical array. They can be digested by proteinase K. However, the fibers cannot be disintegrated even by boiling in the presence of sodium dodecylsulfate at pH 12 (H. König, pers. comm.). The cells are surrounded by a thick envelope of protein subunits, each about 30 nm in diameter (Fig. 28-4), stained by the periodate-Schiff reagent and are, therefore, most likely glycoproteins. Parts of the cells are often extremely flat (Fig. 28-4), possibly in order to increase the surface:volume ratio and, therefore, to enhance metabolism. Uncontrasted cells and fibers were dotted with electron-dense encrustations (Fig. 28-5). Analyses by "Edax" revealed (B. Spray, pers. comm.) that these precipitates consisted of zinc and sulfur, most likely of zinc sulfide, that may have been formed by the reaction of hydrogen sulfide generated by the organisms with the zinc ions present in the culture medium. Therefore, the fibers also seem to participate in metabolism. The encrustations were not observed during growth in the fermenter with constant gassing with H_2/CO_2. The new organisms are archaebacteria due to their possession of phytanyl ether lipids (T. Langworthy, pers. comm.), their lack of murein, and ADP-ribosylation of their ribosomes by diptheria toxin (F. Klinik, pers. comm.). The (still preliminary) 16 S r-RNA catalogue (E. Stackebrandt and C. Woese, pers. comm.) indicates that they represent a third branch of extremely

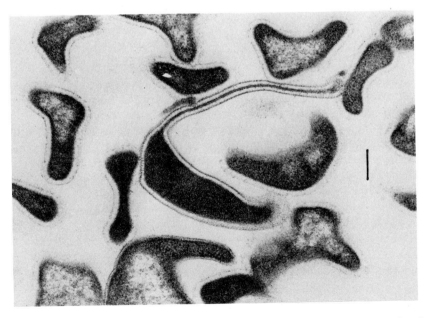

Figure 28–4. *Pyrodictium occultum.* Electron micrograph; thin sections contrasted with lead citrate and uranyl acetate. (Huber et al., 1982). Bar, 0.2 μm.

thermophilic nonmethanogenic archaebacteria and are more closely related to *Sulfolobus* than to *Thermoproteus*. Because of its optimal growth temperature above 100°C and its peculiar shape, we named the new genus *Pyrodictium*, or "fire network." Two species could be distinguished (Stetter, König, and Stackebrandt, 1983): *Pyrodictium occultum*, a strictly autotrophic species with a G + C content of around 62%, and *Pyrodictium brockii*, a mixotrophically growing organism with a G + C content of 52% to 56%. The cell envelopes of *Pyrodictium brockii* did not cross-react serologically with antibodies against those of *Pyrodictium occultum*.

Thermodiscus. *A New Genus of Disc-Shaped Organisms.* Again from the hot sediments of the sea floor off Vulcano, 13 disc-shaped isolates were obtained after enrichment on hydrogen, sulfur, CO_2, and yeast extract at 85°C. Morphologically they are flat discs (Fig. 28-6), highly variable in diameter, and similar to *Pyrodictium*. However, they never show the ultrathin fibers that are characteristic of *Pyrodictium*. Optimal growth occurs at 85°C, and no growth occurs below 70°C. The upper growth limit is around 95°C, the organisms growing either by hydrogen sulfur autotrophy or by sulfur respiration during growth on yeast extract. Some isolates do not form H_2S from yeast

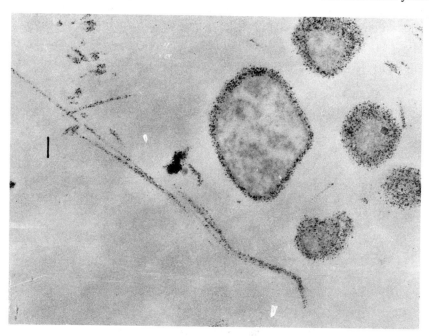

Figure 28-5. *Pyrodictium occultum* grown in serum bottles. Electron micrographs of thin sections, treated as described (Huber et al., 1982), but not contrasted with lead citrate and uranyl acetate. Bar, 0.2 μm.

extract or H_2 in the presence of sulfur, indicating that an electron acceptor other than sulfur is being used. The G + C content of one isolate, S2, is 49%. This isolate grows in a very narrow range of salt concentrations between about 1.5% and 4%. Because of this characteristic, and its flat shape, it was named *Thermodiscus maritimus*. Its preliminary 16 S r-RNA catalogue (E. Stackebrandt, personal communication) shows a relatively close relationship to *Pyrodictium*.

Heterotrophic Sulfur-Oxidizing Microbes. Twenty-four highly irregular lobed or spherical isolates were obtained recently from the hot sea floor off Vulcano and close off shore at Stufe di Nerone, Naples (Fig. 28-7). The isolates grow optimally on protein or yeast extract at 85°C. The temperature maximum is around 92°C. Growth occurs within a salt range between 0.2% and 6% NaCl. Their envelope, consisting of protein subunits, indicates that they too may be archaebacteria. Although differing in their salt requirement, they may belong to the genus *Thermococcus*, which was isolated recently on Vulcano island from a mud hole containing hot sea water (Zillig et al., 1983*b*).

Isolate F1, Fluorescent Nonmethanogenic Archaebacterium. A rod-shaped, extremely thermophilic organism was isolated from the hot sea floor at the

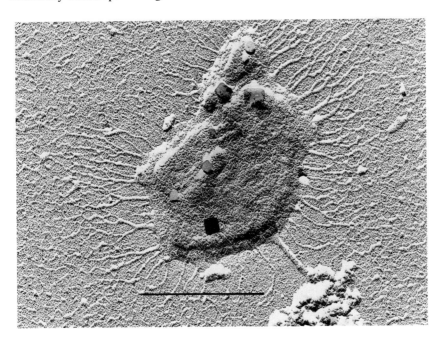

Figure 28-6. *Thermodiscus maritimus.* Electron micrograph; platinum shadowed. Bar, 0.5 μm.

base of the Fossa vulcano close to Faraglione, Vulcano Island. It is flagellate (Fig. 28-8) and has an unusual outer envelope, the tips of which always look empty. Under the UV microscope these organisms show a bright green fluorescence. The lipids (T. Langworthy, pers. comm.) contain phytanylethers, which are typical of archaebacteria. The G + C content of its DNA is only 33%. The isolate grows optimally at around 85°C and shows an upper temperature limit for growth around 97°C. This organism grows either by hydrogen-sulfur-autotrophy or by sulfur respiration on yeast extract.

Isolate MSBB, A Fermentative Organism.   A strictly anaerobic organism growing merely on starch in artificial sea water was isolated very recently. Starch can be substituted with yeast extract or bacterial extracts—for example, from *Methanococcus thermolithotrophicus* or *Pyrodictium occultum*. The rod-shaped cells have an outer envelope that looks as if it is too big (Fig. 28-9). In the stationary growth phase the rods become spheres that are still surrounded by the bag-shaped outer envelope. The G + C content of this organism is 46%. It grows optimally at around 80°C. Growth is obtained at temperatures from about 50°C to 88°C. The taxonomic position of this unusual gram-negative organism is unclear at the moment.

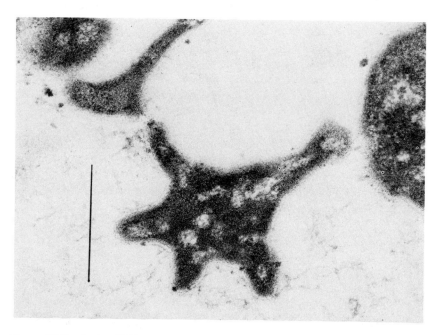

Figure 28-7.   Isolate SDN-1, growing on protein by sulfur respiration. Electron micrograph; platinum shadowed. Bar, 0.5 μm.

The Upper Temperature Limit for Life. Growth of all six *Pyrodictium* isolates at 110°C shows clearly the existence of life about 100°C. The upper temperature limit is still unclear. One prerequisite for the existence of life is the existence of liquid water, which is possible above 100°C under pressures up to the critical point (374.2°C and 225.5 bars at low salt concentration). Above the critical point, water exists in a single phase with about one-third of the density of normal water. At temperatures slightly above 100°C, small molecules like ATP begin to be hydrolyzed rapidly. *Pyrodictium* may have solved this problem with its flat shape and the unusual network-forming fibers that serve to enlarge the cell surface and are possibly a kind of booster power generator used to form additional ATP in order to resynthesize organic matter destroyed by heat.

A paper published recently (Baross and Deming, 1983) claims growth of undefined mixed cultures at up to 300°C and 260 bar. This seems very doubtful, however, since under these conditions various biomolecules are extremely unstable (White, 1983). An extremely thermophilic methanogen isolated from those same geothermal deep-sea vents that were sampled by Baross and Deming (1983) grows only at temperatures up to 95°C and does

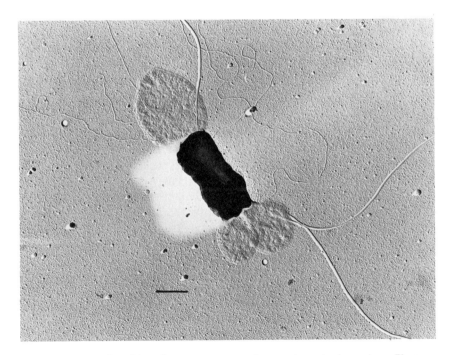

Figure 28–8. Isolate F1, a fluorescent nonmethanogenic archaebacterium. Electron micrograph; platinum shadowed. Bar, 0.5 μm.

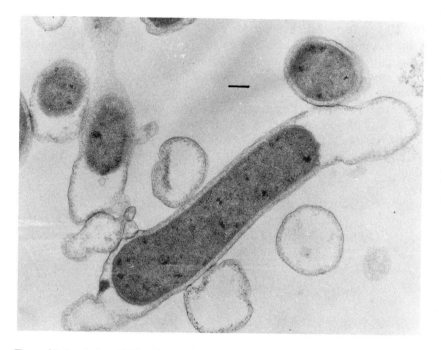

Figure 28-9. Isolate MSB8, a fermentative bacterium. Electron micrograph; thin sections, contrasted with lead citrate and uranyl acetate. Bar, 0.2 μm.

not show any increase of the maximal growth temperature on the application of high pressure (R. Wolfe, pers. comm.).

Macromolecules formed with a negative reaction volume (e.g., DNA) are rapidly hydrolyzed at high pressure and temperature. Preliminary experiments in our laboratory show that, in fact, most of the amino acids are unstable at 250°C and 260 bar, as are the proteins of *Pyrodictium* (Bernhard and König, pers. comm.). Baross and Deming (1983) suggest that high pressure may compensate the destructive force of high temperature on organic material. From the literature, however, it appears that high pressure actually potentiates the destructive effect of high temperatures (Marquis, 1976). Therefore, we believe that the upper temperature limit of life may be much lower, possibly below 150°C, and is dictated not only by the requirement for liquid water, but also and mainly by the stability of the organic components.

ACKNOWLEDGMENTS

I wish to thank Peter Orlean for critical reading of the manuscript.

REFERENCES

Balch, W. E., G. E. Fox, L. J. Magrum, C. R. Woese, and R. S. Wolfe, 1979, Methanogens: Reevaluation of a unique biological group, *Microbiol. Rev.* **43:**260-296.

Baross, J. A., and J. W. Deming, 1983, Growth of "black smoker" bacteria at temperatures of at least 250°C, *Nature* **303:**423-426.

Brock, T. D., K. M. Brock, R. T. Belley, and R. L. Weiss, 1972, *Sulfolobus:* A new genus of sulfur-oxidizing bacteria living at low pH and high temperature, *Arch. Microbiol.* **84:**54-68.

Fischer, F., W. Zillig, K. O. Stetter, and G. Schreiber, 1983, Chemolithoautotrophic metabolism of anaerobic extremely thermophilic archaebacteria, *Nature* **301:**511-513.

Fox, G. E., E. Stackebrandt, R. B. Hespell, J. Gibson, J. Maniloff, T. A. Dyer, R. S. Wolfe, W. E. Balch, R. S. Tanner, L. J. Magrum, L. B. Zablen, R. Blakemore, R. Gupta, L. Bonen, B. J. Lewis, D. A. Stahl, K. R. Luehrsen, K. N. Chen, and C. R. Woese, 1980, The phylogeny of prokaryotes, *Science* **209:**457-463.

Huber, H., M. Thomm, H. König, G. Thies, and K. O. Stetter, 1982, *Methanococcus thermolithotrophicus,* a novel thermophilic lithotrophic methanogen, *Arch. Microbiol.* **132:**47-50.

Jones, J. B., and T. C. Stadtman, 1977, *Methanococcus vannielii:* Culture and effects of selenium and tungsten on growth, *J. Bacteriol.* **130:**1404-1406.

Marquis, R. E.,, 1976, High-pressure microbial physiology, *Adv. Microb. Physiol.* **14:**159-241.

Stetter, K. O., 1982, Ultrathin mycelia-forming organisms from submarine volcanic areas having an optimum growth temperature of 105°C, *Nature* **300:**258-260.

Stetter, K. O., G. Gaag, 1983, Methanogenic bacteria are able to reduce molecular sulfur, *Nature* **305:**309-311.

Stetter, K. O., H. König, and E. Stackebrandt, 1983, *Pyrodictium,* a new genus of submarine disc-shaped sulfur reducing archaebacteria growing optimally at 105°C, *System. Appl. Microbiol.* **4:**535-557.

Stetter, K. O., M. Thomm, J. Winter, G. Wildgruber, H. Huber, W. Zillig, D. Janecovic, H. König, P. Palm, and S. Wunderl, 1981, *Methanothermus fervidus,* a novel extremely thermophilic methanogen isolated from an Icelandic hot spring, *Zbl. Bact. Hyg. I. Abt., Orig.* **C2:**166-178.

Wauschkuhn, A., and H. Gröpper, 1975, Rezente Sulfidbildung auf und bei Vulcano, Äolische Inseln, Italien, *N. Jb. Miner Abh.* **126:**87-111.

White, R. H., 1983, The hydrolytic stability of biomolecules at high temperatures and its implication for life at 250°C or: Life at 250°C—I don't think so!, poster presentation at the 4th International Symposium on Microbial Growth on C$_1$ Compounds, Minneapolis, Minnesota.

Zillig, W., K. O. Stetter, W. Schäfer, D. Janecovic, S. Wunderl, I. Holz, and P. Palm, 1981, *Thermoproteales:* A novel type of extremely thermoacidophilic anaerobic archaebacteria isolated from Islandic solfataras, *Zbl. Bact. Hyg., I. Abt. Orig.* **C 2:**205-227.

Zillig, W., K. O. Stetter, D. Prangishvilli, W. Schäfer, S. Wunderl, D. Janecovic, I. Holz, and P. Palm, 1982, Desulfurococcaceae, the second family of the extremely thermophilic, anaerobic, sulfur-respiring Thermoproteales, *Zbl. Bakt. Hyg., I. Abt. Orig.* **C 3:**304-317.

Zillig, W., A. Gierl, G. Schreiber, S. Wunderl, D. Janecovic, K. O. Stetter, and H. P. Klenk, 1983a, The archaebacterium *Thermofilum pendens* represents a novel genus of thermophilic, anaerobic sulfur-respiring Thermoproteales, *System. Appl. Microbiol.* **4:**79-87.

Zillig, W., I. Holz, D. Janecovic, W. Schäfer, and W. D. Reiter, 1983b, The archaebacterium *Thermococcus celer* represents a novel genus within the thermophilic branch of the archaebacteria, *System. Appl. Microbiol.* **4:**88-94.

29

Modified Sulfur-Containing Media for Studying Sulfur-Oxidizing Microorganisms

James J. Germida

University of Saskatchewan

ABSTRACT: Modified sulfur (S^0)-containing media for the enumeration, isolation, and study of S^0-oxidizing microorganisms are described. Solid or liquid media for chemolithotrophic or heterotrophic microorganisms were prepared with flowable S^0 (FS^0). The addition of different pH indicators increased the usefulness of modified FS^0 media and provided a relative index of microbial S^0-oxidizing ability. Modified liquid FS^0 media were more useful than solid media for the enumeration of S^0-oxidizing populations in agricultural, forest, and S^0-polluted soil samples. Both autotrophic and heterotrophic oganisms grew in the appropriate modified liquid FS^0 media, whereas only heterotrophs and selected autotrophs grew on agar media. Modified solid FS^0 media were very useful for the identification and isolation of S^0-oxidizers. Modified solid FS^0 media also allowed demonstration of synergism between different pairs of microorganisms, which was necessary for them to oxidize FS^0 to SO_4^{-2}. More than 25% of the 56 reference strains tested on modified solid FS^0 media were identified as S^0-oxidizers. Various soil isolates and reference strains of thiobacilli oxidized FS^0 and precipitated S^0 at comparable rates. Based on these results and results from a comparison with other routinely used S^0-containing media, the new modified FS^0 media described should facilitate the study of S^0-oxidizing microorganisms, especially heterotrophs. The advantages and disadvantages of modified FS^0 media are discussed.

INTRODUCTION

Many different microorganisms are capable of oxidizing inorganic sulfur compounds. These include chemolithotrophic organisms such as the thiobacilli

(Vishniac and Santer, 1957; Waksman and Joffe, 1922) and also a variety of heterotrophic bacteria (Swaby and Fedel, 1977; Vitolins and Swaby, 1969), actinomycetes (Wieringa, 1966; Yagi, Katai, and Kimura, 1971) and fungi (Wainwright, 1978b, 1979; Wainright and Killham, 1980). Beneficial uses of sulfur-oxidizing microorganisms include commerical leaching of metals from low-grade ore (Brierley, 1978), reclamation of alkali soils (Rupela and Tauro, 1973), and, when mixed with sulfide minerals, as a fertilizer supplement (Banath, 1969). Detrimental effects of these organisms include acid mine-water pollution of streams (Colmer and Hinkle, 1947) and the deterioration of monument stone (Galizzi and Farrari, 1976).

There is considerable interest in the activity and ecology of S^0-oxidizing microorganisms in nature; however, their study has been hindered by a lack of adequate procedures for their enumeration and isolation from environmental samples. Several interesting ideas, such as enumeration on membrane filters (Tuovinen and Kelly, 1973) or replica plating on thallous sulfide paper (Galizzi and Farrari, 1976), have been proposed for studying thiobacilli. Usually, however, some type of liquid medium containing S^0 or $S_2O_3^{2-}$ is used for the enumeration, enrichment, and isolation of autotrophic and heterotrophic thiobacilli (Postgate, 1966; Vishniac and Santer, 1957; Waksman, 1922). Several new modified media containing precipitated S^0 have recently been developed (Wainright, 1978a; Wieringa, 1966). Nevertheless, more diverse and easily prepared media are needed in order to isolate and identify a wider range of S^0 oxidizers and to study microbial synergism involved in sulfur oxidation.

The present report describes new, modified FS^0-containing media that are easy to prepare and can be adapted for the growth of many different sulfur-oxidizing microorganisms. Of particular interest was the use of these media to study microbial synergism required for complete S^0 oxidation and to identify S^0-oxidizing ability of reference strains and soil isolates.

MATERIALS AND METHODS

Media

Modified S^0 media were prepared by adding flowable, elemental sulfur (FS^0) to a variety of basal media. FS^0 (Stoller Chemical Company, Inc., Houston, Texas) is a creamy liquid suspension (@ 50%-70% wt/vol) of finely divided (i.e., 1- to 2-μm) S^0 that is used as an agricultural sulfur fertilizer (Boone, 1982). FS^0 was washed three times with distilled water to remove the carrier material and other impurities, resuspended in distilled water (approximately 50%-60% S^0 wt/vol), and then sterilized by autoclaving for 15 min at 121°C. Sterilization resulted in some aggregation of the S^0. This aggregate was removed, and hence the concentration of S^0 remaining in suspension was determined by dry weight.

The basal media (without added sulfur) routinely used in this study included: the heterotrophic (HSM) and autotrophic (VSAM) media described by Vishniac and Santer (1957); the American Type Culture Collection (ATCC) media #125, #290, and #528 (ATCC catalog, 1982); Czapek-Dox agar (CzD) at pH 3.5 and 0.3% Trypticase-Soy agar. All media components were from Difco Laboratories, Detroit, Michigan. FS^0 was added to provide a final concentration between 0.1% and 1.0%. For some studies the following pH indicators were added to these media: bromophenol blue (Bpb), 0.005%, pH 4.6-3.0; bromocresol green (Bcg), 0.005%, pH 5.8-3.4; bromothymol blue (Btb), 0.005%, pH 7.0-6.0. The various media were used as liquids or else solidified with Difco Agar (2%).

Enumeration of Sulfur-Oxidizing Populations in Various Soils

The soils used in this study are described in Table 29-2. Autotrophic and heterotrophic sulfur oxidizers were enumerated by spread plate techniques and by MPN determinations. For MPN tests, various selective modified FS^0 media (2.5 ml) were added to 24-well microtiter plates (3.5 ml capacity/well) and inoculated with a 0.1-ml aliquot of the appropriate soil dilution. All samples were incubated at 27°C and checked weekly for sulfur oxidation. Regardless of the technique, a control without FS^0 (or S^0) was used as a check for non-S^0-oxidizing, acid-producing organisms.

Microorganisms

The microorganisms used in this study are listed in Table 29-1. Reference strains were obtained from the American Type Culture Collection (ATCC) or from a departmental culture collection. The bacterial soil isolates BI-2, 3, 4, the fungal isolates FI-1, 6, 8, 20, and the *Streptomyces* spp. A-12, 15, 16, 17 oxidized S^0 and were obtained by plating the soils used in this study on modified FS^0 media. The heterotrophic bacteria and actinomycetes were maintained on glucose, 0.1%, yeast extract, 0.1%; nutrient agar (GYEN) slants; fungi on CzD agar slants; *T. thioparus* and *T. perometabolis* on ATCC #290 and #528 agar slants, respectively; and *T. thiooxidans* in ATCC #125 broth (without S^0) at 5°C.

Sulfur Oxidation Rates in Vitro

The ability of various microoganisms to oxidize different types of S^0 was determined as described by Wainwright and Killham (1980). Stock cultures of selected bacteria and actinomycetes were grown in 50 ml of an appropriate medium, usually HSM (without added S^0). *T. thiooxidans* was grown in

Table 29-1. Microorganisms Tested for Sulfur Oxidation on Modified-Sulfur Media [a]

Bacteria	Fungi
Arthrobacter globiformis ATCC 8010*	Absidia glauca*
Arthrobacter oxydans ATCC 14358*	Aspergillus fumigatus*
Azospirillum brasilense ATCC 29145**	Chrysosporium pannorum
Azospirillum brasilense ATCC 29711**	Chyrsosporium sp.
Azospirillum lipoferum ATCC 29707**	Cladosporium eladum
Azospirillum lipoferum ATCC 29709**	Cladosporium herbarum
Azotobacter chroococcum ATCC 9043**	Cladosporium macrocarpum
Azotobacter vinelandii ATCC 12837**	Fusarium culmorum
Bacillus circulans	Fusarium episphaeria*
Bacillus megaterium	Fusarium roseum
Bacillus polymyxa	Fusarium tricinctum*
Ensifer adhaerens ATCC 33212**	Gliocladium roseum
Ensifer adhaerens ATCC 33499**	Mortierella isabellina*
Enterobacter aerogenes	Mortierella nana
Enterobacter cloacae	Mortierella ramanniana
Micrococcus flavus	Mucor silvaticus
Pseudomonas aeruginosa	Paecilomyces carneus
Psuedomonas fluorescens	Paecilomyces marquandii
Pseudomonas ovalis	Penicillium frequentans
Pseudomonas stutzeri	Penicillium lilacinum
Pseudomonas synxantha	Penicillium nalgiovensis
Pseudomonos sp.	Penicillium pinetorum*
Rhizobium leguminosarum 128C52	Pencillium thomii
Rhizobium meliloti NRG 185	Sporotrichum sp.
Rhizobium trifolii TAl	Trichoderma sp.*
Thiobacillus thiooxidans ATCC 19377	Trichoderma hamatum*
Thiobacillus thioparus ATCC 8158	Trichoderma viride*
Thiobacillus perometabolis ATCC 23370*	Zygorhynchus moelleri*
Soil isolate BI-2*	Zygorhynchus vuilmanii*
Soil isolate BI-3*	Soil isolate FI-1*
Soil isolate BI-4*	Soil isolate FI-6*
	Soil isolate FI-8*
	Soil isolate FI-20*

Actinomycetes

Streptomyces ruber
Streptomyces sp. A-12*
Streptomyces sp. A-15*
Streptpomyces sp. A-16*
Streptomyces sp. A-17*

---

[a] All microorganisms streaked on the appropriate medium containing pH indicator with or without 1.0% FS^0: Media for heterotrophic bacteria and actinomycetes, HSM/Btb; fungi, CzD/Bpb; thiobacilli strains ATCC 19377, 8158, and 23370 streaked on HSM or ATCC media #125, #290, and #528, respectively, each tested with Btb, Bpb, and Bcg as indicators. All plates incubated at 27°C and checked daily for sulfur oxidation.
* Indicates positive test for sulfur oxidation.
**Indicates positive test on media with and without sulfur.

50 ml of ATCC medium #125 with 1% S^0, whereas *T. perometabolis* and *T. thioparus* were grown in ATCC medium #528 or #290 with 1% $Na_2S_2O_3$. All cultures were incubated at 27°C on a gyrotory shaker (150 rpm). Selected fungi were spread onto the surface of CzD agar plates and incubated at 27°C. Stationary-phase broth cultures were harvested by centrifugation, washed three times, and resuspended in sterile water. Fungal spores were harvested by washing plates with 10 ml of sterile water. One ml of washed bacterial or actinomycete cells was added to 50 ml of the appropriate medium (e.g., HSM), whereas 1 ml of fungal spores was added to 50 ml basal medium (BM) (Wainwright and Killham, 1980) containing 1% S^0 or FS^0. All cultures were incubated at 27°C on a gyratory shaker (150 rpm). At various time intervals, cultures were harvested by centrifugation, and the supernate was filtered through a 0.45-μm membrane filter. The filtrates were analyzed for SO_4^{2-} turbidimetrically (Hesse, 1971) and for $S_2O_3^{2-}$ and $S_4O_6^{2-}$ by the colorimetric method of Nor and Tabatabai (1976).

Cross-Streak Experiments

Various soil isolates and named reference strains were cross-streaked on modified sulfur media to study potential synergism necessary for FS^0 oxidation. Each organism was streaked by itself or perpendicular to, but without touching, a second test organism. All plates were incubated at 27°C and examined periodically for growth and FS^0 oxidation.

RESULTS

Flowable Sulfur in Agar Media

Modified solid sulfur media were easy to prepare using FS^0. FS^0 could be sterilized separately by autoclaving, easily diluted to desired concentrations, and added to a variety of basal media. FS^0 was usually added to agar media at concentrations between 0.1% and 1.0%, which would provide an opaque background. This allowed presumptive identification of sulfur-oxidizing microorganisms as they formed a zone of clearing around or beneath the colony (Fig. 29-1). Similar results have been obtained by Wainwright (1978a) and Wieringa (1966) using their precipitated S^0 media. The use of an appropriate basal medium allowed the selection and isolation of a specific microbial population or group of microorganisms capable of S^0 oxidation. There were, however, certain problems with modified FS^0 media. For example, not all organisms capable of oxidizing FS^0 produced distinct clear zones around colony growth because some organisms oxidized FS^0 slowly, and overgrowth of nonoxidizing organisms interfered with observations. This was especially true on spread plates of low soil dilutions, and, in some cases, it was difficult to identify which organism was responsible for the zone of clearing because, as will be shown later, two or more organisms were involved in the oxidation of FS^0.

The addition of various pH indicators to modified solid FS^0 media greatly facilitated the identification of S^0-oxidizing microorganisms. Because of the different sensitivity and pH range of the indicators used, choice of an appropriate indicator allowed one to select or differentiate the degree of oxidation as strong, medium, weak, or none. Unfortunately, those organisms that oxidized FS^0 to $S_2O_3^{2-}$ or $S_4O_6^{2-}$ or only produced small quantities of

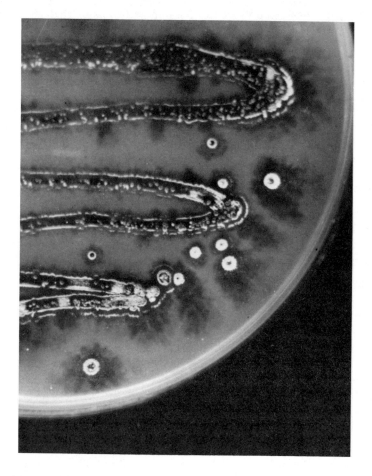

Figure 29-1. *Streptomyces* soil isolate sp. A-12 growing on 0.3% Trypticase-Soy agar containing 0.1% FS^0 after 21 days incubation. Note zones of clearing in sulfur around colonies.

SO_4^{2-} did not change the indicator color and were not identified as S^0 oxidizers. However, in many instances, at least on spread plates of soil dilutions, these organisms were in close association with other organisms that oxidized $S_2O_3^{2-}$ to SO_4^{2-}, and hence the combined action of these two organisms was detected on modified FS^0 media. For example, on several occasions it was observed that when soil dilutions were spread-plated on HSM/Btb/FS^0 agar medium, putative S^0-oxidizing isolates were identified. When these colonies were isolated and then retested on the same isolation medium, they did not cause a change in the pH indicator color, which would have indicated oxidation of FS^0 to SO_4^{2-}. Subsequent cross-streak experiments with several isolates from within or near the zone of sulfur oxidation on the original plate demonstrated that synergism of two or more organisms was required in order to oxidize FS^0 to SO_4^{2-} and change the color of the pH indicator (Fig. 29-2).

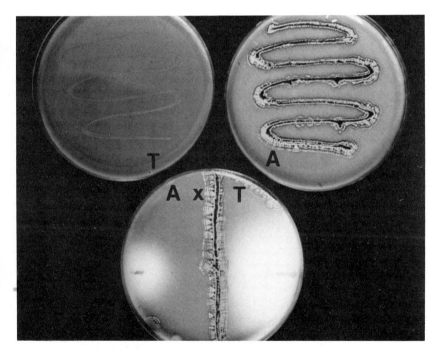

Figure 29-2. Cross-streak of *Streptomyces* soil isolate sp. A-16 and *Thiobacillus perometabolis* on HSM/Bcg/FS^0; plates streaked with A-16 (A); *T. perometabolis* (T) or A-16 x *T. perometabolis* (AXT); Initial medium color was opaque blue (pH 6.8), note the white zone (bright yellow color on plate) on cross-streak (AXT) around *T. perometabolis* growth where pH has dropped to 3.8; 21 days incubation.

Table 29-2. Most Probable Number of Sulfur-Oxidizing Microorganisms in Various Soils

Soil Site[b]	pH	125/Bpb/FS[0]	MPN/g Soil Medium[a] 125/Bcg/FS[0]	HSM/Btb/FS[0]
HS−I	3.4	4×10^4	3.5×10^6	NT[c]
APS/LFH1	1.3	1.2×10^5	15	14
APS/Ae1	3.6	5.2×10^4	5	7
APS/LFH2	3.5	4.1×10^4	7.5×10^4	25
APS/Ae2	4.2	6.5×10^3	2.2×10^3	25
DB-1	7.2	ND[d]	ND	5.6×10^2
GW-1	7.6	ND	ND	1.9×10^3

[a]See materials and methods.
[b]HS-I−Brown Chernozemic Soil polluted with S[0] from University of Saskatchewan field plot; APS 1 and 2, litter forest horizons and Ae horizons next to poured S[0] block in Alberta, 50 m and 250 m from block, respectively; DB-1 and GW-1−wheat fields on Dark Brown Chernozemic and Grey Wooded soils, respectively, in Saskatchewan.
[c]Not tested.
[d]None detected.

Flowable Sulfur in Liquid Media

FS^0 was also used in liquid media for MPN determination of S^0-oxidizing populations in various soil samples (Table 29-2). S^0-polluted soil samples contained a considerable number of autotrophic thiobacilli but only a few heterotrophic S^0-oxidizers. The picture was reversed, however, for agricultural soils (Table 29-2) that contained no detectable autotrophic thiobacilli. FS^0 was not as useful as S^0 when test tubes were used for the MPN determination (because FS^0 settled out of nonagitated broth) but gave results equivalent to, or better than, S^0 when microtiter plates were employed (due to surface-volume ratio). This observation was verified using a pure culture of *T. thiooxidans* ATCC 19377. Similar results were obtained for MPN determinations using heterotrophic modified FS^0 media. Non-S^0-oxidizing microorganisms accounted for about 20% of the positive wells with media such as HSM/Btb/FS^0.

Comparison of Modified Media with Other Sulfur-Containing Media

Modified FS^0 media were compared with the other media (Vishniac and Santer, 1957; Wainwright, 1978a; Waksman 1922; Wieringa, 1966) routinely used for enumeration and isolation of S^0-oxidizers. Modified liquid FS^0 media were comparable to these other media for MPN enumeration of S^0-oxidizers. Similar results were obtained on modified solid FS^0 media. For example, the total microbial population of the soil sample HI-S (Table 29-2) on modified CzD/Bpb/FS_3^0 was 7.2×10^5, and the number of sulfur oxidizers was 6×10^3. On the medium of Wainwright (1978a), the numbers were 2.3×10^7 and 5×10^3, respectively. The lower total count on CzD/Bpb/FS^0 can be explained by the fact that this medium is selective for fungi, based on its low pH (i.e., 3.5), whereas Wainwright's medium supports some bacterial growth.

Modified Sulfur Media for Identification of Sulfur-Oxidizing Ability

A number of reference strains and soil isolates were tested on modified FS^0 media for S^0 oxidation (Table 29-1). None of the reference strains of heterotrophic bacteria tested gave a clear-cut response indicating sulfur oxidation. About 35% of these strains, however, exhibited a positive response on both the control and FS^0-containing media (Table 29-1). When tested in broth culture, only strain NRG 185 oxidized FS^0 to SO_4^{2-}; strains ATCC 29145, 29709, and 33212 produced $S_2O_3^{2-}$. *T. thiooxidans* did not grow on any of the modified FS^0 agar media tested, and *T. thioparus* did not grow on modified solid FS^0 media unless another organism capable of oxidizing FS^0 to $S_2O_3^{2-}$ was present. *T. perometabolis* grew on a variety of solid FS^0 agar media and was able to change different pH indicators (e.g., Bcg) but required a long incubation period (28 days). All soil isolates originally isolated on modified FS^0 medium gave a positive response indicating FS^0 oxidation. The actinomycete soil isolates required very long incubation periods before a slight change in pH was noted. Subsequent studies (Table 29-3) showed that these organisms produced mostly $S_2O_3^{2-}$ and little SO_4^{2-} during the oxidation of FS^0. The majority of the heterotrophic S^0 oxidizers were fungi, although the time required for isolates to change the pH indicators ranged from 6 to 37 days. These fungal isolates produced different amounts of SO_4^{2-} when oxidizing FS^0 (Table 29-3), which would account for the different time intervals required for a positive test.

Sulfur Oxidation Rates in Vitro

The ability of various reference strains and soil isolates to oxidize precipitated (i.e., powder) and flowable S^0 was assessed in broth culture. Most organisms oxidized both types of S^0 at comparable rates (Table 29-3). *T. thioparus* and *T. thiooxidans* oxidized FS^0 in broth, although neither organism grew on modified solid FS^0 media. Many of the heterotrophic soil isolates oxidized either type of S^0 to $S_2O_3^{2-}$ or $S_4O_6^{2-}$ with little production of SO_4^{2-}. Some of the fungal isolates were highly active S^0 oxidizers, but this was not unexpected based on the work of Wainwright (1978b, 1979) and Wainwright and Killham (1980). One soil isolate, FI-1, appeared to prefer FS^0 over precipitated S^0 (Table 29-3), and it is possible that other isolates not tested exhibit similar preferences for either sulfur source or basal medium.

Table 29-3. Oxidation of Precipitated and Flowable Sulfur by Various Microorganisms[a]

Organism	Medium[c]	$S_2O_3^{2-}$	$S_4O_6^{2-}$	SO_4^{2-}
		$\mu g\text{-}S\ ml^{b}$		
T. thiooxidans	125/S^0	0	0	10,110
ATCC 19377	125/FS^0	0	0	8,570
T. thioparus	125/S^0	0	0	135
ATCC 8158	125/FS^0	0	0	80
T. perometabolis	528/S^0	195	19	50
ATCC 23370	528/FS^0	3	0	73
Bacterial soil	HSM/S^0	145	4	0
Isolate BI-2	HSM/FS^0	146	4	0
Bacterial soil	HSM/S^0	16	14	0
Isolate BI-3	HSM/FS^0	12	11	0
Bacterial soil	HSM/S^0	167	5	0
Isolate BI-4	HSM/FS^0	155	9	0
Streptomyces	HSM/S^0	285	9	0
sp. A-15	HSM/FS^0	60	10	0
Streptomyces	HSM/S^0	230	10	14
sp. A-16	HSM/FS^0	135	10	14
Streptomyces	BM/S^0	25	8	0
sp. A-16	BM/FS^0	23	7	2
Fungal soil	BM/S^0	322	247	537
Isolate FI-1	BM/FS^0	410	410	883
Fungal soil	BM/S^0	13	10	0
Isolate FI-6	BM/FS^0	110	85	33
Fungal soil	BM/S^0	595	86	25
Isolate FI-8	BM/FS^0	195	7	1
Fungal soil	BM/S^0	258	148	71
Isolate FI-20	BM/FS^0	220	200	210

[a]50 ml of the appropriate medium containing 1% precipitated (S^0) or flowable (FS^0) sulfur were inoculated with 1 ml of a washed microbial culture and then incubated at 27°C on a gyratory shaker (150 rpm) for 14 days before analysis; T. thiooxidans and T. thioparus cultures incubated 10 and 28 days, respectively, before analysis.
[b]Concentrations of S-ions in excess of noninoculated controls containing appropriate sulfur source; means of triplicate determinations.
[c]See materials and methods.

DISCUSSION

The present study describes new modified FS^0-containing media that were easy to prepare and could be adapted for the growth of a wide range of microorganisms. These media were used to enumerate S^0-oxidizing populations in soil samples, and the results obtained were comparable to those obtained with other routinely used S^0 media (Vishniac and Santer, 1957; Wainwright, 1978a; Wieringa, 1966). The advantages of using FS^0 for modified sulfur media were: (1) FS^0 was easy to manipulate, and known quantities could be added to a variety of basal media that allowed selection or enrichment of specific S^0-oxidizing microorganisms; (2) since FS^0 was sterilized separately, S^0 media did not require Tyndallization (this was especially useful for MPN tests); (3) FS^0 was oxidized by many different microbes at rates comparable to precipitated S^0 powder; (4) FS^0 was readily utilized and, in fact, preferred by some heterotrophs; (5) FS^0-containing media were useful for demonstrating microbial synergism required for S^0 oxidation. The disadvantages of using modified S^0 media were: (1) FS^0 settled out of nonagitated liquid cultures; hence microtiter plates were required for MPN determinations; (2) not all thiobacilli, such as *T. thiooxidans,* grew on modified solid FS^0 media; (3) the surface area of the FS^0 may be altered by autoclaving; (4) modified FS^0 media containing pH indicators may not identify organisms that oxidize FS^0 to $S_2O_3^{2-}$ or $S_4O_6^{2-}$.

The major limitation with modified FS^0 media, which caused problems when testing microbial cultures for S^0-oxidizing ability, involved S^0-oxidizers that produced $S_2O_3^{2-}$ and $S_4O_6^{2-}$. Basically, there were three types of re-actions that could be observed on modified FS^0 media containing pH indicators: (1) a change in the pH of the medium containing FS^0; (2) a pH change occurred for media with and without FS^0; (3) no pH change was observed on either medium. Most of the positive (i.e., S^0-oxidizing) fungal isolates were representative of this first category. Many of the heterotrophic bacteria tested produced some acidic reaction other than from S^0 oxidation (which was typical for the second category); these organisms needed to be checked in broth culture, as many different isolates not only produced acid but also oxidized S^0. Many organisms oxidized S^0 to $S_2O_3^{2-}$ or $S_4O_6^{2-}$, or formed small amounts of SO_4^{2-} (category 3). These organisms would not be readily identified on modified FS^0 media since they did not cause a pH change, or else the change was very small and required a long incubation period. Nevertheless, the S^0-oxidizing ability of these organisms was easily checked by including a simple cross-streak of a $S_2O_3^{2-}$-oxidizing organism such as *T. perometabolis* or *T. thioparus,* or by streaking on modified FS^0 media without pH indicators.

Modified FS^0 media should facilitate the study of S^0 oxidation in a variety of natural habitats because it will allow new S^0 oxidizers, especially heterotrophs, to be isolated. In addition, the use of modified FS^0 media for studying microbial synergism for S^0 oxidation is an advantage that other media do not have. Finally, as more information is obtained concerning the choice of basal media and solidifying agents, the usefulness of modified FS^0 media as proposed in this study will increase.

ACKNOWLEDGEMENTS

This work was supported by grants from the Natural Sciences and Engineering Research Council of Canada and the Sulphur Development Institute of Canada. The technical assistance of N. E. Howse and E. E. Onofriechuk is appreciated. Contribution No. R354, Saskatchewan Institute of Pedology.

REFERENCES

American Type Culture Collection, 1982, *Catalogue of Strains.* I., 15th ed., ATCC, Rockville, Maryland.

Banath, C. L., 1969, Iron pyrites as a sulphur fertilizer, *Aust. J. Agric. Res.* **20:**697-708.

Boone, D. H., 1982, Sulphur-containing products for agriculture, in *Proceedings of the International Sulphur '82 Conference,* A. I. More, ed., London, England: The British Sulphur Corporation Ltd., pp. 519-522.

Brierley, C. L., 1978, Bacterial leaching, *Crit. Rev. in Microbiol.* **6:**207-262.

Colmer, A. R., and M. E. Hinkle, 1947, The role of microorganisms in acid mine drainage: A preliminary report, *Science* **106:**253-256.

Galizzi, A., and E. Ferrari, 1976, Identification of thiobacilli by replica plating on thallous sulfide paper, *Appl. Environ. Microbiol.* **32:**433-435.

Hesse, P. R., 1971, *A Textbook of Soil Chemical Analysis,* Murray, London.

Nor, Y. M., and M. A. Tabatabai, 1976, Extraction and colorimetric determination of thiosulfate and tetrathionate in soils, *Soil Sci.* **122:**171-178.

Postgate, J. R., 1966, Media for sulphur bacteria, *Lab. Pract.* **15:**1239-1244.

Rupela, O. P., and P. Tauro, 1973, Utilization of *Thiobacillus* to reclaim alkali soils, *Soil Biol. Biochem.* **5:**889-901.

Swaby, R. J., and R. Fedel, 1977, Sulphur oxidation and reduction in some tropical Australian soils, *Soil Biol. Biochem.* **9:**327-331.

Tuovinen, O. H., and D. P. Kelly, 1973, Studies on the growth of *Thiobacillus ferrooxidans.* I. Use of membrane filters and ferrous iron agar to determine viable numbers and comparison with $^{14}CO_2$-fixation and iron oxidation as measures of growth, *Arch. Mikrobiol.* **88:**285-298.

Vishniac, W., and M. Santer, 1957, The thiobacilli, *Bacteriol. Rev.* **21:**195-213.

Vitolins, M. I., and R. J. Swaby, 1969, Activity of sulphur-oxidizing microorganisms in some Australian soils, *Aust. J. Soil Res.* **7:**171-183.

Wainwright, M., 1978a, A modified sulphur medium for the isolation of sulphur-oxidizing fungi, *Plant and Soil* **49:**191-193.

Wainwright, M., 1978b, Microbial sulphur oxidation in soil, *Sci. Prog. Oxf.* **65:**459-475.

Wainwright, M., 1979, Microbial S-oxidation in soils exposed to heavy atmospheric pollution, *Soil Biol. Biochem.* **11:**95-98.

Wainwright, M., and K. Killham, 1980, Sulphur oxidation by *Fusarium solani, Soil Biol. Biochem.* **12:**555-558.

Waksman, S. A., 1922, Microorganisms concerned in the oxidation of sulfur in soils: III. Media used for the isolation of sulfur bacteria from soil, *Soil Science* **13:**329-336.

Waksman, S. A., and J. S. Joffe, 1922, Microorganisms concerned in the oxidation of sulfur in soils: II. *Thiobacillus thiooxidans,* a new sulfur-oxidizing organism isolated from soil, *J. Bacteriol.* **7:**239-256.

Wieringa, K. T., 1966, Solid media with elemental sulphur for detection of sulphur-oxidizing microbes, *Antonie van Leeuwenhoek J. Microbiol. Serol.* **32:**183-186.

Yagi, S., S. Katai, and T. Kimura, 1971, Oxidation of elemental sulfur to thiosulfate by *Streptomyces, Appl. Microbiol.* **22:**157-159.

Part V

GROUND WATER

30

Geochemical Conditions in the Ground Water Environment

Georg Matthess
Kiel University

ABSTRACT: Ground water is a part of the hydrological cycle. It occurs below the water table in permeable geologic formations (aquifers). The hydrogeological conditions are defined by the hydraulic conductivity and porosity of the aquifers, the presence of impermeable rocks, the position of the water table, and the hydraulic gradient. The residence time of ground water is controlled by the flow velocities and the dimensions of the aquifer system.

The geochemical properties of ground water are described by temperature, pH values, and general data of the dissolved constituents. They are controlled by the nature of soils and rocks through which the water passes along its subterranean flow path and by various geochemical, physical, biochemical, and biophysical processes. The concentration of dissolved substances is generally due to dissolution, degradation, and hydrolytic processes, which are frequently controlled by pH and Eh. The redox conditions in ground water are dependent on oxygen-consuming reactions involving the breakdown of organic substances and the oxidation of reduced inorganic substances. Microorganisms play an important part in these reactions. The efficiency of the oxygen supply from the atmosphere due to diffusion and flow dispersion and the rate of oxygen-consuming processes control whether there are anaerobic or aerobic conditions in the ground water.

INTRODUCTION

Ground water takes part generally in the hydrologic cycle. It is recharged by infiltration of precipitation water and, when a respective hydraulic gradient is provided, by effluent seepage from surface water. It emerges, after a more-or-less long subterranean flow path, in the form of springs, or it seeps directly into streams, ponds, lakes, or oceans.

In hydrology the subsurface water system is subdivided into saturated and unsaturated zones. In the saturated zone below the water table, the voids of the earth's crust are filled by ground water, the behavior of which is controlled only by gravity. In the unsaturated zone the voids are filled partially by air (called ground air) and partially by water. A part of this water is bound by adhesive forces to the surface of the solid material, and a part is held by surface tension and moves by capillary action. Any excess water— for example, infiltrated water after heavy rainfalls or after snow melt— moves due to gravity and may reach the water table. The thickness of the unsaturated zone depends on the climate and the topography. In lowlands and in humid climates, the thickness may be zero or only centimeters; in arid zones and mountain areas, the thickness may range up to hundreds or even thousands of meters.

Ground water flows in permeable geologic formations known as aquifers, which may be subdivided into porous aquifers, fractured aquifers, and karstic aquifers. In porous aquifers—that is, sands and gravels—ground water flows in pores that form a dense network of interstices. The flow velocities range from fractions of 1 m/day up to a few m/day. Velocities of greater than 10 m/day are rare. In fractured rocks, ground-water movement is restricted to fractures, fissures, and cracks produced by mechanical stresses in any hard rock or by contraction of cooling magmas in magmatites. The fractures are enlarged by solution processes in karstic rocks, such as limestones and dolomites, where solution holes with lengths of several kilometers may occur. In fractured and karstic rocks, ground-water flow velocities range from less than 1 m/day up to tens or hundreds of m/day. Extreme velocities of 8000 m/day and 26,000 m/day have been reported. The velocity decreases generally with increasing depth. The residence time of ground water ranges from a few days to tens of thousands of years depending on the flow velocity and the dimensions of the aquifer system. The distance between recharge and discharge areas and aquifer thickness are also important parameters. Nace (1967) quotes a mean residence time of 5000 yr for ground water between 0 and 4000 m depth. Some deep-lying bodies of water are cut off from the hydrological cycle by practically impermeable rocks. These stagnant waters are called fossil waters if they originate in pre-Holocene times. Some of these waters (the "connate" waters) have been syngenetically incorporated in ancient sediments (Matthess, 1982).

The distribution and the flow velocity are controlled by the hydraulic conductivity and porosity of the rocks and by the local hydraulic gradient of the water table. In addition to the aquifers that generally have appreciable permeabilities and porosities, formations occur that are characterized by low permeabilities. These formations are incapable of transmitting ground water. They may be subdivided according to their porosity into aquicludes, containing water in very minute pores (e.g., clays), and aquifuges, which are practically free of voids (e.g., solid granite).

Ground water, which is in contact with the atmosphere through open spaces in the unsaturated zone, is called unconfined water. Ground water separated from the atmosphere by impermeable layers is referred to as confined water or artesian water (Freeze and Cherry, 1979; Matthess and Ubell, 1983).

THERMAL AND CHEMICAL PROPERTIES OF GROUND WATER

The temperature and chemical composition of ground waters are generally constant; this characteristic makes ground waters distinctly different from surface waters of streams and lakes. The natural quality of ground water depends on the origin of recharge water (i.e., infiltrating precipitation or surface water), the soil surface cover, the nature of the local rocks, and the chemical, physical, and microbial reactions in the unsaturated zone and the aquifer.

Temperature

Ground-water temperature is the result of the heat exchange of the earth's surface controlled by incoming and outgoing radiation, heat conduction, convection, evaporation, and chemical and thermonuclear processes in the rocks. At a depth of about 15–39 m, the temperature variations of the earth's surface disappear. Here the ground-water temperature reflects the mean annual air temperature. Below this neutral zone, water temperature rises with increasing depth according to the geothermal gradient, which is generally 33 m/°C, and ranges between about 10 m/°C in volcanic areas and 100 m/°C in shield areas (Matthess, 1982).

pH of Ground Water

Natural waters usually contain dissolved CO_2 and hydrogen carbonate ions that form a buffered system with the carbon dioxide. Therefore, pH varies in a narrow range between 5.0 and 8.0. Higher pH values (>8.5 pH) are usually associated with high sodium carbonate contents (e.g., pH 11.6 in a Na-Cl-CO_3 water from Mount Shasta, California (Feth, Rogers, and Roberson, 1961). Very low values (<4.0 pH) occur in acid mine waters or in thermal waters that contain H_2S, HCl, and other volatile constituents of volcanic origin (e.g., 0.4 pH in a thermal water from Yakeyama, Niigata Prefecture, Honshu, Japan (White and Waring, 1963)). Low pH values can be attributed to small amounts of sulfuric acid from the oxidation of metal sulfides or to organic acids produced by humidification (Matthess, 1982).

Dissolved Constituents

The total dissolved solids (TDS) comprise dissociated and undissociated substances. TDS of ground water can amount to less than 25 mg/l in water in humid regions with relatively insoluble rocks, and to more than 300,000 mg/l in brines. On the basis of the TDS content, ground water may be classified into fresh water ($\leq$1000 mg/kg), brackish water (1000-10,000 mg/kg), salty water (10,000-100,000 mg/kg), and brines (>100,000 mg/kg) (Davis and DeWiest, 1967).

According to their abundance in potable water, major constituents (1.0-1000 mg/kg; Na^+, Ca^{2+}, Mg^{2+}, HCO_3^-, SO_4^{2-}, SiO_2), secondary constituents (0.01-10.0 mg/kg; Fe^{2+}, Fe^{3+}, Sr^{2+}, K^+, CO_3^{2-}, NO_3^-, F^-, B), and minor and trace constituents (<0.1 mg/kg; e.g., heavy metals, phosphate) may be discriminated (Davis and DeWiest, 1967). More data on the abundance and behavior of these constituents are found in Hem (1970) and Matthess (1982). Dissolved organic substances contribute in only minor amounts to the TDS. Their composition is discussed in the following chapter.

The total dissolved solids do not include suspended sediment, colloids, or dissolved gases. Among these constituents, dissolved oxygen and carbon dioxide are most important for the geochemical conditions in a given ground water.

GEOCHEMICAL AND PHYSICAL PROCESSES

The properties of ground water are the results of processes occurring during its subsurface passage and are listed in Table 30-1. They are discussed in more detail insofar as they are of direct importance to the subject of this symposium. (For further information, see Stumm and Morgan, 1981; and Matthess, 1982.)

Geochemical Processes

Dissolution-Precipitation. Dissolution of solids may occur in the atmosphere (aerosols), on the ground surface, and in the unsaturated and saturated zones. Resulting concentrations of dissolved solids are due to the abundance and the solubility of the constitutent substances, the extent of the solid-water interface, and the contact time. Dissolution includes degradation and hydrolytic processes. The ability of water to dissolve substances is increased by inorganic and organic acids and generally by an increase in temperature. The solubility of numerous substances is dependent on pressure, temperature, pH, and Eh, and co-solutes can be modeled using chemical thermodynamics. The dependence of the solubility of some elements

Table 30-1. Processes Controlling Ground Water Properties

Geochemical Processes	Physical Processes
dissolution-precipitation	dispersion
acid-base reaction	filtration
oxidation-reduction	evaporation
complexation	gas transport
adsorption-desorption	
Biochemical and	
Microbiological Processes	Biophysical Processes
decomposition of organics	transport of microorganisms
cell synthesis	filtration of microorganisms
transpiration	

(e.g., iron and manganese) on pH and Eh is demonstrated by stability field diagrams (Garrels and Christ, 1965; Stumm and Morgan, 1981).

The most important soluble salts occurring in relatively large quantities in rocks include $CaCO_3$, $CaCO_3 \cdot MgCO_3$, $NaCl$, $CaSO_4$ and $CaSO_4 \cdot 2H_2O$. The easily solubilized chlorides and sulfates are leached from soil and rocks by the seepage and ground water. Dissolution of the less-soluble calcium and magnesium carbonates requires the presence of some free acids. Characteristic acids that are involved in forming the typical ground water properties are carbonic acid, sulfuric acid, nitric acid, humic acids, simple fatty acids, naphthenic acids, and tannic acids. The organic components occur mainly in the weathering zone due to biochemical processes. The organic acids are unstable and seem to decay very soon in the subterranean flow path. Hydrochloric acid, carbonic acid, and sulfuric acid are present in volcanic emissions. The inorganic and organic acids attack the less-soluble calcium and magnesium carbonates and other minerals of the rocks that are degraded, thus setting free cations such as Na^+, K^+, Ca^{2+}, Mg^{2+}, Fe^{2+}, and Mn^{2+}. A typical reaction is

$$x CaCO_3 + (x + y)\, CO_2 + x H_2O \rightarrow x Ca(HCO_3)_2 + y CO_2 \qquad \text{(Eq. 30-1)}$$

This reaction demonstrates the necessity of excess CO_2 in this process. The main source of carbon dioxide in ground water is the ground air in which CO_2 originates from root respiration and from microbial degradation of organic matter in the soil. The CO_2 partial pressure in the ground air is 10 to 100 times higher than in the atmosphere (Matthess, 1982).

Hydrolysis is a very important process in the breakdown of silicates under the influence of H^+ and OH^- ions. It causes the formation of new minerals and releases cations such as Na^+, K^+, Ca^{2+}, and Mg^{2+}. A typical reaction is the transformation of albite into kaolinite:

$$2 NaAlSi_3O_8 + 2H^+ + 9H_2O \rightarrow Al_2Si_2O_5 (OH)_4 + 4H_4SiO_4 + 2Na^+ \qquad \text{(Eq. 30-2)}$$

Dissolved materials may be precipitated along subterranean flow paths when their solubility product is exceeded because of concentration by evaporation and transpiration or the admixture of ground water with differing chemical composition, pH, and Eh.

Oxidation-Reduction.   The redox conditions in ground water are dependent on oxygen-consuming reactions involving the oxidation of reduced inorganic substances (e.g., Fe^{2+}, Mn^{2+}, S^{2-}, and NH_4^+) and the breakdown of organic substances present in the aquifer system in dissolved state or as separate phases. Besides the organic acids and humic substances mentioned above, hydrocarbons, phenols, aromatic oxygen compounds, asphaltic compounds, esters, polycyclic hydrocarbons, carbohydrates, amino acids, and various degradation products have been observed in natural ground-water systems (Matthess, 1982). They originate either from the active soil zone or from aquifer materials. The dissolved organic substances are determined mostly as total organic carbon (TOC) or dissolved organic carbon (DOC). In natural ground waters TOC usually is only a few mg/l, but in ground water in contact with oil or peat, much higher concentrations are observed (e.g., hundreds of mg/l). The content of organic substances present as separate phases in aquifers varies greatly with the nature of rocks and usually increases with decreasing grain size. Wedepohl (1978) quotes mean contents of organic carbon in shales between 0.40 wt% and 1.68 wt%, in carbonates between 0.15 wt% and 0.50 wt%, and in sandstones between 0.04 wt% and 0.42 wt%. Measurements in Quaternary sands found concentrations in the order of 0.008 wt%-1.9 wt% (Matthess, 1982, 1983). Microorganisms play an important role in oxidation-reduction processes.

The consumption of free dissolved oxygen in ground water is primarily due to the degradation of organic substances and this gives rise to zonation. Champ, Gulens, and Jackson (1979) proposed a thermodynamic model of the sequential utilization of oxidized species (e.g., O_2, NO_3^-) in aquifers that have an excess of reduced dissolved organic carbon (DOC) rather than dissolved oxygen.

This model assumes that natural waters are recharged through a reactive soil zone into a confined aquifer system containing MnO_2, $Fe(OH)_3$, and excess DOC. If the individual redox reactions are arranged in the order of their decreasing negative value of the free energy change at pH 7, the oxidized species are reduced in this system following a typical sequence: Dissolved oxygen (aerobic respiration), nitrate (denitrification), manganese oxide (Mn IV reduction), ferric iron hydroxide (Fe III reduction), sulfate (sulfate reduction), dissolved CO_2 (methane fermentation), and dissolved nitrogen (nitrogen fixation). Occurrences of such a sequence have been observed in ground water flow systems by Edmunds (1977), Golwer, Matthess, and Schneider, (1970), Golwer et al.(1976), Jackson and Inch (1980), and Schwille (1976).

Complexation.   The solubility may be increased by forming soluble complexes with inorganic or organic compounds (e.g., fulvic acids), or it may be decreased when insoluble complexes (e.g., water-insoluble humic acids) are formed. Organic ligands may be decomposed by microbial activity; solubility is changed due to this degradation.

Adsorption-Desorption.   Many solid substances in the ground coming into contact with subterranean waters tend to release certain constituents into solution and to remove dissociated and nondissociated components from solution. Removal occurs as a result of binding to surfaces of solid particles due to intermolecular interchanges. In the case of exchange between dissolved and adsorbed ions, the process is called ion exchange. Strong adsorbents in rocks include clay minerals, zeolites, iron and manganese hydroxides and hydrates, aluminum hydroxide, and organic substances, especially humic materials. The continuous adsorption-desorption processes, which can be described by Freundlich or Langmuir isotherms, cause a retardation specific for dissolved or suspended substances with respect to the surrounding ground water. Thus a chromatography effect is observed.

Biochemical Processes

Numerous reactions, especially in the carbon, sulfur, nitrogen, iron, and manganese cycles, are accelerated by autochthonic microorganisms that are adapted to local underground environments (Alexander, 1961; Davis, 1967). The direction of the microbial reactions is controlled by the thermodynamic conditions of the respective system; however, these reactions, when proceeding under favorable ecological conditions, occur much faster than the corresponding physico-chemical reactions.

Physical Processes

Gas Transport.   Gas transport in the saturated and unsaturated zone is due to diffusion. However, in ground water diffusion is coupled with flow dispersion, and in the unsaturated zone it is combined with thermal and barometric influences. Oxygen from the atmosphere enters ground water by diffusion or is dissolved in percolating seepage water. The balance between the rate of oxygen entry and the rate of oxygen consumption controls whether anaerobic or aerobic conditions exist in the ground water. Reverse gas movement removes the gaseous decay products (e.g., nitrogen and CO_2) and volatile contaminants from the ground water.

Biophysical Processes

Microorganisms are passively entrained into and within ground water. Extended propagation is likely only in large fissures and solution holes. In porous aquifers, however, filtration controls the transport behavior of microorganisms. The filtration effect is a complex physical and chemical phenomenon. The transport of suspended microorganisms may be limited mechanically by the pore size of the solid medium and the size of the microorganism. Thereafter, the mechanical filter processes in gravelly aquifers cannot be very effective due to the small diameters of bacteria (0.2-5 μm) and viruses (0.25-0.02 μm) (Matthess and Pekdeger, 1981). More important is the particle accumulation on the surfaces of solid substances, due to sedimentation, flow processes, diffusion, and interception (Yao, Habilian, and O'Melia, 1971). Sedimentation is very effective for accumulation of inorganic mineral suspensions (density of about 2.5 g/cm^3), but not for microorganisms (density about 1 g/cm^3). For particles with diameters less than 1 μm (e.g., viruses), diffusion is very important, and this factor becomes increasingly effective in the accumulation process as particle size decreases. The interception process proves to be the most effective means of filtering bacteria. Van der Waals (mass) and Coulomb (electrostatical) forces are primarily involved. The solid particles of an aquifer possess a negative net charge, and the negatively charged bacteria and viruses are usually maintained in suspension as a result of the repulsive electrostatical forces. These forces are stronger than the Van der Waals forces. However, the presence of dissolved cations in water can decrease the repulsive forces of the grain surfaces through adsorption of cations and compensation of the charge deficit (Althaus et al., 1982). Bacteria can attach themselves to surfaces of the solid aquifer materials, where they are protected from adverse influences and are subject to higher nutrient concentrations.

REFERENCES

Alexander, M., 1961, *Introduction to Soil Microbiology*, Wiley, New York and London.

Althaus, H., K. D. Jung, G. Matthess, and A. Pekdeger, 1982, Lebensdauer von Bakterien and Viren in Grundwasserleitern. Materialien 1/82 Umweltbundesamt. Schmidt, Berlin.

Champ, D. R., J. Gulens, and R. E. Jackson, 1979, Oxidation-reduction sequences in ground water flow systems, *Can. J. Earth Sci.* **16**:22-23.

Davis, J. B., 1967, *Petroleum Microbiology*, Elsevier, Amsterdam-London-New York.

Davis, S. N., and R. J. M. DeWiest, 1967, *Hydrogeology*, 2nd ed., Wiley, New York.

Edmunds, W. M., 1977, Groundwater geochemistry—Controls and processes, in *Papers and Proceedings, Groundwater Quality, Measurement, Prediction and Protection*, Water Research Centre, Medmenham, England, pp. 115-147.

Feth, J. H., S. M. Rogers, and C. E. Roberson, 1961, Aqua de Ney, California, a spring of unique chemical character, *Geochim. et Cosmochim. Acta* **22**:75-86.

Freeze, R. A., and J. A. Cherry, 1979, *Groundwater,* Prentice-Hall, Englewood Cliffs, New Jersey.

Garrels, R. M., and C. L. Christ, 1965, *Solutions, Minerals and Equilibria,* Harper & Row, New York.

Golwer, A., G. Matthess, and W. Schneider, 1970, Selbstreinigungsvorgänge im aeroben und anaeroben Grundwasserbereich, *Vom Wasser* **36**:64-92.

Golwer, A., K. H. Knoll, G. Matthess, W. Schneider, and K. H. Wallhaeusser, 1976, *Belastung und Verunreinigung des Grundwassers durch fest Abfallstoffe.* Abh. Hess. Landesamt. Bodenforsch. 73, Wiesbaden.

Hem, J. D., 1970, *Study and interpretation of the chemical characteristics of natural water,* 2nd ed., U.S. Geological Survey Water Supply Paper No. 1473, Washington, D.C.

Jackson, R. E., and K. J. Inch, 1980, *Hydrogeochemical processes affecting the migration of radionuclides in a fluvial sand aquifer at the Chalk River Nuclear Laboratories,* NHRI Paper No. 7, Science Series No. 104, Ottawa.

Matthess, G., 1982, *The Properties of Groundwater,* John Wiley & Sons, New York.

Matthess, G., 1983, Verzögerung des Transports von Halogenkohlenwasserstoffen (HKW) aufgrund von Adsorptions- und Desorptions-Vorgangen im Boden. DVGW-Schriftenr. *Wasser* **36**:79-83.

Matthess, G., and A. Pekdeger, 1981, Survival and transport of pathogenic bacteria and viruses in groundwater, *Sci. Total Environ.* **21**:149-159.

Matthess, G., and K. Ubell, 1983, *Allgemeine Hydrogeologie, Grundwasserhaushalt,* Borntraeger, Berlin Stuttgart.

Nace, R. L., 1967, Water resources: A global problem with local roots, *Environ. Sci. Tech.* **1**:550-560.

Schwille, F., 1976, Anthropogenically reduced groundwater, *Hydrol. Sci. Bull.* **21**:629-645.

Stumm, W., and J. J. Morgan, 1981, *Aquatic Chemistry,* 2nd ed., John Wiley & Sons, New York.

Wedepohl, K. H., ed., 1978, *Handbook of Geochemistry,* Vol. II-1, Springer, Berlin.

White, D. E., and G. A. Waring, 1963, Data of geochemistry, 6th ed. K., Volcanic emanations, *U. S. Geol. Surv. Prof. Papers, No. 440.*

Yao, K. H., M. T. Habilian, and C. R. O'Melia, 1971, *Water and waste water filtration: Concepts and applications,* ESE TUBL. No. 289, Dept. Environ. Sci. Eng., University of North Carolina, Chapel Hill.

31

The Evidence for Zones of Biodenitrification in British Aquifers

S. S. D. Foster

Institute of Geological Sciences

Don P. Kelly

University of Warwick

Richard James

University of East Anglia

INTRODUCTION

Background to Research

Ground water provides more than 30% of the British potable water supply and in excess of 60% in some of the flatter and drier provinces in eastern England. Consolidated sedimentary geological formations give rise to fissured porous aquifers of regional extension. In the unconfined areas of these aquifers, ground waters have high or rising nitrate concentrations, already widely in excess of 10 mg N/l and locally above 15 mg N/l. This problem has been shown to derive largely from increasing nutrient-leaching losses from cultivated soils, subjected to progressively more intensive arable cropping (Foster, Cripps, and Smith-Carington, 1982).

Significant resources of low (or zero) nitrate (and otherwise uncontaminated) ground water exist, however, mainly, although not exclusively, in areas where the aquifers are confined by argillaceous strata or glacial tills. In these zones, ground-water flow is generally slower than in the unconfined zones. However, ground-water circulation has been, and is being, induced by water-supply abstraction. The magnitude, evolution, and security of these low-nitrate ground-water resources is of considerable importance to water-supply interests.

This paper reviews the evidence for biodenitrification (nitrate reduction to nitrogen or nitrous oxide gases or to some intermediate product, such as nitrite or ammonia) being responsible for the low-nitrate ground waters.

Scope of Investigations

The evidence for biodenitrification is derived from investigation of two representative areas, one on the Jurassic Limestone south of Lincoln and the other on the (Cretaceous) Chalk in central Norfolk. Both areas are currently the subject of detailed research. Core samples have been recovered from numerous boreholes and analyzed for pore-water chemical and isotopic concentrations and rock lithologic, hydraulic, and chemical properties. The ground-water flow regime has been established by integrated borehole-flow logging and quality sampling and by hydrogeological pumping tests. Moreover, core samples collected from three boreholes at the Lincoln site and two at the Norfolk site have also been the subject of intensive microbiological investigation. This paper reviews research methods and results that are being published more fully elsewhere (Foster and James, 1984).

Previous Related Microbiological Work

The potential influence of microbiological processes on ground-water quality has been comprehensively reviewed by Matthess (1982), who stresses the relatively widespread subsurface occurrence of bacteria, despite earlier assumptions to the contrary, and their ability to speed up certain chemical reactions already favored by thermodynamic considerations.

In contrast to the (generally chemolithotrophic) nitrifying bacteria, the denitrifiers are mainly chemoorganotrophic, requiring an oxidizable organic carbon substrate for growth. They have been shown to occur widely and abundantly in soils and aquatic sediments (Knowles, 1982), but there are few reports of their occurrence in ground-water systems. Their presence in parts of the Chalk unsaturated zone has been demonstrated (Whitelaw and Rees, 1980), but activity was not proven. The formidable problems associated with the satisfactory enumeration and characterization of ground-water bacteria and the limitations of enrichment culture techniques have been recently discussed (Wilson et al., 1983).

HYDROGEOLOGICAL CONDITIONS IN RESEARCH AREAS

Billingborough (Lincoln) Site

The Jurassic Limestone aquifer in the Billingborough area is highly confined by overlying, predominantly argillaceous, Jurassic strata. Rainfall seasonally recharges the limestone over its outcrop to the west. Ground-water flow is presently governed by abstraction, and active circulation occurs between the outcrop and important water-supply boreholes located 10-15 km downdip in the confined aquifer (Fig. 31-1). Major exploitation of the confined aquifer for public water supply commenced in the 1950s, and the cumulative abstraction to the end of 1982 from the 4-km-wide flow path to the Aslackby source is estimated at 120-150 Mm³.

The limestone comprises a variable succession of oolitic, shelly, more cemented and clayey materials, but pore sizes throughout are very small (mainly in the range 0.05-0.5 μm). This results in very low intergranular permeability, despite quite high interconnected porosity (up to 0.30), and almost all ground-water flow occurs through subhorizontal fissures that give rise to high aquifer transmissivities (in excess of 1000 m²/day) and ground-water flow velocities (10-20 m/day). The limestone is gray in color but alters to buff with mineral (principally sulfide) oxidation by flowing ground water along fissures. This process appears to be occurring on a geological time scale and has not developed only as a result of modern ground-water circulation.

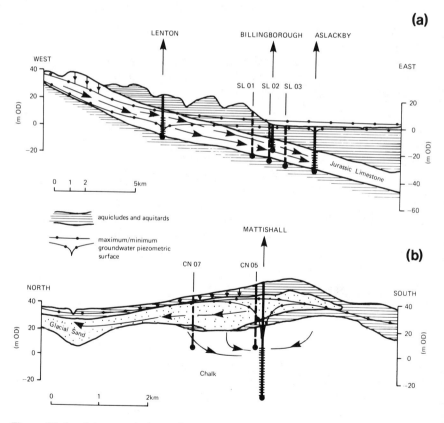

Figure 31-1. Schematic hydrogeological cross-sections of research areas in (a) Lincoln and (b) Norfolk.

Table 31-1. Summary of Investigation Borehole Results from Billingborough (Lincoln) Site

	SL 1		SL 2		SL 3
	Upper Borehole	*Lower Borehole*	*Upper Borehole*	*Lower Borehole*	*All Boreholes*
Pore water					
NO_3-N(mg/l) grey	< 0.2	< 0.2	< 0.2	< 0.2	< 0.2
NO_3-N(mg/l) buff	→ 2	→ 3	→ 1	→ 4	?
NH_4-N(mg/l)	< 0.1	< 0.1	0.1-0.6	< 0.1	0.1-0.3
Fissure water					
NO_3-N(mg/l)	6-7	5-7	0-2	5-7	< 1
DOC (mg/l)	1-2	2-4	?	1-2	?
Dissolved O_2 (mg/l)	1.5-2.0	1.0-1.5	?	1.5-2.5	?
Bacteriological Tests (Scraped Limestone)					
Nitrate consumed from organic media 15°C for 10 days					
sugars	20%-60%	50%-70%	30%-70%	< 20%	< 20%
acids	10%-35%	25%-55%	10%-35%	0%	0%

The mobile ground water currently varies in nitrate concentration from considerably more than 10 mg N/l in the unconfined aquifer (Smith-Carington et al., 1983) to less than 0.1 mg N/l in the confined aquifer at Aslackby (Fig. 31-1). Moreover, there appears to be a rapid change between investigation boreholes SL1/SL2 and SL3 (Table 31-1), the former having concentrations up to 7 mg N/1, depending on the depth from which ground water is abstracted, while the latter has no detectable nitrate throughout. The overall ground-water quality picture appears to have remained relatively stable over a decade or more, and the decrease in nitrate concentration is accompanied by the virtual disappearance of dissolved oxygen and an associated Eh change (+400 mV to +100 mV) at constant pH of about 7.2 in the vicinity of Billingborough (Edmunds and Walton, 1983). The restricted penetration of agriculturally derived nitrate into the confined aquifer could be due to physical retardation, resulting from extensive exchange by aqueous diffusion between mobile fissure water and static pore water (Barker and Foster, 1981), or to bacterial denitrification, or both.

Two semi-independent lines of evidence suggest that significant bacteriological denitrification is, in fact, occurring. First, pore-water nitrate concentrations are generally less than 0.2 mg N/l throughout the cored boreholes SL1, SL2, SL3 (Table 31-1), with the exception of somewhat higher concentrations in the thin bands of buff limestone along the walls of

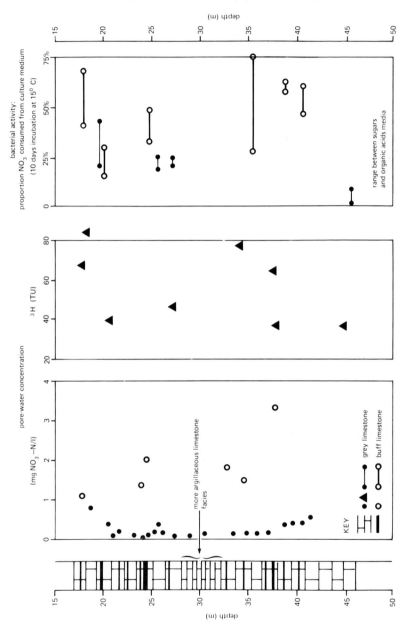

Figure 31-2. Pore-water composition and bacterial activity in investigation borehole SL1 at Billingborough (Lincoln) site.

360

fissures (Fig. 31-2). In contrast, the pore water of borehole SL1 contains appreciable thermonuclear (probably post-1960) tritium, with eight samples having concentrations in the range 38-78 TU. This suggests that denitrification is occurring at the fissured surfaces or in the buff limestone bands, as solutes diffuse from the mobile fissure water into the static pore water. Second, thermonuclear tritium had already penetrated to Aslackby and similarly situated ground-water sources by 1969, albeit at much lower concentrations than those seen in the unconfined aquifer; however, nitrate has never been detected.

Mattishall (Norfolk) Site

The Chalk aquifer of the Mattishall area is covered by a variable suite of glacial sands and clay tills that, to varying degrees, confine its ground water. The superficial sands (Fig. 31-1) are seasonally recharged by rainfall infiltrating locally through more permeable areas of the clayey soil cover, and most natural ground-water circulation is concentrated in these sands and directed under low hydraulic gradient towards neighboring small rivers. Recharge to the Chalk aquifer, in the form of leakage from the overlying sands, can be induced by borehole pumping, as in the case of the small Mattishall source (Fig. 31-1), which commenced pumping in the late 1970s and by the end of 1982 had a cumulative abstraction of almost 2 Mm^3.

The Chalk is also a fissured microporous aquifer, but in the Mattishall area it displays important differences in properties from the Jurassic Limestone at Billingborough. The formation transmissivity is considerably lower, around 100-150 m^2/day. The subhorizontal fissures that conduct most ground-water flow are concentrated in the uppermost 20 m, with velocities increasing from less than 5 m/day to perhaps more than 20 m/day near the pumping borehole. The rock mass is traversed by a high density of microfissures that break it into blocks of small dimension (probably less than 0.1 m). The rock mass also contains essentially static pore water, but the lithology is more uniform, the porosity significantly higher (0.40-0.50), and the pore sizes somewhat larger (0.2-1.0 μ).

The area is one of highly productive arable agriculture, and the ground water of the superficial sands contains nitrate concentrations in the range 2-14 mg N/l together with significant dissolved oxygen (Fig. 31-3). In comparison, the Chalk pore waters have less than 0.1 mg N/l. The mobile Chalk ground waters contain very low levels of dissolved oxygen (less than 0.5 mg/l), and the presence of soluble iron (more than 1.0 mg Fe (II)/l) at mildly alkaline pH confirms the anoxic character of this ground-water system.

Isotopic determinations on pore-water samples (Fig. 31-3) reveal the penetration of thermonuclear tritium not only to the base of the sands

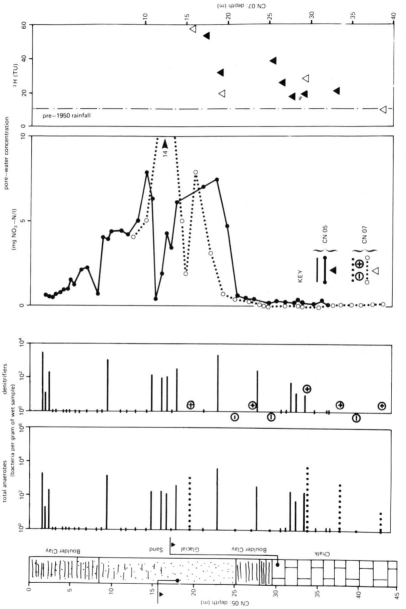

Figure 31-3. Summary of pore-water chemistry and bacteria in Mattishall (Norfolk) investigation boreholes.

362

(32-58 TU), but also into the upper part of the Chalk (9-39 TU). The implication is that a proportion of modern (probably post-1960) recharge has penetrated the Chalk from the overlying sands. Since the soils of the area have been utilized for arable farming since at least 1930, all infiltration since 1960 will have contained significant nitrate. The fact that thermonuclear tritium has been detected in the Chalk, where nitrate levels are negligible, is evidence of active bacteriological denitrification.

MICROBIOLOGICAL INVESTIGATIONS

In the absence of well-established techniques for bacteriological investigation of ground-water systems, two research programs were mounted during 1981-1983 on a parallel, but semi-independent, basis to explore different approaches to the problems of bacteriological sampling and enumeration and to obtain an indication of the corresponding biochemical activity. It was accepted that more sophisticated and expensive field-sampling and laboratory-culturing techniques could be required to evaluate such activity systematically. Publication of the results is intended to stimulate discussion in this new research area, and the methods adopted are not advocated as the only, or the most appropriate, approach.

Billingborough (Lincoln) Site

Field Sampling. The Jurassic Limestone is sufficiently hard for rotary core-drilling techniques to be necessary to obtain samples. A 1- or 3-m triple-core barrel lined with a plastic sheath was employed, producing a 100-mm-diameter core. Air was used as the drilling fluid because it was less likely to contaminate the core than water, foam, or mud. Since the limestone was fully saturated, there was no significant tendency for the core to become overheated by friction. The core was extruded from the core barrel at the surface within its plastic sheath and taped, ready for subsampling for both bacteriological and other investigations. Bacteriological subsamples were selected and maintained at $+4°C$ during transport and prior to handling at the base laboratory. Two methods of sampling were attempted: The first involved scraping the surface of identifiable fissures with a sterile tool in a sterilized environment; the second involved crushing the limestone within a sterile bag, placement in sterile centrifuge liners, and extraction of pore water by spinning at 14,500 rpm (about 25,000 gravity RCF) for 30 min. The water sample obtained was transferred directly into a sterile bottle.

Laboratory Analyses. The basic experimental approach was to assess sample denitrifying bacterial activity by the enrichment culture technique. This procedure is primarily a way of establishing the presence of bacteria, provided the growth of any group is not suppressed as a result of the selected media composition or incubation conditions. Appropriate sterilized media were chosen to detect the presence of three different groups of the denitrifiers: Those that use (a) organic acids and (b) sugars as their carbon and energy sources, and those that (c) oxidize some sulfur compounds to obtain energy. Incubations were carried out in 25-ml microaerophilic screw-topped bottles fitted with an inverted tube to collect any nitrogen gas evolved. The bottles were inoculated with the samples in the form of about 0.5 ml of scraped limestone or 1.0 ml of centrifuged pore water. For a few samples, strictly anaerobic culture conditions were used, the bottles being carefully gassed with nitrogen and completely sealed by rubber stoppers. Two incubation temperatures were chosen, 15°C and 30°C; the lower was close to the *in situ* temperature (10°C), and the upper was to stimulate more rapid bacterial activity. During incubation the enrichment cultures were scored for growth (on the basis of visual turbidity) and for gas accumulation. After 10 days those showing positive growth were assayed for nitrate, nitrite, and, where appropriate, thiosulfate, tetrathionate, and pH.

Summary of Results.   Virtually all the enrichment cultures using scraped limestone samples and one of the two organic media recorded denitrifying activity. It was highly significant in the samples from certain depth intervals (Table 31-1), especially the buff limestones (Fig. 31-2). The consumption of nitrate was generally greater for sugar (Fig. 31-2), but gas evolution and nitrite accumulation were variable in both cases. In those instances where duplicate tests were carried out, the denitrifying activity was significantly greater under microaerophilic conditions than under strictly anaerobic conditions, suggesting that the former best approximates *in situ* conditions. Neither growth, nitrate consumption, nor gas evolution was observed in the media incubated with pore water even after incubation for more than 20 days at 30°C, suggesting that the denitrifying bacteria must be located only close to fissures and probably attached to rock surfaces. Although growth in the thiosulfate medium was observed only in a few enrichments from one borehole, the disappearance of both thiosulfate and nitrate in these cases demonstrated that denitrification could occur in the absence of any organic substrate other than that contained in the limestone inoculum.

Rates of Denitrification.   The amount of nitrate that can be reduced (to nitrogen gas) will be dependent upon the type and quantity of oxidizable organic substrate available and on the establishment of microaerophilic conditions. Different biological oxidations provide different amounts of reducing equivalents (H^+); for example, such oxidizable carbohydrates as glucose would generate 24 H^+ on complete oxidation, and 1 g would be sufficient to reduce almost 0.4 g N of nitrate. To calculate *in situ* denitrification rates it would be necessary to know: (1) the total number of denitrifying bacteria per unit area of active surface; (2) the type and concentration of available organic substrate; and (3) the metabolic rate of bacteria at the prevailing nitrate and substrate concentrations. These are virtually impossible to establish in practice, but the potential range of activity can be approximated from the results of the enrichment cultures. Almost all the scraped limestone samples (of 0.5 ml) exhibited the ability to reduce more than 100 micromoles nitrate in 10 days, and these samples were in effect derived from 8 cm^2 of fissure surface. Crudely extrapolated, this represents a minimum denitrification capacity in excess of 150 mg N/m^2/day. This is compared to an estimated *in situ* rate of about 1.5 mg N/m^2/day, assuming a nitrate reduction of 10 mg N/l over a 2-km flow distance traversed in 100 days with the aquifer having a volume/area ratio of 0.015 m. Thus, even if this extrapolation grossly exceeds the capacity of the standing attached bacterial population, there should still be sufficient denitrifying activity to account for the observed reduction.

The question arises as to the nature and origin of the organic substrate being utilized. Monitoring of water supplies pumped from the Jurassic Limestone (and from the Chalk) shows the mobile fissure water generally to contain less than 2 mg DOC/l, and this, whatever its form, would not be sufficient to support significant denitrification. However, preliminary work suggests that the pore water and the rock matrix of both formations can contain 10-20 mg DOC/l and more than 1000 mg TOC/kg dry weight, respectively. In the case of the Chalk unsaturated zone, it has been demonstrated that the rock matrix together with its pore water generally contains total carbohydrates in the range 30-150 mg/kg dry weight, with up to 5% in the pore-water solution producing concentrations generally in the range 5-10 mg/l (Whitelaw and Edwards, 1980). The apparent correlation with overlying land use is more likely to be the result of a greater rate of organic carbon consumption beneath arable land than grassland. It is considered unlikely that the solid-phase carbohydrates could be derived from the land surface. The pore-water carbohydrates must be derived from the solid phase, with the rate of dissolution being controlled by rate of breakdown or solubility. Further investigations, with careful attention to extraction procedures, will be required before the carbohydrate speciation can be confidently established.

Mattishall (Norfolk) Site

Field Sampling. At this site the strata were softer or unconsolidated, and satisfactory samples could be obtained by the dry-percussion drilling technique utilizing steel sampling tubes of 0.5-m length and 100-mm diameter with cutting shoe and core catcher. This method minimizes possible contamination of core samples since neither water nor air is added. It is possible, though time consuming, to sterilize the entire sampling tool; this process was not generally carried out. Samples for bacteriological analysis were taken from one borehole in each of two drilling phases. The first phase consisted of a survey of the distribution of general bacterial groups in all material from the soil base to 40-m depth. In the second phase a smaller number of samples was subjected to more detailed investigation; attempts were made to identify the major species of denitrifying bacteria present and to study their ability to utilize a range of substrates.

Laboratory Analyses.   Complete sample tubes were taken to the base laboratory daily. A 2-cm layer of surface material was removed aseptically from the base of each tube, and a subsample was extracted from the freshly exposed material using a sterilized apple corer. The plugs of about 10-g weight were transferred to sterile petri dishes to be weighed. Specimens were then transferred to a blender containing a detergent dispersant solution to obtain a homogeneous suspension for subsequent analyses. In the case of the Chalk, it is not generally feasible to identify and scrape specific fissure surfaces, as was done with the limestone samples; however, it is assumed that some microfissures will have transected most of the subsamples selected. For the enumeration of total aerobic and anaerobic bacteria, the homogeneous suspensions in 0.5- or 1.0-ml volumes with duplicates were inoculated on nitrate agar plates and incubated both aerobically at 22°C for 3 days and anaerobically in gas jars at 22°C for 5 days. Indicator strips were included with the latter to confirm that anaerobic conditions were maintained. To identify denitrifying bacteria, representative single colonies were tested for gas evolution, when cultured in nitrate broth. Subsequent tests revealed that the vast majority of these were nonfermenters, thus confirming the presence of denitrifiers. Fermenting isolates were tested chemically to confirm reduction to nitrous oxide or nitrogen gas. Confirmed denitrifying bacteria were characterized by morphological examination, a battery of biochemical tests, and their utilization of a range of organic substrates during aerobic incubations at 22°C in a minimal salts media (Table 31-2).

Summary of Results.   The results of the survey of general bacterial types in the investigation of borehole CN05 proved the existence of potential denitrifiers at some depth below the top of the Chalk aquifer and in the overlying sands (Fig. 31-3). Denitrification rates as low as 0.1 mg N/day/m of aquifer surface area perpendicular to the vertical leakage would be sufficient to explain the observed ground-water nitrate distribution. Both aerobes and anaerobes occurred in the majority of samples, with the former normally being 10 to 100 times more numerous except at depth. However, the complete absence of bacteria from some depths is believed to be associated with the absence of microfissuring in the particular subsamples concerned.

Table 31-2. Characteristics and Substrate Utilization of Selected Representative Colonies from Mattishall (Norfolk) Investigation Borehole CN07

| | Colony Number (Glacial Sand, 16 m) | | | | Colony Number (Chalk, 34 m) |
	1	*7*	*9*	*10*	*2*
Characteristics					
Nitrate broth gas	+	+	+	+	+
Pigment fluorescence	+	+	+	+	+
Gram stain	−	−	−	−	−
Oxidase	−	+	+	+	+
Growth at 4°C	−	+	+	+	+
Growth at 41°C	−	−	−	−	−
Gelatinase	−	+	+	+	+
Catalase	+	+	+	+	+
Arginine dehydrolase	−	+	+	+	+
Motility at 22°C	−	+	+	+	+
Citrate	−	+	+	+	+
Urease	+	−	−	−	−
PHB granules	+	−	−	−	−
Substrate utilization					
Glucose	+	+	+	+	+
Sucrose	+	+	+	+	+
Maltose	+	−	+	−	+
Ethanol	+	−	+	+	+
Sorbitol	+	−	+	+	+
Acetic acid	+	+	+	+	+

The confirmed denitrifying bacteria formed a narrow, specialized, motile, gram-negative group, suggesting selective development during migration with ground water from the soil. Among the representative colonies, four showed characteristics (Table 31-2) typical of *Pseudomonas fluorescens,* and the fifth has proved impossible to identify. Their ability to utilize various organic substrates (a range of sugars, alcohols, and acids) as sole carbon source suggests that different *Pseudomonas fluorescens* biotypes may be present. Attempts to identify the nondenitrifying isolates have also presented problems; unusual profiles of biochemical reactions, not matching any previously described gram-negative organism, perhaps suggest abnormal selective pressures, especially low oxidizable organic carbon, in the subsurface environment. It should be borne in mind that some of these isolates may really be denitrifiers also, but do not maintain this character in the laboratory (Gamble, Betlach, and Tiedje, 1977). Moreover, some colonies, initially producing gas in nitrate broth, subsequently lost this ability during culture storage. The *Pseudomonas* group, like most other denitrifying bacteria, are facultative aerobes that reduce nitrate only when growing under microaerophilic conditions.

CONCLUSIONS

The accumulated data provide convincing evidence for the presence of bacteria at depths up to 50 m, and in one case a large distance downdip, in the British confined aquifers. The potential of these bacteria for nitrate reduction has been demonstrated, and, given the probable microaerophilic conditions, it is reasonable to conclude that denitrification is a likely mechanism for bacterial growth. The absence of nitrate in ground water containing a component of thermonuclear tritium is strong circumstantial evidence for denitrifying activity.

The security of such low-nitrate ground-water resources will depend on the continued generation, presumably from a geological source, of a suitable supply of organic carbon substrate. It will also depend on sustained microaerophilic conditions, a function of possibly related (bio)chemical processes.

The positive identification of *Pseudomonas* species as prominent members of the bacterial population is not surprising in view of this genera's environmental adaptability and its ability to utilize a very wide range of substrates. Their presence could be significant in relation to the biodegradation of organic ground-water pollutants also.

ACKNOWLEDGEMENTS

This paper is published with permission of the Director, NERC Institute of Geological Sciences, with financial support from the U.K. Department of the Environment and E.E.C. Environmental Programme and with practical assistance from the Anglian Water Authority. The authors gratefully acknowledge the painstaking field- and laboratory work of Lionel Bridge, Adrian Lawrence, Judy Parker, Julie Mason, Colin Edge, and William Forbes.

REFERENCES

Barker, J. A., and S. S. D. Foster, 1981, A diffusion exchange model for solute movement in fissured porous rock, *Quart. J. Eng. Geol.* **14:**17-24.
Edmunds, W. M., and N. R. G. Walton, 1983, The Lincolnshire Limestone—hydrogeochemical evolution over a ten-year period, *J. Hydrol.* **61:**201-211.
Foster, S. S. D., A. C. Cripps, and A. K. Smith-Carington, 1982, Nitrate leaching to ground-water, *Royal Soc. London Phil. Trans.* **296:**477-489.
Foster, S. S. D., and R. James, eds., in press, *Hydrogeological and Biochemical Research on Nitrogen Transformations in Groundwater Systems,* IGS Report Series, H.M.S.O. London.

Gamble, T. N., M. R. Betlach, and J. M. Tiedje, 1977, Numerically dominant denitrifying bacteria from world soils, *Appl. Environ. Microb.* **33:**926-939.

Knowles, R., 1982, Denitrification, *Microb. Rev.* **46:**43-70.

Matthess, G., 1982, *The Properties of Groundwater,* John Wiley & Sons, New York.

Smith-Carington, A. K., L. R. Bridge, A. S. Robertson, and S. S. D. Foster, 1983, *The Nitrate Pollution Problem in Groundwater Supplies from Jurassic Limestones in Central Lincolnshire,* IGS Report 83/3, H.M.S.O., London.

Whitelaw, K., and R. A. Edwards, 1980, Carbohydrates in the unsaturated zone of the Chalk, England, *Chem. Geol.* **29:**281-291.

Whitelaw, K., and J. F. Rees, 1980, Nitrate-reducing and ammonium-oxidising bacteria in the vadose zone of the Chalk aquifer of England, *Geomicrob. J.* **2:**179-187.

Wilson, J. T., J. F. McNabb, D. L. Balkwill, and W. C. Ghiorse, 1983, Enumeration and characterisation of bacteria indigenous to a shallow water-table aquifer, *Groundwater* **21:**134-142.

32

Microbial Involvement in Trace Organic Removal during Rapid Infiltration Recharge of Ground Water

Stephen R. Hutchins, Mason B. Tomson, and Calvin H. Ward
Rice University

INTRODUCTION

In recent years population growth and urbanization have led to increased production of municipal and industrial wastes that must be treated and disposed of. One of the management technologies being used to treat these wastes in an environmentally acceptable manner is land application. Land treatment is generally effective for the removal of many wastewater constituents, such as fecal coliforms, heavy metals, phosphorus, and carbonaceous biochemical oxygen demand (Bouwer and Chaney, 1974; Olson, Crites, and Levine, 1980). However, the effects of toxic organic chemicals are the least known among all wastewater constituents (Majeti and Clark, 1981). This is of particular concern in rapid infiltration systems since rapid infiltration depends primarily on percolation for wastewater disposal and thus has the greatest potential for affecting associated ground waters. Most studies indicate that little attenuation of organic chemicals should occur in rapid infiltration systems (Sills, Costello, and Laak, 1978; Jenkins et al., 1983). Furthermore, there is virtually no information on long-term degradation of trace organics in ground water (McCarty, Reinhard, and Rittman, 1981).

Trace organic transport was examined in several rapid infiltration sites in the United States. It was found that many wastewater trace organics are completely or mostly removed by the treatment process, although detectable concentrations of recalcitrant compounds are still present in the associated ground waters. In this study, trace organic transport and fate were evaluated using soil columns in a series of tests designed to address the effects of microbial adaptation and involvement in trace organic removal.

MATERIAL AND METHODS

Test System

A glass-and-Teflon soil column test system was developed to study trace organic fate during rapid infiltration (Fig. 32-1). Amberlite XAD-4 resin and reverse-ion-search gas chromatography/mass spectrometry were used to identify and quantitate selected target compounds. These methods are described in detail elsewhere (Hutchins, Tomson, and Ward, 1983).

Column Test I (CTI): Baseline Information

A newly constructed rapid infiltration site at Fort Polk, Louisiana, was selected as a model for CTI. The purpose of CTI was to establish baseline information on transport of trace organics through unacclimated soil. Topsoil from one of the infiltration basins was obtained prior to the first application of wastewater at the site. Similarly, unchlorinated secondary effluent was collected for feed solution from the sewage treatment plant located there. Four soil columns were packed with soil to a depth of 1 m each, and a 2-day flooding/12-day drying inundation schedule paralleling design operation of the field system was initiated. Nine target compounds were selected for the study: Tetrachlorethylene (TCE), p-dichlorobenzene (PDCB), acetophenone (ACE), diethyl phthalate (DEP), 2-(Methylthio)benzothiazole (2MTBT), benzophenone (BZPN), N-butylbenzenesulfonamide (NBBS), dibutyl phthalate (DBP), and bis(2-ethylhexyl)phthalate (BEHP). During the final two inundation cycles the feed from two of the soil columns was spiked with 0.5% mercuric chloride as a biocide to eliminate microbial activity as a removal mechanism.

Column Test II (CTII): Primary vs. Secondary

The objectives of CTII were twofold: (1) to repeat CTI except using acclimated soil to eliminate reversible adsorption as a removal mechanism, and (2) to evaluate the use of primary effluent instead of secondary effluent in rapid infiltration. Both soil and secondary feed were obtained from the same sources as had been used for CTI. Also obtained was primary feed at a point after the primary clarifiers and before the trickling filters. Eight soil columns were packed to a depth of 1 m each; four columns were conditioned with primary feed and four columns with secondary feed by continuously flooding the soil with approximately 50 pore volumes of the respective feed solution. A 6-day flooding/8-day drying inundation schedule was initiated to more accurately reflect field conditions at the time. Six target compounds were selected for study: ACE, α-terpineol (αTER), skatole (SKA), o-phenylphenol (OPP), p-(1,1,3,3-tetramethylbutyl)phenol (TMBP), and 2MTBT. During one of the inundation cycles, trace organic behavior in the feed

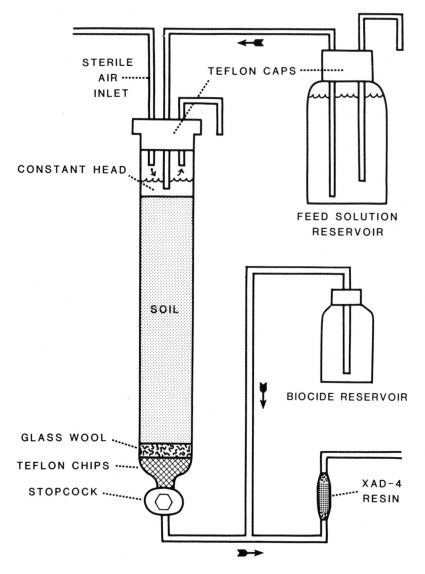

Figure 32-1. Soil column design.

solutions was monitored over the 6-day flooding period to evaluate trace organic removal prior to infiltration.

Column Test III (CTIII):
Microbial Involvement

The rapid infiltration site at Fort Devens, Massachusetts, was selected as a model for CTIII. This site has been operating for more than 40 years with primary effluent and represents an acclimated soil. The objectives of CTIII were again twofold: (1) to evaluate trace organic removal during rapid infiltration using soil from an operational site with an adapted microflora, and (2) to evaluate the extent of trace organic removal during exposure to varying concentrations of selected pollutants. Eight columns were packed with soil from the site to a depth of 1.2 m each, and a 6-day flooding/16-day drying inundation cycle paralleling the operation of the field site was initiated. Primary feed was obtained from the previous source at Fort Polk, Louisiana, and six target compounds were selected: PDCB, 2-methylnaphthalene (2MN), OPP, TMBP, 2MTBT, and BZPN. The feed was divided into four subsamples; each subsample was spiked with increasing concentrations of the target compounds, and the soil columns were separated into four sets of duplicates corresponding to the four feed subsamples.

RESULTS AND DISCUSSION

CTI: Baseline Information

In CTI sandy clay loam that had not been exposed to wastewater application was used, and trace organic transport was evaluated during rapid infiltration. Consistent reductions in concentrations were noted for TCE, PDCB, and 2MTBT only during each inundation cycle (Table 32-1). However, traces of each target compound were detected in the column effluents during all inundation cycles. The only exception was PDCB; no PDCB was detected in the column effluents during the first inundation cycle, but the fractional breakthrough (mass output/mass input) increased with successive inundation cycles. Addition of mercuric chloride during the seventh and eighth inundation cycles resulted in increased fractional breakthroughs of PDCB, ACE, BZPN, and BEHP compared to the other soil columns. Although it is possible that this increase resulted from an inhibition of biodegradation in the soil column, there are other factors to be considered as well. Mercuric chloride has a variety of effects, not the least of which is a dramatic lowering of soil-water pH. This low-pH effect, coupled with the increased ionic strength of soil water due to the biocide salt, could easily disrupt soil integrity and cause release of previously sorbed organics from the soil matrix.

Table 32-1. Trace Organic Concentrations in Feed Solution and Column Effluents. Column Effluent Trace Organic Concentrations Are Expressed as Mean Values with Standard Deviation, Four Columns Total

	First ($\mu g/l$)		Second ($\mu g/l$)		Third ($\mu g/l$)		Fourth ($\mu g/l$)	
Trace Organic	Feed Solution	Column Effluent	Feed Solution	Column Effluent	Feed Solution	Column Effluent	Feed Solution	Column Effluent
				Inundation Cycle				
TCE	4.30	0.46 ± 0.13	1.10	0.18 ± 0.026	0.37	0.30 ± 0.14[a]	0.12	0.048 ± 0.014
PDCB	0.15	0.00 ± 0.00	0.24	0.042 ± 0.015	0.45	0.088 ± 0.015	0.24	0.060 ± 0.010
ACE	0.088	0.12 ± 0.014	0.28	0.095 ± 0.028	0.35	0.085 ± 0.015	0.28	0.042 ± 0.013
DEP	0.17	0.83 ± 0.23	0.19	0.26 ± 0.070[a]	0.66	0.036 ± 0.014[a]	0.019	0.034 ± 0.009[a]
2MTBT	0.65	0.052 ± 0.048	1.30	0.043 ± 0.037	1.0	0.034 ± 0.013	1.40	0.087 ± 0.026
BZPN	0.058	0.049 ± 0.018[a]	0.13	0.077 ± 0.042	0.11	0.060 ± 0.017	0.20	0.12 ± 0.040[a]
NBBS	0.026	0.10 ± 0.087[a]	0.11	0.15 ± 0.033[a]	0.038	0.43 ± 0.25	0.45	0.93 ± 0.22
DBP	0.34	0.22 ± 0.10[a]	0.45	0.12 ± 0.053[a]	0.50	0.22 ± 0.070	0.18	0.15 ± 0.067[a]
BEHP	1.60	7.10 ± 4.30[a]	2.50	1.30 ± 1.50[a]	7.60	0.011 ± 0.022	4.30	0.16 ± 0.16[a]

[a]Difference between column effluent and feed solution trace organic concentrations is not statistically significant at the 5% significance level.

374

Furthermore, it is also possible that the biocide induced a certain degree of cellular lysis resulting in a net release of these compounds into the column effluent.

Results of this column study have shown that several trace organics in unchlorinated secondary effluent are capable of migrating through the soil profile, although concentrations of specific compounds may be significantly reduced by the treatment process. Mechanisms of action on these susceptible trace organics cannot be definitely determined on the basis of the simple design of the study, but results are consistent with the processes of volatilization, sorption, and abiotic and biotic transformations in the soil matrix. There is evidence that compound removal by biodegradation and biosorption may be an important fate for trace organics such as PDCB, ACE, BZPN, and BEHP, although it is unknown if these compounds are permanently removed or merely sequestered and thus subject to possible release.

CTII: Primary vs. Secondary

Conditioned sandy clay loam was used in CTII to eliminate reversible adsorption as a removal mechanism. In contrast to results obtained from the first column test, the trace organic profiles and the target compound studies show that there was little removal of trace organics from either primary or secondary wastewater by the conditioned soil columns (Table 32-2). A soil column study by Wilson also showed that most of the trace organics examined were transported readily through a sandy soil low in organic carbon (Wilson et al., 1981). This is contrary to the results obtained by Bouwer in which many trace organics were removed by rapid infiltration through soil columns that had been in operation for approximately 10 years (Bouwer, McCarty, and Lance, 1981).

The fate of the six target compounds in feed solutions prior to infiltration varied widely for both primary and secondary feeds (Table 32-3). The concentration of 2MTBT was the most stable; it did not change appreciably during the six-day period in either primary or secondary feed. Concentrations of ACE decreased slightly in both solutions during the six-day period, but fluctuations in the measured concentrations indicate that these reductions may not be significant. The behavior of these compounds in the soil columns reflects their stability. The soil-treatment process did not significantly affect 2MTBT levels in the secondary columns (Table 32-2). But in the primary columns, the column effluent concentrations were actually higher than the column feed concentrations. This phenomenon is probably due to leaching of 2 MTBT that had been previously sorbed to the primary column soil during the conditioning phase of the column test. Leaching was not observed for any target compound in any of the secondary column systems, but leaching was consistently observed for the nonlabile compounds in the primary columns.

Table 32–2. Column Test II Summary Results. Fractional Breakthrough (Mass Output/Mass Input) of Trace Organics. Data are Expressed as Fractional Breakthrough (%) and as Mean Fractional Breakthrough ± Standard Deviation (%) for Third and Fourth Inundation Cycles.

Column System	Trace Organic	Inundation Cycle				
		First	Second	Third	Fourth	Fifth
Secondary	ACE	87	120	100 ± 12[b]	67 ± 24	110
	αTER	ND[a]	ND	ND	ND	ND
	SKA	0	0	ND	0 ± 0	ND
	OPP	37	10	46 ± 36	14 ± 17	37
	TMBP	300	110	280 ± 69	140 ± 27	110
	2MTBT	88	25	90 ± 88	54 ± 38	81
Primary	ACE	290	280	290 ± 35	140 ± 50	180
	αTER	9	5	9 ± 4	3 ± 4	0
	SKA	200	34	25 ± 10	4 ± 6	27
	OPP	51	32	54 ± 12	84 ± 21	20
	TMBP	2200	1500	1640 ± 76	3700 ± 400	1940
	2MTBT	2700	720	580 ± 87	660 ± 250	1200

[a]ND denotes that compound was not detected in the feed.
[b]Mean ± standard deviation from four replicates.

Table 32–3. Removal of Target Trace Organics from Primary and Secondary Feeds Prior to Infiltration

Trace Organic	Secondary Feed (µg/l)					Primary Feed (µg/l)					
	Day 0	1	2	4	6	Day 0	1	2	4	6
ACE	0.47	0.27	0.25	0.14	0.27		0.61	0.21	0.32	0.42	0.41
αTER	ND[a]	ND	ND	ND	ND		21.0	ND	ND	ND	ND
SKA	0.10	0.06	0.05	ND	ND		1.3	0.06	ND	ND	ND
OPP	9.4	0.54	0.51	ND	ND		7.8	0.43	3.3	ND	ND
TMBP	1.2	0.76	0.83	0.23	0.35		0.87	0.23	2.0	2.4	2.4
2MTBT	0.25	0.37	0.38	0.11	0.12		0.12	0.19	0.26	0.18	0.12

[a]ND-Not detected.

SKA, OPP, and αTER were very labile in the feed solution and were completely removed before the end of the 6-day feed test (Table 32–3). αTER was by far the most labile trace organic studied; it was completely removed from the primary feed before the first day of the feed test and could not be detected in the secondary feed. SKA disappeared from the secondary feed before the fourth day, and high levels were removed from the primary feed before the second day of the feed test. αTER was not detected in either the secondary feed or the secondary column effluents during the entire column test. Similarly, SKA was detected only sporadically in the secondary feed during the column test, and it was never detected in the secondary column effluents. Primary column effluent concentrations of SKA and αTER were generally less than feed concentrations. The exception occurred with SKA during the first inundation cycle and is again probably related to leaching. Complete removal (less than 10 ng/l) of OPP occurred in both feed solutions before the fourth day.

The occurrence of measurable concentrations of these labile compounds in the primary column effluent has significant consequences. αTER and SKA were reduced to levels below detection in the primary feed before the end of the second day during the feed test (Table 32-3). Yet tracer studies using tritiated water indicated that the hydraulic retention time in the soil columns ranged from 2.6 to 3.1 days. Therefore, if degradation of these compounds continued at the same rate in the soil columns as that observed in the feed test, there should be no detectable concentration of either compound in the primary column effluent during any of the inundation cycles. The fact that measurable concentrations exist in the soil column effluent implies that removal properties inherent in the primary wastewater are inhibited by passage of the feed through the soil matrix. This may indeed occur with OPP as well, but since feed concentrations were not measured on the third day of the feed test, an interpretation of a protective effect is less firm.

Consideration of more refractory compounds highlights the problem of ground-water contamination. The most refractory target compound in this study was 2MTBT, a compound used in fungicide and antimicrobial agents. Results from CTI demonstrated that effluent concentrations of 2MTBT were not affected when a biocide was added to the soil columns during the final inundation cycles. Other researchers have found corresponding resistances of 2MTBT to biodegradation (Chudoba, Tucek, and Zeis, 1977). In the present study, the soil treatment did not affect 2MTBT concentrations in either primary or secondary feed.

The most unusual target trace organic studied was TMBP. Concentrations of TMBP increased with time in primary feed, and column effluent concentrations increased with successive inundation cycles. In secondary feed, concentrations of TMBP decreased with time, and column effluent concentrations remained stable during the course of the test. This anomalous behavior may result from the release of TMBP from an alkylphenol polyethoxylate parent compound. Alkylphenol polyethoxylates are nonionic surfactants that are produced in large quantities in the United States and Europe. These surfactants are considered to be biodegradable, although there have been no studies reporting the formation of the free alkylphenols from their biodegradation. Nevertheless, this type of formation probably does occur, and octylphenol diethoxylate has been tentatively identified in secondary sewage extracts (Stephenou and Giger, 1982). If release of the free alkylphenol from the alkylphenol polyethoxylate is indeed occurring, then this is one case in which the degradation metabolite is more toxic than the parent compound. Alkylphenols, particularly those with alkyl chains ranging from 6 to 12 carbon atoms, are highly toxic to aquatic fauna (McLeese et al., 1981). Therefore, the possible release of these compounds in rapid infiltration systems is of particular concern.

CTIII: Microbial Involvement

Results from the previous test indicated that the soil treatment had little effect in removing trace organics from either primary or secondary effluent. There are several possible failure mechanisms that can account for these results. The most probable cause is that soil conditioning was insufficient to develop an adapted microflora capable of degrading the target compounds under the limiting constraints of short contact time and low concentrations. Therefore, the final column test was designed to use soil from an operational site and to provide information on the dependence of removal efficiency on input concentrations.

It is not possible to present all the results obtained in the context of this short paper. For illustrative purposes the results have been summarized for the six target compounds studied during the first and third inundation cycles. These results demonstrate that the extent of fractional breakthrough is dependent on input concentration and varies with the nature of the target compound (Table 32-4). Each of the target compounds studied, with the possible exception of BZPN, exhibited much higher removals than had been observed for either of the two previous column tests. This treatment efficiency generally increased as higher input concentrations of the target compounds were used, but it was still pronounced at the lowest levels tested. In addition, the infiltration rate for CTIII was about twice that of the two previous tests; thus it is unlikely that failure mechanisms for CTII could be attributed solely to short contact time and low concentrations. Therefore, although alternate failure mechanisms for CTII can be postulated, microbial adaptation is probably a necessary prerequisite for effective trace organic removal during rapid infiltration. Other evidence indicates microbial activity as a fate mechanism in this system. If the columns are operating at steady state, successive inundation cycles should produce similar profiles of fractional breakthrough vs. input concentration for the individual target compounds. A non-steady-state assumption regarding adsorption would imply that fractional breakthroughs would generally increase with successive inundation cycles. This may have been the case with PDCB, since the fractional breakthrough profile increased after the first inundation cycle. The second

Table 32–4. Column Test III. Fractional Breakthrough (FB = Mass Output/Mass Input) as a Function of Input Concentration for First and Third Inundation Cycles

Trace Organic	Inundation Cycle	Column Sets A and B						Column Set C		Column Set D	
		Input µg/l	FB (%)	Input µg/l	FB (%)	Input µg/l	FB (%)	Input µg/l	FB (%)	Input µg/l	FB (%)
PDCB	First	0.28	(0.36)	2.9	(0.03)	11	(0.34)	45	(0.000)	390	(0.047)
	Third			4.2	(0.12)			66	(0.35)	600	(4.1)
2MN	First	0.10	(1.0)	0.75	(0.40)	3.8	(0.00)	34	(0.10)	210	(0.10)
	Third			1.1	(0.00)			13	(0.000)	370	(0.000)
OPP	First	0.78	(9.1)	3.4	(3.5)	14	(2.2)	39	(3.6)	310	(7.2)
	Third			5.9	(1.8)			40	(2.8)	750	(1.4)
TMBP	First	0.37	(9.2)	3.7	(0.86)	7.7	(0.30)	54	(0.41)	360	(0.088)
	Third			4.3	(1.0)			52	(0.067)	660	(0.000)
2MTBT	First	2.1	(6.2)	6.2	(1.6)	12	(0.70)	77	(0.65)	500	(0.74)
	Third	1.6	(4.7)					63	(0.73)	790	(0.91)
BZPN	First	0.33	(52)	4.8	(26)	11	(1.3)	59	(64)	430	(3.2)
	Third			1.5	(8.4)			66	(1.8)	770	(0.34)

and third inundation cycles produced similar profiles for PDCB, implying that a steady-state level may have been attained as regards adsorption. However, fractional breakthroughs for 2MN, TMBP, and BZPN decreased significantly with successive inundation cycles, especially at the higher input concentrations. The most probable explanation for this decrease would be microbial adaptation. Other research indicates that a threshold concentration of 10 μg/l is necessary for adaptation to occur in the degradation of xenobiotics (Spain and Van Veld, 1983). The data for 2MN and BZPN indicate that the threshold may be lower in this system (Table 32-4). Fractional breakthroughs for 2MN decreased from 0.40% during the first inundation cycle to 0.00% during the third inundation cycle upon exposure to concentrations as high as 6.3 μg/l. Similarly, BZPN fractional breakthrough decreased from 52% to 8.4% upon exposure to concentrations as high as 1.5 μg/l.

Microbial activity is further implicated as a fate mechanism for some of the target compounds based upon the relationship of fractional breakthrough to input concentration. Adsorption and volatilization are first-order rate processes. If no limits are imposed on these processes, then fractional breakthrough would be independent of input concentration. There are no discernible lower limits for these processes in this system, although an upper limit for adsorption (i.e., saturation of adsorption sites) may exist. Yet TMBP, 2MTBT, and BZPN show greater fractional breakthroughs at the lower input concentration during the third inundation cycle (Table 32-4) as does 2MN during the second inundation cycle. This indicates that removal mechanisms are operating less efficiently at lower concentrations. This result is consistent with the concept of a minimum substrate concentration below which biodegradation cannot occur (McCarty, Reinhard, and Rittmann, 1981). Column effluent concentrations of 2MTBT and BZPN appeared to approach a minimum substrate concentration of about 0.1 μg/l during the third inundation cycle. Column effluent concentrations of TMBP remained at 0.035 ± 0.008 μg/l during the third inundation cycle regardless of input concentration. Similarly, column effluent concentrations of 2MN remained at 0.014 ± 0.009 μg/l during the second inundation cycle regardless of input concentration, although the concentrations decreased to the detection limit (0.001 μg/l) during the third inundation cycle. Biodegradation of trace organics below the minimum substrate concentration by secondary utilization has been demonstrated (McCarty, Reinhard, and Rittman, 1981), and secondary utilization of trace organics in sewage has been reported as well (Jacobson, O'Mara, and Alexander, 1980). This may be the reason that no minimum substrate concentration effect was observed for PDCB or OPP. Without further information it is not possible to determine if microbial activity is a fate mechanism for these two compounds.

CONCLUSIONS

Trace organic transport through soil columns during rapid infiltration under a variety of conditions has been examined. The results show that trace organic removal is minimal unless an adapted microflora capable of trace organic degradation is present. The extent of trace organic removal is generally dependent not only on the nature of the compound itself but also on its input concentration.

ACKNOWLEDGEMENTS

This work was supported by Cooperative Agreement No. CR806931-0 between the National Center for Ground Water Research and the U.S. Environmental Protection Agency.

REFERENCES

Bouwer, H., and R. L. Chaney, 1974, Land treatment of wastewater, *Adv. Agron.* **26:**133-176.

Bouwer, E. J., P. L. McCarty, and J. C. Lance, 1981, Trace organic behavior in soil columns during rapid infiltration of secondary wastewater, *Water Res.* **15:**151-159.

Chudoba, J., F. Tucek, and K. Zeis, 1977, Biochemischer abbau von benzothiazol-derivaten, *Acta. Hydrochim. Hydrobiol.* **5:**495-498.

Hutchins, S. R., M. B. Tomson, and C. H. Ward, 1983, Trace organic contamination of ground water from a rapid infiltration site: A laboratory-field coordinated study, *Environ. Toxicol. Chem.* **2:**195-216.

Jacobson, S. N., N. L. O'Mara, and M. Alexander, 1980, Evidence for cometabolism in sewage, *Appl. Environ. Microbiol.* **40:**917-921.

Jenkins, T. F., D. C. Leggett, L. V. Farmer, J. L. Oliphant, C. J. Martel, B. T. Foley, and C. J. Diener, 1983, *Assessment of the treatability of toxic organics by overland flow,* U.S. Army Corps of Engineers, Cold Regions Research and Engineering Laboratory (CRREL) Report 83-3.

McCarty, P. L., M. Reinhard, and B. E. Rittmann, 1981, Trace organics in ground water, *Environ. Sci. Technol.* **15:**40-51.

McLeese, D. W., V. Zitko, D. B. Sergeant, L. Burridge, and C. D. Metcalfe, 1981, Lethality and accumulation of alkylphenols in aquatic fauna, *Chemosphere* **10:**723-730.

Majeti, V. A., and C. S. Clark, 1981, Health risks of organics in land application, *Am. Soc. Civ. Eng. Environ. Eng. Div. J.,* **107:**339-357.

Olson, J. V., R. W. Crites, and P. E. Levine, 1980, Ground water quality at rapid infiltration site, *Amer. Soc. Civ. Eng. Environ. Eng. Div. J.,* **106:**885-900.

Sills, M. A., R. Costello, and R. Laak, 1978, Pretreatment techniques and design modification for rapid infiltration land treatment systems, in *State of Knowledge in Land Treatment of Wastewater,* Proceedings International Symposium, August 1978, Hanover, New Hampshire, Vol. 1, pp. 123-131.

Spain, J. C., and P. A. Van Veld, 1983, Adaptation of natural microbial communities to degradation of xenobiotic compounds: Effects of concentration, exposure time, inoculum, and chemical structure, *Appl. Environ. Microbiol.* **45:**428-435.

Stephenou, E., and W. Giger, 1982, Persistent organic chemicals in sewage effluents. 2. Quantitative determinations of nonylphenols and nonylphenol ethoxylates by glass capillary gas chromatography, *Environ. Sci. Technol.* **16:**800-805.

Wilson, J. T., C. G. Enfield, W. J. Dunlap, R. L. Cosby, D. A. Foster, and L. B. Baskin, 1981, Transport and fate of selected organic pollutants in a sandy soil, *J. Environ. Qual.* **10:**501-506.

33

Nutrient Concentrations in Ground Waters from Bermuda: Anthropogenic Effects

J. A. Kent Simmons, Timothy Jickells, and Anthony Knap

Bermuda Biological Station for Research

Wm. Berry Lyons

University of New Hampshire

INTRODUCTION

Fresh water has long been a precious commodity in Bermuda. The island has no freshwater lakes, rivers, or reservoirs, and ever since early colonization Bermudian residents have had to resort to elaborate techniques for collecting fresh water. The primary method of obtaining potable water has been collection of precipitation using specially designed roof and hillside catchments coupled with large storage tanks. Bermuda's annual rainfall is 140 cm (Macky, 1957). For normal domestic usage this meteoric source is quite adequate, but during periods of drought the island may suffer serious shortages. On several occasions in the past the situation has become critical enough to warrant import of fresh water from North America. More recently, the implementation of distillation and reverse osmosis techniques have alleviated the problem, and measures as costly as importation have not been necessary since 1969 (Hayward, Gomez, and Sterrer, 1981).

Since the mid-1940s, freshwater usage has increased from about 30 l per person per day to about 100 l per person per day (Hayward, Gomez, and Sterrer, 1981). The overall rise in consumption can be attributed to increased use of household conveniences such as washing machines and dishwashers. In addition, the growth of tourism, the island's major industry, has contributed to the demand for potable supplies. Typical consumption figures for tourists can run as high as 450 l per person per day (Hayward, Gomez, and Sterrer, 1981).

Development of the ground waters was attempted in the early 1900s, but the occurrence of freshwater lenses does not extend islandwide; most of the early attempts at tapping the resource were abandoned when brackish and salty ground waters were encountered. In 1932, however, the first commercial development of fresh ground-water supplies began with the opening of Watlington Water Works in Devonshire Parish (Hayward, Gomez, and Sterrer, 1981). Although these waters were seldom used for drinking, they conveniently supplemented potable supplies for cleaning and flushing. By 1972 the Bermuda government, in hopes of large-scale development of ground water, inaugurated extensive hydrogeological investigations across the island to locate and map the extent of fresh reserves (Vacher, 1974; Hayward, Gomez, and Sterrer, 1981). The results of this major study were released in a 1974 government publication (Vacher, 1974).

The major problem facing ground-water resources worldwide is contamination by solid, liquid, and chemical waste disposal (Freeze and Cherry, 1979). Although ground water is not as easily polluted as surface waters, once damaged, long periods of time may be necessary for recovery (Freeze and Cherry, 1979).

In 1949 the possibility of unsafe ground water was recognized in Bermuda, and consumption of water from unauthorized wells was forbidden (Hayward, Gomez, and Sterrer, 1981). It is estimated that in the Devonshire area 27% of the freshwater recharge results from septic tank effluent, while another 2% comes from sewage plant input (Vacher, 1974; Hayward, Gomez, and Sterrer, 1981). This implies that chemical compounds such as nutrients (N,P,C) may be mobilized into the aquifer.

If the Bermuda government's policy for the future is to expand usage of this highly significant water resource, it will be necessary to have a better understanding of the chemical, geological, and physical processes that determine the state of the ground water on the island.

The purpose of this study was (1) to investigate some of the primary chemical constituents (such as nutrient compounds, nitrate, nitrite, ammonium, and phosphate) present in Bermuda ground waters, (2) to determine their sources, and (3) to establish whether or not they indicate any potential threat to human health. High levels of nitrogenous compounds have long been an indication of sewage pollution. Virtually all of Bermuda's domestic waste is disposed of via unlined subsurface cesspits that, in order to avoid emptying, are specially designed to promote seepage into the limestone bedrock.

The physical mixing processes controlling these nutrient concentrations in the aquifer and the geochemical mechanisms by which they are removed will also be considered. Phosphate, for example, is the limiting nutrient in the Bermuda marine environment (Morris et al., 1977). This compound reacts quite strongly with calcium carbonate (Ames, 1959; Martens and Harriss, 1970; Kitano, Okumura, and Idogaki, 1978). Phosphate-calcium carbonate reactions may provide a mechanism that inhibits eutrophication processes in Bermuda's inshore waters.

Bermuda Ground Water

In an oceanic island like Bermuda, fresh ground water, which is less dense, floats on top of the more dense saline water. This buoyant freshwater nucleus is called a Ghyben-Herzberg lens and is the result of water percolation through soils (Vacher, 1974; Freeze and Cherry, 1979). Ideally the freshwater-saltwater interface is very sharp, but due to tidal oscillation, storm surges, and diffusion there is some mixing (Vacher, 1974, 1978*a*, 1978*b*).

Five major Ghyben-Herzberg lenses have been located in Bermuda: Somerset, Southampton, Warwick, Devonshire, and St. Georges. These lenses account for 20% of the island's area. The largest is the Devonshire lens; it is centrally located, has a maximum thickness of 15 m and a volume of $7.57 \times 10^6 m^3$, and covers an area of about 6.7 km^2, or twice that of the other four lenses combined (Vacher, 1974, 1978*b*; Plummer et al., 1976).

MATERIALS AND METHODS

Study Area

The major area chosen for study is the Devonshire freshwater lens (Fig. 33-1). The Devonshire lens is centrally located and is the major site of the Bermudian government's commercial extraction. The city of Hamilton lies at the southeastern border of the lens, and the surrounding area is probably the most densely populated area of the island. Also, located just north

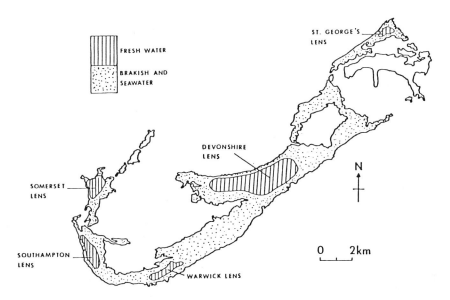

Figure 33-1. Contour map showing the location of freshwater nuclei. (*Source: Vacher, 1974, Fig. 2–6*).

of the city in the Pembroke marsh is the primary landfill for the garbage-pulverization plant.

The lens has been the subject of extensive geological studies, mapping, and pumping tests, and therefore general background information is available (Vacher, 1974). In addition, the Bermuda Department of Public Works maintains approximately 40 observation wells throughout the area. These wells have a diameter of about 20 cm and range in depth from a few meters to a few tens of meters. Some are lined at the top with polyvinyl chloride (PVC) pipe and others with metal pipes, but beyond the top 2 m, the walls are bare rock.

Samples were collected from a total of 12 stations across the lens (Fig. 33-2). Four of these wells were selected for depth-profile studies; however, only one of them will be discussed here. The data from the other stations are discussed in Simmons (1983).

The profiles were obtained from a well in the southeastern corner of the West Pembroke elementary school playing field (WPS). The water depth in this well is 10.6 m, and the top of the well is lined with metal pipe. The location of the school is approximately 200 m north of the well, and the ocean is 200 m downslope beyond this. The exact location of the septic tank that serves the school is not known, but it is probably at least 150 m from the well. To the south and east the houses are located as close as approximately 10 m from the well atop a 6-m limestone cut embankment. To the west the closest houses are approximately 100 m from the well.

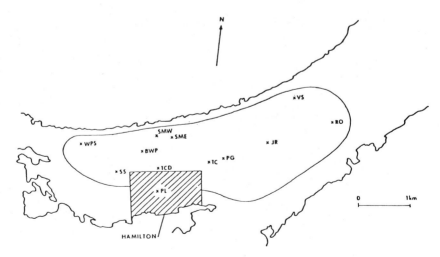

Figure 33-2. A map of the central parishes indicating the 12 stations from which ground water samples were collected.

Sampling and Analysis

The wells were sampled with a flow-through PVC sampling device of our own design and construction (Simmons, 1983). Prior to sample collection the sampler had been soaked in dilute nitric acid and rinsed thoroughly with distilled-deionized water. Blank determinations were conducted on distilled-deionized water samples that had been left in the sampler for a two-hour period. Nitrate concentrations in all cases were less than 0.05 μM (Simmons, 1983).

Samples were immediately transferred to precleaned 120-ml polyethylene bottles while in the field, packed in ice, and rapidly transported to the Bermuda Biological Station (BBS) in St. Georges for analysis. Nutrient determinations were begun within one to three hours of sample collection. The samples were first filtered through prerinsed GF/C glass-fiber filters and analyzed spectrophotometrically using the techniques outlined in Strickland and Parsons (1972). Dissolved oxygen was determined using the modified Winkler technique, also described in Strickland and Parsons (1972). Ammonium, nitrite, and reactive phosphate were analyzed the same day of collection, while nitrate samples were refrigerated overnight and analyzed the next morning. The precision of these measurements reported as coefficients of variation was ±8% for ammonium, ±2% for nitrate, ±0.3% for nitrite, and ±0.1% for reactive phosphate. All samples were collected between July 7 and August 19, 1982.

RESULTS

Assumptions

All ground water samples were collected from wells 20–30 cm in diameter. It is assumed that horizontal water flow through the wells is greater than vertical mixing, and the water in the wells is representative of the water in the surrounding aquifer.

Data describing areal distribution of chemical constituents are derived from water samples taken within 1.5 m of the piezometric surface. The nature of the ground water at depth differs only as a result of oxidation-reduction reactions, physical mixing processes and so forth.

Areal Distribution

Nitrate. The average concentration of nitrate for the Devonshire lens is 749 μmol/l, a value that slightly exceeds the USEPA (1975) and WHO (1970) drinking-water standard of 726 μmol/l (Freeze and Cherry, 1979). Nitrate concentrations range within the lens from 1 μM to just over 2300 μM. The

Table 33-1. Comparison of Mean Nitrate Values from Bermuda with
Background Values from Other Carbonate Areas

	NO_3^- μM	PO_4^{3+} μM	N/P
Bermuda	749	2.04	367
(this work)	<1-2300	<1-~4	
Jamaica	89.5	—	—
(D'Elia, Webb, and Porter, 1981)			
Missouri	89.5	1.93	46.4
(Murray, Rouse, and Carpenter, 1980)			

lowest values are associated with the shallow wells (0-7 m to the water table) in both Pembroke and Devonshire marshes, while the highest concentrations are centered around the deeper wells (15-35 m to the water table) of the densely populated areas near St. Monica's Road, Prospect, and the city of Hamilton. Elevated nitrate levels are also seen at RO and WPS with approximately 1500 μM at the former and 550 μM at the latter. Both areas appear to be slightly less populated than those areas bordering the city. Bermuda data are compared to other carbonate aquifer data in Table 33-1.

It seems very unlikely that nitrate values of this magnitude would occur naturally, for example, from leguminous and nonleguminous terrestrial nitrogen-fixing organisms, plant decomposers, or rainwater (D'Elia, Webb, and Porter, 1981). The fact that Pembroke parish with 2200 persons/km^2 is the most densely populated region of the island (Census Report, 1980), and that cesspits are the predominate method for disposal of domestic wastes, suggests that septic seepage is contaminating the aquifer in this area.

Understanding how nitrate may be transported to the water so that its concentration is one to two orders of magnitude higher than that of rural areas requires discussion of a key mechanism controlling recharge. In undeveloped regions of the island, only 10% of the precipitation reaching the soil contributes to recharge of the aquifer, while 90% returns to the atmosphere via evapotranspiration (Vacher, 1974). However, when a large area is covered by roof catchments, a much greater percentage of the precipitation is channeled past the soil cover and reaches the water table, passing through cesspits en route (Vacher, 1974). Hence, not only does the volume of recharge increase, but since it has passed through a cesspit, its nitrogen concentration could increase markedly. Therefore, the high nitrate values may be due to cesspit leachate development. This mechanism is discussed in detail in Simmons (1983).

Ammonium. Ammonium represents the most highly reduced species of nitrogen and generally results from the decomposition of organic material in oxygen-depleted waters. At 10 of the 12 stations across the lens, ammonium concentrations show the expected inverse relationship to nitrate. That is, sites displaying characteristically high nitrate concentration in the surface waters display negligible ammonium values, and vice versa. Distinctively higher ammonium values are seen at BWP and VS in the Pembroke and Devonshire marshes. Strangely, however, JR, another marsh station, shows fairly low ammonium concentrations. At PG, ammonium values are slightly elevated, although nitrate data still suggest an oxygen rich system.

Phosphate. Phosphate levels in Bermuda are very low in comparison to nitrate. Ordinarily, sewage N:P ratios are on the order of 9:1 (Brandes, 1978; Murray, Rouse, and Carpenter, 1980), and in plant material ratios are approximately 15:1 (forested ecosystem) (Likens et al., 1977). In most samples collected throughout the lens, it was not uncommon to observe N:P ratios as high as 1000:1. The average phosphate concentration was 2.04 μmol/l, with most wells having values less than 1 μM. The adsorption of phosphate by calcium carbonate has been documented by Martens and Harriss (1970), Griffin and Jurinak (1974), and Kitano, Okumura, and Idogaki (1978). Previous studies on phosphate uptake in Bermuda have shown the limestones to provide a significant removal mechanism (Bodungen et al., 1982). Although limestone removal of phosphate appears to be highly efficient, it is not completely removed from the ground water. All wells displayed measurable phosphate concentrations. Sites notable for higher phosphate values were PL, SME, and Tech. These are densely populated areas that also have high values of nitrate. SMW, a station not far from SME, also displayed high nitrate, but no anomalous phosphate was seen. N:P ratios, however, are still on the order of 400-1000:1.

Relatively high phosphate occurs at JR (1.95 μM) and VS (8.69 μM), while low phosphate occurs at BWP and TCD. Morris et al. (1977) report levels as high as 70 μM in the Pembroke dump area, but values this high were not observed in this study. The highest values were seen at VS in Devonshire marsh (3.5-11 μM). This well had water right to the land surface. The water contained much plant debris and lacked direct contact with the limestone.

Depth Profiles. Depth profiles of ΣN (nitrate + nitrite + ammonium) and the various N-species are shown for the WPS well in Figures 33-3, 33-4a, and 33-4b, respectively. The ΣN depth profile (Fig. 33-3) reaches its maximum concentration between 6 and 8 m below the water table. This maximum has been shown to be related to a layer of water that is derived directly from precipitation and percolation (Simmons, 1983). Precipitation that ordinarily reaches the water table via soil percolation may bypass the soil zone and recharge the aquifer through cesspit seepage (Vacher, 1974). Considering the surrounding populace at WPS, it is conceivable that this septic tank recharge mechanism may explain the high ΣN concentrations.

A very high percentage of the ΣN is comprised of the nitrate species, with nitrite and ammonium being very low. Domestic sewage, however, contains little if any nitrate. Most of the nitrogen is tied up as organic-N or ammonium (Benfield, Judkins, and Weand, 1982). Given that the source of ΣN is cesspit leachate, a high degree of nitrification could have occurred within the aerated zone of the aquifer.

On August 10 the maximum concentration of nitrate encountered was 787 μM. Since the organic-N must degrade via ammonium to nitrate in the molar ratio of 1:1, nitrification of 787 μM ammonium was necessary to produce an equivalent amount of nitrate by the equation

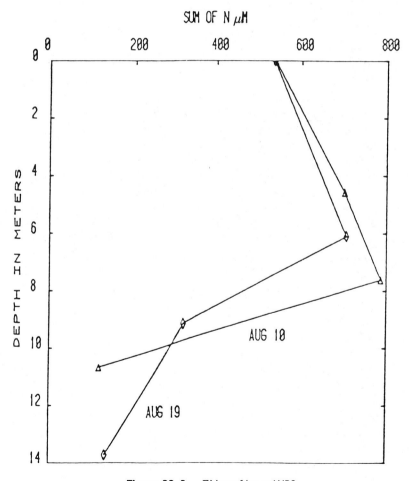

Figure 33–3. ΣN profiles at WPS.

$2NH_4^+ + 4O_2 \rightarrow 2NO_3^- + 4H^+ + 2H_2O$ (Benfield, Judkins, and Weand, 1982).

According to the stoichiometry, the oxygen utilization for this reaction is 1574 μM. For a 1-liter sample this is equivalent to approximately 35 ml of O_2 or 176 ml of air. Assuming Bermuda limestones have a porosity of about 20% (Vacher, 1974, 1978b; Freeze and Cherry, 1979), approximately 900 cm^3 would contain enough O_2 to completely oxidize 787 μmol/l of ammonium.

In addition to nitrification, other reactions that contribute to the decomposition of organic material also utilize considerable amounts of O_2. This suggests that continued large-scale cesspit usage may lead to O_2 depletion within the aerated zone.

Variations in the water table elevations as a means for gas exchange are considered. Oscillation of the entire lens has been documented by Vacher (1974, 1978a). It occurs as a result of tidal action, atmospheric pressure changes, and recharge. This type of upward and downward behavior of the lens may provide a mechanism by which a fairly rhythmic and frequent exchange of air between the land and the atmosphere occurs, maintaining an environment within the rock conducive to aerobic oxidation of organic materials and, therefore, minimizing the possibility of O_2 depletion in the aquifer.

In the WPS well, phosphate concentrations average approximately 0.8 μM at the surface and reach a maximum of 2.1 μM near the bottom and a minimum of 0.4 μM at a depth of 6-8 m (Fig. 33-5). The minimum corresponds with the anomalously high ΣN intrusion (Fig. 33-3). These septic effluents should have the highest phosphate levels. It appears that phosphate is preferentially removed within this layer. These findings are similar to those of Sewell (1982), who found that phosphate was preferentially removed to nitrate in polluted urban ground water in western Australia. Therefore, contrary to the predicted responses, phosphate concentrations show a highly correlated inverse relationship to ΣN. This phosphate removal is due to either: (1) adsorption of phosphate onto calcium carbonate, leading to the formation of the mineral apatite ($Ca_{10}(PO_4)_6F_2$), as suggested by Ames (1959) and Simpson (1964); or (2) the adsorption of phosphate onto iron-oxyhydroxides in the phreatic zone. Phosphorus has been shown to be associated with hydrated ferric oxide coatings and mineral phases in both oxidized marine and freshwater sediments (Strom and Biggs, 1982; Williams, Jaquet, and Thomas, 1976). In addition, terra rosa paleosols in Bermuda have been shown to contain high concentrations of phosphorus (Wilson, 1979).

The higher concentrations of phosphate with depth may be due to the inhibition of phosphorus adsorption or even desorption at the higher salinities in the more saline portion of the ground water, as discussed by Simmons (1983). The N:P ratio in these ground waters decreases with depth and increasing salinity to values approaching that of nearshore Bermuda ocean water (Simmons, 1983).

(a)

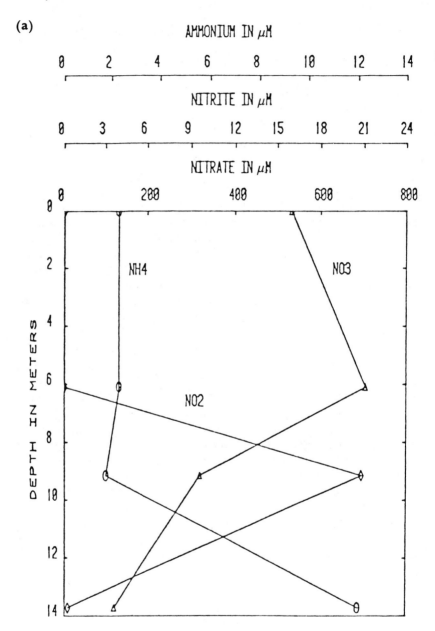

Figure 33–4. (*a*) Nitrate, nitrite, and ammonium profiles at WPS on August 10, 1982. The decrease in nitrate along with the increase in nitrite and ammonium are indicative of suboxic biogeochemical zonation. (*b*) Nitrate, nitrite, and ammonium profiles at WPS on August 19, 1982.

(b)

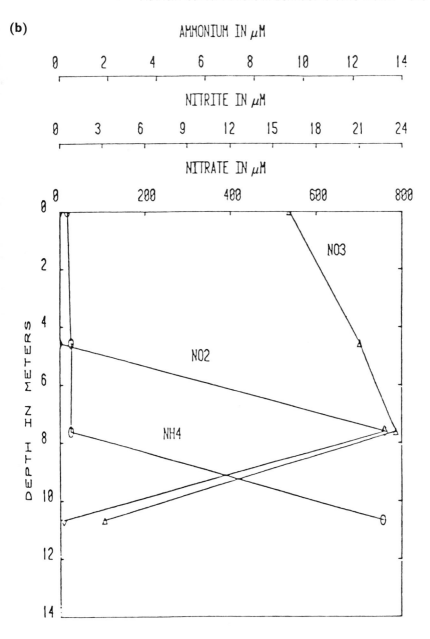

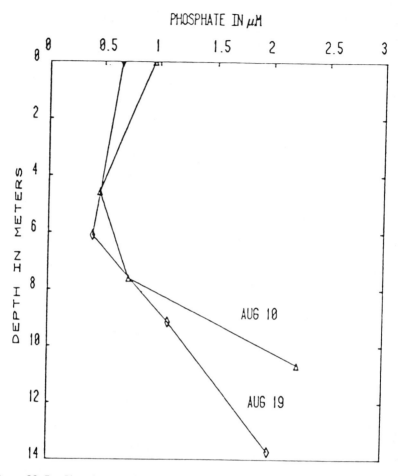

Figure 33-5. Phosphate profiles at WPS. Phosphate is removed from solution between the surface and the septic layer at 6 m while below this depth its removal appears to be either inhibited or possibly desorbed.

DISCUSSION

Biogeochemistry in the WPS Well

Biogeochemical zonation in ground water, analogous to that found in marine sediments, has been discussed by Champ, Gulens, and Jackson (1978) and Baedecker and Back (1979). These zones are induced by the microbial decomposition of organic matter and controlled by the thermodynamic favorability of the electron-transferring oxidation reactions (Froelich et al., 1979; Champ, Gulens, and Jackson, 1978; Berner, 1980). In sediments, the zonal boundaries are dependent on rates of sediment and organic-matter accumulation and the concentration of the different electron acceptors, while in ground-water systems the boundaries are more susceptible to ground-water flow and dispersion within the aquifer (Baedecker and Back, 1979).

Three major biogeochemical zones can usually be distinguished in a vertical profile of marine sediments. The oxic and suboxic zones closest to the sediment-water interface are comprised of the oxygen-reduction zone, followed by the nitrate-reduction zone and the manganese (IV)- and ferric iron-reduction zone. At depth, anoxic conditions bring on the reduction of sulfate, followed by carbonate reduction (Froelich et al., 1979; Champ, Gulens, and Jackson, 1978; Baedecker and Back, 1979; Berner, 1980). Fermentation reactions can occur throughout all these zones (Claypool and Kaplan, 1974).

These reactions that result in the decomposition of organic material are mediated by bacteria. The reaction that yields the greatest amount of free energy proceeds first, and those giving the least occur last in the sequence (Froelich et al., 1979; Berner, 1980).

The chemistry in the WPS well is controlled by three main factors: (1) the input of high concentrations of nitrate; (2) conservative mixing (dilution) of this water mass in the water column, and (3) oxygen depletion below the surface (Simmons, 1983). With some understanding of input and mixing parameters, one can incorporate the dissolved oxygen (DO) data and discuss the biogeochemical reactions within the aquifer.

At the piezometric surface DO appears to be at equilibrium with the atmosphere for the observed temperature and salinity (5.93-6.47 ml O_2/l). Since O_2 is abundant, aerobic respiration (i.e., oxygen reduction) is the dominant decomposition reaction. From the surface to 6-8 m, the ΣN concentration increases, but below that second water layer the ΣN concentration decreases with depth. This is due obviously to mixing with the very low nitrogen (0.2 μM) sea water. But in addition to the dilution process, formation of nitrite and ammonium are shown to play a more significant role (Fig. 33-4). As a result of decomposition, the DO concentration in the deeper water has become diminished, and nitrate becomes the next electron acceptor in the oxidation of organic matter. As stated above, the two possible end products of denitrification are ammonium and dinitrogen. Increases in ammonium are observed, but it may also be possible to determine the amount of ΣN lost to nitrogen (N_2) and other gaseous intermediates, such as nitrous oxide (N_2O).

It is possible to calculate the amount of sea water dilution that has occurred at a specific depth. On August 10 the depth of maximum ΣN occurred at 7.6 m. The water of 10.6 m was determined, by use of salinity data, to have been diluted by a factor of 4.5 (Simmons, 1983). Assuming the 0.2-μM nitrate concentration of sea water to be negligible compared to the

ΣN maximum, a ΣN concentration of 176.6 μmol/l should have been present at this depth; 118 μmol/l were observed. This implies that 58.3 μmol/l could have been lost to gaseous products, most probably N_2. The following week, at depths of 9.1 and 12.7 m, 29.2 μmol/l and 29.7 μmol/l of ΣN were calculated to have been lost by this mechanism.

CONCLUSIONS

The following conclusions can be drawn from this study: (1) High concentrations of nitrogenous compounds, particularly nitrate, occur in areas of higher population density [e.g., Hamilton, St. Monica's Road, Prospect, and near the reverse-osmosis (RO) plant in Devonshire]. Considering that the use of unlined cesspits is the dominant method of domestic waste disposal, there is a strong possibility that septic seepage is the primary source. (2) Either calcium carbonate, or ferric oxyhydroxides, or both appear to play a significant role in the removal of phosphate from the ground water. The salinity of the ground water may exert some control on this process. Phosphate concentrations appear to increase at higher salinities; this may be a result of adsorption inhibition due to competition by sodium and magnesium ions present in sea water, as outlined in Martens and Harriss (1970) and Kitano, Okumura, and Idogaki (1978). Thermodynamic calculations indicate that the major P species in the freshwater zone of the ground water is HPO_4^{2-}, while the major P species in the transition zone of higher salinity are $CaHPO_4^0$ and $MgHPO_4^0$. This change of P speciation with increasing salinity (i.e., depth) may also have an effect on P adsorption/desorption mechanics. (3) Due to the decomposition of organic material, some wells become oxygen deficient below the piezometric surface. As a result, the development of biogeochemical zones much like those seen in marine sediments occurs. However, due to varying velocities and dispersion of ground water, biogeochemical zones may be more transient and exhibit a considerable amount of overlapping.

ACKNOWLEDGEMENTS

This work was funded in part by a Bermuda Junior Service League Scholarship. We are grateful to M. Rowe and J. Conyers of the Department of Public Works in Bermuda for furnishing well access, for providing valuable background information, and for their time. We appreciate the help of M. J. Spencer and T. C. Loder, University of New Hampshire, for their analytical help and data interpretation. We also appreciate the help of numerous BBS volunteers.

REFERENCES

Ames, L. L., 1959, The genesis of carbonate apatite, *Econ. Geol.* **54:**829-841.

Baedecker, M. J., and W. Back, 1979, Modern marine sediments as a natural analog to the chemically stressed environment of a landfill, *J. Hydro.* **14:**393-414.

Benfield, L. D., J. F. Judkins, and B. L. Weand, 1982, *Process Chemistry for Water and Wastewater Treatment,* Prentice Hall, Inc., Englewood Cliffs, New Jersey.

Bermuda Government, 1980, *Report of the Population Census,* Vol. II, Bermuda Government Publication.

Berner, R. A., 1980, *Early Diagenesis: A Theoretical Approach,* Princeton University Press, Princeton, New Jersey.

Bodungen, B. V., T. D. Jickells, S. R. Smith, J. A. D. Ward, and G. B. Hillier, 1982, *The Bermuda Marine Environment,* Vol. II, Bermuda Biological Station for Research, Spec. Pub. No. 18.

Brandes, M., 1978, Characteristics of effluent from gray and blackwater septic tanks, *Wat. Pollut. Control Fed. J.* **50:**2547-2559.

Champ, D. R., J. Gulens, and R. E. Jackson, 1978, Oxidation reduction sequences in ground water flow systems, *Can. J. Earth Sci.* **16:**12-23.

Claypool, G. E., and I. R. Kaplan, 1974, The origin and distribution of methane in marine sediment, in *Natural Gases in Marine Sediments,* I. R. Kaplan, ed., Plenum Press, New York, pp. 99-139.

D'Elia, C. F., K. L. Webb, and N. W. Porter, 1981, Nitrate-rich groundwater inputs to Discovery Bay, Jamaica: A significant source of N to local coral reefs? *Bull. Mar. Sci.* **31:**903-910.

Freeze, R. A., and J. A. Cherry, 1979, *Groundwater,* Prentice Hall, Inc., Englewood Cliffs, New Jersey.

Froelich, P. N., G. P. Klinkhammer, M. L. Bender, N. A. Luedtke, G. R. Heath, D. Cullen, P. Dauphin, D. Hammond, B. Hartman, and V. Maynard, 1979, Early oxidation of organic matter in pelagic sediments of the eastern equatorial Atlantic: Suboxic diagenesis, *Geochim. Cosmochim. Acta* **43:**1075-1090.

Griffin, R. A., and J. J. Jurinak, 1974, Kinetics of the phosphate interaction with calcite, *Soil Sci. Soc. Am. Proc.* **38:**75-79.

Hayward, S. J., V. H. Gomez, and W. Sterrer, 1981, *Bermuda's Delicate Balance: People and the Environment,* Bermuda Biological Station for Research, Spec. Pub. No. 20.

Kitano, J., M. Okumura, and M. Idogaki, 1978, Uptake of phosphate ions by calcium carbonate, *Geochem. J.* **12:**29-37.

Likens, G. E., F. H. Bormann, R. S. Pierce, J. S. Eaton, and N. M. Johnson, 1977, *Biogeochemistry of a Forested Ecosystem,* Springer-Verlag, New York.

Macky, W. A., 1957, *The Rainfall of Bermuda,* Bermuda Meteorological Office Technical Note No. 8.

Martens, C. S., and R. C. Harriss, 1970, Inhibition of apatite precipitation in the marine environment by magnesium ions, *Geochim. Cosmochim. Acta* **34:**621-625.

Morris, B., J. Barnes, F. Brown, and J. Markham, 1977, *The Bermuda Marine Environment,* Vol. 1, Bermuda Biological Station for Research, Spec. Pub. No. 15.

Murray, J. P., J. V. Rouse, and A. B. Carpenter, 1980, Groundwater contamination by sanitary landfill leachate and domestic wastewater in carbonate terrain: Principal source diagnosis, chemical transport characteristics and design implications, *Water Res.* **15**:745-757.

Plummer, L. N., H. L. Vacher, F. T. Mackenzie, O. P. Bricker, and L. S. Land, 1976, Hydrogeochemistry of Bermuda: A case history of groundwater diagenesis of biocalcarenites, *Geol. Soc. Am. Bull.* **87**:1301-1316.

Sewell, P. L., 1982, Urban groundwater as a possible nutrient source for an estuarine benthic algal bloom, *Estuar. Coastal Shelf Sci.* **15**:569-576.

Simmons, J. A. K., 1983, *The Biogeochemistry of the Devonshire Lens, Bermuda,* M. S. thesis, University of New Hampshire, Durham.

Simpson, D. R., 1964, The nature of alkali carbonate apatites, *Am. Mineral.* **49**:363-376.

Strickland, J. D. H., and T. R. Parsons, 1972, *A Practical Handbook of Seawater Analysis,* 2nd ed., Fish. Res. Bd. Can. Bull. 167.

Strom, R. N., and R. B. Biggs, 1982, Phosphorus distribution in sediments of the Delaware River Estuary, *Estuaries* **5**:95-101.

Vacher, H. L., 1974, *Groundwater Hydrology of Bermuda,* Bermuda Public Works Department.

Vacher, H. L., 1978*a*, Hydrology of small oceanic islands—Influence of atmospheric pressure on the water table, *Groundwater* **16**:417-423.

Vacher, H. L., 1978*b*, Hydrogeology of Bermuda—Significance of an across-the-island variation in permeability, *J. Hydrol.* **39**:207-226.

Williams, J. D. H., J. -M. Jaquet, and R. L. Thomas, 1976, Forms of phosphorus in the surficial sediments of Lake Erie, *Can. J. Earth Sci.* **33**:413-429.

Wilson, K. M., 1979, *Diagenesis of Phosphorus in Carbonate Sediments from Bermuda,* M. S. thesis, University of New Hampshire, Durham.

34

In situ Characterization of Microorganisms Indigenous to Water-Table Aquifers

David L. Balkwill

Florida State University

William C. Ghiorse

Cornell University

INTRODUCTION

Water-table aquifers are environmentally significant because they contain ground water reserves that represent most of the fresh water (95% in the United States; Josephson, 1980) available for irrigation or consumption. With the occurrence of organic pollutants in ground waters becoming an increasingly widespread problem (Council on Environmental Quality, 1981), it is important to define and develop an understanding of factors that affect ground-water quality.

Microorganisms could play a major role in maintaining ground-water quality, considering that they can profoundly affect biological and chemical activities in surface soils and other environments (Alexander, 1977). However, the microorganisms in aquifers have been studied only rarely (Dunlap and McNabb, 1973), perhaps because early reports of soil microbiologists (e.g., Waksman, 1916) indicated that the number of microorganisms in soil drops sharply with increasing depth. More recent studies have shown that microorganisms probably are present at considerable depths in the subsurface (Dockins et al., 1980; Dunlap et al., 1972; Whitelaw and Edwards, 1980; Whitelaw and Rees, 1980), but the possibility of contamination from the surface soil has hampered the interpretation of these results. Thus, detailed data on the occurrence and numbers of microorganisms in aquifers remains scant (Dunlap and McNabb, 1973). Even less is known about the *in situ* metabolic activities of such organisms or about how these activities may affect organic contaminants of ground water.

In 1979, efforts were initiated to obtain more detailed information on the aquifer microflora by direct observation of *in situ* microorganisms in subsurface samples. Traditional cultural methods were deemed mostly unsuitable for these initial studies because they were not likely to select for many of the significant organisms in subsurface samples (Ghiorse and Balkwill, in press). Instead, light- and electron-microscopy methods for direct observation of microbial cells in surface soils were modified for application to subsurface materials. The present report reviews the information regarding the characterization of microorganisms in water-table aquifer materials that have been obtained with this approach (Ghiorse and Balkwill, 1981, 1983; Wilson et al. 1983).

MATERIALS AND METHODS

Description of Samples

Samples were collected from a total of four sites in Louisiana, Oklahoma, and Texas, from above and below the water table at each site. Subsurface regions situated above aquifers were sampled because they are likely to affect water that travels from the surface to aquifers below. The samples and their origins are listed in Table 34-1; for more detailed information, see the original references cited in the table. Aquifer and subsurface samples were collected aseptically by using a modification (Wilson et al., 1983) of the procedures developed by Dunlap et al. (1977).

Acridine Orange Direct Counts (AODC)

Epifluorescence light microscopy (LM) of acridine orange (AO)-stained samples was used to determine the morphological characteristics and the total numbers of cells by direct counts (AODC). A modification of Trolldenier's (1973) method was used to determine the AODC as described by Ghiorse and Balkwill (1983).

Respiring Bacteria

The proportion of AODC bacteria capable of reducing 2-(p-iodophenyl)-3-(p-nitrophenyl)-5-phenyl tetrazolium chloride (INT)—that is, the proportion of respiring bacteria (Zimmermann, Iturriage, and Becker-Birck, 1978)—was determined by mixing 2.5 g of subsurface material with 20 ml of filter-sterilized 0.1% sodium pyrophosphate (SPP) in a 125-ml Erlenmeyer flask. The mixture was shaken at room temperature for 15 min at 160 rpm;

Table 34-1. Aquifer and other Subsurface Samples Examined to Date

Sampling Site	Borehole	Depth (m)	Saturation[a]	Reference
Lula, Okla.	—	1.2	U	Wilson et al., 1983
(Sampson Ranch)		3.0	I	
		5.0	S	
Pickett, Okla.	—	5.5	S	Ghiorse and Balkwill, in press
(Walters Ranch)				
Ft. Polk, La.	6B	3.7	U	Ghiorse and Balkwill, 1983
	6B	5.5	I	
	6B	6.7	S	
Ft. Polk, La.	7	1.2	U	Ghiorse & Balkwill, 1983
	7	4.0	I	
	7	5.1	S	
Conroe, Tex.	7G	5.4	I	—

[a]U = unsaturated sample, taken above water table; I = sample taken at interface between saturated and unsaturated zones; S = saturated sample, taken below water table.

2.5 ml of 0.2% aqueous INT was added, and shaking was resumed for an additional 15 min. Excess unreduced INT was removed by centrifuging the entire contents of the flask at 10,000 rpm for 10 min. The supernatant fluid was decanted, and the pellet was washed by centrifugation in 10 ml of 0.1% SPP. The final pellet was resuspended in 22.5 ml of 0.1% SPP, and the AODC procedure (Ghiorse and Balkwill, 1983) was followed.

To count bacteria containing INT-formazan deposits, green fluorescent cells were first identified under epi-illumination. These were then inspected for the presence of INT-formazan employing a 100X bright field objective lens. Care was taken to use bright field illumination conditions that optimized recognition of the red formazan deposits in the cells. This included adjusting the substage iris diaphragm and the illuminator rheostat to the same setting each time, as well as the use of neutral-density filters to reduce brightness of the field.

INT-containing bacteria of two types were counted. One type was characterized by diffuse but distinctly reddish cells with no apparent granules. The second type was characterized by the presence of distinct red granules inside the cell.

Plate Counts

Standard plate counts in triplicate were used to estimate the number of viable microorganisms in subsurface environments. Both nutritionally rich (PYG or 1%-5% PYG agar; Ghiorse and Balkwill, 1983) and low-nutrient (SEA; Wilson et al., 1983) media were used in all cases. All plates were incubated aerobically at 25°C, and colonies were counted after one to two weeks of incubation.

Transmission Electron Microscopy (TEM)

Transmission electron microscopy (TEM) was used to determine the ultrastructural characteristics of subsurface microorganisms. Microbial cells were released and concentrated from subsurface materials prior to TEM examination with the centrifugal washing method described by Ghiorse and Balkwill (1983).

RESULTS

Morphological Characteristics of Aquifer Microorganisms

Epifluorescence light microscopy of AO-stained preparations readily detected microbial cells in all of the subsurface samples examined (Table 34-1). Objects of microbial size that fluoresced bright green, thereby indicating that they contained double-stranded DNA (Daley and Hobbie, 1975), and that possessed appropriate morphological characteristics were considered to be microbial cells. Such cells (mostly bacteria) stood out clearly against a dull orange background of fluorescing abiotic material.

Epifluorescence LM was useful for assessing the range of morphological diversity in each sample and for detecting the occurrence of microcolonies (for illustrations of these results, see Fig. 1 in Ghiorse and Balkwill, 1983, and Fig. 2 in Wilson et al., 1983). Microcolonies (groups of cells with similar morphological characteristics) were present in all samples, but the range of morphological diversity varied considerably in samples from one location to another. Texas and Oklahoma samples contained mostly small, coccoid bacterial cells that were similar in shape to those found in surface soils (Bae, Cota-Robles, and Casida, Jr., 1972; Balkwill and Casida, 1973; Balkwill, Labeda, and Casida, Jr., 1975; Balkwill, Rucinsky, and Casida, Jr., 1977). Few, if any, eukaryotic forms were detected.

In contrast, the samples from Louisiana contained a greater variety of bacterial forms, including small coccoid cells, rod-shaped cells of varying dimensions, and actinomycetes or other filamentous types. Some of these samples contained small numbers of microeukaryotic forms. Very recently, a cyst-forming amoeba and several fungi have been detected in Oklahoma samples by special cultural methods (J. Sinclair, pers. comm.). These microeukaryotes were present in low numbers in comparison to bacteria and, therefore, do not appear to account for a significant portion of the biomass in the sample (see also Hirsch and Rades-Hohkohl, 1983).

Ultrastructural Characteristics of Aquifer Microorganisms

Transmission electron microscopy of aquifer and other subsurface materials confirmed the presence of microorganisms in all samples by revealing objects that possessed ultrastructural features (such as cell walls, membranes, and intracytoplasmic inclusions) unequivocally characteristic of microbial cells (for illustrations, see Figs. 2 and 3 in Ghiorse and Balkwill, 1983, and Fig. 3 in Wilson et al., 1983). The dimensions and shapes of these cells corresponded to those of the green-fluorescing objects considered to be cells in AO-stained preparations for LM (above).

TEM of thin-sectioned microbial cells that were released and concentrated from aquifer or other subsurface samples by blending and centrifugal washing (see Materials and Methods) provided important information on these organisms that could not be obtained readily with other approaches. For example, it was possible to determine the relative proportions of gram-positive and gram-negative bacteria in aquifer environments because thin sectioning revealed the architectural details of their cell walls. Both gram-positive and gram-negative forms were present in all samples, but the former were always clearly predominant (two-thirds or more of the bacteria observed were gram-positive).

The cytoplasm of many subsurface bacterial cells was partially depleted of the intracellular constituents (ribosomes and nuclear material) commonly found in laboratory-cultured cells. Control experiments involving addition of laboratory-cultured cells to subsurface samples established that these cytoplasmic constituents were not lost during preparation of the samples for TEM. Some of the *in situ* bacterial cells with a depleted cytoplasm contained mesosomelike internal membranes and intracellular storage bodies such as polyphosphate granules (Jensen, 1968), or more frequently, poly-β-hydroxy-butyrate (PHB) granules (Dunlop and Robards, 1973). In contrast, other subsurface bacteria lacked such inclusions and contained the normal or healthy-looking cytoplasm that is characteristic of laboratory-cultured cells. A few of these bacteria possessed cross-walls or division septa, implying that they were in the process of dividing when the samples were fixed. This was observed in both coccoid and filamentous forms. A small number of bacterial cells in the samples from Louisiana also contained internal membrane systems that were reminiscent of those found in nitrifying or methane-oxidizing bacteria. Ruthenium red staining indicated that many subsurface bacteria were surrounded by polysaccharide-based capsules and glycocalyx layers.The polysaccharide strands of these structures often extended from the cell surface to surrounding pieces of abiotic materials.

As was true of morphological diversity (above), the internal ultrastructural diversity of the bacteria in samples from Louisiana was greater than that of the bacteria in Oklahoma samples. TEM also confirmed that the overwhelming majority of subsurface microorganisms were prokaryotic.

**Table 34-2. Numbers of Microorganisms in Aquifer and
Other Subsurface Samples**

Sampling Site	Borehole	Depth (m)	AODC[a] (cells/g dry wt)	Plate Counts[b] (CFU/g dry wt) PYG Medium	SEA Medium
Lula, Okla.	—	1.2	6.8×10^6	1.1×10^4	1.4×10^5
		3.0	3.4×10^6	3.5×10^5	6.9×10^5
		5.0	6.8×10^6	1.5×10^3	3.4×10^6
Pickett, Okla.	—	5.5	5.7×10^6	6.7×10^2	3.1×10^6
Ft. Polk, La.	6B	3.7	1.6×10^6	$< 10^2$	$< 10^2$
	6B	5.5	2.2×10^6	2.4×10^2	$< 10^2$
	6B	6.7	2.7×10^6	$< 10^1$	$< 10^2$
	7	1.2	1.5×10^7	5.1×10^3	1.0×10^5
	7	4.0	1.3×10^6	$< 10^2$	$< 10^2$
	7	5.1	8.9×10^6	$< 10^1$	$< 10^2$
Conroe, Tex.	7G	5.4	1.2×10^5	$< 10^2$	$< 10^2$

[a] AODC = acridine orange direct count; see Materials and Methods.
[b] Average from triplicate spread plates.

Numbers of Microorganisms in Aquifer Environments

Epifluorescence LM of AO-stained samples was an effective way to obtain direct counts (AODC) of total microbial cells in aquifers and other subsurface environments. The resulting counts (Table 34-2) were lower, sometimes by two or three orders of magnitude, than those that have been reported for typical surface soils (Alexander, 1977), and they were remarkably consistent from one sampling site to another. Samples from Oklahoma, Louisiana, and Texas typically contained between 1 million and 10 million AODC-cells per gram (dry weight) of subsurface solids. Somewhat surprisingly, the numbers of cells did not decrease appreciably with increasing depth at any site.

Plate counts on media with differing nutrient concentrations were used to estimate the numbers of viable cells present in subsurface samples (Table 34-2). Although traditional cultural methods of this type were of limited value for obtaining meaningful information on aquifer microorganisms (see Discussion), plate counts were included in these studies to provide a basis for comparison with other environmental investigations. Plate counts on nutritionally-rich media like PYG agar were generally lower (sometimes much lower) than on low-nutrient media like SEA (Table 34-2). The highest plate counts, which were usually obtained on SEA, were always lower than the AODCs of the same sample. The magnitude of this discrepancy varied from one sampling site to another. Plate counts for samples from Oklahoma were sometimes as high as 50% of the AODC, but those for samples from Louisiana and Texas were generally much lower (0.01% of the AODC, or less).

Metabolic Activities of Aquifer Microorganisms

Plate counts demonstrated that some of the microorganisms residing in aquifer and other subsurface samples were capable of growth, but this information provided little or no indication of their activities *in situ*. Similarly, direct observation of microbial cells with LM and TEM provides only limited and indirect information on the metabolic activities of these organisms. Therefore, it was of interest to apply methods designed to reveal *in situ* metabolic activity more directly. One such method involves the use of 2-(p-iodophenyl)-3-(p-nitrophenyl)-5-phenyl tetrazolium chloride (INT) as a measure of respiratory activity of microbial cells (Zimmermann, Iturriage, and Becker-Birck, 1978). Preliminary results with the INT method suggest that 1% to 10% of the AODC bacterial cells in samples from Texas were capable of respiration-linked INT reduction. In most cases, however, INT-formazan-containing bacteria contained the diffuse type of deposit. Very few cells contained distinct granules. These results suggest a low level of respiratory activity in the subsurface bacterial population.

An alternate approach to investigating potential metabolic activities of aquifer microorganisms was developed and used by J. T. Wilson (Wilson et al., 1983). This approach involved the use of microcosms constructed from subsurface materials to determine whether the organisms indigenous to those materials could degrade selected organic pollutants. Toluene was degraded rapidly in subsurface samples from above and below the water table at Lula, Oklahoma. Comparison of autoclaved and nonautoclaved samples indicated that the degradation was a biological process. Chlorobenzene was also degraded in these samples, but its degradation rate was considerably slower than that of toluene, and degradation took place only in samples from above the water table. Bromodichloromethane was also degraded slowly, but it was not clear whether this was a direct or indirect result of microbial metabolism. In contrast, there was no detectable degradation of 1,2-dichloroethane, 1,1,2-trichloroethane, trichloroethylene, or tetrachloroethylene in any of the Lula, Oklahoma, samples.

DISCUSSION

The results reviewed here demonstrate conclusively that appreciable numbers (1-10 million per gram) of microbial cells reside in aquifer material. This observation is important for two reasons: (1) It contradicts the traditional belief that such environments are almost devoid of microbial life, and (2) the numbers of cells detected were great enough to potentially affect ground-water quality, provided that these cells were metabolically active.

Morphological and ultrastructural data indicated that, even though the total number of microorganisms was quite consistent from one sampling site to another, the identity of those organisms may have varied considerably. This has important implications with respect to the potential effects of microorganisms on ground-water quality, since different microbial types carry out different metabolic reactions and will respond differently to specific pollutants. Equally important is the fact that the range of microbial types varied widely from one site to another. A pollutant compound might destroy all microbial life in an aquifer that contained only a few types of bacteria, whereas a more diverse microbial community would be more likely to include a species that could survive or even degrade the pollutant compound. In defining the various factors that control ground-water quality, it will probably be necessary not only to consider microorganisms in general, but also to consider the specific microbial population of each aquifer system.

Some of the microorganisms in aquifer and other subsurface environments must be viable because the studies reviewed here showed that they were capable of growth on plates. However, most of the AODC cells in typical samples did not grow on plates. This could mean that these organisms were not viable, but it is more likely that the growth media used for plating simply failed to meet their possibly complex growth requirements (see also Ghiorse and Balkwill, 1983). There is a need, then, to characterize the growth requirements of subsurface microorganisms so that more realistic procedures for enumerating viable cells can be developed. Alternatively, modifications of such direct LM approaches as the INT method might also serve to solve this problem.

The fact that plate counts of aquifer and other subsurface samples usually were higher on nutritionally-rich media than on relatively dilute media implies that the in situ microorganisms in these samples may prefer low levels of nutrients for growth. Morphological and ultrastructural data in the studies reviewed here also point to adaptation by subsurface microorganisms to low-nutrient starvation conditions. The overwhelming predominance of prokaryotes, for example, probably occurred because oligotrophic prokaryotes are much better adapted that eukaryotes to live in environments with very low levels of organic matter (Poindexter, 1981). The PHB granules seen in subsurface bacteria also indicated an adaptation to low-nutrient conditions, since synthesis of these and other storage materials is a common bacterial strategy for surviving periods of nutrient shortage (Poindexter, 1981; Shively, 1974). The depleted cytoplasm of many subsurface bacteria suggests that these cells were actually either nutrient limited or starving at the time of sampling and, therefore, were probably relatively inactive members of the microbial community. On the other hand, the bacteria with a healthy cytoplasm or with division septa must have learned both to survive and to grow actively under low-nutrient conditions.

Although data obtained by direct observation of subsurface microbial cells with LM and TEM allowed reasonable conclusions about the likely physiological characteristics of subsurface microorganisms (above), very little is known about the specific *in situ* or potential *in situ* metabolic activities of these organisms. Specialized techniques such as the INT procedure may prove helpful in this regard, but there is a need to develop more powerful and sophisticated LM and TEM methods for determining *in situ* metabolic activities in subsurface environments. Such information will be critical in order to understand the biology of subsurface microorganisms. It will also be critical in understanding how subsurface microorganisms may affect ground-water quality and how these microorganisms themselves may be affected by pollutants.

ACKNOWLEDGEMENTS

This work was supported by Subcontract No. 6931-5 under U.S.E.P.A. Cooperative Agreement No. CR806931-02.

REFERENCES

Alexander, M., 1977, *Introduction to Soil Microbiology,* 2nd ed. John Wiley & Sons, New York.

Bae, H. C., E. H. Cota-Robles, and L. E. Casida, Jr., 1972, The microflora of soil as viewed by transmission electron microscopy, *J. Bacteriol.* **113:**1462-1473.

Balkwill, D. L., and L. E. Casida, Jr., 1973, Microflora of soil as viewed by freeze-etching, *J. Bacteriol.* **114:**1319-1327.

Balkwill, D. L., D. P. Labeda, and L. E. Casida, Jr., 1975, Simplified procedures for releasing and concentrating microorganisms for transmission electron microscopy and viewing as thin-sectioned and frozen-etched preparations, *Can. J. Microbiol.* **21:**252-262.

Balkwill, D. L., T. E. Rucinsky, and L. E. Casida, Jr., 1977, Release of microorganisms from soil with respect to electron microscopy viewing and plate counts, *Antonie van Leeuwenhoek J. Microbiol. Serol.* **43:**73-81.

Council on Environmental Quality, 1981, *Contamination of Ground Water by Toxic Organic Chemicals,* U.S. Government Printing Office, Washington, D.C.

Daley, R. J., and J. E. Hobbie, 1975, Direct counts of aquatic bacteria by a modified epifluorescence technique, *Limnol. Oceanog.* **20:**875-882.

Dockins, W. S., G. J. Olson, G. A. McFeters, and S. C. Turbak, 1980, Dissimilatory bacterial sulfate reduction in Montana ground waters, *Geomicrobiol. J.* **2:**53-98.

Dunlap, W. J., R. L. Cosby, J. F. McNabb, B. E. Bledsoe, and M. R. Scalf, 1972, Probable impact of NTA on ground water, *Ground Water* **10:**107-117.

Dunlap, W. J., and J. F. McNabb, 1973, *Subsurface Biological Activity in Relation to Ground Water Pollution,* National Environmental Research Center, Office of Research Monitoring, U.S.E.P.A., Corvallis, Oregon.

Dunlap, W. J., J. F. McNabb, M. R. Scalf, and R. L. Cosby 1977, *Sampling for Organic Chemicals and Microorganisms in the Subsurface,* EPA-600/2-77-176, U.S.E.P.A., Washington, D.C.

Dunlop, W. F., and A. W. Robards, 1973, Ultrastructural study of poly-β-hydroxybutyrate granules from *Bacillus cereus, J. Bacteriol.* **114:**1271-1280.

Ghiorse, W. C., and D. L. Balkwill, 1983, Eunmeration and morphological characterization of bacteria indigenous to subsurface environments, *Dev. Ind. Microbiol.* **24:**213-224.

Ghiorse, W. C., and D. L. Balkwill, in press, Microbiological characterization of subsurface environments, in *Ground Water Quality,* C. H. Ward, ed., Wiley and Sons, New York.

Hirsch, P., and E. Rades-Rohkohl, 1983, Microbial diversity in a groundwater aquifer in northern Germany, *Dev. Ind. Microbiol.* **24:**183-200.

Jensen, T. E., 1968, Electron microscopy of polyphosphate bodies in a blue-green alga, *Bostoc pruniforme, Arch. Mikrobiol.* **62:**144-152.

Josephson, J., 1980, Safeguards for ground water, *Environ. Sci. Technol.* **14:**19-44.

Poindexter, J. S., 1981, Oligotrophy, fast and famine existence, *Adv. Microb. Ecol.* **5:**63-89.

Shively, J. M., 1974, Inclusion bodies of prokaryotes, *Ann. Rev. Microbiol.* **28:**167-187.

Trolldenier, G., 1973, The use of fluorescence microscopy for counting soil microorganisms, in *Modern Methods in the Study of Microbial Ecology,* T. Rosswall, ed., Ecological Research Committee, Swedish Natural Science Research Council, Stockholm, pp. 53-59.

Waksman, S. A., 1916, Bacterial numbers in soil, at different depths, and in different seasons of the year, *Soil Sci.* **1:**363-380.

Whitelaw, K., and R. A. Edwards, 1980, Carbohydrates in the unsaturated zone of the Chalk, England, *Chem. Geol.* **29:**281-291.

Whitelaw, K., and J. F. Rees, 1980, Nitrate-reducing and ammonium-oxidizing bacteria in the vadose zone of the Chalk aquifer of England, *Geomicrobiol J.* **2:**179-187.

Wilson, J. T., J. F. McNabb, D. L. Balkwill, and W. C. Ghiorse, 1983, Enumeration and characterization of bacteria indigenous to a shallow water-table aquifer, *Ground Water* **21:**134-142.

Zimmermann, R., R. Iturriage, and J. Becker-Birck, 1978, Simultaneous determination of the total number of aquatic bacteria and the number thereof involved in respiration, *Appl. Environ. Microbiol.* **36:**926-935.

Part VI

NITROGEN OXIDES—
TRANSFORMATION PROCESSES

35

Microbial Transformations as Sources and Sinks for Nitrogen Oxides

Roger Knowles

Macdonald Campus of McGill University

ABSTRACT: Biological and biochemical aspects of the production and consumption of nitric oxide (NO), nitrous oxide (N_2O), nitrite (NO_2^-), and nitrate (NO_3^-) are reviewed. The potential for oxidative and reductive processes in chemolithotrophic, methanotrophic, and heterotrophic nitrifiers, as well as in sulfate reducers, denitrifiers, nitrate respirers, and nitrate utilizers among the bacteria, yeasts, and filamentous fungi, is discussed.

PROCESSES

Oxidative Production of N Oxides

Chemolithotrophic Nitrification. The oxidation of ammonium (NH_4^+) to NO_2^- by members of the genera *Nitrosomonas, Nitrosolobus, Nitrosococcus, Nitrosospira,* and *Nitrosovibrio,* and of NO_2^- to NO_3^- by *Nitrobacter, Nitrospina,* and *Nitrococcus* (process 5, Fig. 35-1) has recently been well reviewed (Schmidt, 1982). In natural systems, NO_2^- does not generally accumulate, except at high pH values in the presence of high NH_4^+ concentrations, when *Nitrobacter* activity is inhibited (Aleem and Alexander, 1960). The process is acid intolerant and occurs mainly above about pH 6.0. However, it has been reported in more acid environments, and a chemolithotrophic NO_2^- oxidizer was isolated from a soil of pH 3.5 (Josserand and Bardin, 1981), although its pH tolerance in culture was not reported. The concentrations of energy substrate (NH$_4^+$) and electron acceptor [oxygen (O$_2$)] are the major limiting factors.

The possession by *Nitrosomonas* of a NH$_2$OH-NO$_2^-$ oxidoreductase (Hooper and Terry, 1979) is presumably responsible for the production of the gaseous oxides NO and N$_2$O (Fewson and Nicholas, 1960; Hooper and Terry, 1979; Lipschultz et al., 1981; Ritchie and Nicholas, 1972; Yoshida and Alexander, 1970, 1971). N$_2$O is produced universally as a very minor product, ranging from 0.05% up to a few percent of the nitrogen oxidized as the O$_2$ concentration is decreased (Blackmer, Bremner, and Schmidt, 1980; Breitenbeck, Blackmer, and Bremner, 1980; Goreau et al., 1980; Hynes, 1983; Lipschultz et al., 1981). In view of the widespread occurrence of nitrification, it is considered that this represents a significant global source of N$_2$O (Breitenbeck, Blackmer, and Bremner, 1980).

Heterotrophic Nitrification.   Many microorganisms (e.g., *Aspergillus flavus, Arthrobacter* sp.) can oxidize NH$_4^+$ and various oximes to produce hydroxyl-amine (NH$_2$OH), NO$_2^-$, or NO$_3^-$, or both (process 5, Fig. 35-1). There is no report of N$_2$O as an oxidative product from NH$_4^+$-grown cells, and it appears that the mechanism of NH$_4^+$ oxidation is different from that in the chemolithotrophic nitrifiers (Focht and Verstraete, 1977). An *Alcaligenes*

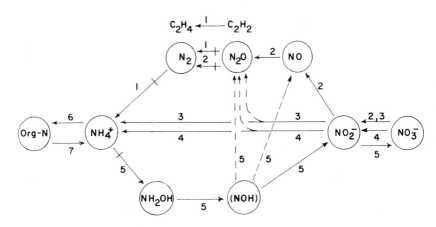

I Nitrogenase activity

2 Denitrification

3 Dissimilatory nitrate reduction

4 Assimilatory nitrate reduction

5 Nitrification

6 Ammonium assimilation

7 Mineralization

Figure 35-1. The role of nitrogen oxides in the nitrogen cycle. Compartments are arranged in order of increasing oxidation number to the right and gaseous components at the top. The crossbars indicate processes that are inhibited by C$_2$H$_2$.

was recently reported to oxidize pyruvic oxime or NH_2OH to NO_2^- and to have the ability also to denitrify (Castignetti and Hollocher, 1982). Heterotrophic nitrification occurs mainly at high NH_4^+ concentrations, but there are no reported comparisons of apparent K_m values for these organisms with those of classical nitrifiers (Table 35-1). Both the specific activity per cell and the maximum product concentration (< 0.1 mM) are at least three orders of magnitude lower than for the chemolithotrophic nitrifiers (Focht and Verstraete, 1977), and there is no strong evidence that heterotrophic nitrification is of any geochemical significance.

Methanotrophic Nitrification.   *Methylosinus* and *Methylococcus* sp., which oxidize methane (CH_4), can also cooxidize NH_4^+ (by the CH_4 monooxygenase) to NH_2OH, which is then oxidized to NO_2^- (Dalton, 1977; Hutton and ZoBell, 1953; O'Neill and Wilkinson, 1977). The detailed mechanism is not known, but there is evidence that the oxidation of NH_2OH to NO_2^- yields ATP that can support reversed electron transport to generate NADH (Malashenko et al., 1979). Apparent K_m values for NH_4^+ appear to be higher than in the classical nitrifiers (Table 35-1, and Dalton, 1977) but, although NH_4^+ and CH_4 seem to be competitive, NH_4^+ oxidation is stimulated by low CH_4 concentrations and inhibited by higher CH_4 concentrations (O'Neill and Wilkinson, 1977). Incidentally, 0.1 mM CH_4 inhibits NH_4^+ oxidation by *Nitrococcus oceanus* by 87%, but up to 1 mM has no effect on *Nitrosomonas europaea* (Jones and Morita, 1983).

Small amounts of N_2O arise during oxidation of NH_4^+ by methanotrophs (O'Neill and Wilkinson, 1977; T. Yoshinari, pers. comm.; E. Topp and R. Knowles, unpub.), but there is no evidence of N_2O reduction by these organisms (Vedenina and Zavarzin, 1979). There is speculation about the possible natural contribution of methanotrophs to NO_2^- accumulation (Hanson, 1980; Knowles, Lean, and Chan, 1981; Schmidt, 1982; Verstraete, 1981) but little direct evidence.

Oxidative Consumption of N_2O.   A mixed consortium of five heterotrophic corynebacteria was reported to consume N_2O under strictly aerobic conditions (Vedenina and Zavarzin, 1977, 1979). This process was not inhibited by ethane (C_2H_2), and the authors suggested that it may be related to hydrogen peroxide (H_2O_2) metabolism. However, the mechanism is not known, and it is conceivable that a reductive process may be involved. In general, it may be said that N_2O is not consumed by strictly aerobic systems.

Reductive Transformation of N Oxides

Chemolithotrophic Nitrifiers.   *Nitrosomonas* and *Nitrobacter* possess NO_2^- and NO_3^- reductases, respectively, but there is no evidence that they can

Table 35-1. Reported Values of Apparent K_m (or K_s) for the Uptake or Transformation of Inorganic Nitrogen, Especially N Oxides

Process	Km (μM)	Organism	Authors
Nitrification			
NH_4^+ oxidation	70–700	Nitrosomonas	Focht and Verstraete, 1977
	480	Nitrosomonas	Suzuki, Dular, and Kwok, 1974
	400	Nitrosomonas	Hynes and Knowles, 1982
	400–7000	Methylosinus	O'Neill and Wilkinson, 1977
NO_2^- oxidation	360–640	Nitrobacter	Focht and Verstraete, 1977
	350	Nitrobacter	Gould and Lees, 1960
N oxide reductions			
NO_3^- reduction, assimilatory	1400	Neurospora (cell free)	Nason, 1962
	110	Ditylum brightwellii	Eppley, Coatsworth, and Solorzano, 1969
	7	Pseudomonas fluorescens	Betlach, Tiedje, and Firestone, 1981
	0.05	Oscillatoria, Skeletonema	Rhee, 1982
denitrification	500	P. aeruginosa (cell free)	Fewson and Nicholas, 1961
	<15	Alcaligenes, Flavobacterium, P. fluorescens	Betlach and Tiedje, 1981

414

dissimilatory	500	*Clostridium*	Caskey and Tiedje, 1980
	100	*Klebsiella*	Van t'Riet and Planta, 1975
	510	*E. coli* (cell free)	Nason, 1962
NO_2^- reduction, denitrification	5–12	*Alcaligenes, Flavobacterium, P.fluorescens*	Betlach and Tiedje, 1981
	51	*Rhodopseudomonas* (pure enzyme)	Sawada, Satoh, and Kitamura, 1978
dissimilatory → N_2O	900	*Citrobacter*	Smith, 1982
→ NH_4^+	40	*Citrobacter*	Smith, 1982
→ NH_4^+	50–70	*Vibrio fischeri* (pure enzyme)	Prakash and Sadana, 1972
NO reduction, denitrification	40	*P. aeruginosa* (pure enzyme)	Fewson and Nicholas, 1960
	10	*Ps. denitrificans* (cell free)	Miyata, Matsubara, and Mori, 1969
N_2O reduction, denitrification	30–60	*Ps. denitrificans*	Matsubara and Mori, 1968
	5	*Paracoccus denitrificans* (crude lysate)	Kristjansson and Hollocher, 1980
	0.44	*Flavobacterium*	Betlach and Tiedje, 1981
Plant root uptake			
NO_3^-	24–600	various	McGill *et al.* 1981
	100	barley (*Hordeum*)	Rao and Rains, 1976
	75	rice (*Oryza*)	Youngdahl *et al.* 1982
NH_4^+	13–200	various	McGill *et al.* 1981

Table 35-2. Relative Rates of Production of N₂O by Cells of *Nitrosomonas europaea* in the Presence and Absence of O₂ and C₂H₂ (Hynes, 1983)

N Ion Present	Aerobiosis		Anaerobiosis	
	$-C_2H_2$	$+C_2H_2$	$-C_2H_2$	$+C_2H_2$
None	0	0	+	+
NH_4^+	+	0	++	+
$NH_4^+ + NO_2^-$	+	0	++++	+++
NO_2^-	0	0	+++	+++

catalyze a significant reduction of these oxides under anaerobic conditions. However, it is clear that *Nitrosomonas* produces much more N_2O from NO_2^- under anaerobic that under aerobic conditions (Ritchie and Nicholas, 1972; and Table 35-2). It is noteworthy that production of N_2O from NH_4^+ under aerobiosis is sensitive to C_2H_2, whereas that produced under anaerobiosis is not so sensitive (Table 35-2). This suggests the participation of the C_2H_2-sensitive NH_4^+, monooxygenase (Hynes and Knowles, 1982) in the aerobic process and of the NO_2^- reductase (or NH_2OH-NO_2^- oxidoreductase) in the anaerobic process (Hynes, 1983). Poth and Focht (1982) termed the latter reaction denitrification, but since the NO_2^- does not act as a major electron acceptor to support anaerobic growth, this view does not appear tenable.

Assimilatory Reductions.   Very many bacteria, fungi, photosynthetic bacteria, cyanobacteria, algae, and lower and higher plants utilize NO_3^- as sole source of nitrogen by means of assimilatory reductases that are repressed by NH_4^+ but unaffected by O_2 (Payne, 1973). Assimilation thus represents a major sink for the ionic N oxides. The reported uptake constants (Table 35-1) are seen to be quite variable, but many organisms, especially the phytoplankton, have extremely low K_m (or K_s) values, and the ability to compete would seem to be in the order phytoplankton > heterotrophic bacteria > plants. However, the outcome of competition for NO_3^- depends not only on the apparent K_m but on the relative V_{max} values and growth rates.

Yoshida and Alexander (1970) reported that some organisms produced N_2O under aerobic conditions when NO_3^--grown but not when NH_4^+-grown. Included were *Bacillus subtilis, Escherichia coli, Aerobacter aerogenes, Aspergillus flavus,* and *Penicillium atrovenetum,* but not *Azotobacter vinelandii, Chlorella, Pseudomonas,* or *Bacillus megaterium.* Subsequently, *A. vinelandii, Hansenula, Rhodotorula, Aspergillus, Alternaria,* and *Fusarium* were shown to produce small amounts of N_2O under similar conditions (Bleakley and Tiedje, 1982). Production of N_2O by fusaria under aerobic conditions was reported to occur from NO_2^-, but not from NO_3^- (Bollag and Tung, 1972), and up to 60% of the NO_2^--N was recovered as N_2O-N (Burth, Benckisev, and Ottow, 1982; Burth and Ottow, 1983). It would be interesting to discover to what extent this process is significant in natural soils.

Dissimilatory Reduction of NO_3^- to NH_4^+. This strictly anaerobic process (process 3, Fig. 35-1) is not NH_4^+-repressible and is a nitrogen cycle pathway that provides an alternative to denitrification (Knowles, 1982). It essentially bypasses the major gaseous compartments, N_2O and N_2, and thus conserves combined nitrogen. The process has been most studied in facultative anaerobes, such as *Aeromonas, Klebsiella, Vibrio fischeri, Citrobacter freundii,* and *E. coli,* in which it is suggested its main function is to reoxidize reduced pyridine nucleotides that would otherwise limit the rate of fermentative growth (Cole and Brown, 1980). It also occurs in obligate anaerobes such as *Clostridium* (Hasan and Hall, 1975; Caskey and Tiedje, 1980) and *Desulfovibrio* (Barton et al., 1983; Keith and Herbert, 1983).

N$_2$O is produced in small quantities during NH_4^+ production by members of the genera *Serratia, Enterobacter, Erwinia, Bacillus, Escherichia, Klebsiella,* and *Citrobacter* (Bleakley and Tiedje, 1982; Smith, 1982, 1983). In *Citrobacter,* high glucose and low NO_3^- levels appear conducive to the formation of NH_4^+ rather than N_2O (Smith, 1982), and in *E. coli* and *K. pneumoniae* the production of N_2O from NO_2^- apparently requires the presence of an active NO_3^- reductase (Satoh, Hom, and Shanmugam, 1983; Smith, 1983). However, the details of these reactions are not known.

Although such dissimilatory N oxide reductions are generally limited to the prokaryotes, it was recently reported that *Loxodes,* a protozoan, possessed a dissimilatory NO_3^- reductase located in the inner mitochondrial membrane and that anaerobically grown cells contained 70% more mitochondria, with more cristae, than did aerobically grown cells (Finlay, Span, and Harmon, 1983). A *Loxodes*-rich layer in a stratified lake was clearly correlated with peaks of both NO_3^- reductase activity and NO_2^- concentration.

Denitrification. The dissimilatory reduction of NO_3^- and NO_2^- to one or more of the gaseous products NO, N_2O, and N_2 (process 2, Fig. 35-1) represents a major sink for the ionic N oxides and a major source of tropospheric N_2O. The process and the bacteria responsible have been extensively reviewed (Firestone, 1982; Knowles, 1982; Payne, 1973, 1981). The reductases involved are repressed and inhibited by O_2 but not by NH_4^+. Assuming availability of organic carbon, denitrification rates are maximum in completely anaerobic microenvironments, and slight exposure to O_2 or to low pH values appears to inhibit first the N_2O reductase, causing accumulation of N_2O, and then the other reductases, reducing the overall denitrification rate. However, regardless of environmental parameters, the transient or permanent accumulation of denitrification intermediates depends also on the state of derepression of the successive reductases and their respective apparent K_m values (Betlach and Tiedje, 1981). Reported K_m values (Table 35-1) vary considerably but suggest that the denitrifying reductases have higher affinities for the N oxides than do the reductases involved in dissimilatory reduction to NH_4^+. However, in highly and permanently anaerobic habitats, denitrifiers, which depend on available O_2 or N oxide electron acceptors, may be at a competitive disadvantage compared to the fermentative dissimilatory NO_3^- reducers that are not dependent on but can nevertheless utilize the greater electron sink capacity of the NO_3^--to-NH_4^+ reduction process (Tiedje et al., 1982).

Since the earlier studies of Stanford et al. (1975), it has become clear that available organic carbon (energy and reductant) in anaerobic soils and sediments tends to decrease the role of denitrification and increase the role of dissimilatory reduction to NH_4^+ as a NO_3^- sink. A model, based on published data, is summarized in Figure 35-2. Unfortunately, there is as yet no

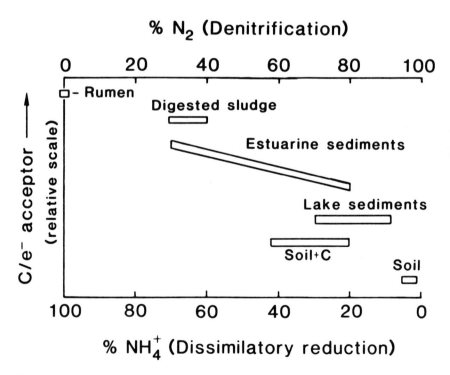

Figure 35-2. Relative importance of denitrification and dissimilatory reduction as sinks for NO_3^- in different habitats as a function of the relative availability of carbon and electron acceptor. *(Source: From Tiedje et al. 1982, p. 574).*

practical way in which the pathway to NH_4^+ may be emphasized in anaerobic agricultural soils.

COMMENTS

Aquatic Systems

Mass-balance studies of lakes have clearly shown the production and consumption of N oxides in surficial epilimnetic sediments and in the water column, usually near the aerobic-anaerobic interface at the top of the hypolimnion. In some studies, denitrification was confirmed using $^{15}NO_3^-$ or C_2H_2 inhibition assays, and nitrification was measured as nitrapyrin-sensitive O_2 uptake or radioactive-labelled bicarbonate ($H^{14}CO_3^-$) incorporation. Regarding the latter measurements, it is noteworthy that CH_4 oxidation frequently occurs at or near the zone of NO_2^- accumulation, suggesting that methanotrophs could be responsible for some cooxidation of NH_4^+, as was discussed earlier. In such methanotrophs, both oxidation of CH_4 and incorporation of carbon dioxide (CO_2) are inhibited by nitrapyrin (Topp and Knowles, 1982), as is the oxidation of NH_4^+ by these organisms (E. Topp and R. Knowles, unpub.), making it more difficult to attribute nitrapyrin-sensitive activities to chemolithotropic nitrifiers or methanotrophs specifically.

The production and subsequent consumption of N oxides at localized depths in lakes (e.g., Brezonik and Lee, 1968; Yoh, Terai, and Saije, 1983) are usually transient, non-steady state processes. Occasionally, however, such localized concentrations are observed to be maintained in the thermocline region for considerable time periods (Fig. 35-3), a phenomenon that suggests the calculation of flux rates based on diffusion coefficients (Table 35-3). The two coefficients used in the calculation were from lakes similar in size to Lake St. George, one low value (Lake 227) being close to nonturbulent diffusion, and the other (Lake 224) being more than an order of magnitude higher. It is assumed that the sum of the upward and downward NO_3^- fluxes represents the nitrification rate necessary to maintain the NO_3^- peak for the 82-day period. Assuming this nitrification to be restricted to a 2-m-depth zone, it represents an activity of at least 89 $\mu g/N/l/day$. Even with one of the highest chemolithotrophic nitrifier activities reported (81 pg N/cell/day) (Schmidt, 1982), this activity would require a population of at least 10^3 nitrifiers per ml. Such a large population has apparently not been recorded in lake water, and it is clear that our understanding of this process in aquatic systems is far from complete. It is interesting that the maximum denitrification rate, assumed to be equal to the downward NO_3^- flux, is approximately equal to the upward flux of NH_4^+ originating in the sediment.

Most evidence suggests that in marine systems, the oceans act as net sources of N_2O to the atmosphere, but the origin of this N_2O remains unresolved (Pierrotti and Rasmussen, 1980). Although in O_2-depleted regions (such as the secondary NO_2^- maximum) denitrification may be implicated (Knowles, 1982), in other waters containing much more O_2 the process is not clear. Ward and Perry (1980) detected *Nitrosococcus oceanus* using fluorescent antibody techniques, but Ward, Olson, and Perry (1982) found no correlation of NH_4^+ oxidizer populations with the NO_2^- maximum. Schropp and Schwarz (1983), on the other hand, showed the existence of potentially denitrifying bacteria in the near-surface particulate maximum, but there was no indication of their *in situ* activity. Recent measurements of $\delta^{15}N$ values of oceanic N_2O by E. Wada and co-workers (Tokyo) may shed some light on this enigma.

Terrestrial Systems

In a limited space it is difficult to make more than a few comments on N oxides in terrestrial systems, a subject that has been extensively reviewed (Knowles, 1982; Rosswall, 1982; Schmidt, 1982). The question of the occur-

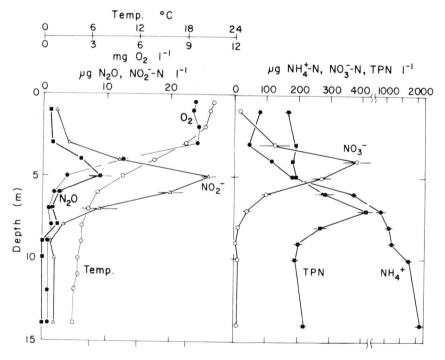

Figure 35-3. Peaks of N oxides in the metalimnion of Lake St. George, Ontario. Data are the means ($\pm$ S.E.) of 11 samplings between 21 June and 11 September, 1979 (from Knowles, Lean, and Chan, 1981, p.864). TPN, total particulate N.

Table 35–3. Concentration Gradients and Flux Rates of Inorganic N Species in Lake St. George, Ontario, During the Steady-State 82-Day Period, 21 June to 11 September 1979 (See Fig. 35–3)

	Concentration Gradients (dc/dz^*)	Flux†	
	$\mu g\ N/l/cm$	$pg\ N/cm^2/s$	$g\ N/m^2/82\ d$
Upward fluxes			
NO_3^-	2.67	133 − 2128	9.4 − 150
NO_2^-	0.14	7.0 − 112	0.5 − 7.9
N_2O	0.024	1.2 − 19	0.08 − 1.4
NH_4^+	1.14	57 − 912	4.0 − 65
Downward fluxes			
NO_3^-	1.48	74 − 1184	5.2 − 84
NO_2^-	0.088	4.4 − 70	0.3 − 5.0
N_2O	0.042	2.1 − 34	0.1 − 2.4
Nitrification = flux ($NO_3^- \uparrow + NO_3 \downarrow$)		=	14.6 − 234
Denitrification (max) = flux $NO_3^- \downarrow$		=	5.2 − 84
(min) = flux $N_2O \downarrow$		=	0.1 − 2.4

*Concentration gradients determined from the data of Figure 35–3.
†The lower and higher values of each range were calculated using diffusion coefficients (Quay et al., 1980) of 5×10^{-5} (Lake 227) and 8×10^{-4} (Lake 224) cm²/s, respectively.

rence of nitrification in climax systems such as forests, especially acid forest soils, is still the subject of discussion. Such systems are very N retentive, and strong assimilatory sinks exist for both NH_4^+ and NO_3^-. When the plant sinks are removed, NO_3^- production frequently occurs and is mainly dependent on NH_4^+ availability from mineralization (Adams and Attiwill, 1982; Robertson, 1982; Vitousek et al., 1982). However, there are reports that are not consistent with this hypothesis (Klein, Kreitinger, and Alexander, 1983). Thus, with small NH_4^+ pools the nitrifier populations are likely to decline, and this decline will be more marked if pH values are decreased (Lodhi, 1982). Forest soils have considerable potential for denitrification, especially after urea fertilization (Pluth and Nömmik, 1981). The potential is highly correlated with NO_3^- concentration, and N_2O appears to be the only significant product under aerobic or anaerobic conditions (Melillo et al., 1983). It is not known what organisms are responsible or to what extent this activity occurs *in situ*.

It is clear from the earlier discussion that N_2O may arise from a variety of sources. Denitrification and nitrification seem to be the major sources, but there is virtually no information on the natural contribution of other dissimilatory or assimilatory processes. In field studies, agricultural soils do not usually act as a net sink for N_2O (e.g., Mosier and Hutchinson, 1981), but may occasionally do so (Ryden, 1983). A great deal more information is required on all of the aspects in this paper. Especially needed are studies that test in the field some of the hypotheses that have been based mainly on laboratory studies.

REFERENCES

Adams, M. A., and P. M. Attiwill, 1982, Nitrogen mineralization and nitrate reduction in forests, *Soil Biol. Biochem.* **14:**197-202.

Aleem, M. I. H., and M. Alexander, 1960, Nutrition and physiology of *Nitrobacter agilis, Appl. Microbiol.* **8:**80-84.

Barton, L. L., J. LeGall, J. M. Odom, and H. D. Peck, 1983, Energy coupling in nitrite respiration in the sulfate-reducing bacterium *Desulfovibrio gigas, J. Bacteriol.* **153:**867-871.

Betlach, M. R., and J. M. Tiedje, 1981, A kinetic explanation for accumulation of nitrite, nitric oxide, and nitrous oxide during bacterial denitrification, *Appl. Environ. Microbiol.* **42:**1074-1084.

Betlach, M .R., J. M. Tiedje, and R. B. Firestone, 1981, Assimilatory nitrate uptake in *Pseudomonas fluorescens* studied using ^{13}N, *Arch. Microbiol.* **129:**135-140.

Blackmer, A. M., J. M. Bremner, and E. L. Schmidt, 1980, Production of nitrous oxide by ammonia-oxidizing chemoautotrophic microorganisms in soil, *Appl. Environ. Microbiol.* **40:**1060-1066.

Bleakley, B. H., and J. M. Tiedje, 1982, Nitrous oxide production by organisms other than nitrifiers or denitrifiers, *Appl. Environ. Microbiol.* **44:**1342-1348.

Bollag, J.-M., and G. Tung, 1972, Nitrous oxide release by soil fungi, *Soil Biol. Biochem.* **4:**271-276.

Breitenbeck, G. A., A. M. Blackmer, and J. M. Bremner, 1980, Effects of different nitrogen fertilizers on emission of nitrous oxide from soil, *Geophys. Res. Lett.* **7:**85-88.

Brezonik, P. L., and G. F. Lee, 1968, Denitrification as a nitrogen sink in Lake Mendota, Wisconsin, *Environ. Sci. Technol.* **2:**120-125.

Burth, I., and J. C. G. Ottow, 1983, Influence of pH on the production of N_2O and N_2 by different denitrifying bacteria and *Fusarium solani,* in *Environmental Biogeochemistry,* R. Hallberg, ed., *Ecol. Bull.* (Stockholm) **35:**207-215.

Burth, I., G. Benckiser, and J. C. G. Ottow, 1982, N_2O-Freisetzung aus Nitrit (Denitrifikation) durch ubiquitäre Pilze unter aeroben Bedingungen, *Naturwissenschaften* **69:**598-599.

Caskey, W. H., and J. M. Tiedje, 1980, The reduction of nitrate to ammonium by a *Clostridium* sp. isolated from soil, *J. Gen. Microbiol.* **119:**217-223.

Castignetti, D., and T. C. Hollocher, 1982, Nitrogen redox metabolism of a heterotrophic, nitrifying-denitrifying *Alcaligenes* sp. from soil, *Appl. Environ. Microbiol.* **44:**923-928.

Cole, J. A., and C. M. Brown, 1980, Nitrite reduction to ammonia by fermentative bacteria: A short circuit in the biological nitrogen cycle, *FEMS Microbiol. Lett.* **7:**65-72.

Dalton, H., 1977, Ammonia oxidation by the methane-oxidizing bacterium *Methylococcus capsulatus* strain Bath, *Arch. Microbiol.* **114:**273-279.

Eppley, R. W., J. L. Coatsworth, and L. Solorzano, 1969, Studies of nitrate reductase in marine phytoplankton, *Limnol. Oceanogr.* **14:**194-205.

Fewson, C. A., and D. J. D. Nicholas, 1960, Utilization of nitric oxide by microorganisms and higher plants, *Nature* **188**:794-796.

Fewson, C. A., and D. J. D. Nicholas, 1961, Nitrate reductase from *Pseudomonas aeroginosa, Biochim. Biophys. Acta* **49**:335-349.

Finlay, B. J., A. S. W. Span, and J. M. P. Harman, 1983, Nitrate respiration in primitive eukaryotes, *Nature* **303**:333-336.

Firestone, M. K., 1982, Biological denitrification, in *Nitrogen in Agricultural Soils,* F. J. Stevenson, ed., Amer. Soc. Agron., Madison, Wisconsin, pp. 289-326.

Focht, D. D., and W. Verstraete, 1977, Biochemical ecology of nitrification and denitrification, *Adv. Microbial Ecol.* **1**:135-214.

Goreau, T. J., W. A. Kaplan, S. C. Wofsy, M. B. McElroy, F. W. Valois, and S. W. Watson, 1980, Production of NO_2^- and N_2O by nitrifying bacteria at reduced concentration of oxygen, *Appl. Environ. Microbiol.* **40**:526-532.

Gould, G. W., and H. Lees, 1960, The isolation and culture of the nitrifying organisms. Part I. *Nitrobacter, Can. J. Microbiol.* **6**:299-307.

Hanson, R. S., 1980, Ecology and diversity of methylotrophic organisms, *Adv. Appl. Microbiol.* **26**:3-39.

Hasan, S. M., and J. B. Hall, 1975, The physiological function of nitrate reduction in *Clostridium perfringens, J. Gen. Microbiol.* **87**:120-218.

Hooper, A. B., and K. R. Terry, 1979, Hydroxylamine oxidoreductase of *Nitrosomonas:* Production of nitric oxide from hydroxylamine, *Biochim. Biophys. Acta* **571**:12-20.

Hutton, W. E., and C. E. ZoBell, 1953, Production of nitrite from ammonia by methane-oxidizing bacteria, *J. Bacteriol.* **65**:216-219.

Hynes, R. K., 1983, *Production of Nitrous Oxide by Nitrification and the Effect of Acetylene on Nitrifying Bacteria,* Ph.D. thesis, McGill University, Montreal.

Hynes, R. K., and R. Knowles, 1982, Effect of acetylene on autotrophic and heterotrophic nitrification, *Can. J. Microbiol.* **28**:334-340.

Jones, R. D., and R. Y. Morita, 1983, Methane oxidation by *Nitrosococcus oceanus* and *Nitrosomonas europaea, Appl. Environ. Microbiol.* **45**:401-410.

Josserand, A., and R. Bardin, 1981, Nitrification en sol acide. I. Mise en évidence de germes autotrophes nitrificants (genre *Nitrobacter*) dans un sol forestier sous résineux, *Rev. Ecol. Biol. Sol* **18**:435-445.

Keith, S. M., and R. A. Herbert, 1983, Dissimilatory nitrate reduction by a strain of *Desulfovibrio desulfuricans, FEMS Microbiol. Lett.* **18**:55-59.

Klein, T. M., T. P. Kreitinger, and M. Alexander, 1983, Nitrate formation in acid forest soils from the Adirondacks, *Soil Sci. Soc. Amer. J.* **47**:506-508.

Knowles, R., 1982, Denitrification, *Microbiol. Revs.* **46**:43-70.

Knowles, R., D. R. S. Lean, and Y.-K. Chan, 1981, Nitrous oxide concentrations in lakes: Variations with depth and time, *Limnol. Oceanogr.* **26**:855-866.

Kristjansson, J. K., and T. C. Hollocher, 1980, First practical assay for soluble nitrous oxide reductase of denitrifying bacteria and a partial kinetic characterization, *J. Biol. Chem.* **255**:704-707.

Lipschultz, F., O. C. Zafiriou, S. C. Wofsy, M. B. McElroy, F. W. Valois, and S. W. Watson, 1981, Production of NO and N_2O by soil nitrifying bacteria, *Nature* **294**:641-643.

Lodhi, M. A. K., 1982, Effects of H ion on ecological systems: Effects on herbaceous biomass, mineralization, nitrifiers, and nitrification in a forest community, *Amer. J. Bot.* **69**:474-478.

McGill, W. B., H. W. Hunt, R. G. Woodmansee, and J. O. Reuss, 1981, Phoenix, a model of the dynamics of carbon and nitrogen in grassland soils, in *Terrestrial Nitrogen Cycles,* F. E. Clark and T. Rosswall, eds., *Ecol. Bull.* (Stockholm) **33:**49-115.

Malashenko, Y. R., I. G. Sokolov, V. A. Romanovskaya, and Y. B. Shkurko, 1979, Elements of lithotrophic metabolism in the obligate methylotroph *Methylococcus thermophilus, Microbiology* **48:**468-474 (1980). transl. of *Microbiologiya* **48:**592-598.

Matsubara, T., and T. Mori, 1968, Studies on denitrification IX. Nitrous oxide, its production and reduction to nitrogen, *J. Biochem.* (Tokyo) **64:**863-871.

Melillo, J. M., J. D. Aber, P. A. Steudler, and J. P. Schimel, 1983, Denitrification potentials in a successional sequence of northern hardwood forest stands, in *Environmental Biogeochemistry,* R. Hallberg, ed., *Ecol. Bull.* (Stockholm) **35:**217-228.

Miyata, M., T. Matsubara, and T. Mori, 1969, Studies of denitrification. XI. Some properties of nitric oxide reductase, *J. Biochem.* (Tokyo) **66:**759-765.

Mosier, A. R., and G. L. Hutchinson, 1981, Nitrous oxide emissions from cropped fields, *J. Environ. Qual.* **10:**169-173.

Nason, A., 1962, Enzymatic pathways of nitrate, nitrite and hydroxylamine metabolisms, *Bacteriol. Revs.* **26:**16-41.

O'Neill, J. G., and J. F. Wilkinson, 1977, Oxidation of ammonia by methane-oxidizing bacteria and the effects of ammonia on methane oxidation, *J. Gen. Microbiol.* **100:**407-412.

Payne, W. J., 1973, Reduction of nitrogenous oxides by microorganisms, *Bacteriol. Revs.* **37:**409-452.

Payne, W. J., 1981, *Denitrification,* Wiley-Interscience, New York.

Pierotti, D., and R. A. Rasmussen, 1980, Nitrous oxide measurements in the eastern tropical Pacific Ocean, *Tellus* **32:**56-72.

Pluth, D. J., and H. Nömmik, 1981, Potential denitrification affected by nitrogen source of a previous fertilization of an acid forest soil from central Sweden, *Acta Agric. Scandinav.* **31:**235-241.

Poth, M. A., and D. D. Focht, 1982, Production of nitrous oxide during denitrification by *Nitrosomonas europaea* ATCC 19718, Abstracts, Annual Meeting of American Society of Microbiology, Atlanta, Georgia, p. 187.

Prakash, O., and J. C. Sadana, 1972, Purification, characterization and properties of nitrite reductase of *Achromobacter fischeri, Arch. Biochem. Biophys,* **148:**614-632.

Quay, P. D., W. S. Broecker, R. H. Hesslein, and D. W. Schindler, 1980, Vertical diffusion rates determined by tritium tracer experiments in the thermocline and hypolimnion of 2 lakes, *Limnol. Oceanogr.* **25:**201-218.

Rao, K. P., and D.W. Rains, 1976, Nitrate absorption by barley. I. Kinetics and energetics, *Plant Physiol.* **57:**55-58.

Rhee, G. Y., 1982, Effects of environmental factors and their interactions on phytoplankton growth, *Microbial Ecol.* **6:**33-74.

Ritchie, G. A. F., and D. J. D. Nicholas, 1972, Identification of the sources of nitrous oxide produced by oxidative and reductive processes in *Nitrosomonas europaea, Biochem. J.* **126:**1181-1191.

Robertson, G. P., 1982, Factors regulating nitrification in primary and secondary succession, *Ecology* **63:**1561-1573.

Rosswall, T., 1982, Microbiological regulation of the biogeochemical nitrogen cycle, *Plant Soil* **67:**15-34.

Ryden, J. C., 1983, Denitrification loss from a grassland soil in the field receiving different rates of nitrogen as ammonium nitrate, *J. Soil Sci.* **34:**355-365.

Satoh, T., S. S. M. Hom, and K. T. Shanmugam, 1983, Production of nitrous oxide from nitrite in *Klebsiella pneumoniae:* Mutants altered in nitrogen metabolism, *J. Bacteriol.* **155:**454-458.

Sawada, E., T. Satoh, and H. Kitamura, 1978, Purification and properties of a dissimilatory nitrite reductase of a denitrifying phototropic bacterium, *Plant Cell. Physiol.* **19:**1339-1351.

Schmidt, E. L., 1982, Nitrification in soil, in *Nitrogen in Agricultural Soils,* F. J. Stevenson, ed., American Society of Agronomy, Madison, Wisconsin, pp. 253-288.

Schropp, S. J., and J. R. Schwarz, 1983, Nitrous oxide production by denitrifying microorganisms from the eastern tropical north Pacific and the Caribbean Sea, *Geomicrobiol. J.* **3:**17-31.

Smith, M. S., 1982, Dissimilatory reduction of NO_2^- to NH_4^+ and N_2O by a soil *Citrobacter* sp., *Appl. Environ. Microbiol.* **43:**854-860.

Smith, M. S., 1983, Nitrous oxide production by *Escherichia coli* is correlated with nitrate reductase activity, *Appl. Environ. Microbiol.* **45:**1545-1547.

Stanford, G., J. O. Legg, S. Dzienia, and E. C. Simpson, 1975, Denitrification and associated nitrogen transformations in soils, *Soil Sci.* **120:**147-152.

Suzuki, I., U. Dular, and S.-C. Kwok, 1974, Ammonia or ammonium ion as substrate for oxidation by *Nitrosomonas europaea* cells and extracts, *J. Bacteriol.* **120:**556-558.

Tiedje, J. M., A. J. Sexstone, D. D. Myrold, and J. A. Robinson, 1982, Denitrification: Ecological niches, competition and survival, *Antonie van Leeuwenhoek, J. Microbiol. Serol.* **48:**569-583.

Topp, E., and R. Knowles, 1982, Nitrapyrin inhibits the obligate methylotrophs *Methylosinus trichosporium* and *Methylococcus capsulatus,* *FEMS Microbiol. Lett.* **14:**47-49.

Van't Riet, J., and R. J. Planta, 1975, Purification, structure and properties of the respiratory nitrate reductase of *Klebsella aerogenes, Biochim. Biophys. Acta* **379:**81-94.

Vedenina, I. Y., and G. A. Zavarzin, 1977, Biological removal of nitrous oxide under oxidizing conditions, *Microbiology* **46:**728-733, transl. of *Mikrobiologiya* **46:**898-903.

Vedenina, I. Y., and G. A. Zavarzin, 1979, Removal of nitrous oxide by a combined bacterial culture, *Microbiology* **48:**459-462, transl. of *Mikrobiologiya* **48:**581-585.

Verstraete, W., 1981, Nitrification, in *Terrestrial Nitrogen Cycles,* F. E. Clark and T. Rosswall, eds., *Ecol. Bull.* (Stockholm) **33:**303-314.

Vitousek, P. M., J. R. Gosz, C. G. Grier, J. M. Melillo, and W. A. Reiners, 1982, A comparative analysis of potential nitrification and nitrate mobility in forest ecosystems, *Ecol. Monogr.* **52:**155-177.

Ward, B. B., and M. J. Perry, 1980, Immunofluorescent assay for the marine ammonium-oxidizing bacterium *Nitrosococcus oceanus, Appl. Environ. Microbiol.* **39:**913-918.

Ward, B. B., R. J. Olson, and M. J. Perry, 1982, Microbial nitrification rates in the primary nitrite maximum off Southern California, *Deep-Sea Res.* **29:**247-255.

Yoh, M., H. Terai, and Y. Saije, 1983, Accumulation of nitrous oxide in the oxygen-deficient layer of freshwater lakes, *Nature* **301:**327-329.

Yoshida, T., and M. Alexander, 1970, Nitrous oxide formation by *Nitrosomonas europaea* and heterotrophic microorganisms, *Soil Sci. Soc. Am. Proc.* **34:**880-882.

Yoshida, T., and M. Alexander, 1971, Hydroxylamine oxidation by *Nitrosomonas europaea, Soil Sci.* **111:**307-312.

Youngdahl, L. J., R. Pacheco, J. J. Street, and P. L. G. Vlek, 1982, The kinetics of ammonium and nitrate uptake by young rice plants, *Plant Soil* **69:**225-232.

36

Influence of Oxygen Aeration on Denitrification and Redox Level in Different Bacterial Batch Cultures

J. C. G. Ottow and W. Fabig
Universitat Hohenheim

INTRODUCTION

In bulk soil or sediment, denitrification seems to begin as soon as the trapped oxygen has been utilized, and the subsequent oxygen (O_2) diffusion is limited by water saturation, physico-chemical soil properties, or both. At the onset of denitrification, the overall redox potential (Eh) of the flooded soil has dropped to approximately +600 to +200 mV (Table 36-1). In the presence of large amounts of easily decomposable organic matter, the demand for external electron acceptors may greatly surpass the supply of O_2 and nitrate, and the microflora will switch to other electron acceptors, such as manganese [Mn(IV)] and iron [Fe(III)] oxides, in their immediate environment (Takai and Kamura, 1966; Ottow and Glathe, 1973; Hammann and Ottow, 1974; Munch and Ottow, 1977, 1980, 1983). Consequently, the gross sequence of reductive transformations proceeds in the succession $O_2 \rightarrow$ nitrate (NO_3^-) $\rightarrow$ manganese dioxide (MnO_2) $\rightarrow$ ferric oxide (Fe_2O_3) $\cdot$ $xH_2O \rightarrow$ sulfate (SO_4^{2-}) $\rightarrow$ carbon dioxide (CO_2) and is generally reflected by a stepwise decrease in redox potential (Table 36-1). These clear successive reductive processes have led to a general view that oxygen, nitrate, Mn(IV), and Fe(III) compounds will act only as subsequent alternate electron acceptors as soon as the appropriate Eh has been established. Oxygen is known to inhibit denitrification, but aerobic denitrification has been observed and claimed before (Delwiche and Bryan, 1976; Krul and Veeningen, 1977). Such reports have been explained by the existence of anaerobic sites in an

Table 36-1. Sequence of Microbial Reductive Transformations in Flooded Soils and Theoretical Equilibrium Potentials of Half-Reactions at pH = 7

Processes	Measured Redox Potential (Eh) During Transformations[a]	rH Level[b]	Redox Systems Involved	$E'o\,(mV)$[c]
Respiration	>400	>26	$O_2 + 4H^+ + 4e^- \leftrightarrows 2H_2O$	+814
NO_3 respiration	+500 – +200	29–19	$2NO_3^- + 12H^+ + 10e^- \leftrightarrows N_2 + 6H_2O$[d]	+741
Formation of Mn(II)	+400 – +200	26–19	$MnO_2 + 4H^+ + e^- \leftrightarrows Mn(II) + 2H_2O$[d]	+410
Production of Fe(II)	+400 – +180	26–18	$Fe(OH)_3^e + 3H^+ + e^- \leftrightarrows Fe(II) + 3H_2O$[d]	−185
S^{2-} production	+100 – −200	16–5	$SO_4^{2-} + 10H^+ + 8e^- \leftrightarrows H_2S + 4H_2O$[d]	−214
CH_4 production	−150 – −280	7–2	$CO_2 + 8H^+ + 8e^- \leftrightarrows CH_4 + 2H_2O$	−244

[a] Redox potentials (Eh) are comparable only if corrected for pH.

[b] rH (= negative log of pH_2) calculated from Eh(mV) at (a) and a pH of 6 (assumed) according to rH = Eh (mV)/29 + 2xpH (30°C). At rH = 0 completely reduced, at rH = 42.6 entirely oxidized, conditions prevail.

[c] E'o = Em = standard electrode potential at 50% reduction at a pH of 7.0.

[d] These electron acceptors are increasingly electromotively sluggish (NO_3^-, MnO_2) or even highly inactive ($Fe_2O_3 \cdot xH_2O$, SO_4^{2-}). In soils their equilibria are shifted to the left, but the relatively high energy of the activation barrier is easily lowered by specific microbial reductases that use the oxidants as electron acceptors (Hammann and Ottow, 1974; Ottow and Munch, 1978; Munch and Ottow, 1977, 1983; Munch and Ottow, 1982; Ottow, 1982). A low redox level (pH-Eh level comparable to completely reduced conditions at rH = 0) does not warrant any significant/reductive transformation (Munch and Ottow, 1982; Ottow, 1982).

[e] Amorphous Fe(III)-oxides are reduced by bacteria in preference to the crystalline forms (Munch and Ottow, 1980).

aerobic microenvironment. However, since the electron flow during energy-conserving respiration and denitrification is not decided until the cytochrome b or c level, there is actually no need to consider respiration and denitrification as entirely alternative processes once the required enzymes have been induced and synthesized. This hypothesis was examined in the present study.

MATERIALS AND METHODS

Test Organisms

Acinetobacter sp. 53b and *Moraxella* sp. 13b were isolated from columnar denitrification and identified as reported elsewhere (Fabig and Ottow, 1976). The strains were subcultured anaerobically in a synthetic mineral salt medium (Fabig and Ottow, 1976) containing glycerol (0.3%) and KNO_3 (1.0% = 1400 μg NO_3^--N/ml). Five-ml samples of a young (48 hr, 30°C), actively denitrifying culture (ca 10^8 cells/ml) were used to inoculate the experimental vessel.

Batch Experiments under
Controlled Conditions

About 500 ml of the separately sterilized (15 min at 120°C) synthetic glycerol-mineral salt stock solution (2 l, pH = 7.4, with or without 1% KNO_3) was pumped into the sterile reaction vessel (Fig. 36-1). Different electrodes for pH, redox potential (Eh) pO_2, and nitrate were air-tight inserted and the whole setup was sterilized with 0.5 ml ethylene oxide (Merck) for 2 hr. Helium (He) gas was flushed through the continuously stirred broth until it was completely free from ethylene oxide (gas chromatography). All parts of the setup were connected with special air-tight rubber tubes, and anoxic conditions in the vessel were ascertained after several days by gas chromatography before inoculating. Experiments were run anaerobically (He gas) as well as aerobically (He/O_2 mixture with 5 ml oxygen/min). Aeration at 5 ml O_2/min corresponds to O_2-saturated conditions with 25-26 mg O_2/l. Each vessel was stabilized for at least 48 hr (30°C), after which it was inoculated by introducing 5 ml of a 4-hr culture together with the remainder of the stock solution (hand pumping). The broth was homogenized continuously (magnetic stirrer), kept at 30°C, and covered entirely with aluminum foil.

Analysis

pH and Ex were measured with a glass silver chloride-Pt-electrode (Ingold Type 405) using a WTW Digi 510 as a recording instrument. The electrodes were adjusted with pH and redox buffer solutions (Ingold 9805, 9807, and 9881). The Eh was calculated from Ex by the addition of 207 mV (30°C) and used to determine the rH values (= negative logarithm of the partial pressure of gaseous hydrogen) according to the formula

$$rH = Eh \ (mV)/29 + 2 \ pH \ (30°C) \qquad (Eq. \ 36-1)$$

At rH = 42.6, completely oxidized conditions exist; at rH = 0, entirely reduced conditions are established (Jacob, 1970). Redox equilibrium ($pO_2 = pH_2$) is obtained at rH = 27.5. Reduced conditions prevail at rH < 17. The rH allows the comparison of the redox level in different systems at any time (Ottow and Glathe, 1973). pO_2 was monitored with a WTW-Oxygen Electrode EO 16 and nitrate using an Orion electrode model 92-67. All data were continuously monitored and collected with a Philips PM 8235 recorder. At regular intervals samples were taken from the culture solution (the outlet is missing in Fig. 36-1) and examined enzymatically for glycerol and color-imetrically for nitrite (Fabig, Ottow, and Muller, 1978). Population densities (*Acinetobacter* sp. 53b and *Moraxella* sp. 13b) were determined by the MPN-technique using the same broth (with and without KNO_3, respectively).

Gas Chromatography

Samples were withdrawn (Fig. 36-1) and analyzed for nitric oxide (NO), nitrous oxide (N_2O), dinitrogen (N_2), CO_2, and O_2 using a Perkin Elmer gas

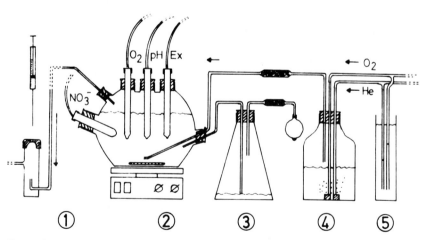

Figure 36-1. Experimental setup to study the effect of aeration on denitrification and redox conditions in batch cultures of *Acinetobacter* sp. 53b or *Moraxella* sp. 13b. The reaciton vessel was continuously homogenized by gas flushing as well as magnetic stirring and kept tat 30°C. 1 = sampling site (gas chromatography; 2 = reaction vessel and magnetic stirrer (30°C); 3 = stock solution and hand pump; 4 = gas mixing device; 5 = gas flow detector

chromatograph F 22 fitted with a hot wire and operating with He gas as a carrier. Porapak Q and R series were used to separate NO, N_2O, and CO_2, while a molecular sieve MS 5A was used to differentiate N_2 and O_2 (Moretti et al., 1974). Results were collected with a Servogor S (Metrawatt) recorder and recorded with a SIP 1 Perkin Elmer Integrator (Fabig, Ottow, and Muller, 1978).

RESULTS

Acinetobacter sp. 53b — Mineralization, Denitrification, and Redox Conditions at Anaerobiosis

In Figure 36-2 the development of the various parameters during the growth of *Acinetobacter* sp. 53b under complete anaerobic conditions (He gas) are given. After a lag phase (24 hr) the population increases while glycerol and nitrate decrease. Nitrite accumulates but decreases slowly as soon as nitrate has been exhausted. Nitrate seems to be the growth-limiting factor. Population development is reflected by CO_2 evolution, as well as N_2 and N_2O production. Both carbon dioxide and denitrification practically cease after 120 hr. As soon as metabolism accelerates, the redox level (rH) drops rapidly and remains at about rH = 18 as long as glycerol and nitrate are available. Gaseous losses of N_2O and N_2 were most intensive at active growth and minimum redox level.

Acinetobacter sp. 53b — Mineralization and Redox Conditions at Continuous Aeration

In Figure 36-3 the same parameters are presented at continuous aeration (5 ml O_2/min). After a prolonged lag phase (48 hr) the population of *Acinetobacter* sp. 53b increases rapidly while glycerol utilization proceeds (up to 168 hr). During the lag phase the pO_2 decreases only slightly, but drops to a complete anaerobic situation within 24 hr as soon as the population enters the lag-growth phase. The redox level is lowered steeply and poises at rH = 13-15 as long as glycerol remains available. Apparently, the redox level of an actively metabolizing *Acinetobacter* sp. 53b population drops much lower with O_2 than with nitrate as the only electron acceptor.

Acinetobacter sp. 53b — Mineralization, Denitrification, and Redox Conditions in the Presence of Oxygen and Nitrate

Figure 36-4 represents the changes in the various parameters of the *Acinetobacter* culture in the presence of both oxygen and nitrate. The following features are obvious. First, glycerol utilization, nitrate dissimilation,

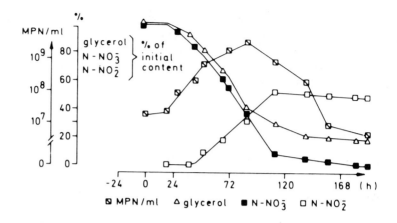

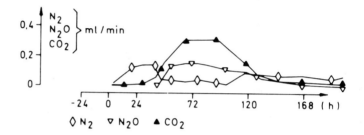

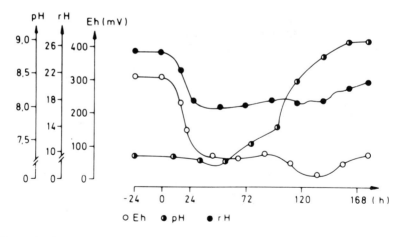

Figure 36-2. Development of *Acinetobacter* sp. 53b population, glycerol utilization, carbon dioxide evolution, denitrification (H_2, N_2O), and ecological conditions (pH, Eh, rH) at anaerobic conditions (He-atmosphere, batch culture, 30°C).

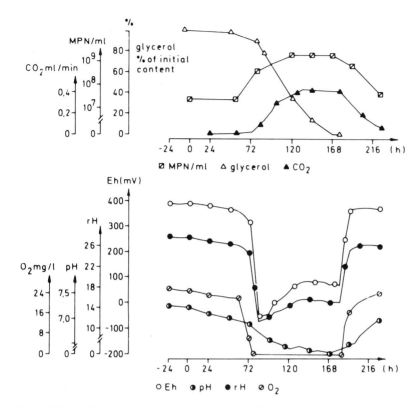

Figure 36-3. Development of *Acinetobacter* sp. 53b population, glycerol utilization, carbon dioxide production, and ecological conditions (pH, Eh, rH) at continuous aeration (5 ml O_2/min, batch culture, 30°C).

and respiration are distinctly more intensive than with either nitrate (Fig. 36-2) or oxygen (Fig. 36-3) alone. Second, the growth of *Acinetobacter* sp. 53b as well as CO_2 production is more intensive in the presence of both electron acceptors than with nitrate or O_2. Third, pO_2 and rH level drop considerably if compared to nitrate or O_2 alone. The minimum rH of 14 poises at this redox level only very briefly. During this most active phase of metabolism, both respiration ($pO_2 = 0$) and denitrification ($N_2O + N_2$) occur simultaneously, indicating that both nitrate and molecular oxygen are used at the same time.

Moraxella sp. 13b—Mineralization, Denitrification, and rH Level at Anaerobiosis

In Figure 36-5 the behavior of the same parameters in the culture of *Moraxella* sp. 13b is given. Although the same conditions exist, the metabolism of

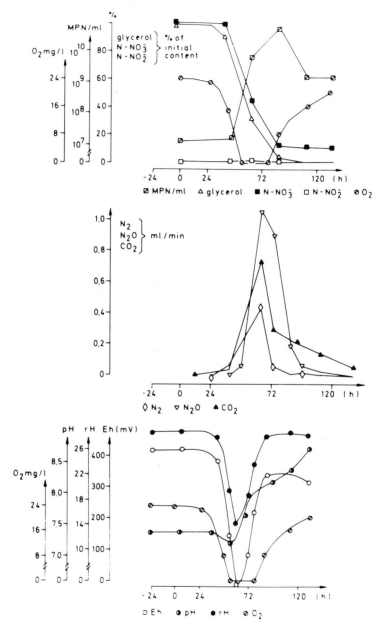

Figure 36–4. Development of *Acinetobacter* sp. 536 population, glycerol utilization, carbon dioxide evolution, denitrification (N₂, N₂O), and ecological conditions (pH, Eh, rH) in the presence of nitrate and O₂ (5 ml/min, batch culture, 30°C).

Moraxella sp. 13b is apparently much less vigorous than that of *Acinetobacter* sp. 53b. Glycerol is exhausted only after 192 days, and this lower rate of mineralization is clearly reflected by delayed population development and nitrate respiration. Again, denitrification becomes most active as soon as the redox level poises at about rH = 14. This level is significantly lower than that of *Acinetobacter* (Fig. 36-2).

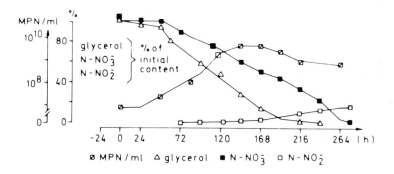

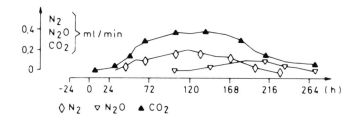

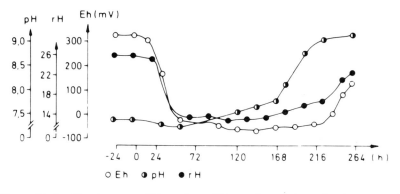

Figure 36-5. Development of *Moraxella* sp. 13b population, glycerol utilization, carbon dioxide evolution, denitrification (N_2, N_2O), and ecological conditions (pH, Eh, rH) at anaerobic conditions (He-atmosphere, batch culture, 30°C).

Moraxella sp. 13b — Mineralization and Redox Conditions at Continuous Aeration

Figure 36-6 shows glycerol mineralization and redox changes during respiration of *Moraxella* sp. 13b at an aeration rate of 5 ml O_2/min. As with *Acinetobacter* sp. 53b, anaerobic conditions ($pO_2 = 0$) are established rapidly and remain until glycerol has been exhausted. The rH level drops and stabilizes at approximately 14 to 10, which is again significantly lower than with nitrate as the only electron acceptor (Fig. 36-5).

Moraxella sp. 13b — Mineralization, Denitrification, and Redox Conditions in the Presence of Nitrate and Oxygen

In Figure 36-7 the changes in the various parameters in the presence of both molecular oxygen and nitrate are presented. The results obtained with *Moraxella* sp. 13b are essentially the same as those observed with *Acinetobacter*

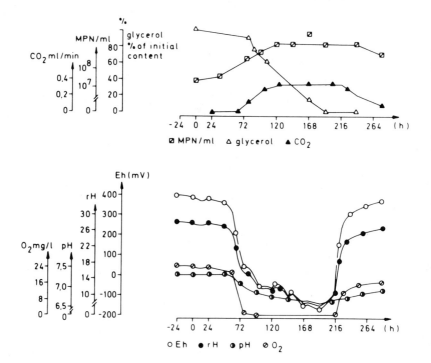

Figure 36-6. Development of *Moraxella* sp. 13b population, glycerol utilization, carbon dioxide production, and ecological conditions (pH, Eh, rH) at continuous aeration (5 ml O_2/min batch culture, 30°C).

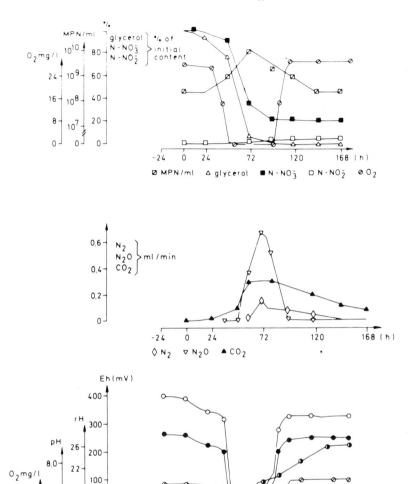

Figure 36-7. Development of *Moraxella* sp. 13b population, glycerol utilization, carbon dioxide evolution, denitrification (N₂, N₂O), and ecological conditions (pH, Eh, RH) in the presence of nitrate and O₂ (5 ml/min, batch culture, 30°C).

sp. 53b (Fig. 36-4). Again, metabolism is much more intensive in the presence of both electron acceptors than with either oxygen or nitrate. Since glycerol is utilized more rapidly, the redox level poises only briefly at rH = 12, which is in between the redox conditions reached with nitrate (Fig. 36-5) and oxygen (Fig. 36-6) alone. Maximum denitrification ($N_2O + N_2$ production) is most active as soon as the rH minimum has been established.

DISCUSSION

From the results obtained, two significant conclusions may be drawn. First, there is no doubt that oxygen and nitrate can be respired simultaneously by an actively metabolizing population of bacteria such as *Acinetobacter* sp. and *Moraxella* sp. This synchronous activity of respiration and denitrification in the same batch culture may be explained by the presence of both respiring and denitrifying bacteria within the same population. This is quite likely, the more so since molecular oxygen could not be detected ($pO_2 = 0$) during the most active phase of metabolism. Consumption of the continuously introduced molecular oxygen must have been fast enough to warrant the induction of denitrification within a part of the flora. On the other hand, energy conservation by respiration and denitrification in the same organism cannot be excluded, since these processes may have occurred in different compartments (inner cell membranes) of one cell. From an ecological viewpoint it is important to stress that respiration and denitrification may, indeed, occur simultaneously even at a relatively high rate of oxygen, if a sufficient amount of easily decomposable organic matter is available. Among the ecological factors that determine denitrification in natural systems, the amount of available organic matter rather than the oxygen diffusion rate (caused by water saturation or small pore size and volume) should be considered as the triggering factor. The higher the demand for electron acceptors, the greater are the chances for denitrification even under aerobic conditions.

Second, it seems as if respiration will lower the redox level of a system much more than denitrification. This is contrary to the sequence predicted by the half reactions involved (Table 36-1). In natural mixed systems such as soils and sediments, the redox conditions (pH-Eh dependent) will poise at the level required to activate the electron acceptor in question (Ottow and Munch, 1978; Munch and Ottow, 1982; Ottow, 1982). Stabilization at a certain level will continue until all oxidants concerned have been transformed into the reduced state. However, if the rate of electron-acceptor availability becomes less than the rate at which electrons are provided by mineralization, the redox level may drop beyond the expected level. This is particularly true if only one electron acceptor is available in the system at a restricted rate. At an aeration rate of 5 mg O_2/min, the demand for O_2 by either *Acinetobacter* sp. or *Moraxella* sp. apparently exceeded the supply ($pO_2 = 0$). This was the opposite with nitrate, which remained in excess even after glycerol had been utilized. The lack of O_2 during aeration was confirmed indirectly by the rise in rH level when nitrate was added. In general, redox potentials provide little diagnostic information on the type and intensity of a process as long as detailed information is lacking on the type and amount of available organic matter, on the composition of the redox systems, and on the ecological conditions (moisture tension, temperature, and so on). In a given situation, the measured Eh depends on the rate of organic-matter turnover (mineralization), the activation energy barrier of the redox system involved, and the proton- and electron-buffering capacity of the whole system, as well as on the oxygen diffusion rate (Ottow and Glathe, 1973; Ottow, 1982). Consequently, the redox level does not provide reliable information on the type of reductive processes that occurred in the past nor at the moment of measuring (Ottow, 1982). Moreover, most relevant redox systems in flooded soils and sediments (nitrate, ferric oxides and hydroxides, sulfate) are electromotively extremely sluggish (their equilibria are shifted to the left). Thus, they do not respond, or respond only weakly, to an increased electron activity (= low Eh) if specific catalysts (enzymes) required to lower the activation energy barrier are missing (Ottow and Munch, 1978; Munch and Ottow, 1977, 1980, 1982; Ottow, 1982).

ACKNOWLEDGEMENTS

This research was supported by the German Research Foundation (DFG) in Bonn, Federal Republic of Germany.

REFERENCES

Delwiche, C. C., and B. A. Bryan, 1976, Denitrification, *Ann. Rev. Microbiol.* **30:**241-262.

Fabig, W., and J. C. G. Ottow, 1976, Isolierung und Identifizierung neuer denitri-fizierender Bakterien aus einer Modell-Kläranlage mit anaeroben Festbettreak-toren, *Arch. Hydrobiol.* **85:**272-391.

Fabig, W., J. C. G. Ottow, and F. Muller, 1978, Mineralisation von ^{14}C-markiertem Benzoat mit Nitrat als Wasserstoff-Akzeptor unter vollständig anaeroben Ben-dingungen sowie bei vermindertem Sauerstoffpartialdruck, *Landw. Forsch.* **35:**441-453.

Hammann, R., and J. C. G. Ottow. 1974. Reductive dissolution of Fe_2O_3 by saccharolytic clostridia and *Bacillus polymyxa* under anaerobic conditions, *Z. Pflanzenernaehr. Bodenkd.* **137:**108-115.

Jacob, H. E., 1970, Redox potentials, in *Methods in Microbiology,* Vol. 2, Chap. 4, J. R. Norris and D. W. Ribbons, eds., Academic Press, New York, pp. 91-123.

Krul, J. M., and R. Veeningen, 1977, The synthesis of the dissimilatory nitrate reductase under anaerobic conditions in a number of denitrifying bacteria, isolated from activated sludge and drinking water, *Water Res.* **11:**39-43.

Moretti, E., G. Leofanti, D. Andreazza, and N. Giordano, 1974, Gas chromatographic separation of effluent from the ammonia oxidation reaction: O_2, N_2, N_2O, NO_2, NH_3, and H_2O, *J. Gaschrom. Sci,* **12**:64-66.

Munch, J. C., and J. C. G. Ottow, 1977, Modelluntersuchungen zum Mechanismus der bakteriellen Eisenreduktion in hydromorphen Böden, *Z. Pflanzenernaehr. Bodenkd.* **140**:549-562.

Munch, J. C., and J. C. G. Ottow, 1980, Preferential reduction of amorphous to crystalline iron oxides by bacterial activity, *Soil Sci.* **129**:15-21.

Munch, J. C., and J. C. G. Ottow, 1982, Einfluß von Zellkontakt und Eisen-(III)-Oxidform auf die bakterielle Eisenreduktion. *Z. Pflanzenernaehr. Bodenkd.* **145**:66-77.

Munch, J. C., and J. C. G. Ottow, 1983, Reductive transformation mechanisms of ferric oxides in hydromorphic soils, *Ecol. Bull.* (Stockholm) **35**:383-394.

Ottow, J. C. G., 1982, Bedeutung des Redoxpotentials für die Reduktion von Nitrat und Fe(III)-Oxiden in Böden, *Z. Pflanzenernaehr. Bodenkd.* **145**:91-93.

Ottow, J. C. G., and H. Glathe, 1973, Pedochemie und Pedomikrobiologie hydromorpher Böden: Merkmale, Voraussetzungen und Ursachen der Eisenreduktion, *Chem. Erde* **32**:1-44.

Ottow, J. C. G., and J. C. Munch, 1978, Mechanisms of reductive transformations in the anaerobic microenvironment of hydromorphic soils, in *Environmental Biogeochemistry and Geomicrobiology,* Vol. 2, *The Terrestrial Environment,* W. E. Krumbein ed., Ann Arbor Science, Ann Arbor, Michigan, pp. 483-491.

Takai, Y., and T. Kamura, 1966, The mechanisms of reduction in water-logged paddy soil, *Fol. Microbiol.* **11**:304-313.

37

Denitrification in a Shortgrass Prairie: A Modeling Approach

Arvin R. Mosier

U.S. Department of Agriculture—ARS

William J. Parton

Colorado State University

INTRODUCTION

Nitrogen (N) input and losses in shortgrass prairie ecosystems are not well understood. Analogously, many aspects of N transport and transformations in native systems have not been investigated. As part of an overall project to quantitate N additions, losses, and transformations in a native and a grazed shortgrass prairie, the importance of denitrification as an N-loss mechanism has been assessed. From preliminary observations it was concluded that nitrous oxide (N_2O) is the primary nitrification-denitrification product. As a result, only N_2O-N losses in this shortgrass prairie system will be discussed. Early work in this study evaluated N_2O losses from the native soil and from simulated cattle urine patches (Mosier, et al., 1981). The initial research suggested that, although N_2O losses from the native prairie were small (about $2 \text{ g N/ha} \cdot \text{d}^{-1}$), these losses accounted for about 10% of total N inputs into the system. Nitrous oxide losses from urea-treated sites were larger than from untreated soils but amounted to only 0.6% of the added N. Large temporal variability in N_2O-emission rates was evident from this study. Initial efforts to develop a model to predict daily N_2O loss rates showed that N_2O temporal variability from the grassland soil was related to a complex combination of factors, such as soil nitrate, ammonium, and water content (Mosier, Parton, and Hutchinson, 1983). This study was continued to better quantify these relationships, to look at effects of carbon availability, and to determine if total denitrification (conversion of nitrate to N_2) is important in the shortgrass

441

prairie N cycle. This paper will discuss the results of these studies and the improvement of the initial model to explain the temporal variability in N_2O losses (denitrification) from a shortgrass prairie site.

MATERIALS AND METHODS

The study site was located in the Central Plains Experimental Range, 56 km northeast of Fort Collins, Colorado. Annual precipitation at the site averages about 30 cm, 80% of which occurs between May and September. Typically, more than one-half of the vegetation is blue grama (*Bouteloua gracilis* Lag.). At this virgin site individual miniplots were established by driving open-ended steel cylinders that were 10 cm in diameter and 40 cm long into the soil. The cylinders were placed in blue grama swards that were as similar as possible in cover, exposure, and soil. Urea (99 atom % ^{15}N) equivalent to 450 kg N/ha in 1.5 cm H_2O (0.76 g urea in 120 ml H_2O) was added uniformly to the surface of each treated plot. Nothing was added to the control plots. Four replicate treated and untreated plots were sampled at each sampling period. The miniplots were established at midslope (July 1981) and the bottom (May 1982) of a northwest-facing slope having a 6° slope. The soil for the midslope is classified as an Ustollic Haplargid with a solum depth of 25 cm. The bottom soil is an Aridic Argiustoll with a solum depth greater than 1 m. Adjacent to each of the two plots, a 5-m² area was treated with natural ^{15}N abundance urea at the same rate as the miniplots to allow rigorous soil sampling to monitor soil water (H_2O), ammonium (NH_4^+), nitrate (NO_3^-), and nitrite (NO_2^-) at each gas-flux-measuring event. Soil water content was determined gravimetrically. Each soil sample was extracted with 2 N KCl, and the extracts were analyzed for NH_4^+ and $NO_3^- + NO_2^-$ by distillation (Bremner and Keeney, 1965) and for NO_2 as described by Bremner (1965). Meteorological data were collected at a site about 300 m from the plots. Soluble organic carbon was determined by persulfate oxidation of a 5:1 (w/w) 1 N Na_2SO_4 extract of freshly collected soil. Another estimate of carbon availability was made by incubating 10 g of previously frozen soil in 125-ml serum vials at 25°C, monitoring CO_2 production from the soil at 1, 3, 5, and 7 hr, and calculating the rate of CO_2 production.

Vertical N_2O and N_2 flux density was determined by measuring the increase in N_2O or $^{30}N_2$ concentration beneath vented cylindrical enclosures placed over a miniplot for 60 min (Mosier and Hutchinson, 1981). Thirty-ml samples of the enclosure air were withdrawn through a fixed orifice with 60-ml polypropylene syringes equipped with two-way valves, 0, 30, and 60 min after each enclosure was installed. After sampling, the gas samples were transported to the laboratory for same-day analysis for N_2O by gas chromotography (GC) (Mosier and Mack, 1980) and dinitrogen by a VG-903 mass spectrometer. Dinitrogen-flux estimates were calculated by a method described by Siegel, Hauek, and Kirtz, (1982). Sensitivity of the N_2 flux measurements is limited to about 10 ± 10 g N/ha $\cdot$ d^{-1} (analytical error decreases to <10% for flux rates >100 g N/ha $\cdot$ d^{-1}), with detection limits of about 5 ± 10 g N/ha $\cdot$ d^{-1}.

Laboratory experiments were conducted as follows: Ten g of surface-to-15-cm-depth soil, freshly collected from the nonlabeled urea-treated plot, were placed in 125-ml serum bottles. Water was added to the desired soil moisture content, and the bottles were sealed with rubber serum stoppers and incubated 1 to 7 days at 25°C. The flask air atmosphere was amended with 10 atmosphere percent acetone-free acetylene when desired. Headspace N_2O concentration was determined by GC (Mosier and Mack, 1980), and total N_2O production was corrected for soil H_2O.

RESULTS AND DISCUSSION

Vertical N_2O flux density from the midslope site for the period 14 July 1981 to 15 October 1982 is shown in Figure 37-1a for urea-treated plots. The N_2O flux data for the bottom site are similar to those of the midslope site in that N_2O flux density varied in relation to soil water content and nitrate concentration, relationships that will be discussed later in this paper. The highest N_2O emission occurred during the last two weeks of August 1981, following two rains that totaled about 4 cm, and during mid-May and mid-June 1982, following rain events of greater than 2 cm each. No data are shown for dinitrogen flux density because the events during which N_2 flux was detectable were limited to only 8 g N/ha $\cdot$ d^{-1} at the midslope site on 19 August 1981 and to 18, 28, and 22 g N/ha $\cdot$ d^{-1} at the bottom site on 14 June, 24 June, and 18 August 1982, respectively. Dinitrogen flux events of less than 5 g N/ha $\cdot$ d^{-1} were not detectable.

It was of interest that N_2O and N_2 emissions from the treated plots, which contained at times as much as 75 μg NO_3^--N/g soil, were so low following rain events. Several rain events were large enough to raise the surface 0- to 15-cm soil moisture content to 12% bar (Fig. 37-1b), which corresponds to -0.1 bar. During these times temperature and evapotranspiration were high, and soil water content quickly decreased after each rain.

To investigate these observations further, soil samples were collected from the nonlabeled urea-treated areas from both midslope and bottom sites and returned to the laboratory for immediate study. Experiments were conducted to determine the effect of soil moisture content on N_2 and N_2O production. Sufficient water was added to provide soil water potentials of -0.3 and -0.1 bars and saturation with a few-millimeters-thick standing-water surface. Soil nitrate content ranged between 40 and 70 μg for all of the laboratory studies. The data presented in Table 37-1 represent the mean of three separate laboratory incubations (soils collected on different days), with each study

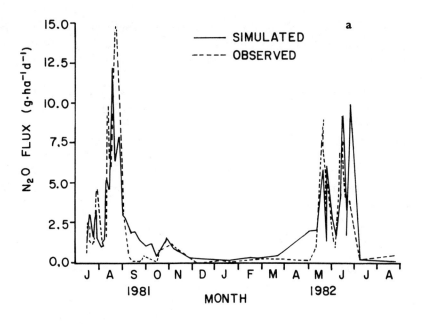

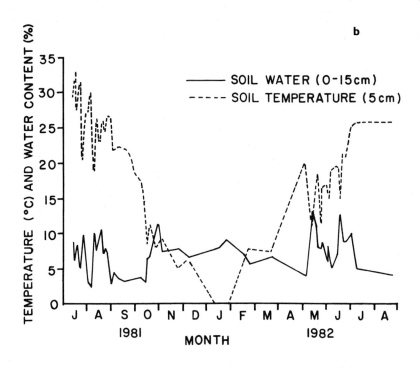

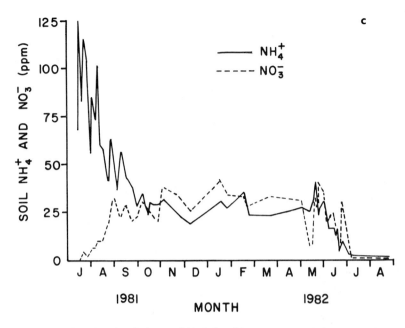

Figure 37-1. Simulated and observed N_2O flux (*a*), gravimetric soil water and average soil temperature (*b*), and soil NH_4^+ and NO_3^- levels for the midslope experiment (*c*).

replicated three times. These data show that at both -0.3 and -0.1 bars soil water potential, little N_2O and no N_2 was produced after 24 hr or 7 days of incubation. Other data in the study using acetylene blockage indicate that the small amounts of N_2O from the -0.3 and -0.1 bar water potential arise from nitrification, not denitrification. Only when a layer of water was maintained over the soil surface were large amounts of N_2 and N_2O produced. These data support conclusions drawn from field measurements that very little N_2 is produced in grassland soil under the climatic conditions normally encountered. Because the study area is in a semiarid climate, soil moisture is infrequently above 10% (Fig. 37-1*b*), and soils normally dry rapidly after precipitation events, when conditions are otherwise conducive to denitrification. From these observations it can be concluded that N_2O is the primary nitrification-denitrification product. As a result, in this shortgrass prairie system only N_2O-N losses will be discussed. The soils do, however, have the potential (Table 37-1) to denitrify nitrate to N_2 if saturated conditions are maintained for more than a day. However, such events are probably rare.

These data and observations were used as a background to develop a simple model that allows prediction of the temporal variability of N_2O-N

Table 37–1. Effect of Soil Water Content on N₂O and N₂ Production in Urea-Fertilized Grassland Soil

Soil Water Potential	N₂O Production $(ng\ N_2O\text{-}N\ g^{-1}soil \cdot day)$		N₂ Production $(ng\ N_2\text{-}N\ g^{-1}soil \cdot day)$	
	Incubation Period (days)			
	1	*7*	*1*	*7*
Midslope site				
0.3	8a*	14a	0a	0a
0.1	14a	14a	0a	0a
Submerged	380b	770b	230b	1600b
Bottom site				
0.3	7a	10a	0a	0a
0.1	50b	57b	0a	0a
Submerged	300c	320c	300b	5900b

*Numbers in each column, for each plot site, followed by the same letter are not significantly different at the 0.05 level.

losses (denitrification) from the shortgrass prairie ecosystem. In this model the fertilized sites (simulating cattle urine deposition) are considered separately from the unfertilized native soil system. The best-fit estimates of parameters in the model were determined using the nonlinear data-fitting technique of Powell (1965).

Mosier, Parton, and Hutchinson (1983) showed that N₂O loss can be predicted as a function of soil water content and the soil NO_3^- and NH_4^+ levels. The new data set collected during 1981 and 1982 was much more comprehensive and included observations of soil water content (at 0-5, 5-10, and 10-15 cm), soil temperature (at 5 cm), and soil NO_3^- and NH_4^+ levels (at 0-5, 5-10, and 10-15 cm) for each day the N₂O flux was measured. For selected dates, estimates of available carbon were also measured. The new data allowed quantification of the effect of soil temperature on N₂O flux and determination of which soil depth (0-5, 5-10, or 10-15 cm) gave the best model for predicting N₂O flux. The new data set includes N₂O-loss data from two sites (bottom and midslope) from fertilized and unfertilized treatments and from different seasons.

The data from the fertilized sites were used to develop a model for N₂O flux:

$$^E N_2O = a \cdot E_\psi \cdot CNO_3 \cdot E_T + b \cdot E_\psi \cdot E_T \cdot E_N, \qquad \text{(Eq. 37-1)}$$

where $^E N_2O$ is the flux of N₂O (g N/ha · d⁻¹). E_ψ is the effect of soil moisture on N₂O flux (see Fig. 37-2a), E_T is the effect of soil temperature (at 5 cm) on N₂O flux (see Fig. 37-2b), CNO_3 is the NO_3^- concentration in the soil, E_N is the effect of soil NH_4^+ on N_2O flux (see Fig. 37-2c), and a and b are empirical constants. The three major changes to the model are alteration of the soil water term (E_ψ), addition of the soil temperature term (E_T), and simplification of the soil-NO_3^- term to a linear function of NO_3 concentration. The denitrification model of Rolston et al. (1980) uses a temperature term with constant Q_{10} of 2, while these results (see Fig. 37-2b) show that Q_{10} changes as a function of soil temperature and ranges from 5 at 10°C to 1.5 at 30°C. It is assumed that the first term on the right side of the equation is N_2O flux from denitrification, while the second term is the nitrification term. Data from the untreated site did not include data on the soil-NO_3^- and -NH_4^+ concentrations because they were generally below the 2-μg N/g soil detection limit of the analytical method. Thus, Equations (37-2) and (37-3) predict N_2O flux from the midslope (sandy loam) and bottom (clay loam):

$$^{E^1}N_2O = 2.67\ E_\psi\ \cdot\ E_T, \qquad\qquad \text{(Eq. 37-2)}$$

$$^{E^2}N_2O = 5.17\ E_\psi\ \cdot\ E_T, \qquad\qquad \text{(Eq. 37-3)}$$

where $^{E^1}N_2O$ is the N_2O flux (g N/ha $\cdot$ d^{-1}) from the midslope and $^{E^2}N_2O$ is the N_2O flux (g N/ha $\cdot$ d^{-1}) from the bottom. The N_2O flux from the bottom soil is twice as large as the midslope and is probably a result of higher NO_3^- and NH_4^+ levels in bottom soil.

Table 37-2 shows a summary of the values of a and b for Equation (37-1) and the error terms for different versions of Equation (37-1). The data input into the equation were average daily soil temperature at 5 cm and soil water and soil NO_3^- and NH_4^+ levels for the 0- to 15-cm soil layer. A plot of observed vs. simulated data for the midslope is shown in Figure 37-1a. The results for Equation (37-1) show that the absolute mean error (γ_a) is 3.51 and square of the correlation coefficient (γ^2) is 0.57. This compares with a mean of the observed data of 6.23 g N/ha $\cdot$ d^{-1}. Dropping the soil temperature term and soil water term causes a substantial drop in γ^2 (0.28 and 0.16, respectively), while dropping the nitrification term $(E_N = 0)$ causes only a slight drop in γ^2 (0.55 vs. 0.57). These results suggest that the soil temperature term and soil water term have a substantial effect on the model and that the nitrification term contributes relatively little to the model. Combining the data from the fertilized and unfertilized treatments (Equations (37-1), (37-2), and 37-3)) results in γ^2 of 0.62 and γ_a of 1.95, while the mean of observed data is 3.44 g N/ha $\cdot$ d^{-1}. In general, γ_a is approximately equivalent to 50%-60% of the mean of the observed data.

Although the nitrification term contributes relatively little to the data fit, under certain conditions (high NH_4^+ and low NO_3^-) it is the major source of N_2O. Laboratory data show that nitrification plays an important role in the

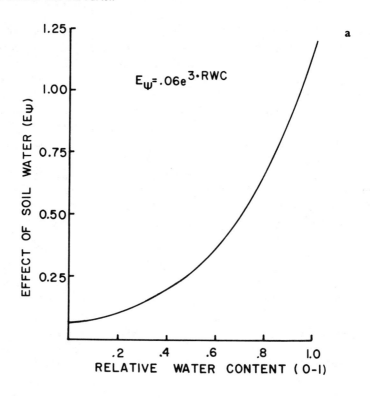

$$E_\psi = .06e^{3 \cdot RWC}$$

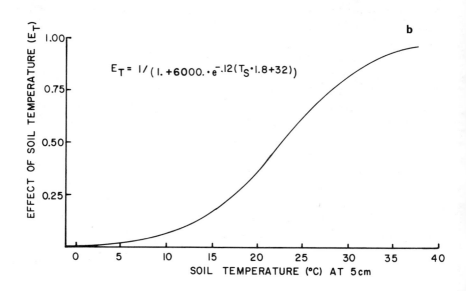

$$E_T = 1/(1. + 6000. \cdot e^{-.12(T_S \cdot 1.8 + 32)})$$

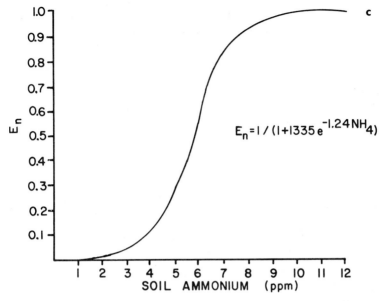

Figure 37-2. Effect of soil water (E_4) on N_2O loss (a), effect of soil temperature on N_2O loss (b), and the effect of soil NH_4^+ level on N_2O loss. (c).

total N_2O flux at specific times. Figure 37-3 shows model estimates of the ratio of nitrification N_2O flux to total N_2O flux from midslope. During the first month of the experiment, most of the N_2O flux ($>50\%$) came from nitrification, while denitrification was the dominant term for the remainder of the experiment (80%-90%). During the entire experiment, 26% of the N_2O flux resulted from nitrification. For the bottom soil, only 10% of the flux resulted from nitrification. The difference was that the nitrification rate was much more rapid in the bottom soils.

Soil NO_3^-, NH_4^+, and water content from the 0- to 5-, 5- to 10-, and 10- to 15-cm depths were used as inputs into Equation (37-1), and the results showed that the best model fit was obtained using the 0- to 5-cm data ($\gamma^2 = 0.58$), while the model fit decreased for the 5- to 10- and 10- to 15-cm layers ($\gamma^2 = 0.44$ and 0.51, respectively). The results also show that the coefficient for the denitrification term (a) decreased from 2.0 for the 0- to 5-cm layer to 1.69 for the 10- to 15-cm layer, while the coefficient for the nitrification term increased from 1.14 for the 0- to 5-cm depth to 20.1 for the 10- to 15-cm depth. This suggests that the effect of the denitrification term decreased with depth, while the nitrification term increased with depth. The reason for the change is unclear at this time. The model fit was compared using the 0- to 5-, 0- to 10- and 0- to 15-cm data, and the results showed that the γ^2 were quite similar (0.58, 0.55, and 0.57, respectively).

Table 37-2. Summary of the Error Terms and Values of *a* and *b* for Different Versions of Equation 37-1

Model	a	b	γ_a $(g\,N\,ha^{-1}$	γ^2 $\cdot\,d^{-1})$
Equation 37-1	1.83	8.72	3.52	0.57
Equation 37-1 with $E_T = 1$	0.63	2.67	4.82	0.28
Equation 37-1 with $E_\psi = 1$	0.83	3.2	5.69	0.16
Equation 37-1 with $E_N = 0$	2.07	0	3.65	0.55
Combined model				
[Equations (37-1), (37-2), and (37-3)	1.83	8.72	1.95	0.62

$$\gamma_a = \frac{\sum\limits_{i=1}^{N} ABS\,(x_i^s - x_i^o)}{N},$$

where N = number of observations, x_i^s = simulated data, and x_i^o = observed data.

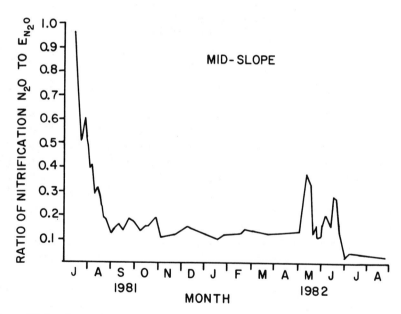

Figure 37-3. Ratio of the N_2O produced by nitrification to the total N_2O produced (E_{N_2O}).

The N_2O-loss model developed in a previous paper (Mosier, Parton, and Hutchinson, 1983) was tested using the new data set, and the results show that the γ^2 were equal to 0.24. The γ^2 for the previous model and new model without the temperature term are quite similar (0.24 vs. 0.28) and suggest that the major improvement in the new model is the inclusion of the soil temperature term. The new model was used to analyze the previous N_2O

data set, and data fit was quite similar ($\gamma^2 = 0.79$ vs. 0.80). This result suggests that adding the soil temperature term did not improve the fit to the old data set. This is consistent with the fact that soil temperature variation for the old data set was small (21°C to 30°C) compared with that of the new data set (-2°C to 32°C).

In a previous discussion of factors that influence N_2O production (Mosier, Parton, and Hutchinson, 1983), it was suggested that knowing the carbon available for denitrification would benefit the predictive model. In the present study estimates of available carbon were made by two different methods. Neither estimate of available carbon improved the model. These results indicate that either available carbon is not a limiting factor for denitrification in this shortgrass steppe or that the carbon fraction that influences denitrification could not be measured.

Loss of nitrogen from native soil and simulated cattle urine patches in a shortgrass prairie resulting from denitrification is not large. Both field and laboratory studies suggest that N_2O is the primary nitrification-denitrification product; therefore, by quantifying N_2O alone, denitrification can be characterized in this system. The model developed by Mosier, Parton, and Hutchinson (1983) to predict temporal variation in N_2O emissions from soils has been modified to include the effect of temperature on denitrification. Work with the model indicates that, for data collected during the warm season, including temperature does not improve the model. However, when the data set included data collected during cold periods, including temperature was essential. The data suggest that Q_{10} ranges from 5 at 10°C to 1.5 at 30°C. Soil data taken from the 0- to 5-cm depth is best correlated to N_2O loss, but integration over the 0- to 15-cm soil depth gives similar results. When comparing the relative importance of denitrification and nitrification to N_2O production, denitrification is most important overall. Nitrification, however, was the principal N_2O production process when NH_4^+ concentrations were high and NO_3^- concentrations low during the first month after urea application. Including a measure of soluble carbon in the model did not improve the prediction of N_2O emissions from the shortgrass prairie ecosystem.

ACKNOWLEDGEMENTS

We thank Ms. S. J. Crookall, Dr. Luis Amoros, and Brad Walker for their capable technical assistance. This work was supported in part by National Science Foundation Grant DEB-7906009.

REFERENCES

Bremner, J. M., 1965, Inorganic forms of nitrogen, in *Methods of Soil Analysis, Part 2,* C. A. Black et al., eds., *Agronomy* **9:**1219-1224, American Society of Agronomy, Madison, Wisconsin.

Bremner, J. M., and D. R. Keeney, 1965, Steam distillation methods for determination of ammonium, nitrate, and nitrite, *Anal. Chim. Acta* **32:**475-495.

Mosier, A. R., and G. L. Hutchinson, 1981, Nitrous oxide emissions from cropped fields, *J. Environ. Qual.* **10:**169-173.

Mosier, A. R., and L. Mack, 1980, Gas chromatographic system for precise, rapid analysis of N₂O, *Soil Sci. Soc. Am. J.* **44:**1121-1123.

Mosier, A. R., M. Stillwell, W. J. Parton, and R. G. Woodmansee, 1981, Nitrous oxide emissions from a native shortgrass prairie, *Soil Sci. Soc. Am. J.* **45:**617-619.

Mosier, A. R., W. J. Parton, and G. L. Hutchinson, 1983, Modelling nitrous oxide evolution from cropped and native soils, in *Environmental Biogeochemistry,* R. Hallberg, ed., *Ecol. Bull.* (Stockholm) **35:**229-241.

Powell, M. J. D., 1965, A method for minimizing a sum of squares of nonlinear function without calculating derivation, *Comput. J.* **7:**303-307.

Rolston, D. E., A. N. Sharpley, D. W. Toy, D. L. Hoffman, and F. E. Broadbent, 1980, Denitrification as affected by irrigation frequency of a field soil, EPA-600/2-80-006, National Technical Information Service, Springfield, Virginia.

Siegel, R. S., R. D. Hauek, and L. T. Kurtz, 1982, Determination of $^{30}N_2$ and application to measurements of N₂ evolution during denitrification, *Soil Sci. Soc. Am. J.* **46:**68-74.

38

The Effect of Land Use on the Nitrogen Biogeochemical Cycle in Central Italy

Paolo Nannipieri
Istituto Chimica Terreno

Fabio Caporali
Tuscia University

P. G. Arcara
Istituto per lo Studio e la Difesa del Suolo

ABSTRACT: The main nitrogen (N) inputs and outputs in some soil-plant systems localized in central Italy are reported. The nitrogen input by rainfall was found to range from 12 to 33 kg N/ha on an annual basis. The dinitrogen (N_2) fixation by legume *Rhizobium* symbiosis was responsible for a nitrogen input of about 100 kg N/ha in grass-legume association. Nitrogen fixation was markedly inhibited by the addition of nitrogen fertilizers.

Nitrate-N (NO_3^--N) concentrations were higher in agricultural than in forested watersheds. In both watersheds, however, monthly nitrogen losses (ammonia, nitrate, and organic N), calculated as the product of the average flow and the respective average, were more dependent upon water flow than upon concentration and land use.

In a clay-loam soil, which is, together with clay, the predominant type of soil in central Italy, about 60% of the labeled urea applied to a grass-legume association in early spring was present as a residual form at the end of the cropping season. Only 27% of the total labeled urea was removed with the harvests.

Further research on denitrification, volatilization losses of nitrogen, and the fate of fertilizer nitrogen in other common crops is needed.

INTRODUCTION

Nitrogen is one of the major factors limiting primary production in many ecosystems. Nitrogen is also closely associated with carbon in the process of life. The measurement of the types and the extent of nitrogenous components, abiotic and biotic transfers, and chemical and biological transformations is therefore receiving extensive coverage in the detailed analyses of ecosystems. In addition, there has been an increased interest in the construction of nitrogen flow charts at a regional level. Since there are large regional differences in human activities, and the turnover of substances in air is often small, regional flow charts give a more accurate picture of the impact of human activities than do global flow charts.

Recently it has been emphasized that there is a particular need for studies on nitrogen inventories and on the dynamics of nitrogenous components in terrestrial ecosystems where soils constitute a considerable nitrogen storage capacity (Eriksson and Rosswall, 1976). Such inventories may reveal geographic patterns that will aid our understanding of biogeochemical processes and their dependence on the physical and chemical environment. Despite the importance of such work, the biogeochemistry of nitrogen has been studied in Italy only since the late 1970s. The purpose of this paper is to review recent studies in Italy.

NITROGEN INPUTS

Nitrogen in Rainfall

The inorganic nitrogen content in rainfall over a three-year study period (from April 1964 to March 1967) was found to depend more on sampling site than on the year. Large nitrogen inputs were found near urban areas where nitrogen deposition was more than 20 kg/ha (Nucciotti and Rossi, 1968).

Large depositions of inorganic nitrogen in rainfall were found at two different sites: a mountainous area near Borgo Taro (900 m above sea level) and a hilly area near Volterra (about 150 m above sea level). The rainfall for a two-year period at Borgo Taro was 31% below the 10-yr annual mean of 700 mm (Table 38-1). At Volterra the rainfall was 25% higher than the 10-yr mean. Monthly precipitation in both areas is characterized by a summer minimum, which is more marked near Volterra, and an autumn maximum. Generally, monthly nitrate-N and ammonium-N concentrations were lower than 1.0 ppm, while nitrite-N was either absent or present in traces. Only in some cases did monthly nitrate-N and ammonium-N (NH_4^+-N) concentrations reach high values (up to 6 ppm).

Over the two-year study period at the mountainous station, the annual inorganic-N deposition associated with precipitation did not change significantly (14 kg/ha, see Table 38-1), even though the relative contributions of nitrate-N and ammonium-N changed markedly. Near Volterra in 1979 the total inorganic-N deposition was quite similar to that observed in the mountainous area, but in 1978 the nitrogen input reached 33 kg/ha. Wet deposition values, noted on a monthly basis, in samples collected at the same station from June to December 1977 and from January to March 1980, agree with the lower

**Table 38–1. Precipitation and Associated Inorganic-N
for the Two-Year Study Period**

	1978	*1979*
Borgo Taro (mountain)		
Precipitation (mm)	600	802
Precipitation		
NH$_4$-N (kg/ha)	3.3	8.3
NO$_3$-N (kg/ha)	11.2	6.5
Volterra (hill)		
Precipitation (mm)	848	812
Precipitation		
NH$_4$-N (kg/ha)	8.7	8.8
NO$_3$-N (kg/ha)	24.8	3.4

values obtained in 1979, thus indicating that the input observed in 1978 was unusually high (Nannipieri et al., 1983).

Nitrate in precipitation is basically manmade (Soderlund, 1977), and the very pronounced increase in nitrate-N observed in central Europe in the 1950s has been ascribed to the large emission of NO$_x$ from various industries, including high-temperature engines and oil-based power plants (Soderlund, 1977). Assuming an annual value of 0.5 kg of nitrate-N/ha as a natural deposition rate (Soderlund, 1977), the influence of anthropogenic emission on wet deposition is considerable even for the low nitrate-N inputs (Table 38-1).

The increase, observed in central Europe during the last decades, of wet deposition of ammonium-N also reflects human influence. The maximum deposition is generally found in areas of agricultural activity (Soderlund, 1977). Our results, with the exception of rainfall input in the mountains in 1978, showed an annual deposition of about 8 kg N/ha, which is more than the highest input of about 6 kg N/ha observed in central Europe (Soderlund, 1977).

Nitrogen Fixation

The sampling station in the mountains at Borgo Taro is mainly a mixed grass-legume pasture used for cattle breeding. Symbiotic nitrogen fixation, measured *in situ* by acetylene reduction assay, was 139 kg N/ha in 1978 (Table 38-2). In both the mountains and hills, symbiotic nitrogen fixation occurred primarily during spring and early summer, accounting for 75% of the total annual input (Arcara and Sparvoli, 1983). The decrease in nitrogen-fixing activity during the late summer was related to lack of water that reduced the growth of plants. A positive correlation was found between nitrogenase activity and soil moisture in meadows located in both areas from the end of June to the beginning of November. On the other hand, no

**Table 38-2. Nitrogen Fixation in
Permanent Meadows**

Year	Taro Valley	Era Valley
1978	139	78
1979	121	87

Note: Values are expressed as kg N/ha on an annual basis.

significant correlation was found between nitrogen fixation and soil temperature (Arcara and Sparvoli, 1983). This is contrary to previous reports (Vlassak, Paul, and Harris, 1973; Sloger et al., 1975). Nitrogenase activity correlated significantly with the cumulative solar radiation during the whole year (Arcara and Sparvoli, 1983), thus confirming that a close relationship exists between nitrogen fixation and photosynthetic activity (Balandreau et al., 1973).

The application of nitrogen fertilizer, which is normally carried out in Italy not only as starter dressing when sowing but also as top dressing, has a detrimental effect on legumes. This effect is likely due to inhibition of nitrogen fixation. When urea was added as fertilizer to a grass-legume association, the symbiotic nitrogen fixation decreased by 27% compared to the unfertilized control (Arcara and Sparvoli, 1981).

NITROGEN OUTPUTS

Leaching and Surface Runoff Losses

Extensive research was carried out in the late 1970s to investigate the effect of land use on nitrogen concentrations in watersheds in central Italy. The selected watersheds are large in area with a human population density four times lower than the natural mean (180 person/km^2). This population is distributed in small villages. In the mountainous watershed the human population works mostly in centers localized outside the basin. The geology of the selected watershed was unique (Table 38-3), having forested basins associated with sandstone. The forests of the other watersheds are associated with either Pliocenic clays or sands (Caporali, 1980). The mountainous watersheds are located in the central North Appennine chain and are largely covered by forests (Table 38-3), containing chestnut, beech, and few conifers (Caporali, 1980; Caporali, Nannipieri, and Pedrazzini, 1981). The agricultural basins of Sterza watershed are equally forested with oak and Mediterranean bush; however, the Fossa Nuova watershed is completely cropped with rotations of winter cereals and forages (Table 38-3). Rates of applications of nitrogen fertilizer are often 100 kg N/ha and up to 150 kg N/ha for cereal crops, vineyards, and olive groves. Animal manure is often applied to the latter two crops.

Table 38–3. Characteristics of Watersheds Studied

	Mountain Forested Watersheds			Agricultural Watersheds			
	Reno	*Orsigna*	*Faldo*	*Era*	*Ragone*	*Sterza*	*Fossa Nuova*
			(*height above sea level*)				
Sea level (m)	1640-609	1732-600	1269-605	603-98	591-70	675-57	14-5
Size (ha)	4090	1500	320	9630	5500	7360	900
Land use (%)							
Forest	97	90	100	36	28	47	0
Winter cereals and rotated forages	0	0	0	60	70	50	100
Vineyard and olive grove	0	0	0	4	2	3	0
Pasture	3	10	0	0	0	0	0

Source: Data from Caporali, 1980.

Table 38–4. Annual Range of Nitrate Concentrations in Water Samples

	Streams	NO_3-*N ppm*
Mountain forested watersheds	Reno	1.0-2.0
	Orsigna	0.5-1.3
	Faldo	0.5-1.5
Agricultural watersheds	Era	2.5-6.3
	Ragone	3.5-6.0
	Sterza	1.0-4.0
	Fossa Nuova	3.0-11.0

Source: Data from Caporali, Nannipieri, and Pedrazzini, 1981.

As a rule, nitrate-N concentrations were higher in the agricultural watersheds (Table 38-4), especially in samples collected from December to March. The Fossa Nuova basin, which is completely cropped, showed the highest nitrate-N concentrations. However, the concentration considered to be a health hazard for drinking water was reached only at the end of January 1979. The Sterza basin, which has the largest forested surface, showed the lowest nitrate-N concentrations (Table 38-4) among the agricultural watersheds.

The greatest nitrate concentrations observed in the agricultural watersheds during the winter agree with the behavior of average monthly concentrations in English rivers and in some rural North American basins. These concentrations depend on the factors that control the release of nitrate from the soil reserves. A common agricultural practice in Italy is to plow

Table 38–5. Linear Equations between Nitrogen Losses (Y) and Flow (X)

	Forested Watershed (Reno)		Agricultural Watershed (Era)	
NO_3-N	$Y = 0.360 + 1.181X$	$r = 0.99$	$Y = -1.884 + 4.36X$	$r = 0.98$
NH_4^+-N	$Y = 0.023 + 0.074X$	$r = 0.95$	$Y = 0.030 + 0.072X$	$r = 0.85$
Organic-N	$Y = -1.437 + 2.519X$	$r = 0.96$	$Y = -0.494 + 2.007X$	$r = 0.93$

Note: Nitrogen losses and flows are expressed as g/sec and m^3/sec, respectively.
Source: Data from Caporali, Nannipieri, and Pedrazzini, 1981.

the soil during summer months and leave it bare until mid-autumn, when sowing of winter cereals and nitrogen fertilization are carried out. Therefore, the rise in nitrate concentrations in autumn and winter, when rainfall is particularly heavy, may be due to mineralization of soil organic matter and nitrogen fertilization.

Ammonium-N and organic-N contents were not significantly different in the forested watershed and in a larger agricultural watershed that collects waters from Era, Ragone, Sterza, and Fossa Nuova basins. Nitrogen losses, expressed as grams of N per second, were essentially a linear function of stream flows, and equations for organic-N and ammonium-N from the two watersheds gave similar slopes. This result indicated that different land use and watershed characteristics do not influence these nitrogen losses (Table 38–5). On the other hand, the agricultural watershed showed greater nitrate-N losses due to unitary increase of flows (Caporali, Nannipieri, and Pedrazzini, 1981).

Organic-N and inorganic-N losses amounted to 58.64 kg/ha in the mountainous basin in the 1979 water year. Much of the nitrogen loss (about 30 kg/ha) was due to organic forms and was probably derived from the litter layer. Clear-cutting of vegetation is a common practice in the forested areas in central Italy. A wide range of nitrate losses from disturbed forests has been reported (Vitousek et al., 1979). Four processes affect nitrate loss: nitrogen uptake by regrowing vegetation; nitrogen immobilization; lag in nitrification; and lack of water for nitrate transport. However, the net effect of these processes, with the exception of uptake by regrowing vegetation, is insufficient to prevent nitrate losses (Vitousek et al., 1979). Although disturbed forests have the potential for very high nitrate losses, such losses occur in a limited way in forested watersheds in central Italy.

In the agricultural watershed, which receives waters from Era, Sterza, Ragone, and Fossa Nuova basins, nitrogen losses amounted to only 15.65 kg/ha, and nitrate-N content accounted for 67% of the total loss (Caporali, Nannipieri, and Pedrazzini, 1981). In general, nitrogen losses from the Reno watershed are nearly four times as high as losses from the larger agricultural watershed. However, runoff from the Reno watershed is about six times that from the Era (1,921 and 294 mm, respectively, in the 1979 water year). The highest runoff in the forested basin is due to higher rainfall, topographic characteristics (Table 38-3), presence of hardwood forests, cutting practices, and absence of plowing.

CONCLUSIONS

There are serious deficiencies in the knowledge of the biogeochemical cycle of nitrogen in central Italy; these gaps need to be bridged by further research at local and regional levels. Studies on the fate of ^{15}N-enriched fertilizer have been carried out with maize (Cervelli, Ciardi, and Perna, 1982) and pasture (Nannipieri, Ciardi, and Palazzi, in press) and are in progress with apple orchards. However, studies need to be extended to include more common crops, such as wheat. Nitrogen losses (denitrification, ammonia volatilization, and outputs due to leaching and surface runoff) have been determined indirectly by the proportion of ^{15}N unaccounted for at the end of the experiment. Only 12% of the total ^{15}N applied as urea fertilizer to a grass-legume association in a clay loam soil, which characterizes the hilly areas of central Italy, was lost from the soil-plant system at the end of the first cropping season (Nannipieri, Ciardi, and Palazzi, in press). In a sandy soil, nitrogen losses have been more than 50% when ^{15}N-enriched urea fertilizer was applied to apple trees.

Denitrification rates are poorly characterized. The influence of nitrate concentrations, soil water content, oxygen concentration, and different management practices on both reaction rates and on the N_2O/N_2 ratio of the gaseous end products should be investigated in terrestrial habitats.

Ammonia volatilization rates, especially in agricultural soils where ammonia and urea fertilizers are applied, have to be estimated, and the influence of soil temperature, pH, moisture, and plant cover needs to be investigated.

Forested areas, where both vegetation and soils constitute large N storehouses, require more detailed investigations. Large increases in N content of the organic horizon of forest soils have been reported (Tchagina, Vishnyakova, and Vedrova, 1968). Such increases cannot be attributed to nitrogen fixation alone. Nitrogen may be redistributed through the soil profile by trees or translocated in upper organic horizons by fungal hyphae.

REFERENCES

Arcara, P. G., and E. Sparvoli, 1981, Nitrogen-fixing activity and microbial biomass of a mixed hay field treated with three different nitrogen fertilizers, in *Colloque humus azote,* P. Dutil and F. Jacquin, eds., I.S.S.S. and I.N.R.A., Reims, France, pp. 118-124.

Arcara, P. G., and E. Sparvoli, 1983, Nitrogen fixation by *in situ* open system assay (C_2H_2) in grass-legume meadows in Central Italy, *Ist. Sper. Stud. Dif. Suolo Ann.* **14:**60-70.

Balandreau, J., C. R. Millier, P. Weinhard, P. Ducerf, and Y. Dommergues, 1973, A modelling approach of acetylene reducing activity of plant-rhizosphere diazotroph system, in *Recent Developments in Nitrogen Fixation,* W. Newton, J. R. Postgate, and C. Rodriquez-Barrueco, eds., Academic Press, New York, pp. 523-530.

Caporali, F., 1980, Nitrogen contents of streams draining agricultural and forested watersheds in Tuscany, in *Ecologia,* A. Moroni, O. Ravera, and A. Anelli, eds., S.I.T.E., First Congress, Salsomaggiore, Italy, pp. 109-116.

Caporali, F., P. Nannipieri, and F. Pedrazzini, 1981, Nitrogen content of streams draining an agricultural and a forested watershed in Central Italy, *J. Environ. Qual.* **10:**72-76.

Cervelli, S., C. Ciardi, and A. Perna, 1982, Effect of atrazine on soil transformations of nitrogen and uptake by corn, *J. Environ. Qual.* **11:**82-86.

Eriksson, E., and T. Rosswall, 1976, Man and biogeochemical cycles: Impacts, problems and research needs, in *NPS Global Cycles,* B. H. Svensson and R. Soderlund, eds., Ecological Bulletins, Stockholm, pp. 11-16.

Nannipieri, P., S. Gori, P. G. Arcara, and E. Sparvoli, in press, Nitrogen balance of unfertilized grass-legume meadow in Central Italy, *Agrochimica.*

Nannipieri, P., C. Ciardi, and T. Palazzi, in press, Plant uptake, microbial immobilization and residual soil fertilizer of spring-applied, labeled urea-nitrogen in a grass-legume association, *Soil Sci. Soc. Am J..*

Nucciotti, F., and N. Rossi, 1968, Chemical composition of precipitation in Emilia region (Italy), *Agrochimica* **12:**540-548.

Sloger, C., D. Bezdicek, R. Milbberg, and N. Boonkerd, 1975, Seasonal and diurnal variations in $N_2(C_2H_2)$ fixing activity in field soybeans, in *Nitrogen Fixation by Free-Living Microorganisms,* W. D. P. Stewart ed., International Biological Programme, Cambridge University Press, Cambridge, pp. 271-284.

Soderlund R., 1977, NO_x pollutants and ammonia emissions — A mass balance for the atmosphere over NW Europe, *Ambio* **6:**118-122.

Tchagina, E. C., Z. V. Vishnyakova, and E. F. Vedrova, 1968, Fixation of nitrogen during litter decomposition in pine forests of the western Sayan mountains, *9th Int. Cong. Soil Sci. Trans.* **2:**163-171.

Vitousek, P. M., J. R. Gosz, C. C. Grier, J. M. Melillo, W. A. Reyners, and R. L. Todd, 1979, Nitrate losses from disturbed ecosystems, *Science* **204:**469-474.

Vlassak, K., E. A. Paul, and R. E. Harris, 1973, Assessment of biological nitrogen fixation in grassland and associated sites, *Plant and Soil* **38:**637-649.

Part VII

SOILS

39

The Biogeochemistry of Soil Phosphorus

Holm Tiessen and John W. B. Stewart
University of Saskatchewan

INTRODUCTION

Weathering of Mineral Phosphorus

The phosphorus (P) contained in most soils is entirely inherited from the parent materials. With the exception of fertilizer additions, no significant inputs of P into soils occur once the parent material is laid down. A variety of primary minerals contain P, since phosphate can isomorphously substitute for silicate in many crystal structures (McConnell, 1979). The major forms of primary rock phosphate in all types of parent material are the apatites— calcium phosphates containing varying amounts of carbonate, fluoride, sulfate, hydroxide, and a number of cations. Secondary P minerals are formed from these phosphorites during the weathering and formation of soils.

The action of chemical weathering on primary phosphorites has been studied in the major phosphate deposits of marine origin (Lucas et al., 1980; McArthur, 1980). Leaching preferentially removes carbonate from the rock phosphate; at the same time, smaller amounts of sulfate, sodium, strontium, and other trace elements are lost. The loss of these components increases the resistance of apatite to further weathering. Many soils of relatively young pedological age, therefore, contain a wide range of calcium phosphates from complex primary phosphorites such as francolite $[(Ca, Na)_5 (PO_4, CO_3)_3(F, OH)]$ to simpler, more stable forms such as fluor- or hydroxy-apatite $[Ca_5(PO_4)_3 \text{ F or OH}]$.

With the exception of the podzols of the boreal zones, soils are not exposed to acid weathering regimes in the early stages of their development (Pedro and Sieffermann, 1979), but become acid as a result of prolonged weathering and leaching of all carbonates and accompanying bases. In an acidified soil, apatites dissolve rapidly, releasing their phosphates into the soil solution. The phosphate ion can then enter other iron or aluminum minerals (substituting for silicate), or it can react with solution aluminum and iron, resulting in the neoformation of stable secondary P minerals such as phosphosiderite or strengite ($FePO_4 \cdot 2H_2O$).

The heterogeneity of the soil system makes analysis of the different P mineral species difficult, and indirect approaches must be taken. Each P mineral maintains characteristic P solution concentrations under defined conditions of pH and Eh, and so forth. For many compounds this solubility is known, and typical solubility diagrams can be constructed plotting P activity in solution against pH (Lindsay and Vlek, 1977) (Fig. 39-1). Knowing the solution concentration of P in a specific soil at a known pH, one can infer which P minerals contribute to the solution P in that soil.

Alternatively, soil P may be characterized in terms of its extractability in certain reagents that are capable of preferentially extracting specific forms of calcium, aluminum, or iron phosphates (Chang and Jackson, 1958). With many modifications to accommodate different species of metal phosphates, this method has frequently been used to characterize the P status of soils (Williams and Walker, 1969; Williams et al., 1971; Syers, Shah, and Walker, 1969). Westin and Buntley (1967) observed an increase in alkali and reductant soluble, iron-associated phosphates along a gradient of increasing moisture and temperature in South Dakota soils. This was accompanied by a corresponding decrease in acid-soluble calcium-P. Sadler and Stewart (1975) similarly showed a decrease of acid-extractable calcium-P downslope along a catenary sequence. The more strongly developed soils at the bottom slope positions were enriched in secondary P forms extracted in ammonium fluoride or sodium hydroxide (Fig. 39-2). The relative stability of P within a landscape, together with its slow transformations during pedogenesis, has been used by Smeck (1973) to infer the relative age of soils and their weathering history.

Organic P Transformations

At the same time as the physical and chemical weathering processes transform primary inorganic P (Pi) to secondary forms, organisms in the soil or on the parent rock take up P from the solution or actively dissolve P for their consumption. Phosphorus is incorporated into the biomass and supports carbon assimilations and nitrogen fixation. As more P becomes available through the weathering of the parent material, larger amounts become immobilized in organic matter, and in many soils the maximum amounts of organic matter accumulated depend directly on the amounts of Pi available for biological consumption (Walker and Syers, 1976; Cole and Heil, 1981). At later stages of soil development, Pi is progressively transformed into less-soluble iron- and aluminum-associated forms, and organic P (Po) contents of the soil decline. In a strongly leaching environment, total P con-

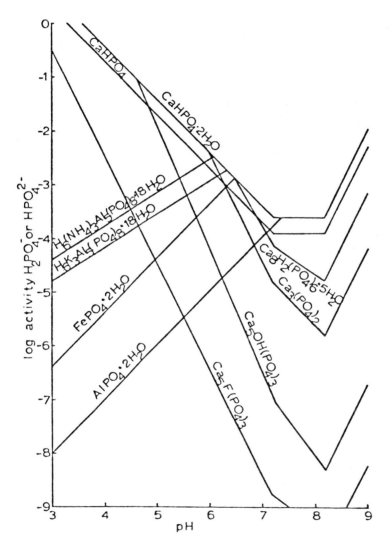

Figure 39-1. A unified solubility diagram for various calcium, iron, and aluminum phosphate minerals in soils *(Source: Reproduced from Lindsay and Vlek, 1977, p. 658, by permission of the Soil Science Society of America.)*

tents will similarly decline as soil development progresses (Walker and Syers, 1976).

The primary step in the accumulation of soil Po is its synthesis in plants or soil organisms. Many plant materials are extensively modified by microbial action before entering the soil organic-matter pool, and the microbial population is an important source for soil Po as well as a significant component of the total soil Po (Brookes, Powlson, and Jenkinson, 1982; Hedley and Stewart, 1982).

Changes in the nutritional state of the microbial population cause major changes in its P composition (Karl, 1980; Beever and Burns, 1980) through processes such as the accumulation of polyphosphates (Pepper, Miller, and Ghonsikar, 1976; Harold, 1966). Such changes can significantly affect the composition of labile soil P (Chauhan, Stewart, and Paul, 1979, 1981). Release of P from the microbial population is controlled by environmental factors such as freezing (Biederbeck and Campbell, 1971) or drying (Alexander, 1977), or by predation by soil fauna (Coleman, Reid, and Cole, 1983).

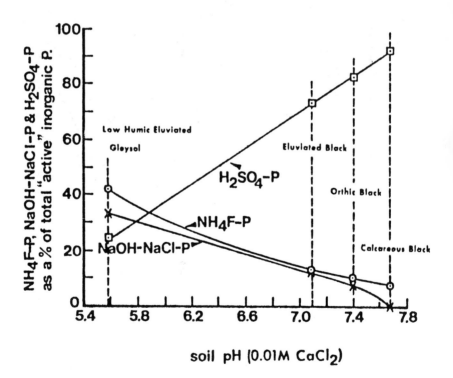

Figure 39–2. Relationship of NH_4P-P, $NaOH/NaCl$-P, and H_2SO_4-P to soil pH and soil type *(Source: From Sadler and Stewart, 1975, p. 155).*

Organic P released into the soil environment can be stabilized chemically or biochemically as part of a larger humic molecule, or it may adsorb to or precipitate with soil mineral components, particularly sesquioxides. Because of its high charge density, the phosphate group is highly reactive and may contribute greatly to the overall stability of organic matter by sorption and precipitation reactions. The compositions of biomass Po and the Po found in soils are very different as a result of the preferential stabilization of specific Po compounds in the soil environment. For instance, inositol phosphates that contain a large number of phosphate ester groups in close proximity are rapidly adsorbed to soil minerals, interacting extensively with sesquioxides, and are therefore very persistent in soils. The inositols that are protected from degradation depend entirely on the contents of active aluminum and iron present for sorption or precipitation reactions (Anderson, Williams, and Moir, 1974). Phosphodiesters have lower charge densities, and their phosphate groups are largely shielded from ionic interactions. This leaves them accessible to microbial or enzyme attack and explains the small proportion present in the soil system. The small amounts of diesters found in soils (Dalal, 1977) are probably stabilized with polysaccharides that envelope most microbial cells in soils (Foster, 1981). Co-adsorption of iron or other metal ions and phosphates to microbial polysaccharide flocs is recognized as an important part of waste purification by activated sludge processes (Brown and Lester, 1979), and similar reactions may account for the stability of soil P esters towards degradation and extraction (Moucawi, Fustec, and Jambu, 1981).

Several specific Po compounds, such as phospholipids (Kowalenko and McKercher, 1970), nucleic acids (Anderson, 1970), glycerol phosphates, choline (Hance and Anderson, 1963), and others (Dalal, 1977), have been identified in soil extracts. Most of these compounds, however, are found in only small quantities in soils, with the exception of inositol phosphates, which may constitute a third of the soil Po and occur in various combinations with humic and fulvic acids or as metal complexes (McKercher and Anderson, 1968; Baker, 1977; Omotoso and Wild, 1970). Less than half the soil Po can usually be identified as specific chemical compounds, while the remainder remains unextractable or in close association with humic and fulvic acids.

Similar to the work on soil Pi, specific extraction techniques have been developed that attempt to characterize soil Po by its extractability in various reagents such as sodium bicarbonate (Bowman and Cole, 1978a) or sodium hydroxide (Bowman and Cole, 1978b). Such extraction techniques, together with experiments examining the biological availability of the various Po fractions, are used to study the transformations of soil Po (Stewart and McKercher, 1982).

DISCUSSION

The Soil Phosphorus Cycle

The previous sections show that the transformations of organic and inorganic forms of soil P are closely interrelated, since Pi is a source of P uptake for soil organisms, and Po may hydrolyze to replenish solution Pi or may be stabilized with the mineral phase of the soil. To follow the interactions of various portions of the soil P cycle, Hedley, Stewart, and Chauhan (1982) developed a sequential extraction procedure that characterizes inorganic and organic P forms in terms of their extractability in specific reagents and their bioavailability (Fig. 39-3). This extraction procedure has been applied to soils (1) that actively transformed P in laboratory incubations (Hedley, Stewart, and Chauhan, 1982), (2) that underwent P depletions in the vicinity of plant roots (Hedley, White, and Nye, 1982), (3) that underwent long-term changes in their P status during cultivation (Tiessen, Stewart, and Moir, 1983), and (4) that exhibited widely differing P compositions as a result of different pedogenesis (Tiessen, Stewart, and Cole, 1984).

Plants appear to take up exclusively Pi from the soil solution. The depleted solution P pool is immediately replenished from exchangeable and labile Pi forms (resin, bicarbonate, and to some extent hydroxide-extractable P) (Hedley, White, and Nye, 1982). When these labile pools are exhausted and less soluble species determine the equilibrium P concentration in the soil,

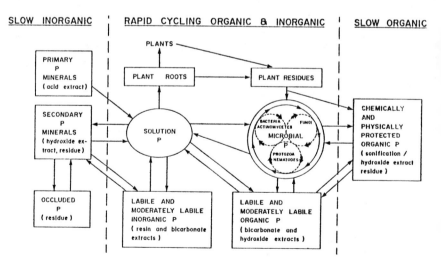

Figure 39-3. The soil phosphorus cycle: Its components and measurable fractions (Source: Adapted from Chauhan, Stewart and Paul, 1981; extracts from Hedley, Stewart, and Chauhan, 1982).

some plants can lower the pH in the vicinity of their roots (Grinsted et al., 1982) and dissolve portions of the acid-soluble calcium P minerals (Hedley, White, and Nye, 1982).

Consumption of labile Pi by the microbial population can be stimulated by additions of microbial substrates, such as litter or crop residues, to the soil (Hedley Stewart, and Chauhan, 1982). Microbes and their metabolic products are extensively stabilized by adsorption to soil minerals (Marshall, 1971) and frequently act as nucleation points for soil aggregate formation (Tisdall and Oades, 1982).When solution P supply to the growing microbial population is abundant, a net immobilization of Pi into organic forms will therefore result. On the other hand, inadequate Pi supply will stimulate the production of phosphatases (McGill and Cole, 1981) and the mineralization of labile organic forms of P for microbial uptake (Hedley, Stewart, and Chauhan, 1982).

A continuous drain on the soil P pools by cultivation and crop removal will rapidly deplete labile Pi and Po forms (Tiessen, Stewart, and Moir, 1983). Reduced inputs of plant litter and accelerated rates of degradation under cultivation will cause the net mineralization of organic matter accompanied by substantial carbon and nitrogen losses (Haas, Evans, and Miles, 1957; Tiessen, Stewart, and Bettany, 1982). Since the concomitantly mineralized P cannot be lost to the atmosphere or easily leached (like carbon or nitrogen), it is free to interact with various components of the P cycle (Fig. 39-3). If plant uptake is less than the amount of P mineralized, precipitation of calcium-P occurs in soils of high base saturation. Apatites of low solubility can thus act as scavengers of newly mineralized P (Tiessen, Stewart, and Moir, 1983; Hooker et al., 1980; Fares, Fardeau, and Jaquin, 1974).

Labile Pi forms (resin and bicarbonate-extractable) are maintained by the mobilization of secondary P minerals in such base-saturated soils, since iron- and aluminum-associated P has relatively high solubilities at pH values above neutrality (Fig. 39-1). Across 60 different grassland soils from the USDA-SCS collection of benchmark soils, the exchangeable (resin) Pi contents were, therefore, closely related to bicarbonate- and hydroxide-extractable Pi (Tiessen, Stewart, and Cole, 1984). In more highly weathered soils that are characterized by higher activities of aluminum and iron in solution and by lower solubilities of iron and aluminum phosphates, the sizes of the labile Pi pool correlated well with labile forms (bicarbonate-extractable) of Po. This indicated that mineralization of Po was the major contributor to plant-available P in such soils.

The effects of soil development on the more stable portions of soil P could also be shown in the comparison of these soil groups. Stable P was in the form of acid-soluble, primary minerals in the grassland soils (Mollisols) and existed as occluded and secondary P forms as well as organic P in association with active iron oxides in the more weathered Alfisols and Ultisols. The important role of active iron oxides in association with organic matter and

Po was shown in all major soil orders. No similar importance was evident for extractable forms of aluminum.

It can be seen that a conceptual model of the soil P cycle such as the one illustrated in Figure 39-3 aids the understanding of the short-term transformations of soil Po and Pi that lie at the basis of long-term changes in the chemistry and biochemistry of P during the development of soils.

ACKNOWLEDGEMENTS

We thank Dr. C. V. Cole for helpful discussions during the preparation of this manuscript published as Journal Paper R 352 of the Saskatchewan Institute of Pedology. Research was supported by the Natural Science and Engineering Research Council of Canada.

REFERENCES

Alexander, M., 1977, *Introduction to Soil Microbiology,* 2nd ed., John Wiley & Sons, New York.

Anderson, G., 1970, The isolation of nucleoside diphosphates from alkaline extracts of soil, *J. Soil Sci.* **21**:96-104.

Anderson, G., E. G. Williams, and J. O. Moir, 1974, A comparison of the sorption of inorganic orthophosphate and inositol hexaphosphate by six acid soils, *J. Soil Sci.* **25**:51-62.

Baker, R. T., 1977, Humic acid associated organic phosphate, *N.Z.J. Sci.* **20**:439-441.

Beever, R. E., and D. J. W. Burns, 1980, Phosphorus uptake, storage and utilization by fungi, *Adv. Bot. Res.* **8**:128-219.

Biederbeck, V. O., and C. A. Campbell, 1971, Influence of simulated fall and spring conditions on the soil system. I. Effect on soil microflora, *Soil Sci. Soc. Am. Proc.* **35**:374-379.

Bowman, R. A., and C. V. Cole, 1978a, Transformations of organic phosphorus substances in soils as evaluated by $NaHCO_3$ extraction, *Soil Sci.* **125**:49-54.

Bowman, R. A., and C. V. Cole, 1978b, An exploratory method for fractionation of organic phosphorus from grassland soils, *Soil Sci.* **125**:95-101.

Brookes, P. C., D. S. Powlson, and D. S. Jenkinson, 1982, Measurement of microbial biomass phosphorus in soil, *Soil Biol. Biochem.* **14**:319-330.

Brown, M. J., and J. N. Lester, 1979, Metal removal in activated sludge—the role of bacterial extracellular polymers, *Water Res.* **13**:817-837.

Chang, S. C., and M. L. Jackson, 1958, Soil phosphorus fractions in some representative soils, *J. Soil Sci.* **9**:109-119.

Chauhan, B. S., J. W. B. Stewart, and E. A. Paul, 1979, Effect of carbon additions on soil labile inorganic, organic, and microbially held phosphate, *Can. J. Soil Sci.* **59**:387-396.

Chauhan, B. S., J. W. B. Stewart, and E. A. Paul, 1981, Effect of labile inorganic phosphate status and organic carbon additions on the microbial uptake of phosphorus in soils, *Can. J. Soil Sci.* **61**:373-385.

Cole, C. V., and R. D. Heil, 1981, Phosphorus effects on terrestrial nitrogen cycling, in

Terrestrial Nitrogen Cycle, Processes, Ecosystem and Management Impact, F. E. Clark and T. Rosswall, eds., *Ecol. Bull.* (Stockholm) **33:**363-374.

Coleman, D. C., C. P. P. Reid, and C.V. Cole, 1983, Biological strategies of nutrient cycling in soil systems, in *Advances in Ecological Research,* Vol.13, A. MacFadyrn and E. D. Ford, eds., Academic Press, New York, pp. 1-55.

Dalal, R. C., 1977, Soil organic phosphorus, *Adv. in Agron.* **29:**83-117.

Fares, F., J. C. Fardeau, and F. Jaquin, 1974, Etude quantitative du phosphore organique dans differents types de sols, *Phosphore et Agric.* **63:**25-41.

Foster, R. C., 1981, Polysaccharides in soil fabrics, *Science* **214:**665-667.

Grinsted, M. J., M. J. Hedley, R. E. White, and P. H. Nye, 1982, Plant-induced changes in the rhizosphere of rape seedlings. I. pH change and the increase in P concentration in the soil solution, *New Phytol.* **91:**19-29.

Haas, H. J., C. E. Evans, and E. F. Miles, 1957, Nitrogen and carbon changes in Great Plains soils as influenced by cropping and soil treatments, USDA Tech. Bull. 1164.

Hance, R. J., and G. Anderson, 1963, Extraction and estimation of soil phospholipids, *Soil Sci.* **96:**94-98.

Harold, F. M., 1966, Inorganic polyphosphates in biology: Structure, metabolism and function, *Bacteriol. Rev.* **30:**772-794.

Hedley, M. J. and J. W. B. Stewart, 1982, A method to measure microbial phosphorus in soils, *Soil Biol. Biochem.* **14:**377-385.

Hedley, M. J., J. W. B. Stewart, and B. S. Chauhan, 1982, Changes in inorganic and organic soil phosphorus fractions induced by cultivation practices and by laboratory incubations, *Soil Sci. Soc. Am. J.* **46:**970-976.

Hedley, M. J., R. E. White, and P. H Nye, 1982, Plant-induced changes in the rhizosphere of rape seedlings. III. Changes in L-value, soil phosphate fractions and phosphatase activity, *New Phytol.* **91:**45-56.

Hooker, M. L., G. A. Peterson, D. H. Sander, and L. A. Daigger, 1980, Phosphate fractions in calcareous soils as altered by time and amounts of added phosphate, *Soil Sci. Soc. Am. J.* **44:**269-277.

Karl, D. M., 1980, Cellular nucleotide measurements and applications in microbial ecology, *Microb. Rev.* **44:**739-796.

Kowalenko, C. G., and R. B. McKercher, 1970, An examination of methods for extraction of soil phospholipids, *Soil Biol. Biochem.* **2:**269-273.

Lindsay, W. L., and P. L. G. Vlek, 1977, Phosphate minerals, in *Minerals in Soil Environments,* J. B. Dixon and S. D. Weed, eds., Soil Science Society of America, Madison, Wisconsin, pp. 639-672.

Lucas, J., R. Flicoteaux, Y. Nathan, L. Prevot, and Y. Shahar, 1980, Different aspects of phorphorite weathering, in *Symp. 10th Int. Congr. Sediment,* Society for Economic Paleontology and Mineralogy, Special Publication, Vol. 20, pp. 4-51.

McArthur, J. M., 1980, Post depositional alteration of the carbonate-fluorapatite phase of Moroccan phosphates, in *Symp. 10th Int. Congr. Sediment.* Society for Economic Paleontology and Mineralogy, Special Publication, Vol. 20, pp. 53-60.

McConnell, D., 1979, Biogeochemistry of phosphate minerals, in *Biogeochemical Cycling of Mineral-Forming Elements (Studies in Environmental Science 3),* P. A. Trudinger and D. J. Swaine, eds., Elsevier Scientific Publishing Co., Amsterdam, pp. 163-204.

McGill, W. B., and C. V. Cole, 1981, Comparative aspects of C, N, S, and P cycling through soil organic matter during pedogenesis, *Geoderma* **26:**267-286.

McKercher, R. B., and G. Anderson, 1968, Content of inositol penta- and hexa-phosphates in some Canadian soils, *J. Soil Sci.* **19:**47-55.

Marshall, K. C., 1971, Sorptive interactions between soil particles and microorganisms, in *Soil Biochemistry,* Vol. 2, A. D. McLaren and J. Skujins, eds., Marcel Dekker, New York, pp. 409-445.

Moucawi, J., E. Fustec, and P. Jambu, 1981, Biooxidation of added and natural hydrocarbons in soil: Effect of iron, *Soil Biol. Biochem.* **13:**335-342.

Omotoso, T. I., and A. Wild, 1970, Occurrence of inositol phosphates and other organic phosphate components in an organic complex, *J. Soil Sci.* **21:**224-232.

Pedro, G., and J. Sieffermann, 1979, Weathering of rocks and formation of soils, in *Review of Research on Modern Problems in Geochemistry,* Vol. 16, F. R. Siegel, ed., Earth Sciences (UNESCO), Paris, pp. 39-55.

Pepper, I. L., R. H. Miller, and C. P. Ghonsikar, 1976, Microbial inorganic polyphosphates: Factors influencing their accumulation in soil, *Soil Sci. Soc. Am. J.* **40:**872-875.

Sadler, M. J., and J. W. B. Stewart, 1975, Changes with time in form and availability of residual fertilizer phosphorus in a catenary sequence of chernozemic soils, *Can. J. Soil Sci.* **55:**149-159.

Smeck, N. E., 1973, Phosphorus: An indicator of pedogenic weathering processes, *Soil Sci.* **115:**199-206.

Stewart, J. W. B., and R. B. McKercher, 1982, Phosphorus cycle, in *Experimental Microbial Ecology,* R. G. Burns and J. H. Slater, eds., Blackwell Scientific Publications, Oxford, pp. 221-238.

Syers, J. K., R. Shah, and T. W. Walker, 1969, Fractionation of phosphorus in two alluvial soils and particle size separates, *Soil Sci.* **108:**283-289.

Tiessen, H., J. W. B. Stewart, and J. R. Bettany, 1982, Cultivation effects on the amounts and concentration of carbon, nitrogen and phosphorus in grassland soils, *Agron. J.* **74:**831-835.

Tiessen, H., J. W. B. Stewart, and J. O. Moir, 1983, Changes in organic and inorganic phosphorus composition of two grassland soils and their particle size fractions during 60-90 years of cultivation, *J. Soil Sci.* **34:**815-823.

Tiessen, H., J. W. B. Stewart, and C. V. Cole, 1984, Pathways of phosphorous transformations in soils of differing pedogenesis, *Soil Science Society of America J.* **48:**853-858.

Tisdall, J. M., and J. M. Oades, 1982, Organic matter and water-stable aggregates in soils, *J. Soil Sci.* **33:**141-163.

Walker, T. W., and J. K. Syers, 1976, The fate of phosphorus during pedogenesis, *Geoderma* **15:**1-19.

Westin, F. C., and G. J. Buntley, 1967, Soil P in South Dakota: III. P fractions of some Borolls and Ustolls, *Soil Sci. Soc. Am. Proc.* **31:**521-528.

Williams, J. D. H., and T. W. Walker, 1969, Fractionation of phosphate in a maturity sequence of New Zealand basaltic soils profiles, *Soil Sci.* **107:**22-30.

Williams, J. D. H., J. K. Syers, R. F Harris, and D. E. Armstrong, 1971, Fractionation of inorganic phosphate in calcareous lake sediments, *Soil Sci. Soc. Am. Proc.* **35:**250-255.

40

Controls of Soil-Solution Chemistry in Lodgepole Pine Forest Ecosystems, Wyoming

Timothy J. Fahey

Cornell University

Joseph B. Yavitt, Alex E. Blum, and James I. Drever

University of Wyoming

ABSTRACT: The soil solution in four contrasting lodgepole pine forest ecosystems was sampled at three depths (forest floor, 0.4 m, 1.5 m) to evaluate biotic and geochemical controls on solute fluxes. Maximum concentrations of all solutes were observed in the early stages of snowmelt, with rapid declines being noted as melting progressed. Dissolved organics derived from the forest floor were the major mobile anions associated with cation transport in the surface-soil horizons, but these were lost from solution deeper in the profile, possibly by adsorption in the mineral soil. This process was more dramatic in a finer-textured soil. Bicarbonate and the anions of strong acids, especially sulfate, also were present at high concentrations in the surface horizons, the former being contributed by dissolution of respiratory carbon dioxide and the latter by wet and dry deposition of neutral salts during the previous summer period. Hydrogen ions derived from biotic processes within the ecosystem appeared to control the mobilization of metallic cations. Soluble aluminum concentrations in the soil solution (up to 0.4 mg/l) were highly correlated with undissociated organic acid levels, suggesting that soil weathering reactions also are controlled by biotic processes.

INTRODUCTION

The factors controlling the ionic chemistry of soil solutions in forest ecosystems vary among sites and geographic regions (Cronan et al., 1978). The role of external and internal sources of free hydrogen ions (H^+) and more-or-less mobile anions in cation mobilization and transport processes has been firmly established (Wiklander, 1971; Johnson et al., 1977; Cronan, 1980;

Sollins et al., 1980). In areas subjected to highly acidic precipitation, soil weathering and cation transport are controlled by external sources of strong mineral acids (Gjessing et al., 1976; Likens et al., 1977), but biotic assimilation of nitrate (Cronan, 1980) and soil sulfate adsorption (Johnson et al., 1980) may limit the mobility of these anions. This will affect total cation flux from these systems. In some less-polluted areas, internal sources of weak acids, especially carbonic and various organic acids, appear to control the ionic composition of the soil solution (Sollins et al., 1980). Johnson et al. (1977) suggested that because of lowered soil solution pH, the role of carbonic acid in leaching of forest soils diminishes as mean annual temperature declines, and they suggested increasing importance of organic acids in high-elevation and northern ecosystems. Ugolini et al. (1977) noted the importance of mobile fulvic acid, derived from the forest floor, in cation transport and soil weathering of a subalpine podzol in Washington State, and a similar role has been noted for a variety of organic acids in other forested ecosystems (Graustein, Cromack, and Sollins, 1977; Cronan, 1980; Antweiler and Drever, 1983).

Soil leaching in lodgepole pine (*Pinus contorta* ssp. *latifolia*) forests of the central Rocky Mountains is usually restricted to the snowmelt interval (May–June) because summer rains are rarely sufficient to cause significant movement of soil water (Knight, Fahey, and Running, in press). Despite some recent evidence for changes in cation output from forested watersheds resulting from acid precipitation in this region (Lewis and Grant, 1979, 1980), it was hypothesized that large internal sources of weak acids are the principal agents of cation flux. Variation among sites within the lodgepole pine ecosystem was expected to result mostly from differences in soil physical-chemical characteristics, particularly the capacity for adsorption of organic anions and sulfate and weatherability of soil minerals. To test these hypotheses, soil-solution chemistry was measured during snowmelt in three contrasting lodgepole pine sites using tension and tension-free lysimeters.

MATERIALS AND METHODS

Study Area

This research was conducted in three contrasting *Pinus contorta* forest ecosystems in southeastern Wyoming (41°N, 106°W), ranging in elevation from about 2900 m (Albany and Nash Fork stands) to 3050 m (French Creek stand). The forest at Nash Fork was even-aged, originating following a fire in about 1870, whereas the French Creek and Albany stands were uneven-aged, and the oldest trees exceeded 200 yr. The typic Cryoboralf soil (Soil Survey Staff, 1975) at the French Creek and Nash Fork sites developed from bouldery glacial till, whereas the lithic Cryochrept at the Albany stand was derived from a highly weathered granitic regolith. Moderately high accumulations of forest floor organic matter (2000-3000 g/m^2) are typical of these ecosystems (Fahey, 1983). The climate in the study area has been described in detail elsewhere (Fahey, 1983; Knight, Fahey, and Running, in press) and is dominated by a long winter period, during which nearly two-thirds of the average annual precipitation of about 60 cm falls as snow.

Field Sampling

Bulk snow samples were collected from three depths in each stand and from large forest openings adjacent to the French Creek and Nash Fork stands in each year of the study (1979-1982). Forest floor leachate was sampled with alundum plate lysimeters (Cole, 1958) evacuated to 10 KPa tension, and collection bottles were emptied every two to three days during the snowmelt interval. The soil solution was sampled at 40-cm depth (beneath the major rooting zone) and at 1.2- to 1.8-m depth (subsoil) in each stand using porous-cup lysimeters (Parizek and Lane, 1975) evacuated to 12 KPa; cup lysimeters were collected at two- to five-day intervals throughout the snowmelt drainage interval. In 1983, zero-tension lysimeters (Jordan, 1968) were installed at 40-cm depth at Nash Fork and Albany, and some results from these collectors will be reported. Soil samples were obtained from the faces of soil trenches in each stand in midsummer 1980.

Laboratory Analyses

Soil samples were air dried and passed through a 2-mm sieve prior to analysis. Free iron and aluminum were extracted from a (<2-μm fraction by the citrate-dithionate-bicarbonate method (Mehra and Jackson, 1960), with analysis of aluminum and total iron by plasma atomic absorption spectrophotometry. Soil-particle size distribution was measured by the hydrometer method (Day, 1965), soil pH on a 1:1 soil-water slurry, and organic matter percentage by dry-ashing at 600°C in a muffle furnace. Soil cations were extracted with 0.1 M ammonium acetate and analyzed by atomic absorption spectrophotometry.

Water samples were returned to the laboratory and analyzed for pH, alkalinity, and acidity within 8 hr of collection. In 1979-1981, total alkalinity was estimated by titration to pH 4.5 with 0.01 N HCl (Golterman, Chimo, and Ohnstad, 1978). Thereafter, total alkalinity was estimated by Gran titration with 0.01 N HCl to pH 3.5 (Stumm and Morgan, 1981) and nonvolatile alkalinity by back-titration with 0.01 N NaOH to the original sample pH after purging samples of carbon dioxide. Nonvolatile acidity was measured by titration of purged samples to pH 8.3. Cations (calcium, magnesium, sodium, and potassium) were analyzed by atomic absorption spectrophotometry on a

Table 40-1. Selected Soil Physical-Chemical Features for Three Lodgepole Pine Ecosystems, Southeastern Wyoming

Stand	Depth	pH	% Organic Matter	% of <2mm Fraction Silt	Clay	% of 2μ Fraction[a] Al	Fe	meq/g[b] Total Cations
Albany	0-10	5.50	5.8	15.8	12.4	0.97	2.97	.049
	25	5.34	4.2	19.8	12.8	0.80	2.85	.056
	50	5.58	2.7	5.8	8.4	0.32	0.75	.084
	100	5.66	2.3	—	—	0.32	—	.066
Nash	0-10	5.54	7.4	37.8	15.4	0.79	3.02	.076
	25	5.58	3.3	29.8	15.4	0.49	2.61	.048
	50	5.55	1.9	23.8	15.4	0.38	2.44	.040
	100	5.62	2.3	27.6	16.4	0.29	3.19	.064
French	0-10	5.32	9.3	40.8	22.4	0.68	2.36	.162
	25	5.12	4.8	38.8	25.4	0.66	2.52	.075
	50	4.93	2.8	28.6	19.8	0.67	2.45	.075
	100	4.85	2.4	16.8	22.4	0.48	2.04	.092

[a]Citrate-dithionate-bicarbonate extractable.
[b]Ammonium acetate extractable.

Perkin-Elmer model 560 and sulfate and chloride by colorimetric methods using a Scientific Instruments autoanalyzer. Aluminum and iron in leachate were measured by colorimetric methods on filtrate from a 0.1-μm membrane filtration (Dougan and Wilson, 1974; Stookey, 1970). Dissolved organic carbon was measured in 1982 by coulometric titration using a Coulometrics, Inc., CO_2 coulometer following a sealed ampoule persulfate digestion. Nitrate and ammonium were measured colorimetrically, the former following cadmium reduction, and the latter by a phenolhypochlorite method with the autoanalyzer.

RESULTS AND DISCUSSION

Soil Features

Physical and chemical features of the soils differed markedly among stands (Table 40-1). High clay and organic matter content and low pH were observed in the soil at French Creek, while low clay and organic matter levels were noted at Albany; the Nash Fork soil was intermediate in these features. Free iron and aluminum content generally decreased with soil depth, but some evidence of subsoil accumulation was observed at French Creek and Nash Fork. Total extractable base content of the soils ranged from 16 meq/100 g in the surface soil at French Creek to 4-5 meq/100 g in the subsoil at Nash Fork (Table 40-1).

Snowpack Chemistry

The bulk snowpack was extremely dilute (total cation concentration = 0.05 meq/l), and only slight chemical enrichment was observed beneath the forest canopy in the study area (0.06 meq/l). Moreover, free acidity levels in snow were low (mean pH = 5.2), indicating that hydrogen ion loading to the ecosystem from melting snow was minor.

Chemistry of Forest Floor Leachate

Snow water was altered dramatically by passage through the forest floor. Because no significant differences were observed in the chemistry of forest floor leachate among three stands and three years of collection, these data were pooled for the present analysis (Fig. 40-1). Maximum concentrations of most chemical species occurred at the beginning of the snowmelt period, probably as a result of heterotrophic activity beneath the winter

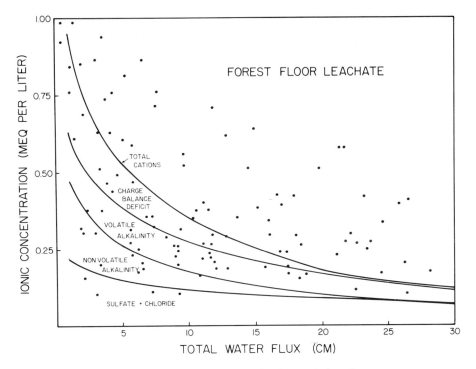

Figure 40-1. Ionic concentrations in forest floor leachate in lodgepole pine ecosystems as a function of total water flux. Curves represent the best-fit linear regressions for the entire data set (240 data points), for which R^2 values ranged from about 0.25 to 0.40. Data points are for total cation concentrations.

snowpack (Fahey, 1983), and ionic strength declined logarithmically as snowmelt progressed.

In the early stages of snowmelt, organic anions (detected as nonvolatile alkalinity) and sulfate were the dominant anions in forest floor leachate, with lesser contributions from bicarbonate (HCO_3^-) and chloride (Cl^-). The significant charge-balance deficit was probably accounted for by organic anions that were not titrated as alkalinity because of the low pKa values of their acids (e.g., acetate, oxalate; Yavitt and Fahey, 1984b).

Despite the addition of organic carbon to snowmelt water, the pH of forest floor leachate was high (mean pH = 5.70), allowing dissociation of carbonic acid and bicarbonate leaching of the lodgepole pine forest floor (Fig. 40-1). Apparently, hydrogen ions generated in the forest floor were neutralized by bases leached from decaying tissues and from the forest floor exchange complex (Yavitt and Fahey, 1984a). Thus, the forest floor was not an important source of free hydrogen ions for weathering/exchange reactions in the mineral soil.

Relatively high total iron and aluminum concentrations were noted in forest floor leachate, indicating weathering of soil minerals mixed with the 02 horizon (nonvolatile matter in the 02 ranged from 30% to 35% in the three stands). This might also be explained by leaching of these elements from decaying tissue (Cronan, 1980); however, aluminum and total iron concentrations have not been measured in lodgepole pine leaf litter. Concentrations of aluminum and total iron in forest floor leachate were highly correlated with levels of undissociated organic acids (detected as nonvolatile acidity), suggesting that these rather insoluble elements were transported from the forest floor as organic ligands (Fig. 40-2).

Soil-Solution Chemistry

The chemistry of the soil solution collected at the base of the major tree rooting zone (35-40 cm) varied both among stands and among years within stands (Fig. 40-3). Moreover, chemical concentrations varied among the porous-cup samplers within each stand. Nevertheless, the patterns of change in element concentrations as snowmelt progressed usually were similar for all the samplers in each stand. In general, total ionic strength declined rapidly during the initial stages of snowmelt (Fig. 40-4); this pattern probably reflects in part a similar trend in forest floor leachate (Fig. 40-1), but other factors may be involved as well.

Sulfate was a major component of the soil solution in all the stands, and high atmospheric inputs are indicated because S-bearing minerals are uncommon in these soils (T. Fahey, unpub. data). Weighted-mean sulfate concentrations in summer through fall in the study area were about 0.1 meq/l, the relatively high levels probably being attributed to regional pollution from

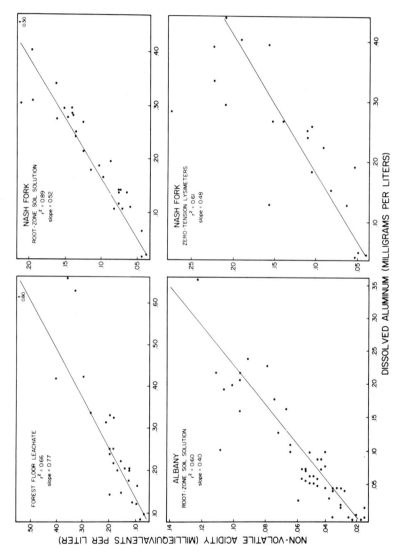

Figure 40-2. Linear correlations between nonvolatile acidity, (estimated by titration to pH 8.3) and dissolved aluminum concentrations in forest floor leachate and root-zone soil solution in lodgepole ecosystems, southeastern Wyoming.

coal-fired power plants and to deflation from adjacent dry basin areas. Yavitt and Fahey (1984*b*) have demonstrated that sulfate transport to the mineral soil by forest floor leaching in summer is a major flux pathway, but the leaching rainwater rarely penetrates beyond the upper 30 cm of mineral soil (Knight, Fahey, and Running, in press). Thus, solution sulfate flux appears to occur in two stages: (1) input in summer precipitation and leaf wash, and (2) mobilization by melting snow the following spring. The typically rapid decline in sulfate concentrations in the root-zone solution during early snowmelt supports this hypothesis, as do the extremely high concentrations observed after heavy rains in late summer 1979 (Fig. 40-3). Because of high spatial variability, few significant differences in sulfate concentrations were detected between root-zone and subsoil solutions, but soil solution sulfate levels far exceeded those in forest floor leachate (Figs. 40-1 and 40-3).

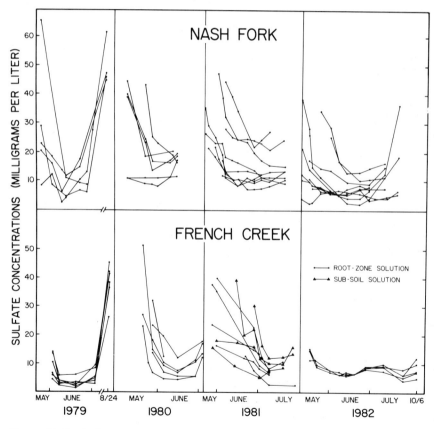

Figure 40–3. Sulfate concentrations in root-zone soil solution and sub-soil solution (French Creek 1981, only) in two lodgepole pine ecosystems, southeastern Wyoming. Each line represents values from one porous-cup lysimeter.

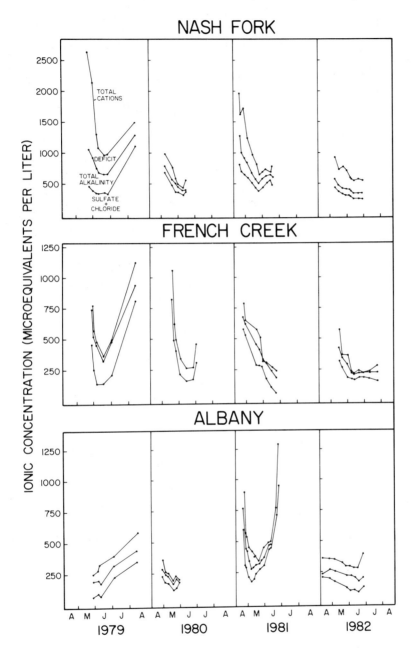

Figure 40–4. Concentrations of total cations (H, Ca, Mg, K, Na, Mn) and various anions in root-zone soil solution for four years in three lodgepole pine ecosystems, southeastern Wyoming. Each data point represents the mean of five to nine porous-cup lysimeters.

At the two stands with coarser soils (Nash Fork and Albany), organic anions were an important component of the root-zone solution charge-balance (Fig. 40-4), but at French Creek nonvolatile alkalinity was very low (< 0.02 meq/l), as were nonvolatile acidity and dissolved organic carbon (DOC). Adsorption of DOC by this soil was postulated and has been demonstrated in the laboratory (Yavitt and Fahey, 1984b). Bicarbonate was an important anion in the root-zone solution in all the stands. Measurements of soil atmosphere pCO_2 during snowmelt leaching at Nash Fork and Albany in 1983 indicated that equilibrium probably was not established between soil-solution and soil-atmosphere carbon dioxide as pCO_2 ranged from 0.003 to 0.005 atm (T. Fahey, unpub. data). Root-zone solution pH values generally were high, and carbonic acid dissociation was not severely depressed. However, no evidence was obtained for high carbon dioxide accumulations beneath the winter snowpack, as summer and winter levels usually were not significantly different. Nevertheless, biotic processes that generate the "carbon anions" (i.e., HCO_3^- and organics; Sollins et al., 1980) obviously play an important role in soil cation flux in lodgepole pine ecosystems.

Annual variation in total cation concentration in the root-zone solution was related both to variation in strong-acid anion levels (especially at French Creek and Albany) and to variation in weak-acid carbon anions (Nash Fork). Observations from 1979 may have been influenced by lysimeter installation that was carried out in summer/fall 1978 and observations in 1981 by un-usually low snowpack and heavy spring rains.

High correlations were noted between undissociated organic acids and dissolved aluminum concentrations in the root-zone solution (Fig. 40-2). The slope of the correlation indicated higher aluminum concentrations relative to organics in root-zone solution than in forest floor leachate, as might be expected because of the higher mineral mass in the soil. The slope of the correlation differed significantly between stands, with higher aluminum levels relative to organics at Albany than at Nash Fork. Moreover, the same pattern was observed for zero-tension as for tension lysimeters (Fig. 40-2). These observations may indicate that the granitic Albany soil is more readily weathered than the predominantly gneiss-derived soil at Nash Fork. Preliminary investigations also indicated that iron concentrations in soil solutions at Albany were similar to aluminum concentrations, with the same slopes and correlations in regressions with nonvolatile acidity.

SUMMARY

Cation leaching in lodgepole pine forest soils appears to be mediated mainly by biotic processes, particularly respiratory carbon dioxide production and generation of soluble organic compounds. The latter compounds form soluble complexes with aluminum and total iron that are transported through the soil in leaching water, enhancing soil development. Thus, the ion-transport phenomenon in central Rocky Mountain forest soils is similar to that in other unpolluted forest ecosystems (Johnson et al., 1977; Ugolini et al., 1977; Sollins et al., 1980). Although sulfate was a dominant anion in forest floor and soil solutions, it probably only slightly influences net cation leaching because it both enters and leaves the ecosystem predominantly as a neutral salt. Soil characteristics, especially the capacity for adsorption of dissolved organic carbon and sulfate, influence ion mobility and result in spatial variation in the leaching characteristics of lodgepole pine ecosystems. Understanding of ion flux and soil formation in this ecosystem could profit from detailed studies of the nature and synthesis of soluble organic acids in detritus and the adsorption processes that limit their mobility.

REFERENCES

Antweiler, R. C., and J. I. Drever, 1983, The weathering of a late-Tertiary volcanic ash: Importance of organic solutes, *Geochim. Cosmochim. Acta* **47**:623-629.

Cole, D. W., 1958, An alundum tension lysimeter, *Soil Sci.* **85**:293-296.

Cronan, C. S., 1980, Solution chemistry of a New Hampshire subalpine ecosystem: A biogeochemical analysis, *Oikos* **34**:272-281.

Cronan, C. S., W. A. Reiners, R. C. Reynolds, and G. E. Lang, 1978, Forest floor leaching: Contributions from mineral, organic and carbonic acids in New Hampshire subalpine forests, *Science* **200**:309-311.

Day, P. R., 1965, Particle fractionation and particle-size analysis, in *Methods of Soil Analysis.* Part 1. *Physical and Mineralogical Properties, Including Statistics of Measurement and Sampling,* C. A. Black, ed., Society of Agronomy, Madison, Wisconsin, pp. 545-566.

Dougan, W. K., and A. L. Wilson, 1974, The absorptiometric determination of aluminum in water. A comparison of some chromogenic reagents and the development of an improved method, *Analyst* **99**:413-430.

Fahey, T. J., 1983, Nutrient dynamics of aboveground detritus in lodgepole pine *Pinus contorta* ssp. *latifolia*) ecosystems, southeastern Wyoming, *Ecol. Monogr.* **53**:51-72.

Gjessing, E. T., A. Henricksen, M. Johannessen, and R. F. Wright, 1976, Effects of acid precipitation on freshwater chemistry, in *Impact of Acid Precipitation on Forest and Freshwater Ecosystems in Norway,* F. H. Braekke, ed., SNSF Project, SNNS, Oslö, Norway, pp. 64-85.

Golterman, H. L., R. S. Chimo, and M. A. M. Ohnstad, 1978, *Methods for Physical and Chemical Analysis of Fresh Waters,* 2nd ed., Blackwell, Oxford.

Graustein, W. C., K. Cromack, and P. Sollins, 1977, Calcium oxalate: Occurrence in soils and effect on nutrient and geochemical cycles, *Science* **198**:1252-1254.

Johnson, D. W., D. W. Cole, S. P. Gessel, M. J. Singer, and R. V. Minden, 1977, Carbonic acid leaching in a tropical, temperate, subalpine and northern forest soil, *Arc. Alp. Res.* **9**:329-343.

Johnson, D. W., J. W. Hornbeck, M. J. Kelly, W. T. Swank, and D. E. Todd, 1980, Regional patterns of soil sulfate accumulation: Relevance to ecosystem sulfur budgets, in *Atmospheric Sulfur Deposition: Environmental Impact and Health Effects,* D. S. Shriver, C. R. Richmond, and S. E. Lindberg, eds., Ann Arbor Science Publishers, Ann Arbor, Michigan, pp. 507-520.

Jordan, C. F., 1968, A simple tension-free lysimeter, *Soil Sci.* **105:**81-86.

Knight, D. H., T. J. Fahey, and S. W. Running, in press, Water and nutrient outflow from contrasting lodgepole pine forests in Wyoming, *Ecol. Monogr.*

Lewis, W. M., Jr., and M. C. Grant, 1979, Changes in the output of ions from a watershed as a result of the acidification of precipitation, *Ecology* **60:**1093-1097.

Lewis, W. M., Jr., and M. C. Grant, 1980, Acid precipitation in the western U.S., *Science* **207:**176-177.

Likens, G. E., F. H. Bormann, R. S. Pierce, J. S. Eaton, and N. M. Johnson, 1977, *Biogeochemistry of a Forested Ecosystem,* Springer-Verlag, New York.

Mehra, O. P., and M. L. Jackson, 1960, Iron oxide removal from soils and clays by a dithionate-citrate system buffered with sodium bicarbonate, *Clays and Clay Miner.* **7:**317-327.

Parizek, R. R., and B. E. Lane, 1975, Soil water sampling using pan and deep pressure-vacuum lysimeters, *J. Hydrology* **11:**1-21.

Soil Survey Staff, 1975, *Soil Taxonomy,* Agricultural Handbook No. 436. Soil Conservation Service, U.S. Dept. of Agriculture, Washington, D.C.

Sollins, P., C. C. Grier, F. M. McCorison, K. Cromack, Jr., R. Fogel, and R. L. Fredricksen, 1980, The internal element cycles of an old-growth Douglas-fir ecosystem in western Oregon, *Ecol. Monogr.* **50:**261-285.

Stookey, L. L., 1970, Ferrozine—A new spectrophotometric reagent for iron, *Anal. Chem.* **42:**779-791.

Stumm, W., and J. J. Morgan, 1981, *Aquatic Chemistry,* 2nd ed., John Wiley & Sons, New York.

Ugolini, F. C., R. Minden, H. Dawson, and J. Zachara, 1977, An example of soil processes in the *Abies amabilis*-zone of Central Cascades, Washington, *Soil Sci.* **124:**291-302.

Wiklander, L., 1971, The role of neutral salts in ion exchange between precipitation and soil, *Geoderma* **14:**93-105.

Yavitt, J. B., and T. J. Fahey, 1984*a*, An experimental analysis of solution chemistry in a lodgepole pine forest floor, *Oikos* **43:**222-234.

Yavitt, J. B. and T. J. Fahey, 1984*b*, Organic chemistry of the soil solution during snowmelt leaching in *Pinus contorta* forest ecosystems, Wyoming, in *Planetary Ecology: Selected Papers from the Sixth International Symposium on Environmental Biogeochemistry* D. E. Caldwell, J. A. Brierley, and C. L. Brierley, eds., Van Nostrand Reinhold Co., Inc., New York (Paper 4).

41

Organic Chemistry of the Soil Solution During Snowmelt Leaching in *Pinus contorta* Forest Ecosystems, Wyoming

Joseph B. Yavitt

University of Wyoming

Timothy J. Fahey

Cornell University

ABSTRACT: Dissolved organics were analyzed in forest floor, root-zone, and subsoil solutions during snowmelt leaching to evaluate the role of these compounds in nutrient transport, soil weathering, and soil organic matter formation. Dissolved organic carbon (DOC) concentrations in forest floor leachate decreased with increasing water flux during the spring snowmelt period (from >70 mg/l to <20 mg/l), suggesting that most of the dissolved organics were decomposition products that accumulated beneath the winter snowpack until being leached by melting snow in the spring. DOC concentrations in the root-zone soil solution were initially low, then rose to a maximum of 40 mg/l during mid-snowmelt before decreasing to <15 mg/l by late snowmelt. Several classes of compounds—that is, low-molecular-weight organic acids, phenolic acids and aldehydes, free amino acids, carbohydrates, and large phenolic polymers—were quantified in solution. Carbon fractionation by liquid chromatography indicated that about 85% of the organics were acidic compounds, whereas the remaining compounds were primarily neutral. The relative importance of the different organic fractions changed as snowmelt progressed, with an increasing proportion of dissolved humics. This was reflected by an increase in the C:N ratio from 25:1 to >75:1 during the snowmelt period. Most of the dissolved organic compounds were apparently immobilized in the soil rooting zone, as DOC levels in the subsoil solution were low (< 9 mg/l). Some of the dissolved compounds are known precursors of soil organic matter, and others form soluble complexes with aluminum (Al) and iron (Fe), thereby enhancing soil weathering reactions.

INTRODUCTION

Dissolved organic compounds rarely have been measured in soil solutions of forest ecosystems, even though they have been shown to play an important role in ion transport (Cronan et al., 1978), mineral weathering reactions (Graustein, Cromack, and Sollins, 1977; Antweiler and Drever, 1983), and soil organic matter formation (Whitehead, Dibb, and Hartley, 1981). These compounds are supplied to the soil solution by leaching of the forest canopy (Zinke, 1962; Gersper and Holowaychuk, 1970) and forest floor (Cronan, 1980; Sollins and McCorison, 1981), and in most temperate-zone forests they represent a significant proportion of the anionic charge facilitating cation mobilization (Johnson et al., 1977). Specific organic anion-cation complexes, such as calcium oxalate crystals (Graustein, Cromack, and Sollins, 1977), have been identified in soil profiles, and the importance of such complexes in mineral weathering and subsequent iron and aluminum transport in podzols has been demonstrated (Ugolini, Dawson, and Zachara, 1977). Most of the organics that enter the soil apparently are immobilized there because concentrations of organics in rivers and ground water are very low (Malcolm and Durum, 1976; Mulholland and Kuenzler, 1979). Consequently, dissolved organics probably contribute to the accumulation of soil organic matter.

The objective of the present study was to characterize and quantify dissolved organic carbon in forest floor, root-zone, and subsoil solutions during snowmelt leaching in two contrasting lodgepole pine (*Pinus contorta* ssp. *latifolia*) ecosystems. It was hoped to provide evidence of spatial and temporal variation in the concentration and speciation of dissolved organics as a first step towards understanding the mechanisms of organic-mediated weathering and ion-transport processes in this widespread Rocky Mountain forest type.

MATERIALS AND METHODS

The study area, in the Medicine Bow Mountains of southeast Wyoming, U.S.A., was underlain by Quaternary glacial till. The climate, soils, and vegetation of the study sites (Nash Fork, French Creek) are described in detail in a companion paper (Paper 40) and in other papers from ongoing studies of the biogeochemistry of lodgepole pine ecosystems (Yavitt and Fahey, 1982; Fahey, 1983; Knight, Fahey, and Running, in press). The most critical difference between the sites was a finer soil texture and higher soil organic matter content at the French Creek stand than at the Nash Fork stand.

Water samples were obtained from beneath the forest floor, the major soil-rooting zone (40 cm), and the subsoil (1.5-1.8 m) in each stand during snowmelt drainage using tension lysimetry. Methods of sample collection, storage, and inorganic analysis are described in Paper 40. Beginning in 1982 a variety of organic analyses were also performed. Samples from selected dates were pooled for carbon fractionation by a liquid chromatography procedure (Leenheer and Huffman, 1976). Several wet chemical analyses were performed on all water samples, including dissolved carbohydrates (Handa, 1966), soluble polyphenolics (Rand, 1976), free amino acids (Satake et al., 1960), and soluble proteins by a protein-dye method (Bradford, 1976). Organic acids were analyzed by gas-liquid chromatography on both non-derivatized samples (C_2-C_5 acids; DiCorcia and Samperi, 1974, and trimethyl-silyl derivatives (phenolic acids; Kaminsky and Muller, 1977).

The retention of organics in soils was investigated in laboratory studies using sorption isotherms. Isotherms were calculated by shaking soil samples with forest floor leachate of varying dissolved organic carbon (DOC) concentration, measuring initial and final DOC levels in solution, and then calculating the concentration at which organics were neither sorbed nor desorbed from the soil (i.e., the equilibrium concentration). Studies were done on soils collected from four depths (0-15 cm, 15-30 cm, 45 cm, and 100 cm) at both sites. To calculate the equilibrium concentration, four concentrations of forest floor leachate were prepared (2 mg/l, 6 mg/l, 12 mg/l, and 24 mg/l) and 25 mls of each solution were shaken with 2.5 g of air-dried soil (< 2 mm fraction) for a measured time interval (Taylor and Kunishi, 1971). Isotherms were calculated for several incubation times ranging from 5 min to 4 hr.

RESULTS AND DISCUSSION

Forest Floor Leachate

Maximum concentrations of all organic compounds in forest floor leachate were observed in the first increment of water flux, with rapid declines being noted as snowmelt proceeded. Dissolved organic carbon concentration decreased more than 3-fold during the snowmelt period, whereas nitrogen (N) concentration decreased by 8-fold (Fig. 41-1). The result was an increase in the C:N ratio in forest floor leachate from 20:1 to 75:1. Most of the dissolved organics were acidic in nature (Table 41-1). Hydrophobic acids accounted for 49% of the DOC, whereas hydrophilic acids and neutral species comprised 34% and 11%, respectively. The majority of these acidic compounds were undissociated at the pH values that occurred in forest floor leachate, as indicated by much higher values of nonvolatile acidity than nonvolatile alkalinity throughout the snowmelt interval (Table 41-1). Maximum concentrations of dissolved carbohydrates, polyphenolics, proteins, and free amino acids were observed in early snowmelt, with 3- to 5-fold concentration decreases occurring by late snowmelt. Several free organic acids were detected in forest floor leachate, including acetic, propionic, n-butyric, 2-methylbutyric, and 3-methylbutyric acids. Concentrations of most of these organic acids were less than 2 mg/l, and they were detected only in the early stages of snowmelt. However, high concentrations of acetic acid were observed throughout the snowmelt period. Free phenolic acids and aldehydes were also detected in forest floor leachate, including p-hydroxybenzoic, vanillic, quinic, and protocatechuic acids, and p-hydroxybenzaldehyde and vanillan. The concentration of p-hydroxybenzaldehyde in forest floor leachate decreased from 4.2 mg/l in early snowmelt to 0.2 mg/l by mid-snowmelt, whereas vanillic acid increased from 0.13 mg/l to 0.3 mg/l during the same period. Most other free phenolics were present at low concentrations (<0.05 mg/l) throughout the course of snowmelt leaching.

The apparent logarithmic declines in organic concentrations in forest floor leachate during the snowmelt period suggest that most of the dissolved

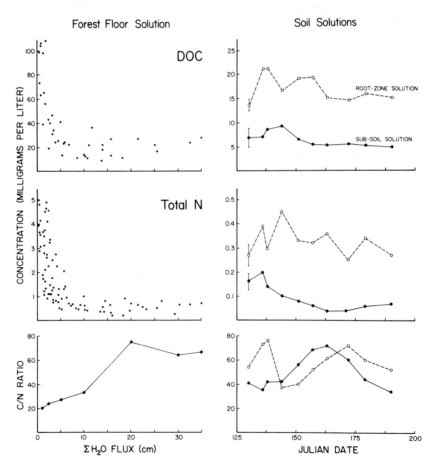

Figure 41–1. Concentrations of DOC and total-N and C:N ratio of forest floor leachate and soil solutions at two depths at the Nash Fork site in *Pinus contorta* ecosystem, Wyoming. Nitrogen values from 1981 and 1982 snowmelt years; carbon values from 1982 only.

Table 41-1. Summary of Organic Chemistry in Forest Floor Leachate from a *Pinus contorta* Ecosystem

Snowmelt Period	pH	Acids (mg/l)[a] Hydrophobic	Acids (mg/l)[a] Hydrophilic	Neutrals (mg/l)[b] Hydrophobic	Neutrals (mg/l)[b] Hydrophilic	Nonvolatile(meq/l) Alkalinity	Nonvolatile(meq/l) Acidity
Early	5.70	20.6	14.6	5.9	2.0	0.140	0.340
Mid	5.68	16.5	12.6	2.0	0.9	0.060	0.190
Late	5.68	11.5	7.9	1.4	1.4	0.030	0.110

Snowmelt Period	Sample Size	Dissolved Carbohydrates[a]	Concentration (mg/l) Soluble Polyphenolics[a]	Concentration (mg/l) Free Amino Acids[b]	Soluble Proteins[b]	Acetate Anion
Early	16	24.9	9.3	0.19	1.07	35.0
Mid	28	9.5	3.4	0.10	0.61	11.2
Late	17	4.3	2.4	0.06	0.29	1.5

[a]Expressed as mg/l of carbon.
[b]Expressed as mg/l of nitrogen.

compounds were decomposition products that accumulated beneath the winter snowpack. Soluble organics were probably released continuously throughout the winter by decomposers that were active beneath the insulating snowpack. Fahey (1983) found that leaf litter decay rates did not differ between summer and winter in the study area. Alternatively, organics could have become soluble in a rapid flush in spring just prior to snowmelt, as detrital heterotrophs responded to the initial wetting front reaching the forest floor. Because many of the dissolved organics in forest floor leachate were compounds that are readily metabolized (e.g., carbohydrates and free amino acids), they would not be expected to accumulate over the long winter period. It is likely that both long-term accumulation of more resistant fulvic acids and short-term buildup of labile organics led to the high DOC levels at the onset of snowmelt. Labile components could have accumulated in the forest floor through microbial-fungal interactions in which fungal grazing of microbial biomass accelerated the rate of release of organics. For example, preliminary studies indicated that active fungal hyphae decreased between pre-snowmelt and just following the onset of leaching, presumably due to grazing by ciliates or protozoans (E. R. Ingham, C. P. P. Reid, and D. C. Coleman, pers. comm.). More resistant compounds could have been produced during the long winter period as a result of microbial or chemically mediated polymerization of low-molecular-weight compounds.

Soil Solution

In contrast to forest floor leachate, maximum concentrations of organics in the root-zone solution at Nash Fork occurred at mid-snowmelt. Both DOC and total-N concentrations in this leachate were highest in the second and third water collections following onset of leaching (Fig. 41-1), with lower concentrations in early and late stages of snowmelt. As for forest floor leachate, DOC in the root-zone solution at Nash Fork was acidic in nature (Table 41-2), with relative concentrations of hydrophobic and hydrophilic acids accounting for about 40% and 35% of the DOC, respectively. Nonvolatile alkalinity in root-zone solution was only slightly lower than the values in forest floor leachate following the initial stages of snowmelt; however, nonvolatile acidity was 6-fold lower. Thus it appears that undissociated organics were retained to a greater degree in surface soil than the dissociated organics.

Concentrations of various soluble organics were 3- to 10-fold lower in leachate from the root-zone than from the forest floor (Tables 41-1 and 41-2). Peak concentrations also occurred at mid-snowmelt for all the measured organics except acetic acid, which declined rapidly during the course of snowmelt leaching. Some phenolic acids and aldehydes were measured in root-zone solution at early snowmelt, but thereafter concentrations were below detection limits.

Average DOC and total-N concentrations were much lower in the subsoil solution than in forest floor or root-zone solutions at Nash Fork (Fig. 41-1). A significant peak in concentration of DOC and total-N in the subsoil solution was observed at a later time than the peak in the root-zone solution (Fig. 41-1). Organic acids, free amino acids, and proteins were not detected in the subsoil solution at Nash Fork, and the low levels of carbohydrates and polyphenolics showed no changes during the snowmelt interval.

Very low concentrations of dissolved organics were observed in the root-zone and subsoil solutions in the French Creek stand (Table 41-3). Concentrations of DOC and total-N did not change significantly as snowmelt progressed, and, consequently, a constant C:N ratio of 30:1 in the soil solutions was calculated. Concentrations of DOC and total-N were not significantly different between root-zone and subsoil leachate at French Creek, and organic concentrations were slightly lower than subsoil leachate at the Nash Fork site.

Differences in organic chemistry between forest floor leachate and soil solutions probably may be attributed to several physical, chemical, and biological properties of soil. A general pattern of lower organic concentrations with increasing depth of water percolation was observed, indicating that organics were retained in the soil profile. Although the mechanisms that regulate retention of organics could vary between sites, it is likely that physical-chemical adsorption on organic and inorganic soil colloids plays a major role. For example, higher retention was observed on the sites with relatively high clay and soil organic matter (SOM) contents. The role of clays in organic-inorganic interactions has been well documented (Theng, 1979;

Table 41-2. Summary of Organic Chemistry of Soil Solutions from Two Depths at Nash Fork Site in Pinus contorta Ecosystem, Southeastern Wyoming

Leachate	Snowmelt Period	pH	Acids (mg/l)[a]		Neutrals (mg/l)[b]		Nonvolatile (meq/l)	
			Hydrophobic	Hydrophilic	Hydrophobic	Hydrophilic	Alkalinity	Acidity
Root-zone	Early	6.12	7.6	5.8	2.2	0.6	0.037	0.053
	Mid	5.35	16.9	17.7	3.4	0.0	0.040	0.093
	Late	6.03	7.3	7.2	0.6	0.5	0.038	0.022
Subsoil		6.55	2.4	2.0	1.6	0.5	0.024	0.022

Leachate	Snowmelt Period	Sample Size	Dissolved Carbohydrates[a]	Soluble Polyphenolics[a]	Free Amino Acids[b]	Soluble Proteins[b]	Acetate Anion
Root-zone	Early	26	3.4	1.15	0.019	0.0	1.10
	Mid	36	4.5	1.38	0.072	0.0	0.10
	Late	30	2.8	1.07	0.030	0.0	0.00
Subsoil		42	1.3	0.22	0.0	0.0	0.00

[a]Expressed as mg/l of carbon.
[b]Expressed as mg/l of nitrogen.

Table 41-3. Summary of Organic Chemistry of Soil Solution from Two Depths at French Creek Site in Pinus contorta Ecosystem, Southeastern Wyoming

Leachate	Sample Size	Concentration (mg/l)		Nonvolatile (meq/l)		Concentration (mg/l)	
		DOC	Total-N	Alkalinity	Acidity	Dissolved Carbohydrates[a]	Soluble Polyphenolics[a]
Root-zone	36	4.4	0.14	0.010	0.026	0.97	0.17
Subsoil	31	4.4	0.14	0.013	0.026	1.12	0.18

[a] Expressed as mg/l of carbon.

Greenland and Hayes, 1981), and several specific mechanisms have been proposed to explain the sorption of both negatively charged and uncharged organics on clays.

Some aspects of organic sorption in soils that have rarely been studied include the roles of soil structure and residence time of leachate in the soil profile. These aspects were indirectly studied in a laboratory investigation of equilibrium concentrations (EC)—that is, the solution concentration at which organics are neither sorbed nor desorbed from soil (Table 41-4). Much higher EC values were observed for Nash Fork than for French Creek soils, and higher sorption capacity was noted for subsoil than for surface soils at both sites. Furthermore, investigation of the Nash Fork soils indicated the importance of solution residence time, as higher EC values were observed for 10- to 20-min incubations than for longer incubations. Thus, the high DOC values in the root-zone solution at Nash Fork during spring snowmelt may indicate that rapid flux of water prevented complete sorption of dissolved organics. Concentrations of DOC in this solution also may have exceeded those in forest floor leachate because of rapid desorption of organics from SOM in surface soils. Apparently, some of the compounds forming SOM at this site are water soluble, allowing redistribution of SOM within the soil profile.

Soil structure may also be important in controlling the concentration of organic solutes. For example, in the French Creek soils EC values were higher than the actual DOC concentrations in soil solution. It seems possible that small structural voids and dead-end pores, which occur in field soils, supported large populations of soil heterotrophs (Elliott et al., 1980; Wilkinson, Miller, and Millar, 1981) that could metabolize labile organics and

Table 41-4. Measured Equilibrium Concentrations for Air-Dried Soils (< 2 mm Fraction) Collected at Two Sites in the *Pinus contorta* Ecosystem, Southeastern Wyoming

| Site | Residence Time | Equilibrium Concentration (mg/l) | | | |
		0-15 cm	15-30 cm	45 cm	100 cm
Nash Fork	5 min	71	22	21	12
	10-20 min	44	52	36	13
	40-60 min	55	38	17	25
	1.5-2 hr	28	34	15	12
	>4 hr	20	30	17	14
French Creek	5 min	12	9	13	8
	10-20 min	16	11	12	8
	40-60 min	24	13	11	10
	1.5-2 hr	17	16	13	8
	>4 hr	—	15	9	12

thus reduce DOC concentrations below values resulting from purely physical sorption processes.

Synthesis

Results from the present study indicate that a diverse mixture of predominantly acidic and neutral compounds comprise DOC in forest floor and soil solutions during snowmelt in lodgepole pine ecosystems. High concentrations of all dissolved organics in forest floor leachate at the beginning of snowmelt probably can be attributed to heterotrophic activity beneath the winter snowpack, possibly including a rapid flush of activity at the onset of snowmelt. In contrast, the peak in DOC concentrations in the soil solution, which occurred at mid-snowmelt, may be attributed to changes in chemical characteristics of DOC. In particular, the higher C:N ratio and lower ionic charge (indicated by a higher proportion of undissociated organic acidity, Table 41-2) of organics from mid-snowmelt could inhibit microbial and physical-chemical immobilization processes leading to higher solution DOC levels. Alternatively, soil sorption sites could be partially saturated by the early, organic-laden forest floor leachate, decreasing the capacity of the soil for retention of DOC. Certainly the factors influencing the sorptive capacity of soils are highly complex, but soil physical-chemical properties, soil structure, and residence time of percolating water are probably involved. In the lodgepole pine ecosystem, variations in the rate of snowmelt and the occurrence of extended cold periods during the snowmelt interval could also influence dissolved organic concentrations in the soil solution. For example, in laboratory column studies of solute flux through sample soils, significant declines in solution DOC levels have been observed following several days without leaching (J. Yavitt and T. Fahey unpub. data), probably as a result of microbial metabolism of flocculated organics that opens previously saturated sites.

REFERENCES

Antweiler, R. C., and J. I. Drever, 1983, The weathering of a late-Tertiary volcanic ash: Importance of organic solutes, *Geochim. Cosmochim. Acta* **47**:623-629.

Bradford, M. M., 1976, A rapid and sensitive method for the quantitation of microgram quantities of protein utilizing the principle of protein-dye binding, *Anal. Biochem.* **72**:248-254.

Cronan, C. S., 1980, Solution chemistry of a New Hampshire subalpine ecosystem: A biogeochemical analysis, *Oikos* **34**:272-281.

Cronan, C. S., W. A. Reiners, R. C. Reynolds, and G. E. Lang, 1978, Forest floor leaching: Contributions from mineral, organic, and carbonic acids in New Hampshire subalpine forests, *Science* **200**:309-311.

DiCorcia, A., and R. Samperi, 1974, Determination of trace amounts of C_2-C_5 acids in aqueous solutions by gas chromatography, *Anal. Chem.* **46:**140-143.

Elliott, E. T., R. V. Anderson, D. C. Coleman, and C. V. Cole, 1980, Habitable pore space and microbial trophic interactions, *Oikos* **35:**327-335.

Fahey, T. J., 1983, Nutrient dynamics of aboveground detritus in lodgepole pine (*Pinus contorta* ssp. *latifolia*) ecosystems, southeastern Wyoming, *Ecol. Monogr.* **53:**51-72.

Fahey, T. J., J. B. Yavitt, A. E. Blum, and J. I. Drever, 1984, Controls of soil solution chemistry in lodgepole pine forest ecosystems, Wyoming, in *Planetary Ecology: Selected Papers from the Sixth International Symposium on Environmental Biogeochemistry,* D. E. Caldwell, J. A. Brierley, and C. L. Brierley, eds., Van Nostrand Reinhold Co., Inc., New York (Paper 40).

Gersper, P. L., and N. Holowaychuk, 1970, Effects of stemflow water on a Miami soil under a beech tree: I. Morphological and physical properties, *Soil Sci. Soc. Amer. Proc.* **34:**779-786.

Graustein, W. C., K. Cromack, Jr., and P. Sollins, 1977, Calcium oxalate: Occurrence in soils and effect on nutrient and geochemical cycles, *Science* **198:**1252-1254.

Greenland, D. J., and M. H. B. Hayes, 1981, *The Chemistry of Soil Processes,* Wiley-Interscience, Chichester.

Handa, N., 1966, Examination of the applicability of the phenol sulfuric acid method to the determination of dissolved carbohydrate in sea water, *J. Oceanogr. Soc. Japan* **22:**79-86.

Johnson, D. W., D. W. Cole, S. P. Gessel, M. J. Singer, and R. V. Minden, 1977, Carbonic acid leaching in a tropical, temperate, subalpine, and northern forest soil, *Arc. Alp. Res.* **4:**329-343.

Kaminsky, R., and W. H. Muller, 1977, The extraction of soil phytotoxins using a neutral EDTA solution, *Soil Sci.* **124:**205-210.

Knight, D. H., T. J. Fahey, and S. W. Running, in press, Water and nutrient outflow from contrasting lodgepole pine forests in Wyoming, *Ecol. Monogr.*

Leenheer, J. A., and E. W. D. Huffman, Jr., 1976, Classification of organic solutes in water by using macroreticular resins, *J. Res. U.S. Geol. Survey* **4:**737-751.

Malcolm, R. L., and W. H. Durum, 1976, *Organic carbon and nitrogen concentrations and annual organic load of six selected rivers of the United States,* U.S. Geol. Surv. Water-Supply Paper 1817-F.

Mulholland, P. J., and E. J. Kuenzler, 1979, Organic carbon export from upland and forested wetland watersheds, *Limnol. Oceanogr.* **24:**960-966.

Rand, M. C., 1976, *Standard Methods for the Examination of Water and Wastewater,* 14th ed. American Public Health Association, Washington, D.C.

Satake, K., T. Okuyama, M. Ohashi, and T. Shinada, 1960, The spectrophotometric determination of amine, amino acid, and peptide with 2,4,6-trinitrobenzene 1-sulfonic acid, *J. Biochem.* **47:**654-660.

Sollins, P., and F. M. McCorison, 1981, Nitrogen and carbon solution chemistry of an old growth coniferous forest watershed before and after cutting, *Water Resour. Res.* **17:**1409-1418.

Taylor, A. W., and H. M. Kunishi, 1971, Phosphate equilibria on stream sediment and soil in a watershed draining an agricultural region, *J. Agr. Food Chem.* **19:**827-831.

Theng, B. K. G., 1979, *Formation and Properties of Clay-polymer Complexes,* Elsevier Scientific Publ. Co., Amsterdam.

Ugolini, F. C., H. Dawson, and J. Zachara, 1977, Direct evidence of particle migration in the soil solution of a podzol, *Science* **198:**603-605.

Whitehead, D. C., H. Dibb, and R. D. Hartley, 1981, Extractant pH and the release of phenolic compounds from soils, plant roots and leaf litter, *Soil Biol. Biochem.* **13:**343-348.

Wilkinson, H. T., R. D. Miller, and R. L. Millar, 1981, Infiltration of fungal and bacterial propagules into soil, *Soil. Sci. Soc. Amer. J.* **45:**1034-1039.

Yavitt, J. B., and T. J. Fahey, 1982, Loss of mass and nutrient changes of decaying woody roots in lodgepole pine forests, southeastern Wyoming, *Can. J. Forest. Res.* **12:**745-750.

Zinke, P. J., 1962, The pattern of influence of individual forest trees on soil properties, *Ecology* **43:**130-133.

42

The Nature of Precipitation, Soil, and Surface-Water Chemistry in a Subalpine Ecosystem

Jill Baron and P. Mark Walthall

Colorado State University

INTRODUCTION

The nature and similarities of precipitation, soil solutions, and surface waters were examined in 1982-1983 to determine the processes that control the chemical composition of a stream drainage in Rocky Mountain National Park, Colorado. This is a study of the long-term capability of the watershed to neutralize incoming acidic deposition.

Loch Vale is a calibrated watershed located east of the Continental Divide. The 860-ha study area begins at approximately 4000 m at the Continental Divide and terminates at the outlet of the lowest of three lakes at 3100 m. Sky Pond, Glass Lake, and the Loch (Fig. 42-1) form a string of cirque lakes linked by a perennial stream, Icy Creek. The geology of the northeast-facing drainage is composed of metasediment of biotite gneiss and schist with intrusions of Silver Plume granite (Cole, 1977). Alpine glaciation is responsible for the present geomorphology: U-shaped valleys, cirque lakes, and morainal deposits. Vegetation in the drainage ranges from alpine tundra and fell field to subalpine spruce-fir forests and wet-sedge meadows.

Average precipitation is 75 cm/yr, with approximately 50% occurring as snow (Marr, 1967). While freezing and thawing may occur during any of the winter months, snowmelt generally begins in April or May, causing the hydrograph to rise until June or July. A permanent snowfield at the top of the watershed supplies meltwater to Icy Creek throughout the summer months.

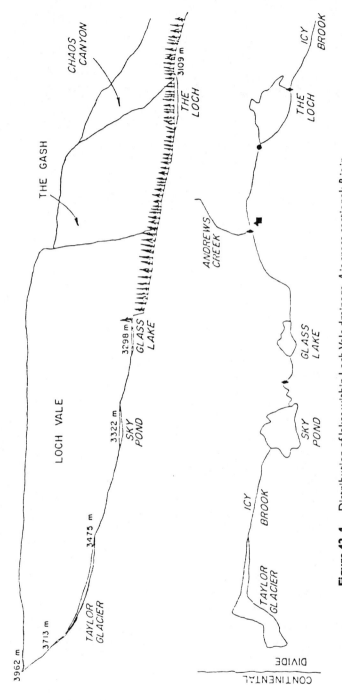

Figure 42-1. Distribution of lakes within Loch Vale drainage. A is cross-sectional; B is in plan view.

498

MATERIALS AND METHODS

Surface Waters

Surface samples (depth of 0.3 m) were collected monthly from the center of each lake in the Loch Vale drainage—Loch Lake, Glass Lake, and Sky Pond—for the period May 1982-May 1983. Samples were collected in acid-washed 250-ml polyethylene bottles and preserved according to the analyses conducted. Conductance and pH values (Fisher conductivity bridge, Corning pH meter, and probe model 401) were obtained immediately during the summer months and within two hours during the winter months. Dissolved silica and the cations calcium [Ca(II)], magnesium [Mg(II)], potassium [K(I)], and sodium [Na(I)] were analyzed with atomic absorption spectroscopy. All analyses were conducted using standard methods (APHA, 1981). Ten percent of the samples analyzed were replicates or blanks as a quality-assurance measure. There was good agreement between samples and replicates. Alkalinities were periodically determined with a Gran titration (Gran, 1952) to an endpoint pH of 3.3.

Precipitation

Precipitation chemistry was obtained from the National Atmospheric Deposition Program (NADP) collector located at Rocky Mountain National Park headquarters. Samples were collected weekly according to standard NADP protocol (NADP, 1982). Values reported here are volume-weighted monthly averages.

Soils

Major landforms and soils regimes occurring in the Loch Vale watershed were characterized and their areal distribution determined. Soil regimes expected to have a potential influence on surface-water chemistry were identified. Typical pedons of these soils were selected and described according to the profile nomenclature of the National Cooperative Soil Survey (Soil Survey Staff, 1981) and classified in accordance with Soil Taxonomy (Soil Survey Staff, 1975). Bulk samples of genetic horizons were collected for laboratory analyses.

The following chemical parameters were determined to characterize the existing soil environment with respect to its potential buffering capacity. Soil pH determinations were made on 1:1, soil:water suspensions. Base saturation (BS), cation exchange capacities (CEC), and exchange acidity (EA) were determined by the sum-of-cations method recommended by Chapman (1965) for acid soils. Organic-matter content was determined by the wet-oxidation method described by Prince (1955). Water-extractable aluminum was determined by atomic absorbtion spectroscopy. Mineralogical analysis of coarse (2 to 0.2 μm) and fine (< 0.2 μm) clay fractions was determined by X-ray diffraction spectroscopy. Particle size analysis was by the pipette method described by Day (1965).

RESULTS

Characterization of Lakes

The lakes of Loch Vale comprise 1% of the total area of the drainage. Sky Pond and Glass Lake are cirque lakes; they have a maximum depth of 7.1 m and 4.9 m respectively. The Loch is also a glacial basin, but the bottom is primarily shallow (1.5 m) and flat. There is a deeper channel extending from the inlet to the outlet with a depth of about 5 m. The lakes remained frozen from December 1982 until May 1983, with a maximum ice thickness of 80-90 cm. The lakes did not stratify during the summer months.

Surface-Water Chemistry

These lakes are characterized as being extremely dilute, oligotrophic water bodies with conductivities ranging between 9 and 25 μmhos/cm (Table 42-1). Average pH values for the Loch, Glass Lake, and Sky Pond are 6.2, 6.5, and 6.6, respectively. A yearly cycle of pH was observed, with lower values occurring in late-summer months, September and October, and the highest values for all lakes occurring immediately before and during ice breakup in late winter, April and May. A similar cycle was observed in yearly cation concentrations (Fig. 42-2). All three lakes exhibited a decline in Ca(II) plus Mg(II) from June through August with values rising through the fall. The highest cation values were found in December and April, both during the period of ice cover. While the months June, July, and August exhibited an inverse relationship of concentration with elevation, that pattern is not consistent throughout the year. Of the cations present in the waters, Ca(II) was always most abundant, Mg(II) and Na(I) were less abundant, and K(I), a major nutrient, was least common. Dissolved silica was seen to be comparable in concentration to Ca(II). Alkalinities ranged between 20 and 120 μeq/l among the lakes.

Precipitation

Precipitation solutions were also rather dilute, with conductivities ranging between 6.0 and 40.4 μmhos/cm (Table 42-1). The average pH of 5.13 matches closely the long-term NADP average for this site (NADP, 1983) of 5.10 and also agrees with results obtained for this area and time period by Lewis, Grant and Saunders (1983). Their reported average pH was 5.0. Of the four cations measured, Ca(II) was found in the highest concentration,

Table 42-1. Inorganic Constituents of Surface Samples of Three Lakes and Precipitation in Loch Vale, May 1982–May 1983

Month	pH	Conductivity (μmhos/cm)	Ca(II)	Mg(II)	Na(I) (mg/l)	K(I)	SiO₂
			Loch				
M	5.89	—	1.20	.34	.60	.26	—
J	—	—	1.50	.34	.54	.25	—
J	6.46	11.8	1.19	.22	.40	.11	—
A	6.37	10.7	.70	.17	.33	.09	—
S	6.32	10.1	.83	.19	.35	.14	—
O	5.78	18.5	1.17	.25	.48	.16	—
N	5.45	16.7	1.64	.24	.49	.13	2.55
D	6.20	18.0	2.20	.36	.90	.24	3.70
J	—	—	—	—	—	—	—
F	6.00	20.0	2.30	.27	1.00	.24	2.20
M	—	—	—	—	—	—	—
A	6.90	24.0	3.00	.50	1.40	.36	3.10
M	6.40	17.0	2.20	.33	1.00	.33	4.00
			Glass				
J	6.93	16.2	1.50	.30	.53	.26	—
J	6.73	11.1	.99	.18	.32	.11	—
A	6.81	9.4	.70	.13	.25	.08	—
S	6.13	15.2	.75	.16	.29	.12	—
O	6.25	14.5	1.47	.22	.44	.16	—
N	6.57	18.9	1.82	.26	.48	.18	2.55
D	6.10	21.0	2.70	.39	.90	.25	3.20
J	—	—	—	—	—	—	—
F	6.60	15.0	2.10	.27	.70	.20	1.10
M	—	—	—	—	—	—	—
A	6.60	18.0	2.20	.30	.70	.20	1.00
			Sky				
J	6.96	13.4	1.96	.34	.48	.21	—
J	6.71	9.3	.92	.15	.21	.08	—
A	6.70	1.7	.59	.11	.23	.06	—
S	6.62	—	.72	.13	.29	.11	—
O	6.52	—	1.27	.19	.39	.13	—
N	6.32	17.0	1.75	.22	.61	.32	2.33
D	6.10	17.0	2.10	.33	.70	.24	2.80
J	—	—	—	—	—	—	—
F	6.20	15.0	1.90	.16	.60	.22	1.50
M	—	—	—	—	—	—	—
A	6.80	14.0	1.80	.20	.60	.17	0.40
			Precipitation				
M	5.26	11.9	.41	.25	.08	.34	—
J	4.57	17.4	.88	.15	.21	.10	—
J	4.36	26.3	.51	.11	.11	.33	—
A	5.30	25.3	.32	.22	.81	1.00	—
S	4.67	12.3	.19	.03	.05	.03	—
O	—	—	—	—	—	—	—
N	5.37	40.4	1.16	.28	.76	.20	—
D	5.13	6.85	.18	.07	.47	.31	—
J	6.41	20.9	1.02	.54	1.56	.16	—

Mg(II) and Na(I) in similar, but lower, concentrations, and K(I) the lowest. Silica was not measured. Seven months of Ca(II) plus Mg(II) concentrations are shown in Figure 42-2. For the most part, the values can be seen to closely resemble the monthly concentrations reported for lake surface waters.

Distribution of Soils

Six major landforms and soil regimes were identified within the watershed. The areal and proportionate extent of these units and surface water are given in Table 42-2. The following discussion characterizes each of these units.

Forest Soils. This soil regime comprises 5% of the watershed. It is characteristic of the soil environment occurring below treeline in a forest of spruce and fir. These are the most highly weathered soils occurring in the study area. Drainage from these soils is expected to have an effect upon the surface-water chemistry of the Loch. Sky Pond and Glass Lake both occur above treeline and would not be affected by drainage from these soils. Two major soil types were recognized within the forest soil regime. The most dominant, a Cryoboralf, is characteristic of the forest floor. Occurring to a lesser extent in this unit is a Cryumbrept that generally occurs where talus slopes merge into glacial till deposits.

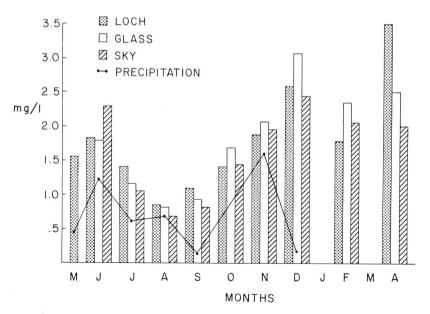

Figure 42-2. Ca(II) plus Mg(II) concentrations for lakes and precipitation.

Table 42-2. Approximate Areal and Proportionate Extent of Soil Regimes

Soil Regime	Hectares	Percent
Forest soils	45	5
Alluvial soils	4	0.5
Rock outcrop-organic soils	15	1.5
Rock outcrop	465	54.0
Rock slides	230	27
Alpine ridge	91	11
Surface water	10	1.0
Total	860	

Alluvial Soils. This soil regime comprises only 0.5% of the watershed but, because of its position adjacent to stream channels, it is expected to have a high potential for influencing the composition of soil drainage entering the surface-water environment. Two major soil types were found to dominate. Organic soil, Cryohemists, occur in open areas where relief is minimal and conditions are optimal for bog development. Where stream channels flow through forested areas with thick canopy covers, bog development is minimized and soils more characteristic of a depositional environment, Cryofluvents, occur.

Rock Outcrop and Cryohemists. This regime comprises 1.5% of the watershed and occurs primarily below Sky Pond and surrounding Glass Lake. It consists of glacially scoured bedrock with approximately 20% of the area covered by thin organic soils ranging from 5 to 40 cm in thickness. Wet conditions necessary for the development of the organic soils come primarily from poorly channelized lake-discharge zones and slow-melting snowpacks from the surrounding area.

Rock Outcrop. This landform comprises the largest area in the watershed (54%), and it is expected to have the least influence on surface-water chemistry, as it consists of more than 90% barren rock. It includes masses of bedrock that form the mountain peaks, the nearly vertical walls of cirques, and the steep canyon walls. These areas are essentially devoid of vegetation except for lichens and mosses on rocks that have been exposed for some time.

Rock Slides. This landform covers 27% of the study area and consists of talus areas of coarse gravel and angular rock fragments at the bases of steep slopes. The rock fragments range from 10 cm to over 2 m in diameter. These slopes have the form of fans or aprons at the bases of steep rock escarpments. Most areas are relatively unstable, with gradients of 65% or greater. Rock slides have little vegetation except where material weathered from rocks has accumulated in sufficient quantities to support grasses, forbs, and trees.

Alpine Ridge.   This soil regime comprises approximately 11% of the watershed. It consists of two inaccessible ridgetops that are situated approximately 600 m above the valley floor. Although these areas have not been characterized, similar soils from more accessible areas are being investigated.

Chemical Characterization of the Soil Environment

Chemical data characterizing the soil environment is given in Table 42-3 for typical pedons of the major soil types occurring in the forest and alluvial regimes. Extremely acid soil pH values were found to occur in all of the pedons, ranging from a maximum of 4.8 to a minimum of 3.7. Such extreme acidity would indicate that some source other than carbonic acid (H_2CO_3) dominates the soil system with respect to pH. Organic acids originating in surface litter and preserved by the cool temperatures of this alpine environment are most likely responsible for this extreme acidity (Ugolini et al., 1977). Two distinct patterns exist within the soil profiles of the forest and alluvial soil regimes. In both of the forest soils, the Cryoboralf and the Cryumbrept, pH values are highest in the organic layers and decrease with depth. In the alluvial soils, the Cryohemist and the Cryofluvent, the opposite trend occurs, with a slight increase in pH with depth.

Cation exchange capacities (CECs) are relatively high and originate from two primary sources: (1) the organic component of these soils, and (2) the mineralogy of the clay fraction that was found to be dominated by smectite. The exchange complex in organic matter is highly pH dependent and, for the humic fraction, is expected to have an exchange capacity of approximately 90 meq/100 g at a pH of 4.0 (Brady, 1974). It is important to point out that this value is for the humic fraction and not for the entire organic mass as reflected in CEC values found in the organic surface layers that have an average value of approximately 50 meq/100 g. The organic surface of the Cryoboralf (pH 4.8) was noted to be more highly decomposed than the other surface layers and was found to have a value of 95.5 meq/100 g.

The clay mineral group of smectites has a characteristic CEC value of 100 meq/100 g that is largely independent of pH. X-ray analysis and CEC determinations of the coarse- and fine-clay fractions indicate a dominance of this material. This mineral is believed to be a weathering product of biotite and chlorite commonly occurring in the parent rock of this environment. Smectites are not generally considered to be stable weathering products in acidic environments. However, their occurrence in such environments has been noted when accompanied by high silica levels (Borchardt, 1979). The

Table 42-3. Chemical Properties of Soils

Horizon	Depth (cm)	pH	Ca(II) (meq/100g)	Mg(II) (meq/100g)	K(I) (meq/100g)	Na(I) (meq/100g)	Exchange Acidity (meq/100g)	Cation Exchange Capacity (meq/100g)	Base Saturation (%)	Organic Matter (%)
				Exchangeable Bases						
				Cryoboralf						
Oe	9–0	4.8	53.1	4.3	2.3	0.2	35.7	95.5	62.6	—a
E	0–19	3.8	6.7	0.6	0.3	0.1	12.3	20.0	38.7	2.7
B'	19–32	3.7	5.2	0.4	0.3	0.1	22.8	28.9	21.0	3.1
BC	32–56	3.7	1.7	0.2	0.3	0.1	14.3	16.5	13.7	1.6
R	56+									
				Cryumbrept						
Oi	1–0	4.8	20.1	2.7	1.3	0.1	26.7	50.9	47.6	11.9
A	0–15	4.3	7.2	0.9	0.4	0.1	28.4	36.9	23.2	2.1
2Bw1	15–46	4.2	1.9	0.2	0.1	0.1	13.7	16.0	14.3	1.3
2Bw2	46–74	4.1	1.1	0.1	0.1	0.1	12.9	14.3	9.9	
2R	74+									
				Cryohemist						
Oi	0–4	3.8	8.0	1.2	1.2	0.6	38.5	49.4	22.1	—
Oe1	4–15	4.0	2.9	0.2	0.3	0.3	38.1	41.8	8.8	—
Oe2	15–25	4.1	4.3	0.2	0.2	0.2	34.9	39.8	12.3	—
Oa	25–33	4.3	5.6	0.2	0.1	0.2	35.1	41.3	14.9	—
Cg	33–43	4.4	5.7	0.4	0.1	0.2	19.0	25.4	25.2	7.0
R	43+									
				Cryofluvent						
Oi	9–0	3.8	10.0	2.5	1.9	0.4	34.1	48.9	30.2	—
A	0–10	3.8	2.4	0.2	0.2	0.2	34.9	37.9	7.9	11.5
Cal	10–24	4.0	2.5	0.2	0.1	0.2	26.4	29.4	10.1	6.8
Cg2	24–30	4.0	4.0	0.5	0.1	0.3	25.1	30.0	16.4	12.7
Cg3	30–37	4.2	5.0	1.0	0.1	0.3	28.9	35.4	18.3	26.9
Ob	37–47	4.4	7.5	1.3	0.1	0.4	33.1	42.4	21.9	—
R	47+									

a —: Organic matter percentages were not determined for organic horizons.

highest clay percentages were found in the sediments of the alluvial soils, ranging from 19.7% to 24.6%. Clay distribution in the Cryoboralfs was highest in the argillic horizon at 15.7%, with 9.1% and 5.6% occurring in the E and BC horizons, respectively. The Cryumbrept had a maximum of 13.4% clay in the A horizon, dropping to 4.2% and 4.4% in the 2Bw1 and 2Bw2 horizons, respectively.

Base saturation percentages were extremely low for the soil environment as a whole, with maximums occurring in the organic surface layers ranging from 22.1% to 62.6%. Extremely low values, ranging from 7.9% to 38.7%, were found in underlying horizons. The sum of the exchangeable bases Ca(II), Mg(II), K(I), and Na(I), plus the exchangeable acidity, equals the exchange capacity of the soil. Calcium was found to be the dominant basic cation, followed by lesser amounts of Mg(II), Na(I), and K(I). Concentration of these cations in surface layers is apparently a result of nutrient cycling. This process seems quite effective in recovering K(I) from the soil solution, as evidenced by its significantly higher concentrations in the organic surface layers compared to Na(I). In underlying horizons these two cations occur at similarly low levels. Total aluminum values determined on saturated-paste extracts ranged from 6.5 mg/l to 1.4 mg/l. Aluminum [Al(III)] activities speciated from total concentrations were found to be extremely high, approaching a level that would be supported by amorphous aluminum hydroxide [Al(OH)$_3$]. These activities ranged from $10^{-4.46}$M to $10^{-5.36}$M.

DISCUSSION

The minimal influence of soils and the close comparison between surface waters and precipitation suggest the lakes of Loch Vale are extremely vulnerable to changes in precipitation chemistry.

The slightly acidic surface waters and precipitation found in Loch Vale indicate that they are presently dominated by H_2CO_3. In contrast, the extremely acidic soil environment suggests a dominance of organic acids characteristic of coniferous forests and tundra environments. Considering that the effective soil environment comprises approximately 7% of the watershed, it appears that the acidifying effect of soil drainage into surface waters is being minimized by dilution from precipitation bypassing the soil environment.

The relative basic cation concentrations between soils and surface waters are similar. This suggests a possible linkage of the two environments. In both soils and the lakes, the most abundant cation is Ca(II), followed by Mg(II), Na(I), and K(I). If this relationship is shown to be more than coincidence, however, one would expect increasing concentrations of cations in surface waters with increasing influence of soils. The chemical compositions of Sky Pond and Glass Lake are expected to have little linkage with the soil environment as they are surrounded primarily by rock slides and rock outcrop. The Loch, on the other hand, can be expected to receive significant inputs of soil drainage with respect to its location in the watershed and the surrounding soil environment. The soil-surface water relationship, then, can be expected to increase with decreasing elevation. This was not found in the surface-water cation data, with the exception of the months of July, August, and September. These are the months that correspond to the falling limb of the stream hydrograph and possibly to a greater relative contribution of soil waters. This is supported by seasonal increases in silica levels in the Loch compared to Glass Lake and Sky Pond (Table 42-1). These data suggest that during periods of minimal input from precipitation and snowmelt, soil drainage may become significant in affecting the chemistry of the Loch.

Average pH values for the three lakes did show a slight relationship between elevation and acidity that might reflect increasing soil influence with lower elevations. That this gradient of pH with elevation is not consistent probably reflects different sampling days for lake chemistry and the influence of storm events.

A close comparison can be made between cation concentrations in precipitation and in surface waters (Fig. 42-2). The relationship occurs during all seasons, although the greater concentrations found in surface waters point toward some input from the soil environment or weathering of parent material. This data suggests that the surface-water chemistry of Loch Vale is presently dominated by precipitation, with some minimal, perhaps seasonal, input from the soil environment. Similarities between cation concentrations in the lakes and precipitation, as well as the low areal extent of the soil component, support this conclusion. Comparison of pH values between precipitation, surface waters, and soils also shows greater similarity between the more neutral values of precipitation and surface waters. The very acidic soil environment bears no resemblance to surface waters; in fact, its high aluminum values would be extremely toxic to aquatic organisms if soil were allowed into the lake environment. In the event of acid precipitation, the soil environment will have little, if any, neutralizing effect, but that environment could possibly serve as a toxic source of aluminum, depending on the volume of soil drainage relative to the volume of precipitation bypassing the soil environment.

ACKNOWLEDGEMENT

We would like to acknowledge the assistance of David R. Beeson and Katherine P. Bricker in the preparation of this manuscript.

REFERENCES

American Public Health Association (APHA) 1981, *Standard Methods for Examination of Water and Wastewater,* 15th ed., R. R. Donelley and Sons, Co., Springfield, Virginia.

Borchardt, G. A., 1979, Montmorillonite and other smectite minerals, in *Minerals in Soil Environments,* J. B. Dixon and S. B. Weed, eds., Soil Science Society of America, Madison, Wisconsin, pp. 293-325.

Brady, N. C., 1974, *The Nature and Properties of Soils,* MacMillan Publishing Co., Inc., New York.

Chapman, H. D., 1965, Cation-exchange capacity, in *Methods of Soil Analysis,* Part 2, C. A. Black, ed., American Society of Agronomy, Madison, Wisconsin, pp. 891-900.

Cole, J. C., 1977, *Geology of East-central Rocky Mountain National Park and Vicinity, with Emphasis on the Implacement of the Precambrian Silver Plume Granite in the Longs Peak-St. Vrain Batholith,* Ph.D. dissertation, University of Colorado, Boulder.

Day, P. R., 1965, Particle fractionation and particle-size distribution, in *Methods of Soil Analysis,* Part 1, C. A. Black, ed., American Society of Agronomy, Madison, Wisconsin, pp. 545-566.

Gran, G., 1952, Determination of the equivalence point in potentiometric titrations. Part II, *International Congress on Analytical Chemistry* **77:**661-671.

Lewis, W. M., M. C. Grant, and J. F. Saunders, III, 1983. *A precipitation chemistry network study of the State of Colorado, 1982-1983,* paper presented at Conference on Acid Rain in the Rocky Mountain West, Golden, Colorado.

Marr, J. W., 1967, *Ecosystems of the East Slope of the Front Range in Colorado,* University of Colorado Studies Series in Biology #8, University of Colorado Press, Boulder.

National Atmospheric Deposition Program, 1982, in *NADP Instruction Manual: Site Operation,* D. S. Bigelow, ed., Natural Resource Ecology Laboratory, Colorado State University, Fort Collins.

National Atmospheric Deposition Program, 1983, *NADP Report: Precipitation Chemistry; First Quarter 1981,* Natural Resource Ecology Laboratory, Colorado State University, Fort Collins.

Prince, A. L., 1955, Methods in soils analysis, in *Chemistry of the Soil,* F. E. Bear, ed., Reinhold Publishing Co., New York, pp. 328-362.

Soil Survey Staff, 1975, *Soil Taxonomy,* U.S. Department of Agriculture Handbook No. 436, U.S. Govt. Printing Office, Washington, D.C.

Soil Survey Staff, 1981, Procedure guide: Application of survey information, in *Soil Survey Manual,* U.S. Department of Agriculture, U.S. Govt. Printing Office, Washington, D.C.

Ugolini, F. C., R. Minden, H. Dawson, and J. Zachara, 1977, An example of soil processes in the *Abies amabilis* zone of Central Cascades, Washington, *Soil Sci.* **124:**291-302.

43

Acidic Deposition Influences on Biogeochemistry of Four Forest Ecosystems in Northwestern Wisconsin

Edward A. Jepsen and James G. Bockheim

University of Wisconsin

ABSTRACT: Hydrogen (H^+), ammonium (NH_4^+), calcium (Ca^{2+}), magnesium (Mg^{2+}), potassium (K^+), sodium (Na^+), sulfate (SO_4^{2-}), nitrate (NO_3^-), and bicarbonate (HCO_3^-) were measured in bulk precipitation, throughfall, stemflow, and in soil solutions at 75 mm and 600 mm from July 1981 to August 1982 in four forest ecosystems in northwestern Wisconsin: *Pinus banksiana, Quercus ellipsoidalis-Q. macrocarpa, Betula papyrifera,* and *Populus tremuloides-P. grandidentata.* The dominant cation and anion in precipitation were hydrogen ion and sulfate, respectively. Whereas hydrogen, ammonium, and nitrate were attenuated by the canopies of all forest types, calcium, mangesium, potassium, sulfate, and bicarbonate were released by the canopies. Significantly greater amounts of hydrogen ion were retained by the aspen canopy than by the other canopies during the growing season (ca. April 21-October 31). During the dormant season significantly greater amounts of magnesium and potassium were released by the jack pine canopy than by the deciduous canopies. Despite the differences in throughfall + stemflow loadings beneath the various canopies, there were not significant differences in loadings of ions in the soil water solutions. From 88% to 93% of the leaching loss volume at 600 mm occurred during the dormant season (ca. November 1-April 20), primarily as a result of spring snowmelt. Whereas hydrogen, calcium, ammonium, and nitrate were conserved by the ecosystems, there were net losses in sodium, magnesium, and bicarbonate. The four ecosystems were able to neutralize acidic deposition during the study period.

INTRODUCTION

There is widespread concern that acidic deposition may reduce forest productivity (Dochinger and Seliga, 1976; Hutchinson and Havas, 1980;

Overrein, Seip, and Tollan, 1980). There are no long-term studies documenting a decline in forest site quality as a result of sustained acidic deposition. Evidence for detrimental effects of acidic deposition on forest soils is based on (1) studies of nutrient fluxes at varying distances from point-source emission centers (e.g., Baker, Hocking, and Nyborg, 1977); (2) studies along precipitation pH gradients (e.g., Hanson, Norton, and Williams, 1982); and (3) laboratory studies dealing with simulated acid rain (e.g., Lee and Weber, 1982).

The concern over acidic deposition influences on forest productivity has stimulated biogeochemical studies in various forest ecosystems (e.g., Nihlgard, 1970; Heinrichs and Mayer, 1977; Likens et al., 1977; Mollitor and Raynal, 1982; Richter, Johnson, and Todd, 1983). These studies have examined the interaction between acidic deposition, the forest canopy, and the mineral soil.

Johnson and Cole (1980) used anion mobility as an index in determining the sensitivity of forest ecosystems to acidic deposition. Forest ecosystems where nitrate and sulfate are mobile, such as the Hubbard Brook ecosystem, were considered to be most sensitive to acidification. However, Krug and Frink (1983) criticized the anion mobility model and proposed that atmospheric additions of sulfuric acid would increase the flux of sulfate but would decrease the flux of organic anions from humic materials with little or no measurable change in pH of the soil solution.

The following study was initiated to evaluate the sensitivity to acidic deposition of four forest types growing on the same soil in northwestern Wisconsin. Specific objectives of the study are as follows: (1) to characterize bulk precipitation; (2) to examine seasonal changes in throughfall and stemflow chemistry under deciduous and coniferous vegetation; and (3) to determine seasonal changes in soil water chemistry beneath the various forest types.

STUDY SITE

The experimental site is located in the 235-ha East Eightmile Lake watershed in northwestern Wisconsin (46° 25' N, 91° 27' W). The watershed consists of gently rolling landscapes derived from deep (200+ m), pitted, glacial outwash sands of late Wisconsin age (ca. 10,000 yr B.P.). The predominant soil is the Omega loamy sand (sandy, mixed, frigid, typic Udipsamment).

The climate is continental but is modified by the influence of Lake Superior and local topography. Solon Springs, 29 km southwest of the study area, has a mean annual temperature of 5.2°C. January is the coldest month (−11.9°C), and July is the warmest (20.2°C). Mean annual precipitation is 836 mm.

Dominant forest communities and their relative abundance are jack pine (*Pinus banksiana* Lamb.), 51%; northern pin oak (*Quercus ellipsoidalis* E. J. Hill)-bur oak (*Q. macrocarpa* Michx.), 20%; trembling aspen (*Populus*

Table 43-1. Stand Characteristics of the Forest Types at East Eightmile Lake Watershed

Stand Characteristics	Jack Pine E1	E4	Mixed Oak E5	Paper Birch E2	Aspen E3
Stand age (yr)	54	59	29	68	20
Basal area (m^2/ha)	31.2	23.6	5.7	19.8	13.1
Density (stems/ha)	1450	3475	4925	1825	4725
Site index (m at 50 yr)	18.8	14.5	8.8	14.6	24.8
Biomass (t/ha)	138	84	13	117	61
Maximum leaf area index (m^{2}m^{-2})	10.2	9.3	0.4	1.1	1.1

tremuloides Michx.)-bigtooth aspen (*P. grandidentata* Michx.), 11%; red pine (*Pinus resinosa* Ait.), 9%; and paper birch (*Betula papyrifera* Marsh.), 6%. Stand characteristics for the various forest types are given in Table 43-1.

MATERIALS AND METHODS

After a comprehensive vegetation survey, five 400-m^2 circular plots were located in the major forest types of the watershed. Two plots were located in jack pine (E1, E4) and one each in paper birch (E2), aspen (E3), and mixed oak (E5). Three each of four types of collectors—bulk throughfall, stemflow, shallow soil water (75 mm depth), and deep soil water (600 mm depth) —were randomly located on each of the plots. Bulk precipitation was collected in three areas free of overtopping vegetation within the watershed.

Bulk precipitation during the growing season (April 21-October 31) was collected with polyethylene funnels and during the dormant season with polyethylene buckets (Likens et al., 1977). Throughfall collectors were the integrating-trough type (Leyton, Reynolds, and Thompson, 1968) over the growing season. During the dormant season polyethylene buckets were used to collect throughfall under jack pine (E1), paper birch (E2), and aspen (E3). Stemflow was sampled from three trees representative of the diameter distribution and species composition of the plot. Stemflow collectors consisted of plastic garage door gaskets wound in a spiral fashion around the tree at a height of 1.4 m (Foster and Morrison, 1976). Soil water from both depths was collected with porous ceramic cup lysimeters that were evacuated to a suction of 50 kPa. All solutions, except bulk precipitation, were composited by collector for each plot, yielding one sample per collector type per visit. Samples were collected during 15 visits at 14- to 45-day intervals over the study period (July 25, 1981, to August 18, 1982).

Drainage at the 600-mm depth was estimated using the water balance equation and a modification of the Thornthwaite and Mather (1957) method for calculating potential evapotranspiration. The Thornthwaite and Mather (1957) method was modified so as to consider interception by the forest canopy, understory vegetation, and the forest floor (Rowe, 1955; Helvey and Patric, 1965; Verry, 1976). Dormant season, deep lysimeter, ionic loadings were estimated by averaging the last fall and first spring visits' concentrations and then multiplying by snowmelt deep-percolate volume.

Hydrogen ion concentration was measured with a digitized pH meter. Precipitation, throughfall, and stemflow samples were filtered to remove particulate matter. Solutions were stored in the dark at 4°C until additional analyses could be made. Nitrate and ammonium were detected on an autoanalyzer; calcium and magnesium were detected on an atomic absorption spectrophotometer; sodium and potassium were detected with a flame photometer; and sulfate was measured on a nephelometer. All methods are given in *Standard Methods for the Examination of Water and Wastewater* (APHA et al., 1975). Bicarbonate was estimated from pH measurements and equations by Reuss (1975).

Differences in loadings in throughfall + stemflow and precipitation and interspecies differences for a given collector type were tested using Duncan's Multiple Range Test and the Statistical Analysis System (SAS Institute, 1979).

RESULTS

Bulk Precipitation

Bulk precipitation had an average pH of 4.6. The dominant anions in bulk precipitation were sulfate and nitrate. Sulfate contributed 60% of the total inorganic negative charges over the growing season (April 21–October 31) (Table 43-2) and 52% during the dormant season (November 1–April 20) (Table 43-3). The annual loading of cations (mol/ha) was hydrogen > ammonium > calcium > potassium > sodium > magnesium.

The cation:anion ratios for bulk precipitation were 1.7:1 for the growing season (Table 43-2) and 1.2:1 during the dormant period (Table 43-3).

Throughfall + Stemflow Chemistry

Growing Season.   The dominant anions in throughfall + stemflow during the growing season were sulfate and nitrate for all species (Table 43-2). Whereas nitrate was attenuated by the canopies of all forest types, sulfate was released in throughfall + stemflow of all species. Bicarbonate was present in small quantities and was greatest in throughfall + stemflow of aspen. The order of loading of cations (mol/ha), excluding hydrogen, in throughfall + stemflow was potassium > calcium > ammonium > magne-

Table 43-2. **Loadings of Ions in Bulk Precipitation and Throughfall + Stemflow Beneath the Various Forest Types during the Growing Season (in mol/ha/yr) ***

Ion	Bulk Precipitation	Throughfall + Stemflow				
		Jack Pine		Mixed Oak	Paper Birch	Aspen
		E1	E4	E5	E2	E3
SO_4^{2-}	123	160	137	132	110	171
NO_3^-	134	85	107	108	85	97
HCO_3^-	2 b	2 b	3 b	7 b	9 b	30 a
Ca^{2+}	73 c	118 b	104 b	124 b	123 b	232 a
Mg^{2+}	17 c	53 b	49 b	27 bc	53 b	91 a
K^+	19 b	181 a	219 a	128 a	207 a	222 a
Na^+	15	18	16	12	13	18
NH_4^+	150	67	77	73	70	90
H^+	182 ab	107 ab	82 ac	64 c	66 c	28 c
Σcations: Σanions	1.8	2.2	2.3	1.7	2.6	2.2
mm H_2O	632	527	428	641	540	596

*Lower-case letters represent the results of the Duncan Multiple Range Test among plots. Loadings for a given ion with the same letter are not statistically different ($\alpha = 0.05$). Ion loadings that showed no significant differences are not lettered.

sium > sodium, except for aspen (calcium > potassium > magnesium > ammonium > sodium). Whereas hydrogen was ranked before ammonium beneath jack pine, hydrogen was ranked after ammonium for all deciduous forest types. Hydrogen and ammonium were attenuated by the canopies of all species, and calcium, potassium, and magnesium were released from all canopies. Calcium loadings were significantly greater under aspen than the other forest types. There were no differences in sodium loadings in bulk precipitation and throughfall + stemflow.

Stemflow contributed less than 15% of the total throughfall + stemflow loadings for all ions in the jack pine and oak forest types. Stemflow provided 10% to 42% of the total throughfall + stemflow loadings of calcium, magnesium, potassium, and bicarbonate in the aspen and paper birch forest types and less than 10% of the loadings of the other ions.

Throughfall volume as a function of forest type, which relates to differences in canopy interception, decreased in the following order: oak > aspen > paper birch > jack pine. Stemflow volumes accounted for less than 10% of net precipitation and decreased in the following order: aspen > paper birch > jack pine > oak.

The cation:anion ratio of throughfall + stemflow during the growing season ranged from 1.7:1 for oak to 2.6:1 for paper birch (Table 43-2).

Dormant Season. The order of cation loading (mol/ha) in throughfall during the dormant season was hydrogen > calcium > ammonium >

Table 43-3. Loadings of Ions in Bulk Precipitation and Throughfall + Stemflow Beneath the Various Forest Types during the Dormant Season (in mol/ha/yr)*

Ion	Bulk Precipitation	Jack Pine E1	Paper Birch E2	Aspen E3
SO_4^{2-}	44	97	43	52
NO_3^-	80	80	89	96
HCO_3^-	<1	<1	<1	<1
Ca^{2+}	20 b	66 a	26 ab	45 ab
Mg^{2+}	5 b	21 a	8 b	11 ab
K^+	5 b	35 a	7 b	12 b
Na^+	11	12	9	10
NH_4^+	51	50	42	45
H^+	58	76	55	55
Σcations: Σanions	1.2	1.5	1.1	1.2
mm H_2O	193	167	186	193

*Lower-case letters represent the results of the Duncan Multiple Range Test among plots. Loadings for a given ion with the same letter are not statistically different ($a = 0.05$). Ion loadings that showed no significant differences are not lettered.

potassium > magnesium > sodium, which differed from the growing-season sequence. Canopy attenuation of hydrogen and enrichment of potassium were reduced during the dormant season (Table 43-3).

The only statistically significant canopy-precipitation interactions during the dormant season were greater calcium, magnesium, and potassium loadings under jack pine than in bulk precipitation (Table 43-3). Potassium was significantly greater in throughfall during the dormant season beneath jack pine than under aspen and paper birch.

Throughfall volumes for aspen and paper birch during the dormant season were similar to precipitation values, while throughfall volume beneath jack pine was slightly lower, reflecting snow interception due to the presence of a winter canopy.

Cation:anion ratios of throughfall during the dormant season ranged between 1.1:1 for paper birch and 1.5:1 for jack pine (Table 43-3). These ratios are similar to those for precipitation during the dormant season but are substantially lower than values for throughfall + stemflow during the growing season.

Shallow Soil Water Chemistry

The predominant anion in the shallow (75 mm) soil water in all forest types during the growing season was sulfate, with relatively small amounts of nitrate and bicarbonate (Table 43-4). Cation loadings (mol/ha) in the shallow soil water were predominately sodium and potassium, followed by magnesium and calcium, with lesser amounts of hydrogen and ammonium. Magnesium,

Table 43–4. Loadings of Ions in the Shallow Soil Water (75 mm) Beneath the Various Forest Types during the Growing Season (in mol/ha/yr) *

Ion	Jack Pine E1	E4	Mixed Oak E5	Paper Birch E2	Aspen E3
SO_4^{2-}	160	150	164	160	198
NO_3^-	0	14	11	1	1
HCO_3^-	9	11	19	8	16
Ca^{2+}	169	109	63	100	121
Mg^{2+}	168	109	136	146	168
K^+	193	288	357	175	280
Na^+	285	161	347	263	316
NH_4^+	4	2	6	7	4
H^+	10	5	10	14	9
Σcations: Σanions	3.5	2.5	3.1	2.9	2.8
mm H_2O	367	291	526	411	458

*Results of the Duncan Multiple Range Test indicated no significant differences in shallow leaching losses among plots.

sodium, calcium, and potassium constituted 98% of the cations in the shallow soil water during the growing season. No significant interspecies differences in ionic fluxes were observed during the growing season.

The volume of shallow soil precolate decreased in the following order: oak > aspen > paper birch > jack pine. Cation:anion ratios for the shallow soil water during the growing season were the highest for any ecosystem component monitored, ranging between 2.5:1 and 3.5:1 (Table 43-4). No estimates were made of shallow soil ionic fluxes during the dormant season.

Deep Soil Water Chemistry

Growing Season. The dominant inorganic anion in deep soil water was sulfate, followed by bicarbonate and nitrate for all plots. The relative abundance of cations in the deep percolates for all plots was sodium > magnesium ≥ calcium ≥ potassium > hydrogen > ammonium (Table 43-5).

There were no significant differences between species in ion loadings of deep soil water during the growing season (Table 43-5). Loadings of all ions were significantly lower in the deep percolates than in the shallow soil water. Losses of hydrogen, ammonium, and nitrate were negligible beyond 600 mm (depth of rooting). Sulfate, calcium, magnesium, sodium, and potassium loadings decreased from 10- to 100-fold from shallow to deep soil water.

The volume of leachate as a function of forest type decreased: oak > aspen > paper birch > jack pine, which is the same sequence observed for shallow soil water. The cation:anion ratio ranged between 2.0:1 for aspen to

Table 43–5. Loadings of Ions in the Deep Soil Water (600 mm) Beneath the Various Forest Types during the Growing (A) and Dormant (B) Seasons (in mol/ha/yr) *

| | Jack Pine | | | | Mixed Oak | | Paper Birch | | Aspen | |
| | E1 | | E4 | | E5 | | E2 | | E3 | |
	A	B	A	B	A	B	A	B	A	B
SO_4^{2-}	6	105	7	107	7	72	13	122	15	124
NO_3^-	<1	4	<1	0	<1	2	<1	0	<1	0
HCO_3^-	1	34	2	34	6	69	4	52	3	43
Ca^{2+}	3	63	4	66	6	43	4	56	6	57
Mg^{2+}	4	52	4	54	8	45	6	64	6	65
K^+	2	23	2	19	6	16	5	23	5	31
Na^+	10	188	9	194	26	275	19	265	18	216
NH_4^+	<1	<1	<1	0	<1	20	<1	4	<1	2
H^+	<1	3	<1	3	<1	<1	<1	<1	<1	4
Σcations:Σanions	2.9	2.3	2.2	2.4	3.6	2.1	2.1	2.4	2.0	2.2
mm H_2O	13	134	11	143	24	178	23	166	25	175

*Results of the Duncan Multiple Range Test indicated no significant differences in leaching losses among plots.

3.6:1 for oak, which is slightly lower than for the shallow soil solution during the growing season.

Dormant Season.   The dominant anions and cations in order of loading during the dormant season follow the same sequence observed during the growing season (Table 43-5). Estimated leaching losses for most ions were considerably greater during the dormant than during the growing season. Potassium, magnesium, calcium, and sodium leaching losses potentially increased 2- to 10-fold, and sulfate and bicarbonate may have increased 5- to 25-fold compared to losses during the growing season.

Deep percolate volumes during the dormant season decreased in the following order: oak > aspen > paper birch > jack pine, which is similar to the sequence noted during the growing season. Cation:anion ratios for the dormant season ranged narrowly between 2.1:1 and 2.4:1.

DISCUSSION

Bulk Precipitation Chemistry

The abundance of sulfate and nitrate in bulk precipitation is characteristic of areas receiving acidic deposition (Likens et al., 1979). The cation:anion imbalance was partially related to the lack of data on chloride and organic anions. Factors that could account for the narrower cation:anion ratio during the dormant season are changes in seasonal weather patterns and reduced biological and agricultural activities during the dormant period (Gorham, 1976).

Canopy Effects

Both the coniferous (jack pine) and deciduous (mixed oak, birch, aspen) canopies attenuated hydrogen, ammonium, and nitrate and released calcium, potassium, magnesium, sulfate, and bicarbonate. Similar trends were reported in deciduous forest types by Nihlgard (1970), Verry and Timmons (1977), Heinrichs and Mayer (1977), Mollitor and Raynal (1982), and Richter, Johnson, and Todd (1983). The attenuation of hydrogen and release of potassium, magnesium, and calcium by forest canopies has been attributed to (1) the exchange of potassium, magnesium, and calcium on and within the leaf for precipitation hydrogen (Tukey, Mecklenburg, and Morgan, 1965; Wood and Bormann, 1975), and (2) the washing of impacted dryfall from leaf surfaces (Schlesinger and Reiners, 1974; Mayer and Ulrich, 1977). An additional mechanism for the attenuation of hydrogen may be the exchange of weak acids for strong acids within the canopy (Hoffman, Lindberg, and Turner, 1980). Coniferous canopies generally release hydrogen (Nihlgard, 1970; Olson et al., 1981; Mollitor and Raynal, 1982); however, in this study jack pine attenuated hydrogen.

Whereas deciduous canopies often attenuate inorganic nitrogen (ammonium, nitrate) and release organic nitrogen (Verry and Timmons, 1977) and coniferous canopies generally release inorganic nitrogen (Nihlgard, 1970; Heinrichs and Mayer, 1977; Mollitor and Raynal, 1982), in this study canopies of both types attenuated inorganic nitrogen. The attenuation of inorganic nitrogen by forest canopies has been attributed to (1) direct foliar absorption (Wittwer and Teubner, 1969); (2) microbial uptake of nitrogen on leaf surfaces (Carroll, 1978); and (3) absorption of nitrogen by epiphytic lichens on boles and branches (Lang, Reiners, and Heier, 1976).

The aspen canopy returned significantly greater amounts of calcium and magnesium to the forest floor in throughfall + stemflow than did the other species. Verry and Timmons (1977) compared throughfall and stemflow loadings beneath black spruce [*Picea mariana* (Mill.) B.S.P.] and trembling aspen in Minnesota and reported considerably greater quantities of calcium and magnesium beneath the aspen.

Soil Effects

There were no significant differences in loadings of ions in the shallow (75 mm) and deep (600 mm) soil percolates beneath the various forest types. Whereas loadings of sulfate, bicarbonate, calcium, magnesium, potassium, and sodium increased substantially as solutions passed through the upper 75 mm of soil, loadings of hydrogen, ammonium, and nitrate declined. These trends can be explained on the basis of ion exchange, mineral weathering, mineralization of litter and soil organic matter, and plant uptake.

Loadings of all of the chemical constituents measured declined as the solution passed from 75 mm to a depth of 600 mm in the soil. These trends likewise can be explained on the basis of ion exchange and plant uptake. The greatest proportion of the losses in the deep soil percolate occurred during the dormant season. From 88% to 93% of the drainage volume occurred during this period, primarily as a result of spring snowmelt. Studies in Norway (Seip, 1980) and the Adirondack Mountains, U.S.A. (Galloway et al., 1980), have shown the importance of snowmelt in seasonal modification of surface water chemistry.

Net Ecosystem Effects

A comparison of precipitation inputs and deep leaching outputs indicates there were net losses of bicarbonate, magnesium, and sodium, net gains in sulfate, nitrate, calcium, ammonium, and hydrogen, and no change in potassium in the four ecosystems. The magnesium and sodium may have been provided by mineral weathering and probably left the system as bicarbonate and sulfate salts. The nitrate and ammonium were retained primarily by the vegetation on leaf surfaces and within organic tissues (Wittwer and Teubner, 1969). Hydrogen was retained on exchange sites in the soil and vegetation. Calcium was retained in the soil and in the perennial tissues of the vegetation following uptake from the soil. Sulfate was retained by oxyhydroxides of iron, aluminum, and manganese within the soil and to a lesser extent by organic tissues following uptake.

The four ecosystems at East Eightmile Lake appear capable of buffering existing acidic deposition. The ecosystems retained both sulfate and nitrate and, according to the anion mobility model of Johnson and Cole (1980), are relatively insensitive to acidic deposition. Cronan et al. (1978) characterized forest floor leachates in various forest ecosystems as indicative of the degree of atmospheric pollution. Forest floor leachates in unpolluted ecosystems were dominated by bicarbonate or organic anions. In contrast, leachates from forest floors receiving acidic deposition are dominated by sulfate. Although sulfate was the dominant inorganic anion in the soil solution at 75 mm, the cations exceeded the anions by a factor of 2.5 to 3.5. If the excess cations were compensated for by organic anions, the organic anions would be dominant in the forest floor and the deep soil leachates. According to the model of Cronan et al. (1978), the ecosystems at East Eightmile Lake are not receiving sufficient atmospheric sulfate to dominate the chemistry of soil leachate.

Krug and Frink (1983) questioned whether acidic deposition has acidified soil and water. They proposed that atmospheric additions of sulfuric acid would increase the flux of sulfate but would decrease the flux of organic anions with little or no measurable change in pH of the soil solution. Studies conducted over the past 20 years at Hubbard Brook Experimental Forest have shown no trends toward increasing acidification of streamwater following sustained acidic deposition (Johnson, 1979).

Although our studies indicate that the canopy and soils were able to neutralize acidic deposition in the short term, it is not known how long the ecosystems will be able to function in this manner. Long-term biogeochemical studies are needed in various forest ecosystems to document effects of sustained acidic deposition on forest productivity, soil nutrient levels, and surface-water and groundwater chemistry.

ACKNOWLEDGEMENTS

This study was funded by the Wisconsin Utilities and the Electric Power Research Institute (RP2174.2) and supported by the College of Agricultural and Life Sciences and the School of Natural Resources, University of Wisconsin-Madison.

REFERENCES

American Public Health Association (APHA), American Water Works Association, and Water Pollution Control Federation, 1975, *Standard Methods for the Examination of Water and Wastewater,* 14th ed., APHA, Washington, D.C.

Baker, J., D. Hocking, and M. Nyborg, 1977, Acidity of open and intercepted precipitation in forests and effect on forest soils in Alberta, Canada, *Water Air Soil Pollut.* **7**:449-460.

Carroll, G. C., 1978, Canopy subsystems of western coniferous forests, National Science Foundation proposal DEB 78-03583.

Cronan, C. S., W. A. Reiners, R. C. Reynolds, Jr., and G. E. Lang, 1978, Forest floor leaching: Contributions from mineral, organic, and carbonic acids in New Hampshire subalpine forests, *Science* **200**:309-311.

Dochinger, L. S., and T. A. Seliga, eds., 1976, *Proceedings of the First International Symposium on Acid Precipitation in Forest Ecosystems,* USDA Forest Service, Northeast Forest Experimental Station, Gen. Tech. Rep. NE-23.

Foster, N. W., and I. K. Morrison, 1976, Distribution and cycling of nutrients in a natural *Pinus banksiana* ecosystem, *Ecology* **57**:110-120.

Galloway, J. N., C. L. Schofield, G. R. Hendrey, N. E. Peters, and A. H. Johannes, 1980, Lake acidification during spring snowmelt: Processes and causes, in *The Integrated Lake-Watershed Acidification Study: Proceedings of the ILWAS Annual Review Conference,* Electric Power Research Institute, Sagamore Lake, New York, pp. 11-1 to 11-21.

Gorham, E., 1976, Acid precipitation and its influences upon aquatic ecosystems—an overview, in *Proceedings of the First International Symposium on Acid Precipitation in Forest Ecosystems,* L. S. Dochinger and T. A. Seliga, eds., USDA, Forest Service, Northeastern Forest Experimental Station, Gen. Tech. Rep. NE-23, pp. 425-458.

Hanson, D. W., S. A. Norton, and J. S. Williams, 1982, Modern and paleolimnological evidence for accelerated leaching and metal accumulation in soils in New England, caused by atmospheric deposition, *Water Air Soil Pollut.* **18**:227-239.

Heinrichs, H., and R. Mayer, 1977, Distribution and cycling of major and trace elements in two central European forest ecosystems, *J. Environ. Qual.* **6**:402-407.

Helvey, J. D., and J. H. Patric, 1965, Canopy and litter interception of rainfall by hardwoods of eastern United States, *Water Resour. Res.* **1**:193-206.

Hoffman, W. A., Jr., S. E. Lindberg, and R. R. Turner, 1980, Precipitation acidity: The role of the forest canopy in acid exchange, *J. Environ. Qual.* **9**:95-100.

Hutchinson, T. C., and M. Havas, eds., 1980, *Effects of Acid Precipitation on Terrestrial Ecosystems,* Plenum Press, New York.

Johnson, D. W., and D. W. Cole, 1980, Anion mobility in soils: Relevance to nutrient transport from forest ecosystems, *Environ. Int.* **3**:79-90.

Johnson, N. M., 1979, Acid rain: Neutralization within the Hubbard Brook ecosystem and regional implications, *Science* **204**:497-499.

Krug, E. C., and C. R. Frink, 1983, Acid rain on acid soil: A new perspective, *Science* **221**:520-525.

Lang, G. E., W. A. Reiners, and R. K. Heier, 1976, Potential alteration of precipitation chemistry by epiphytic lichens, *Oecologia* **25**:229-241.

Lee, J. J., and D. E. Weber, 1982, Effects of sulfuric acid rain on major cation and sulfate concentrations of water percolating through two model hardwood forests, *J. Environ. Qual.* **11**:57-64.

Leyton, L., E. R. C. Reynolds, and F. B. Thompson, 1968, Interception of rainfall by trees and moorland vegetation, in *The Measurement of Environmental Factors in Terrestrial Ecology,* R. M. Wadsworth, ed., Blackwell Scientific Publishing Co., London, pp. 97-108.

Likens, G. E., F. H. Bormann, R. S. Pierce, J. S. Eaton, and N. M. Johnson, 1977, *Biogeochemistry of a Forested Ecosystem,* Springer-Verlag, New York.

Likens, G. E., R. E. Wright, J. N. Galloway, and R. J. Butler, 1979, Acid rain, *Sci. Am.* **241**:42-51.

Mayer, R., and B. Ulrich, 1977, Acidity of precipitation as influenced by the filtering of atmospheric sulfur and nitrogen compounds—Its role in the element balance and effect on soil, *Water Air Soil Pollut.* **7**:409-416.

Mollitor, A. V., and D. J. Raynal, 1982, Acid precipitation and ionic movements in Adirondack forest soils, *Soil Sci. Soc. Am. J.* **46**:137-141.

Nihlgard, B., 1970, Precipitation, its chemical composition and effect on soil water in a beech and a spruce forest in south Sweden, *Oikos* **21**:208-217.

Olson, R. K., W. A. Reiners, C. S. Cronan, and G. E. Lang, 1981, The chemistry and flux of throughfall and stemflow in subalpine balsam fir forests, *Holarctic Ecol.* **4**:291-300.

Overrein, L. N., H. M. Seip., and A. Tollan, 1980, *Acid precipitation—Effects on forest and fish,* Final report of the SNSF Project, 1972-1980, Oslo-As, Norway.

Reuss, J. O., 1975, *Chemical/biological relationships relevant to ecological effects of acid rainfall,* EPA-660/3-75-032, National Environmental Resource Center, U.S. Environmental Protection Agency, Corvallis, Oregon.

Richter, D. D., D. W. Johnson, and D. E. Todd, 1983, Atmospheric sulfur deposition, neutralization, and ion leaching in two deciduous forest ecosystems, *J. Environ. Qual.* **12**:263-270.

Rowe, P. B., 1955, Effects of the forest floor on disposition of rainfall in pine stands, *J. Forestry* **53**:342-348.

Schlesinger, W. H., and W. A. Reiners, 1974, Deposition of water and cations on artificial foliar collectors in fir krummholz of New England mountains, *Ecology* **55**:378-386.

Seip, H. M., 1980, Acid snow—Snowpack chemistry and snowmelt, in *Effects of Acid Precipitation on Terrestrial Ecosystems,* T. C. Hutchinson and M. Havas, eds., Plenum Press, New York, pp. 77-94.

Statistical Analysis System Institute, 1979, *SAS Users' Guide,* SAS, Cary, North Carolina.

Thornthwaite, C. W., and J. R. Mather, 1957, Instructions and tables for computing potential evapotranspiration and the water balance, *Publ. in Climatol.* x(3):185-311.

Tukey, H. B., Jr., R. A. Mecklenburg, and J. V. Morgan, 1965, A mechanism for the leaching of metabolites from foliage, in *Isotopes and Radiation in Soil-Plant Nutrition Studies,* International Atomic Energy Commission, Vienna, pp. 371-385.

Verry, E. S., 1976, *Estimating water yield differences between hardwood and pine forests: An application of net precipitation data,* USDA, Forest Service, North Central Forest Experimental Station, Research Paper NC-128.

Verry, E. S., and D. R. Timmons, 1977, Precipitation nutrients in the open and under two forests in Minnesota, *Can. J. For. Res.* 7:112-119.

Wittwer, S. H., and F. G. Teubner, 1969, Foliar absorption of mineral nutrients, *Ann. Rev. Plant Physiol.* 10:13-32.

Wood, T., and F. H. Bormann, 1975, Increases in foliar leaching caused by acidification of an artificial mist, *Ambio* 4:169-171.

44

The Biogeochemistry of Boron and Its Role in Forest Ecosystems

Bo Wikner

Chalmers University of Technology and University of Göteborg

ABSTRACT: An extensive investigation into the distribution of boron (B) in rocks, soils, water, and forest trees using a modified curcumin method gave the following conclusions:

1. Different parts of Scandinavia receive different amounts of boron from precipitation. The west coast of Sweden annually receives 25–40 g/ha, central Sweden around 12 g/ha, and the northern part of the country less than 2 g/ha. Boron lost by rivers is much larger than the input by precipitation; Uppsala (central Sweden) receives around 12 g/ha annually from precipitation, while at the same time 100 g/ha is leached out from the system.
2. The boron concentration in one-year-old needles of Scotch pine and Norwegian spruce in mature stands in Sweden ranges from 10–40 ppm along the west coast, gradually falling to less than 7 ppm in the north of the country. Levels less than 3 ppm were frequently associated with dieback and other growth disorders typical of boron deficiency. Nitrogen fertilization, liming, or both, cause a drastic decrease in the boron levels in needles, which sometimes leads to boron-deficiency injuries.
3. Forest soils are generally low in boron (total concentration) and extremely poor in water-soluble boron. The main part of the available boron is found in the organic top layer of podzolic soils. During drought periods boron becomes unavailable in the organic layer, and the trees have to compensate with boron from deeper horizons. Consequently, in forestry, boron-deficient areas are found mainly on peatlands or on fine-textured freshwater sediments where the trees are not able to develop deep roots. Boron determination in forest soils is therefore of

limited value for defining areas susceptible to boron deficiency. Needle analysis is preferred.

INTRODUCTION

Deficiencies of boron are recognized for a wider range of crops, soils, and climatic conditions than are deficiencies of any other micronutrients. Boron toxicity is an ever-increasing problem in parts of the world where irrigation is necessary. Salt-affected areas are usually boron toxic. Since the gap between boron-toxic levels and boron-deficient levels is very narrow, problems could easily develop. It is obvious that difficulties in determining boron in biological samples have led to the omission of boron from micronutrient studies. In the forests of Scandinavia and Finland, dieback injuries often develop on trees that have been fertilized with NPK or urea. The cause of these injuries was unknown in the past, but presently they are recognized as boron-deficiency injuries. This is a clear example of antagonism. Although new analytical methods have made it possible to determine boron in a large number of samples, this is not enough to understand the behavior of boron in soils and plants. Other factors in addition to the boron concentration must be taken into consideration. In forestry, such factors are climatic conditions, unbalanced nutrient supply, antagonism, and root depth.

DISCUSSION

Boron in Precipitation

Boric acid and the majority of the borates are, like sodium (Na) and chloride (Cl^-) ions, very soluble in water and pass rapidly through most soils. Thus, the oceans have become the main reservoir for boron, and an accumulation of boron is steadily going on in the marine sediments on the ocean floors. The mean boron concentration in the sea water is about 4.6 ppm according to Bowen (1966). Boron is carried back to land via precipitation and distributed in almost the same manner as is chloride. The ratio of B:Cl is not constant. Since boron is carried in the air more easily than chloride, this ratio increases with the distance from the coast—that is, with decreasing maritime influence. Iodine (I) and boron have similar distribution patterns. The annual deposition of boron is 30 g/ha in Denmark (Jensen, 1962), 24-42 g/ha in Poland (Chojnacki, 1968), 10-170 g/ha in Germany (Riehm, Quellmalz, and Kraus, 1965), 13-30 g/ha in France (Chabannes, 1959), and 12 g/ha in Sweden (Uppsala, a few km north of Stockholm) (Philipson, 1953). Measurements carried out according to Wikner (1981) show that the west coast of

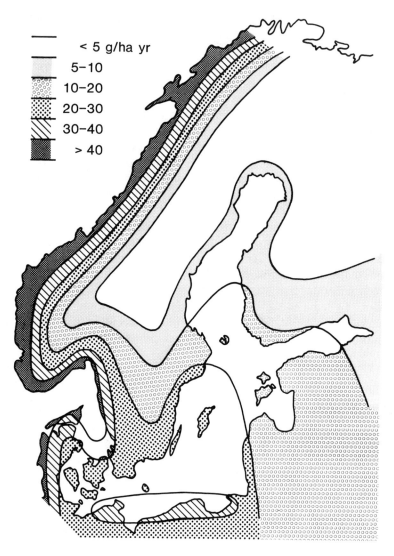

Figure 44-1. The annual deposition of boron in the north of Europe.

Sweden annually receives 25-40 g/ha, while areas of northern Sweden have annual depositions of boron of less than 8 g/ha. Suddesjaur, in the province of Lappland, northern Sweden, receives only 1-3 g boron/ha annually. In Garpenberg, in the province of Dalecarlia, the total boron deposition during 1976 was 6 g/ha and during 1977, 8 g/ha (Fig. 44-1). The boron deposition is larger during the summer than during the winter, and the boron concentration is higher in rain than in snow. Industrial regions have a higher boron deposition due to an extensive use of fossil fuels (Chojnacki, 1968). During stormy years, such as 1969, the coastline may receive several hundred grams per hectare. This is probably the reason why Stevenage in Great Britain during 1969 received around 600 g/ha (Statens Naturvårdsverk, 1976). Also, Japanese figures on boron in snow and rain are surprisingly high (Muto, 1956).

Boron in Fresh Water

The boron concentration in rivers and watercourses has tended to increase — partly, it is believed, because of increasing use of sodium perborate as a bleach in household washing powders. Very little of the boron can be removed by conventional sewage treatment. Boron is present in fresh water as boric acid, and the main sources are leaching of rocks and soils, human activities, and saltwater-affected ground water. In particular, water for irrigation must contain low concentrations of boron, otherwise boron toxicity could be expected in arid areas.

According to Scofield and Wilcox (1931), irrigation water should not exceed 0.67 ppm boron. This is, of course, an approximate figure. In very arid areas, even lower concentrations of boron are necessary due to rapid evaporation. The problems with boron toxicity do not appear until after 10-15 years of irrigation due to permanent accumulation of boron in the top soil layers. Boron toxicity in arid areas due to irrigation with unsuitable water is, therefore, a problem whose significance has yet to be fully appreciated.

An extensive survey on boron in Scandinavian fresh waters done by Ahl and Jönsson (1972) shows that, with a few exceptions, the concentrations are lower than 0.1 ppm (Fig. 44-2). From Table 44-1, which shows boron annually added by precipitation compared with the annual loss of boron via river water, it is seen that the loss of boron is much larger than the input by precipitation. Leaching of soil is the main reason, but perborates in washing powder and boron from fertilizers contribute to a small extent. The total consumption of boron in Sweden in 1969 was only 600,000 kg (IVA, 1970), and the boron released by burning fossil fuels was 190,000 kg (Statens Naturvårdsverk, 1976). The leaching process has probably been accelerated by acid precipitation and also by clear cutting. The concentration in world average river water is estimated as 0.013 ppm (Bowen, 1966).

Boron in Rocks

The geochemistry of boron is complicated. Unsatisfactory analytical methods have made it impossible to form a clear picture of the behavior of boron in rocks and during rock-forming processes. Thanks to the work of Landergren

(1945), Harder (1959a, 1959b), Sahama (1945), and Goldsmidt and Peters (1932a, 1932b), it is today possible to understand the main pathway in the geochemistry of boron. Boron is found in rocks in many different minerals, among which tourmaline is the most important. Tourmaline is a very

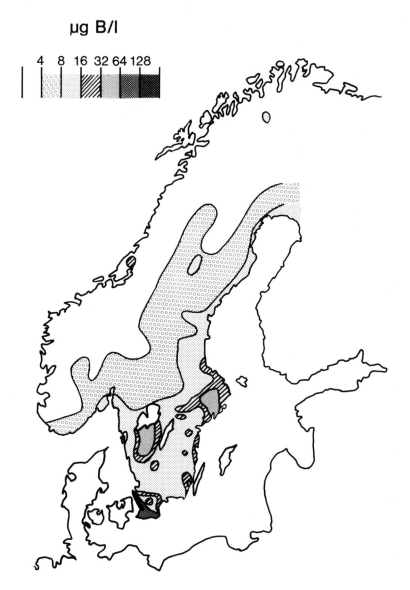

Figure 44-2. The concentration of boron in Scandinavian fresh water according to Ahl and Jönsson (1972).

Table 44-1. Boron Annually Added by Precipitation in Relation to the Annual Loss of Boron with River Water Recalculated as a Loss from the Area and Expressed in g/ha

River	Station	Boron Added by Precipitation (g/ha)	Boron Lost with River (g/ha)	Remarks
Ume älv	Stornorrfors	4-8	21	
Ångermanälven	Sollefteå	4-8	28.1	
Dalälven	Näs bruk	5-9	41.3	
Fyrisån	Kuggebro	10-18	99.7	Sediments of
Göta älv	Trollhättan	15-35	41.5	maritime origin

Note: See Wikner and Uppström (1980) for boron-determination methods. See Ahl and Jönsson (1972) for boron lost with river.

resistant mineral that cannot be dissolved even in hydrofluoric acid. Also, during many rock-forming processes tourmaline remains unchanged. The mineral is thus useless as a boron source for plants. A better boron source is the large amount of boron that is found as impurities in many common minerals. In general, igneous rocks are poor in boron, while rocks of maritime origin contain more. Basic, calcareous, and volcanic rocks are usually very poor in boron. The distribution of boron in igneous rocks is uneven. In massives of granite, boron seems to be enriched in the parts that were coldest during the metamorphosis—that is, in the margins of the granitic bodies.

The igneous rocks of Scandinavia have, in general, very low boron concentrations. Granites in Roslagen contain about 4 ppm (Lundegardh, 1947), granites from southern Norway 0.2-0.8 ppm (Landmark, 1944), and granites from Finland 1-3 ppm (Sahama, 1945). According to Harder (1959a, 1959b), a probable mean value for granites of the world is 10 ppm. Boron in gneisses varies between 1 ppm and 50 ppm. The gneisses seem to be somewhat richer than the granite in Scandinavia. Paragneisses in particular often reach 50 ppm. Quartzites are richer than granites in boron. Unfortunately, a major part is found as tourmaline.

The Boron Geocycle

The ocean is the starting point for the formation of boron-containing rocks. Boron is strongly adsorbed by clay minerals, especially illite, on the ocean floor. Boron adsorption by clays is dependent on (1) the boron concentration in the solution, (2) the particle size of the clay, (3) the salinity, and (4) the temperature. The boron-adsorbing capacity of illite is much higher in saline water than in fresh water. A high temperature hastens the reaction. Marine sediments, with their high boron concentrations (around 220 ppm), could turn to sedimentary rocks and slates (mean boron concentration 100 ppm).

Scandinavian sandstones usually have low concentrations of boron (around 30 ppm). There are many exceptions, however. Since sand does not adsorb boron during sedimentation, most of the boron is found as tourmaline. Boron is lost upon metamorphosis and palingenesis, especially at high temperatures. It can be transported by water vapor or as volatile boric acid. The circumstances determine whether boron will be trapped and forced into tourmaline formation or lost to other phases in the surroundings. On the whole, the formation of igneous rocks leads to a further drop in the boron concentration (around 10 ppm). The boron content will decrease further if igneous rocks are remetamorphosed. Thus, gneiss granites are extremely poor in boron (around 2 ppm). Gneisses with a marine clay sediment origin (paragneisses) sometimes contain more than 100 ppm of boron.

Boron in Soils

Boron is found in soils as boric acid in the soil solution. A minor part is found as complexes between boric acid and different organic substances. Sugars and hydroxy acids are the strongest complexing agents. From a nutritional point of view, boron in soil could be divided into four fractions: (1) totally unavailable boron (mainly tourmaline); (2) unavailable boron (impurities in igneous rocks); (3) temporarily unavailable boron (boron in micas, illite, sesquioxides, and undecomposed matter); and (4) available boron, usually regarded as equal to hot-water-soluble boron, according to Berger and Truog (1939).

In agriculture there is a fairly good agreement between the concentration of boron in crops and the concentration of available boron. Thus, this fraction is usually the only one of special interest. In forestry, unfortunately, there are no such good relationships between needle/leaf concentration and any of the soil boron fractions. Forest soils are very heterogeneous, and the top soil layers differ markedly from the deeper horizons. Sampling is therefore problematic, and the analytical results are difficult to interpret. A typical podzol profile and the distribution of boron throughout the profile are illustrated in Figure 44-3. The different layers are: Ao, 0-4 cm; A_2, 4-8 cm; B_2, 8-18 cm; and B_3, 18-30 cm. Note that B_2 horizons usually contain very little water-soluble boron. Stronger extractions reveal that the B_2 layers have a certain boron-sequestering ability. Boron is adsorbed on the sesquioxides and is only to some extent available to plants.

It is, however, very hard to judge from soil analyses how much boron is actually available. Available boron sometimes becomes unavailable because of drought. On the other hand, with an increased microbiological activity and a steady production of acid extractants from roots, temporarily unavailable boron could be released. If these favorable conditions are combined with a steady supply of water, the boron uptake by plants will be better than could be expected from the analytical figures. It is obvious from fertilizer experiments that large applications of nitrogenous fertilizers, through antagonism, make it difficult for the roots/trees to take up boron. The result is decreasing boron concentrations in the needles, sometimes leading to boron deficiency and dieback injuries. The fractions of boron available to plants do not change significantly after nitrogen fertilization. If the roots no longer retain boron, this portion of mobile boron could be lost from the biocirculation if not fixed by sesquioxides in the iron-rich B_2 horizon of a podzol profile.

Liming acid soils causes a fixation of boron in the soils. This could be a serious consideration since there are plans to lime large forest areas in order to compensate for increased acidification. There are reasons to believe that

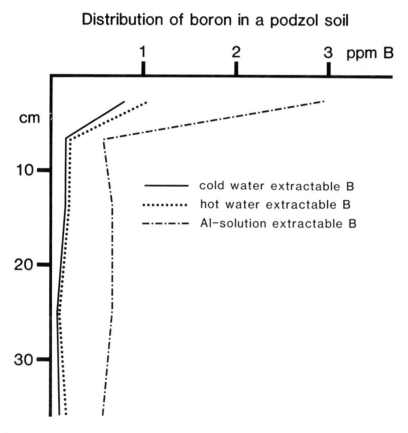

Figure 44-3. The distribution of boron throughout a typical podzol profile. _____Cold-water-extractable boron (soil:water 1:5, extraction time 6 hours). ...... Hot-water-soluble boron according to (3). ......Ammonium lactate–acetic acid-extractable boron. (Soil: Extractant 1:5, extraction time 6 hours).

boron deficiency could be expected in certain areas due to decreased mobility of boron in the soil. The mechanism of boron fixation in soil is not fully understood, but most likely the hydroxyl groups in the organics and in the sesquioxides provide adsorption sites. To be able to fix boron, these hydroxyl groups should be present in a diol of cisconfiguration, and the environment should not be too acidic. The hydroxyl groups of sesquioxides seem to be favorable for boron binding.

Citrate and oxalate (good complexing agents for boron, aluminum, and iron) hinder the adsorption completely. This ability of some organic hydroxy acids to form soluble complexes with boric acid is probably of greater significance in plant nutrition than previously assumed. Citric acid and oxalic acid are found in exudates produced by roots. Since these acids also form soluble complexes with aluminum, iron, and many other metals, they are very important in the process of dissolving sesquioxides—the main reservoir of boron during droughts, when the organic layer can no longer supply the plants/trees with boron. Thus, boron deficiency in natural forests could be expected mainly on soils lacking a well-developed B_2 horizon. However, liming or nitrogen fertilizing could easily change the mechanisms of boron uptake. There are reasons to suspect that nitrogen fertilization will give rise to decreased production of root exudates (sugars, citric and oxalic acid, vitamins, and so on), which is reflected in a drop in microbiological activity in the root zone.

Another question regarding the biocirculation of boron in forests is the speed of circulation. A very limited amount of boron can, if the circulation is fast, do the same amount of biochemical work as a larger amount of boron. Rapid turnover is favored by a good water factor, high microbiological activity, a well-established root system, a high temperature, good aeration in the soil, and active roots producing exudates. This increased mobility can, however, lead to losses by leaching if the root systems die or are nonfunctional. This could happen in connection with clear cutting and possibly with nitrogen fertilization.

Boron in Trees and Plants

There have been some attempts to find a limit below which deficiency symptoms occur. Such a deficiency level is hard to define precisely, because of clonal differences and seasonal variation in concentrations of boron in needles (Fig. 44-4). The deficiency level in general is higher on nutrient-rich calcareous soils due to antagonism and on drought-susceptible sites due to unusually great seasonal fluctuation in the boron concentration of needles. Pines and spruces are more susceptible to boron deficiency than previously assumed. Species with a maritime origin, such as *Pinus radiata* and *P. caribaea,* are very susceptible. The same holds true for all coastal provenances within a species. In Sweden, provenances from the north of the country are less susceptible than those from the south.

An extensive investigation into the boron concentration in pine and spruce needles in mature stands on podzolic moraine soils in different parts of Sweden clearly shows that the concentrations of boron in one-year-old needles are gradually falling from 10–40 ppm along the west coast to less than 7 ppm in the north of the country. Levels less than 3 ppm were frequently associated with dieback and other growth disorders typical of boron deficiency. The difference between Scotch pine, *Pinus sylvestris,* and Norwegian spruce, *Picea abies,* was not great nor significant. On average, the pines are 10% lower in their boron concentrations than the spruces. The Swedish results are much lower than the German results given by Ahrens (1964).

As mentioned above, determination of boron in soil is of limited value as a tool to define boron-deficient areas; needle analysis is preferred. Boron determination in needles must, however, be combined with good knowledge of the distribution of boron in different parts of the tree and seasonal changes in the boron concentration. Figure 44-4 shows the mean seasonal changes of 50 pines and 50 spruces from mature stands in southern Sweden. Apart from the ordinary drop in needle boron concentration in connection with elongation of needles, it was found that the boron concentration in spruce needles increased continuously with needle age. In pines the reverse was true, indicating a certain ability to translocate boron from old to young tissues. It

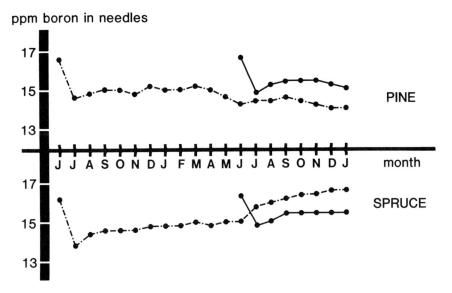

Figure 44-4. The variation of boron concentration in pine and spruce needles with time.

Table 44-2. The Distribution of Boron in a Forest Stand Indicates the Water-Soluble Fraction of Boron

	Biomass ton/ha	Boron ppm	Boron g/ha	Nitrogen kg/ha
Wood	110	2	220	64
Bark	9	8	72	33
Twigs, 20 mm	7	3	21	11
Twigs, 5-20 mm	8	4	32	22
Twigs, 5 mm	6	6	36	43
Needles	12	12	144	122
Dry twigs	1	4	4	3
˙ Total	153		529	298
Stumps	17	2	34	13.9
Roots, 20 mm	3	5	15	6
Roots, 20-5 mm	3.5	5	17.5	12.9
Roots, 5-2 mm	1.4	7	9.8	8.1
Roots, 2 mm	1.3	9	11.7	10.1
Roots in mineral soil	2.4	6	14.4	14
Total	28.6		102.4	65
Vaccinium Myrtillus	0.187	15	2.81	1.97
Vaccinium Vitis-idaea	0.004	10	0.04	0.04
Deschampsia flexuosa	0.060	4	0.24	1.07
Luzula pilosa	0.003	5	0.02	0.06
Other herbs	0.016	20	0.32	0.31
(Mosses)	(0.524)	(6)	(3.14)	
Total, ex mosses	0.27		3.46	3.45
Soil: S-layer	8.9	11 (1.2)	98 (11)	112
F-layer	25.1	9 (1.0)	226 (25)	366
H-layer	69.8	9 (1.1)	628 (77)	681
Total	103.8		952 (113)	1159
0-5 cm	327	15 (0.3)	4905 (26)	722
5-10 cm	361	13 (0.04)	4693 (14)	469
10-15 cm	354	10 (0.1)	3540 (35)	454
15-20 cm	366	12 (0.01)	4392 (37)	384
20-25 cm	390	13 (0.15)	5070 (59)	426
25-30 cm	455	15 (0.15)	6825 (68)	483
30-35 cm	400	12 (0.1)	4800 (40)	284
35-40 cm	417	12 (0.1)	5004 (42)	259
40-45 cm	426	10 (0.05)	4260 (21)	209
Total	3496		43489 (342)	3690

Note: See Wikner and Uppström (1980) for boron-determination methods.

is generally considered that translocation of boron from old leaves to buds is impossible. Pines seem to be an exception to this rule, while the spruces behave according to the rule. Further, seasonal and yearly fluctuations in needle boron concentrations were more pronounced in trees growing on ridges with coarse-textured soils and on plains with peat or fine-textured sediments than in trees on slopes and hillsides, regardless of soil type.

The distribution of boron in a typical forest stand in Garpenberg is found in Table 44-2. Biomass and nitrogen values were earlier calculated by Nyqvist (pers. comm.). Both total and hot-water-soluble boron in soil are accounted for. The stand is made up mainly of Norwegian spruce *Picae abies*, 70 years of age.

Occurrence of Boron Deficiency

In general, boron-deficient areas are characterized by (1) extensively leached soils, (2) coarse-textured soils, (3) soils poor in organic matter, (4) bog soils (acid peats), (5) soils susceptible to drought, (6) overlimed acid soils, (7) soils with a minimum of maritime influence, and (8) soils derived from volcanic rocks.

Mapping for the occurrence of boron-deficiency symptoms in pine and spruce in Sweden gave the following conclusions. Spontaneously occurring deficiency symptoms were found only within the areas receiving less than 10 g of boron/ha annually (Fig. 44-1). Boron-deficiency symptoms were associated with acid peats, freshwater sediments, waterlogging, and drought susceptibility. Topographically, deficiency symptoms were associated with plains. Slopes and hillsides were usually free from deficient trees. Coarse-textured soils were usually covered by healthy trees. Dieback injuries on such soils were usually caused by drought alone. Boron-deficiency dieback was frequently found in connection with nitrogen fertilization or liming. In particular, refertilized plots were affected. Stands having a boron concentration in the needles below 3 ppm were commonly associated with dieback injuries. Within an affected stand there was no correlation between needle concentration of boron and severity of injury. In fact, severely injured trees often had higher concentrations of boron than less severely affected trees. This was probably due to sampling artifacts. At high altitudes, dieback phenomena were very common and were probably caused by drought and frost in combination. These trees and plants were usually low in boron (around 4 ppm), and there is reason to suspect that a higher boron concentration in the needles would be beneficial in order to improve the frost and drought resistance. Boron-deficient soils were often covered by grasses.

Forest trees and agricultural crops cannot be directly compared since the root systems differ significantly. Coarse-textured soils, usually boron deficient when arable, seldom develop boron deficiency under silviculture.

Conversely, fine-textured freshwater sediments more often cause boron deficiency in pine and spruce than in agricultural crops.

Boron-Deficiency Symptoms

No visual symptoms are evident in the early stages. Raitio and Rantala (1977), however, have shown that cell disorders could be found in needles from boron-deficient trees before any striking symptoms appeared. The main disorders found in needles were large cavities, irregular cell sizes, and irregular, sometimes broken, cell walls. In the later stages the symptoms are more striking, with shoot- or bud-dieback injuries as the most common features. Additionally, white resin patches are common, abundant new shoots from lateral or fascicle buds cause greater crown density or bushiness, lateral shoots often grow faster than apical ones, and needles are irregular in shape, size, and position, with yellow tips. In contrast to drought-dieback injuries, the dead leaders are comparatively thick after a dieback caused by boron deficiency. All gradients exist, from pure drought-dieback injuries to pure boron-deficiency dieback. The picture is further complicated by the fact that boron-deficient trees are very susceptible to frost damage. The roots fail to grow and are short and brittle, with swollen ends.

The Role of Boron in Forest Ecosystems

The primary role of boron in plants has not yet been established. Secondary effects are, however, well known. A shortage of boron leads to numerous defects and abnormalities. Shoot and root meristems, in particular apical meristems, die or become extremely moribund, the apical dominance is lost, and cell division does not proceed satisfactorily. New cells fail to differentiate. Disorganization of the vascular tissues with cell collapse occurs, leading to cavities. Cell walls break down or become abnormally thick and are non-functional. The middle lamella becomes disorganized with vesicle formation.

A common feature in boron-deficient plants is the build-up of sugar, free amino acid, and/or phenolic substances. The enrichment of sugars was the first effect observed, and it led to the assumption that boron in one way or another was responsible for the translocation of sugars. Today it is thought that these accumulations of simple substances could be ascribed to a nonfunctional vascular system and a reduced utilization of these substances. The chaotic cell situation in deficient plants is believed to be caused by a supra-optimal concentration of indoleacetic acid (IAA). Nevertheless, from literature available today it is impossible to distinguish between cause and effect. These defects and abnormal concentrations of certain compounds cause, in turn, reduced frost and drought resistance, probably reduced resistance to fungi and insects due to enhanced sugar and amino acid concentrations in needles, and all the striking boron-deficiency symptoms memtioned earlier. The importance of boron for good root growth is probably underestimated. Further studies are highly desirable since the roots are affected before any visible symptoms appear on the aboveground parts of a plant.

REFERENCES

Ahl, T., and E. Jönsson, 1972, Boron in Swedish and Norwegian fresh waters, *Ambio* **1:**66-70.

Ahrens, E., 1964, Investigations on the content of copper, zinc, boron, molybdenum and manganese in leaves and needles of different trees, *Allg. Forst und Jagdztg.* **135:**8-16.

Berger, K. C., and E. Truog, 1939, Boron determination in soils and plants using the quinalizarin reaction, *Ind. Eng. Chem., Anal. Ed.* **11:**540-545.

Bowen, H. J. M., 1966, *Trace Elements in Biochemistry,* Academic Press, London.

Chabannes, J., 1959, Boron and sulfur in rainwater at different points in France, *Chim. and Ind.* **82:**835-837.

Chojnacki, A., 1968, Results of investigations on chemical composition of atmospheric precipitations in Poland, *Pamiet. Pulawski* **35:**163-172.

Goldsmidt, V. M., and C. Peters, 1932a, The geochemistry of boron, *Nachr. Ges. Wiss. Gottingen Math. -physik. Kl.* **1:**402-407.

Goldsmidt, V. M., and C. Peters, 1932b, The geochemistry of boron, *Nachr. Ges. Wiss. Gottingen Math.-physik. Kl.* **2:**528-545.

Harder, H., 1959a, Geochemistry of boron. I. Boron in minerals and igneous rocks, *Nachr. Akad. Wiss. Gottingen, II. Math. physik. Kl.* **5:**67-122.

Harder, H., 1959b, Geochemistry of boron. II. Boron in sediments, *Nachr. Akad. Wiss. Gottingen, II. Math. physik. Kl.* **6:**123-183.

IVA (Royal Swedish Academy of Engineering Sciences), 1970 Environmental aspects on boron, Report IVA Number 33, Stockholm.

Jensen, J., 1962, Boron contents in Danish crops, *Tidsskr. Planteavl.* **65:**894-906.

Landergren, S., 1945, Geochemistry of boron. II. Distribution of boron in some Swedish sediments, rocks, and iron ores. Boron cycle in the upper lithosphere, *Arkiv. Kemi, Mineral, Geol.* **26:**1-31.

Landmark, K., 1944, Boron content of rocks from Vestlandet and Vest-Jotunheimen (in Norway), *Bergens Mus. Årsbok* **5:**1-20.

Lundegårdh, P. H., 1947, Rock composition and development in central Roslagen, Sweden, *Arkiv. Kemi, Mineral, Geol.* **9:**1-160.

Muto, S., 1956, Distribution of boron in natural waters, *Chem. Soc. Japan Bull.* **29:**532-536.

Philipson, T., 1953, Boron in plant and soil, with special regard to Swedish agriculture, *Acta. Agr. Scand.* **3:**121-242.

Raitio, H., and E. M. Rantala, 1977, Microscopic and macroscopic symptoms of a growth disturbance in Scotch pine, *Commun. Inst. For. Fenniae* **91:**5-30.

Riehm, H., E. Quellmalz, and M. Kraus, 1965, Results from atmospheric chemical investigations in central Europe, *Zentalbl. Biol. Aerosol Forsch* **12**(5):434-454.

Sahama, T. G., 1945, Trace elements in the rocks of southern Finnish Lapland, *Comm. Geol. Finlande Bull.* **135**:5-86.

Scofield, C. S., and L. V. Wilcox, 1931, Boron in irrigation waters, *U.S. Dept. Agr. Tech. Bull.* **264**:1-65.

Statens Naturvårdsverk, (Swedish Environmental Protection Agencies), 1976, *Metals, Publikationer 7.*

Wikner, B., 1981, Boron determination in natural waters with curcumin using 2, 2-dimethyl-1, 3-hexanediol to eliminate interferences, *Soil Sci. Plant Anal. Commun.* **12**:697-710.

Wikner, B., and L. Uppström, 1980, Determination of boron in plants and soils with a rapid modification to the curcumin method utilizing different 1, 3-diols to eliminate interferences, *Soil. Sci. Plant Anal. Commun.* **11**:105-126.

45

Silver Contamination of Soils by Windblown Spoil from a Derelict Mine

Kevin C. Jones
Chelsea College, University of London

Brian E. Davies
University College of Wales

Peter J. Peterson
Chelsea College, University of London

INTRODUCTION

In Wales, United Kingdom, there are more than 300 abandoned lead mines that were in active production between 1750 and 1900. These abandoned workings are accompanied by sterile waste heaps and dressing floors that have persisted to the present day as sources of metal contamination. Colonization of mine tailings by vegetation is limited by their physical instability, paucity of major nutrients, and high concentrations of toxic heavy metals. The failure of vegetation to become established on these sites makes them susceptible to erosion (by fluvial, gravitational, and atmospheric processes), so that the hazard presented by the waste's metal-rich materials can spread to surrounding land. Runoff and leaching may contaminate fluvial ecosystems and sediments (Alloway and Davies, 1971; Jones, Peterson, and Davies, 1983), and gravitational movement (downslope creep) may affect surrounding agricultural land (Davies, Jones, and Peterson, 1983). Mobilization by windblow becomes important where the tips are composed of fine material. Johnson and Roberts (1978) investigated the environs of a Welsh mine, Y Fan, near Llanidloes, Powys, and concluded that wind erosion contributed to low-level contamination close to the heaps. Davies and White (1981) studied windblown contamination at the derelict lead-silver mine at Cwmsymlog, Dyfed, and concluded that most of the <2-mm fraction of the spoil is of sufficiently small particle diameter to move by saltation. Tailings dust had been transported as far as 1800 m downwind. Movement was caused by dry, northerly winds that were funnelled in an easterly direction by the valley. This has led to subsequent reclamation of the site by the Welsh Development Agency to remove the potential health risk of lead contamination.

The conditions for windblown dispersal at Cwmsymlog are not unique and apply to other waste heaps in the district. The Graig Goch mine, for example, is also located in a narrow east-west valley, and when observed from up-valley a marked yellowing of the pasture grass is seen near the mine. This yellowing decreases westward down the valley.

Lead and zinc are the elements that have received most attention in the literature on contamination from Welsh mines. However, other elements, notably silver, have been mined in the area. Despite knowledge of silver as an element of known toxicity (Sax, 1975), there is a scarcity of information regarding the environmental significance of silver contamination. The increasing use of more sensitive analytical techniques, notably graphite-furnace atomic absorption spectrometry, enables this imbalance to be redressed. The argentiferous-rich nature of much of the galena in west Wales was frequently the incentive for exploitation, and this association makes the region an interesting study area for an investigation of the environmental behavior of silver. The location of the Graig Goch mine makes it an appropriate site for examining the movement of silver and other metals by windblow.

MATERIALS AND METHODS

Study Area

The mine at Graig Goch (Ordnance Survey grid reference SN 704741) is located on the southern slope of a narrow east-west valley (see Fig. 45-1). Spoil tips and slime pits occupy part of the south side of the study area. Land above 500 ft on the north slope is covered by Forestry Commission conifer plantations; the lower land and the north-facing slope around the mine site are rough pasture for cattle and sheep grazing. A small stream, Nant Cwm-newydion, flows through the valley, and this stream is a tributary of the river Ystwyth, which empties into Cardigan Bay, at Aberystwyth, 15 km to the west.

The solid geology of the Aberystwyth mining district comprises shales and grits of Silurian age, and these are overlain by Pleistocene deposits. Galena (PbS) and sphalerite (ZnS) with occasional chalcopyrite (CuFeS$_2$), in a gangue of quartz or sometimes calcite, are the chief minerals. Silver in its sulfide form, argentite (Ag$_2$S), is also present, but more frequently occurs as a guest element in galena and sphalerite (El Shazly, Webb, and Williams, 1956). In north Cardiganshire (now Dyfed), silver mining centered on the

Goginan and Daren lodes where the concentration of silver associated with galena varied between 8 and 100 oz per ton (200-2800 μg/g). The Graig Goch mine is part of the less argentiferous Frongoch formation (approximately 100 μg silver/g lead ore). It was productive in the mid-nineteenth century and yielded 3282 tons of lead concentrates and 25 tons of zinc blend. The probable amount of silver contained in the whole ore had been estimated at approximately 12,000 oz (340 kg), although only 815 oz (23 kg) are actually known to have been recovered (Jones, 1922).

Field Techniques

The study area was defined as an area 800 x 370 m in which samples of soil were collected on a regular rectilinear grid. Mine spoil or sediment from

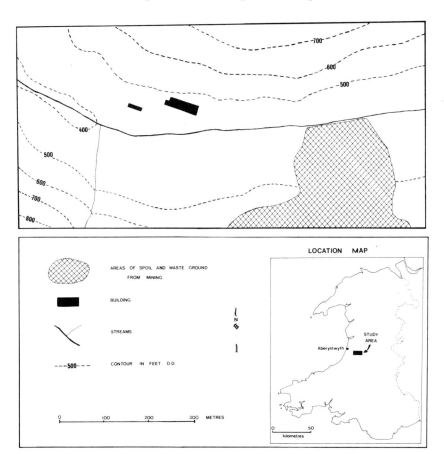

Figure 45-1. Map of the study area.

the slime pits was collected wherever the sampling grid covered areas of spoil or waste ground from the mine.

Analytical Methods

Full details of analytical procedures are given by Jones, Peterson, and Davies (1984). In brief, after air drying, disaggregation, and sieving (2-mm-aperture nylon sieve), heavy metals were extracted from soils and tailings by refluxing with concentrated nitric acid. Silver (Ag), cadmium (Cd), and copper (Cu) were determined by flameless atomic absorption spectrometry, while lead (Pb) and zinc (Zn) were analyzed by conventional flame spectrometry. Soil organic content was estimated gravimetrically by the loss on ignition at 430°C, and soil reaction (pH) was determined electrometrically following equilibration with distilled water (1:2.5 soil:solution).

Bulk samples of mine waste were collected from the spoil tips at Graig Goch and, for comparison, from another mine at Cwm Rheidol (SN 730783). The spoil samples were dried in the same way as soil samples, but not ground, and the <2000-μm fraction was sequentially passed through nylon sieves of apertures 1000, 200, 140, and 64 μm. The choice of sieves was decided on the basis of the relationship between wind erodibility and particle size according to Bagnold (1941). The sieved fractions were weighed and then analyzed for silver as described above.

RESULTS

The analytical results for 76 soil samples are summarized in Table 45-1. The data for all the variables, except pH, were found to be log-normally distributed, so the correlation coefficients (r) performed on log-transformed data where applicable. These data were subsequently plotted on cumulative frequency curves, and values for the 50th, 60th, 70th, 80th, 90th, and 95th percentiles were obtained for each metal (Table 45-1). These were used as contour intervals for the preparation of contour maps using the computer program SYMAP. Isoline maps interpreting the data for lead and silver are shown in Figures 45-2a and 45-2b for the area defined in Figure 45-1. Areas containing metal levels less than the 50th percentile (<575 μg lead/g and <0.76 μg silver/g, respectively) have been left unshaded; to draw attention to higher levels of contamination, the shading has been intensified with successive contours. The distribution of silver anomalies is perhaps better appreciated by looking at Figure 45-3, a perspective block diagram prepared by SYMVU from the contours in Figure 45-2b. It depicts the study area viewed from the southwest corner and at 45° to the horizontal plain. High silver levels are represented as though they are hills to produce a geochemical landscape. Table 45-2 presents data on graded spoil from two mines—the main study area at Graig Goch and a lead-zinc mine at Cwm Rheidol.

Table 45-1. Summary of the Analytical Data (76 Samples)

	pH	% Ignition Loss	Heavy Metals µg g⁻¹ Dry Soil				
			Ag	Pb	Zn	Cu	Cd
Minimum	3.8	0.50	0.25	195	70	10	0.11
Maximum	5.65	28.70	37.7	30500	3800	415	10.3
Mean	4.98	8.81	2.74	2422	709	42.8	1.10
Standard deviation	0.39	4.04	5.96	5158	631	45.7	1.54
50th percentile of log transformed data			0.76	575	501	35	0.61
60th percentile of log transformed data			0.98	912	562	40	0.78
70th percentile of log transformed data			1.38	1445	631	43	0.90
80th percentile of log transformed data			2.29	2455	741	48	1.51
90th percentile of log transformed data			6.31	4467	1202	59	2.30
95th percentile of log transformed data			10.00	10000	1778	76	3.70

Correlation Matrix of pH and $\log_{10}$ Organic and Metal Contents

	pH	% Ignition Loss	Ag	Pb	Zn	Cu	Cd
pH	—	−0.147	−0.365**	−0.120	−0.117	−0.168	−0.140
% Ign. Loss		—	−0.392**	−0.126	−0.177	−0.087	−0.338**
Ag			—	0.604**	0.198*	0.657**	0.353**
Pb				—	0.214*	0.955**	0.118
Zn					—	0.195*	0.420**
Cu						—	0.121
Cd							—

*Indicates significant at the 95% level ($r > 0.190$).
**Indicates significant at the 99.5% level ($r > 0.294$).

DISCUSSION

Metal Concentration in Soils and Spoil

Silver concentrations in soil are ordinarily very low. Swaine (1955) quotes an average value of <1 μg silver/g dry soil, while Boyle (1968) reported that the range of silver in normal soils varies from <0.01 to 5 μg silver/g and estimated an average value of about 0.10 μg silver/g (0.3 μg silver/g in mineralized zones). In a previous paper Jones, Peterson, and Davies (1983) examined uncontaminated soils from a number of locations to establish a normal or background concentration for silver in Welsh soils. Soils derived from black shales (typically 0.5 μg silver/g) or rich in organic matter (0.5 μg silver/g) were inherently richer in silver than limestone-derived soils. Soils taken from the Aeron valley (an area devoid of derelict mines) have formed on similar geological material to those sampled in the present work at Graig Goch and, therefore, serve as a useful guide to the extent of contamination. A normal range of 0.01-0.17 μg silver/g with a mean of 0.046 μg silver/g was calculated for the Aeron samples. The 76 soil and spoil samples from Graig Goch were in the range 0.25-37.7 μg silver/g (see Table 45-1), all above the regional background.

Elevated silver concentrations may be found in soils near areas of mineralization, or near sources of contamination such as smelters, mines, or industrial complexes. Ragaini, Ralston, and Roberts (1977) found silver enriched in surface soils (0-2 cm) by aerial fallout to 20.0 ± 10.2 μg silver/g near a lead-smelting complex in Kellogg, Idaho, U.S.A. Ward, Brooks, and Roberts (1977) investigated soils near a silver mine and treatment plant at Maratoto, New Zealand, and found levels an order of magnitude higher (0.75-6.00 μg silver/g) than background. Fluvial contamination of soils from mine effluent in the Ystwyth Valley, Wales, have elevated silver to between 0.11 and 7.0 μg silver/g (Jones, Peterson, and Davies, 1983; Alloway and Davies, 1971).

Davies (1983) concluded that lead concentrations greater than 100 μg lead/g will not arise naturally in uncontaminated British soils and that values greater than this should be regarded as anomalous. As for silver, all the samples from Graig Goch are above this limit and should, therefore, be regarded as contaminated with lead. Zinc background concentrations can be taken as being <175 μg zinc/g in this area (Alloway and Davies, 1971), copper <62 μg copper/g (Jones, Peterson, and Davies, 1983), and cadmium <0.35 μg cadmium/g (Bowen, 1979). The number of samples that exceeded the relevant thresholds were zinc = 74 (97% of sample population), copper = 7 (9%), and cadmium = 60 (79%). Thus the order of contamination at Graig Goch is silver = lead > zinc > cadmium > copper.

Support for the hypothesis that metals have arisen from losses during ore processing is found in correlations within the data (Table 45-1). High and significant correlations are found between silver:lead, silver:zinc, silver:copper, silver:cadmium, lead:zinc, lead:copper, zinc:cadmium, and zinc:copper. This is consistent with their derivation from the common source of mixed lead-zinc ores.

Patterns of Metal Dispersal

Figures 45-2a and 45-2b show contour maps produced by the computer program SYMAP. They show clearly, for lead and silver respectively, the markedly elevated concentrations of metals associated with the spoil tips and mine workings. Figure 45-3 indicates that silver concentrations are highest in the slime pits, where fine materials produced by ore crushing and processing were allowed to settle before discharge of effluent into the Nant Cwm-newydion stream.

The <2-mm spoil fraction contained about 25,000 μg lead/g, 30 μg silver/g, 1600 μg zinc/g, 400 μg copper/g, and 10 μg cadmium/g. A westwards transect (X $-$ X on Fig. 45-2) was taken to 600 m from the spoil, traversing the area from no grass through yellow grass to apparently healthy grass. At 600 m the soil contained 255 μg lead/g, 1.5 μg silver/g, and 300 μg zinc/g. The eastwards transect (X $-$ Y) measured 200 m, and the farthest soil sample contained 650 μg lead/g, 0.8 μg silver/g, and 670 μg zinc/g. Scatter-grams of metal values against distance for the longer transect indicated an exponential distance-decline effect similar to that observed for fallout from point sources (Franzin, McFarlane, and Lutz, 1979). The following exponential regressions were fitted to the data. They are of the form:

$$Y = Ae^{BX}$$

where Y = metal concentration and X = distance

silver	$A = 7.60$	$B = -0.007$	$R^2 = 0.825$
lead	$A = 4988$	$B = -0.006$	$R^2 = 0.746$
zinc	$A = 962$	$B = -0.002$	$R^2 = 0.525$

The Department of the Environment (DOE) has issued tentative guide-lines for the redevelopment of industrially contaminated land (DOE, 1980), and the implications of these metal concentrations are considered in this context by Davies, Jones, and Peterson (1983).

Table 45-2 indicates that the silver content of spoil increases as particle size decreases. A similar trend has been observed for lead in graded spoils at Cwmsymlog (Davies and White, 1981). It is the finer particles that are more susceptible to windblow; broadly speaking, particles with an equivalent

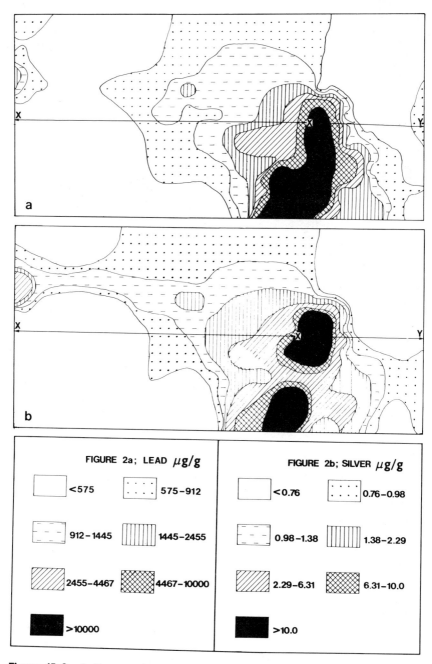

Figure 45-2. Isoline map for soil metal contents in the area around the Graig Goch mine: *a*, lead; *b*, silver.

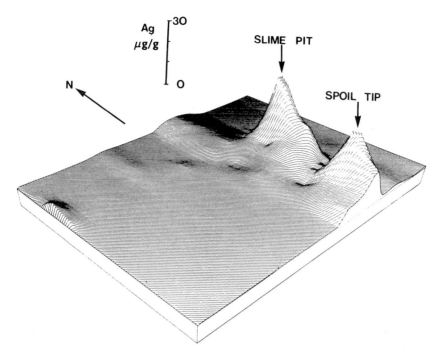

Figure 45-3.   Perspective block diagram derived from the silver contours of Figure 45-2*b*.

Table 45-2. Percent Size Fractions and Silver Contents of Waste from Spoil at Graig Goch and Cwm Rheidol

	Graig Goch Mine Spoil		*Cwm Rheidol Mine Spoil*	
Particle Size (μm)	*% of Sample in Each Size Fraction*	*Ag μg g^{-1} dry weight*	*% of Sample in Each Size Fraction*	*Ag μg g^{-1} dry weight*
>2000	2.5	16.1	—	—
1000-2000	24.2	18.8	8.7	12.7
200-1000	36.9	22.4	41.5	21.8
140-200	17.6	44.3	27.9	23.3
<64-140	11.6	75.4	19.0	37.3
64	7.2	143.5	2.9	72.3

spherical diameter of $<10\,\mu$m can be carried in suspension by air currents, whereas those with diameters between 100 and $500\,\mu$m are likely to travel by saltation (Bagnold, 1941). The strong association between silver and lead suggests that dispersal of silver from Graig Goch may be controlled by processes similar to those described at Cwmsymlog by Davies and White (1981). There, dry conditions that make spoil more susceptible to aerial transportation are most often accompanied by northeasterly winds funnelled down the valley westwards. Since Graig Goch lies in a similarly oriented valley, it seems reasonable to suggest that spoil will move predominantly down the valley. This hypothesis is supported by the isoline contour map in Figure 45-2.

Biological Implications of Silver in Contaminated Soils

The toxicity of silver to plants is reviewed and discussed by Girling and Peterson (1979), but it is the effects of the silver ion (Ag^+) on bacteria and fungi that are likely to be of most concern in a consideration of the biological implications of silver in contaminated soils. Silver is known to be one of the most toxic metal ions to microorganisms and fungi (Porter, 1946; Horsfall, 1956); 6×10^{-9} g Ag^+/ml, for example, kills *Escherichia coli* in 2 to 24 hours, depending on numbers of bacteria (Lawrence and Block, 1968), and the heterotrophic activity of bacteria is reduced by 0.1×10^{-9} g silver /ml (Albright, Wentworth, and Wilson, 1972). Silver nitrate at a concentration of 10×10^{-9} g silver/ml killed 50% of the fungi *Alternatia tenuis,* and, of the metals tested, only osmium was more toxic to the fungus (Somers, 1961).

The toxic action of Ag^+ is related to the binding potential of these ions to enzymes and other active molecules at cell surfaces. Consequently, concern has been expressed about the possible harmful effect of Ag^+ on the soil microbial population, and its subsequent influence on soil biochemical processes. Drucker, Garland, and Wildung (1979) showed that 1 μg silver/g decreased the total number of aerobic soil bacteria and the soil respiration rate of cultures; silver was the most toxic of the 17 metals investigated. Hoffman and Hendrix (1976) showed that soluble silver had an adverse effect on *Thiobacillus ferrooxidans* at $0.1\,\mu$g silver/ml, and solutions containing 1.0 μg silver/ml or more restricted growth of the bacterium. Adverse effects of silver on various microbial metabolic processes, such as glucose respiration (Molise and Klein, 1974), carbon dioxide release (Cornfield, 1977), nitrogen mineralization (Liang and Tabatabai, 1977), and arylsulphatase activity (Al-Khafaji and Tabatabai, 1979) have been reported under laboratory conditions with spiked soils. Tyler (1974) obtained approximately straight regression lines between enzymatic activity or respiration and log copper + zinc concentration in soils polluted from a brass foundry in Sweden, but no attempt has been made to assess the importance of silver on metabolic processes under field conditions. The impact of silver on soil biochemistry will, however, be limited by a number of factors. First, only a fraction of total soil silver is biologically available (Jones, Peterson, and Davies, 1984); second, Ag^+ is readily reduced to form insoluble compounds, such as silver chloride (AgCl), or compounds with a low ionization potential, such as $Ag_2 S$, which will restrict the presence of the silver ion in some soils (Lindsay, 1979); third, other authors have demonstrated the ability of some species of bacteria and fungi to accumulate silver without any apparent adverse affect (Charley and Bull, 1979; Pooley, 1982; Byrne, Dermelj, and Vakselj, 1979).

ACKNOWLEDGEMENTS

The authors are grateful to the Welsh Office for financial support.

REFERENCES

Albright, L. J., J. W. Wentworth, and E. M. Wilson, 1972, Technique for measuring metallic salt effects upon the indigenous heterotrophic microflora of a natural water, *Water Res.* **6:**1589-1596.

Al-Khafaji, A. A., and M. A. Tabatabai, 1979, Effects of trace elements on arylsulfatase activity in soils, *Soil Sci.* **127:**129-133.

Alloway, B. J., and B. E. Davies, 1971, Trace element content of soils affected by base metal mining in Wales, *Geoderma* **5:**197-208.

Bagnold, R. A., 1941, *The Physics of Blown Sand and Desert Dunes,* Chapman Hall, London.

Bowen, H. J. M., 1979, *Environmental Chemistry of the Elements,* Academic Press, London.

Boyle, R. W., 1968, The geochemistry of silver and its deposits, with notes on geochemical prospecting for the element, *Geol. Surv. Canada Bull. 160.*

Byrne, A. R., M. Dermelj, and T. Vakselj, 1979, Silver accumulation by fungi, *Chemosphere* **10:**815-821.

Charley, R. C., and A. T. Bull, 1979, Bioaccumulation of silver by a multispecies community of bacteria, *Arch. Microbiol.* **123:**239-244.

Cornfield, A. H., 1977, Effects of addition of 12 metals on carbon dioxide release during incubation of an acid sandy soil, *Geoderma* **19:**199-203.

Davies, B. E., 1983, A graphical estimation of the normal lead content of some British soils, *Geoderma* **29:**67-75.

Davies, B. E., and H. M. White, 1981, Environmental pollution by windblown lead mine waste: A case study in Wales, U.K., *Sci. Tot. Environ.* **20:**57-74.

Davies, B. E., K. C. Jones, and P. J. Peterson, 1983, Metalliferous mine spoil in Wales: A toxic and hazardous waste, in *Proceedings of the International Conference on Heavy Metals in the Environment,* Heidelberg, Sept., 1983, vol. 2, pp. 984-987.

Department of the Environment, 1980, *Tentative guidelines for acceptable levels of contaminants in soils,* Inter Departmental Committee on Reclamation of Contaminated Land ICRCL, Paper 38/80.

Drucker, H., T. R. Garland, and R. E. Wildung, 1979, Metabolic response of microbiota to chromium and other metals, in *Trace Metals in Health and Disease,* N. Kharasch, ed., Raven Press, New York, pp. 1-25.

El Shazly, E. M., J. S. Webb, and D. Williams, 1956, Trace elements in sphalerite, galena and associated minerals from the British Isles, *Inst. Min. Metall. Trans.* **66:**241-271.

Franzin, W. G., G. A. McFarlane, and A. Lutz, 1979, Atmospheric fallout in the vicinity of a base metal smelter at Flin Flon, Manitoba, Canada, *Environ. Sci. Technol.* **13:**1513-1522.

Girling, C. A., and P. J. Peterson, 1979, Other trace metals, in *Effects of Heavy Metal Pollution on Plants,* N. W. Lepp, ed., Applied Science, London, pp. 213-279.

Hoffman, L. E., and J. L. Hendrix, 1976, Inhibition of *Thiobacillus ferrooxidans* by soluble silver, *Biotechnol. Bioeng.* **18:**1161-1165.

Horsfall, J. G., 1956, *Principles of Fungicidal Action,* Chronica Botanica Co., Waltham, Massachusetts.

Johnson, M., and D. Roberts, 1978, Lead and zinc in the terrestrial environment around derelict metalliferous mines in Wales (U.K.), *Sci. Tot. Environ.* **10:**61-78.

Jones, K. C., P. J. Peterson, and B. E. Davies, 1983, Silver concentrations in Welsh soils and its dispersal from derelict mine sites, *Miner. Environ.* **5:**122-127.

Jones, K. C., P. J. Peterson, and B. E. Davies, 1984, Extraction of silver from soils and its determination by atomic absorption spectrometry, *Geoderma* **33:**157-168.

Jones, O. T., 1922, The mining district of North Cardiganshire and West Montgomeryshire, Memoirs of the Geological Survey, Special Reports on the Mineral Resources of Great Britain, HMSO, London.

Lawrence, C. A., and S. S. Block, 1968, *Disinfection, Sterilization and Preservation,* Lea and Febiger, Philadelphia.

Liang, C. N., and M. A. Tabatabai, 1977, Effects of trace elements on nitrogen mineralization in soils, *Environ. Pollut.* **12:**141-147.

Lindsay, W. L., 1979, *Chemical Equilibria in Soils,* Wiley, London.

Molise, E. M., and D. A. Klein, 1974, Effects of silver on radio respiration of glucose by soil microorganisms, *Am. Soc. Microbiol. Annu. Meet. Abstr.* **74:**2.

Pooley, F. D., 1982, Bacteria accumulate silver during leaching of sulphide ore minerals, *Nature* **296:**642-643.

Porter, J. R., 1946, *Bacterial Chemistry and Physiology,* John Wiley & Sons, New York.

Ragaini, R. C., H. R. Ralston, and N. Roberts, 1977, Environmental trace metal contamination in Kellogg, Idaho, near a lead smelting complex, *Environ. Sci. Technol.* **11:**773-781.

Sax, N. I., 1975, *Dangerous Properties of Industrial Materials,* 4th ed., Van Nostrand Reinhold Co., New York.

Somers, E., 1961, The fungi toxicity of metal ions, *Ann. Appl. Biol.* **49:**246-253.

Swaine, D. J., 1955, The trace element content of soils, Commonwealth Bureau of Soil Science, Tech. Comm. No. 48, Harpenden.

Tyler, G., 1974, Heavy metal pollution and soil enzymatic activity, *Plant Soil* **41:**303-311.

Ward, N. I., R. R. Brooks, and E. Roberts, 1977, Silver in soils, stream sediments, waters and vegetation near a silver mine and treatment plant at Maratoto, New Zealand, *Environ. Pollut.* **13:**269-280.

46

Waste Oil Biodegradation and Changes in Microbial Populations in a Semiarid Soil

John Skujiņš and S. O. McDonald

Utah State University

ABSTRACT: Biodegradation of acidic waste oil in a semiarid soil, disposed by landfarming technology, was studied for four years at a 16-ha site near Ogden, Utah. Following soil preparation and additions of nitrogen (N), potassium (P), calcium (Ca), and oil (62-$110 \, m^3$/ha), hydrocarbon-utilizing bacteria increased $>10^8$/g of soil. Coryneforms, pseudomonads, *Flavobacterium*, and *Bacillus* were the most prominent genera. *Cunninghamella, Penicillum, Aspergillus,* and *Alternaria* were the principal fungal genera showing significant relative increases, although the total mycofloral numbers were suppressed. Total aerobic-soil normal flora and anaerobic bacteria increased during the active oil-degradation period, whereas the actinomycete population decreased. Considerable nitrification of applied urea occurred. Decrease of applied nitrogen was significantly correlated with oil-degradation rates, but there was no significant increase in oil-degradation rates in presence of N as compared with rates in absence of N. In the semiarid climate (annual precipitation 303 mm) the biodegradation of oil occurred during seasonal periods of moisture availability associated with elevated air temperatures. During four years 91% of the applied oil was degraded.

INTRODUCTION

In the spring of 1974 the U.S. Environmental Protection Agency (EPA) disposed of approximately $6500 \, m^3$ of waste industrial oil stored in an open pond near Ogden, Utah, using the landfarming method (Huddleston, 1979). The operation was an environmental protection measure including fertilization and seeding of the landfarming area (Snyder, Rice, and Skujiņš, 1976).

As the disposal site was in a semiarid region, where little was known about the oil biodegradation in soil, it provided an opportunity to obtain information on the degradation process and microbiological changes taking place.

The waste oil emulsion was acidic and contained appreciable amounts of metal ions. To counteract the effects of acidity and to promote oil degradation, the disposal site was treated with calcium hydroxide, phosphate, and nitrogen fertilizers. In this report the process of oil degradation and the microbiological changes in the soil during the degradation process are described. The effects of these treatments on biological availability of metal ions to plants during the initial revegetation of the site have been described elsewhere (Skujiņš, McDonald, and Knight, 1983).

MATERIALS AND METHODS

Site

The 16.4-ha oil disposal site was located 20 km west of Ogden at the base of Little Mountain, Weber County, Utah, elevation 1300 m, on a 5% to 15% slope. The soil was a coarse loamy mixed mesic type Argixeroll (Barton series). Parent material was derived from a massive tillite; depth to the bedrock pre-Cambrian slate varied from exposed outcroppings to below 1 m. The soil (Skujiņš, McDonald, and Knight, 1983) contained 2.25% organic carbon and 0.19% total nitrogen and had a pH value of 7.2.

The semiarid site had an average annual air temperature of 10°C, with the average daily maximum of 33°C occurring in July and the average daily minimum of −9°C in January. Average annual precipitation was 303 mm, with snow contributing about 40% of the total precipitation input (Fig. 46-1).

The dominant vegetation at the site included *Artistida longiseta* Steud., *Asclepias speciosa* Torr., *Bromus tectorum* L., *Euphorbia glyptosperma* Engelm., *Helianthus annus* L., *Opuntia polyacantha* Haw., and *Sporobolus cryptandrus* (Torr.) Gray.

Treatments

The disposal site was cultivated to an average depth of 12 cm with a disc and a barber-shank harrow in May 1974 and was subdivided into 15 plots receiving defined amounts of fertilizers, waste-oil emulsion, and water. The waste-oil emulsion and the oil-free water were obtained from a 4-km-distant 2-ha pond where the material had separated into three layers: the surface oil emulsion (45.7% oil, pH 2.7) with an oil-free water phase (pH 3.5) below and sludge on the bottom. Following addition of each amendment, the plots were tilled with an agricultural rototiller or barber-shank harrow. The effective mixing depth was about 10 cm. The plots received defined amounts of waste-oil emulsion ranging from 62 to 110 m³/ha, resulting in 0.05 to 0.09 ml oil/g of soil, calcium hydroxide (lime) from 1050 to 2130 kg/ha, $Ca_3 (PO_4)_2$ from 56 to 170 kg P/ha, and urea from 420 to 1250 kg N/ha. All sites also received the oil-free aqueous phase ranging from 103 to 140 m³/ha. The total amount of water added with the treatments was equivalent to 19 mm additional precipitation (Fig. 46-1). The aqueous phase was treated with 0.85 kg $Ca(OH)_2$ m⁻³ before spreading. One site received waste oil, 42 m³/ha (0.035 ml oil/g soil) and aqueous phase (103 m³/ha) without N, P, and Ca amendments. A nontreated control site was established.

A second nitrogen (urea) addition was made in April 1975 (370 to 1360 kg N/ha) to the previously fertilized plots, followed by tilling. In October 1975, all plots were tilled and seeded with crested wheatgrass, *Agropyron cristatum* (L.) Gaertn.

The oil emulsion contained dissolved metal ions (Skujiņš, McDonald, and Knight, 1983). Calcium hydroxide was added to the soil to counteract the acidity introduced by the application of oil emulsion and by the water phase. Both calcium hydroxide and calcium phosphate were expected to induce some insolubilization of the added excess metal ions.

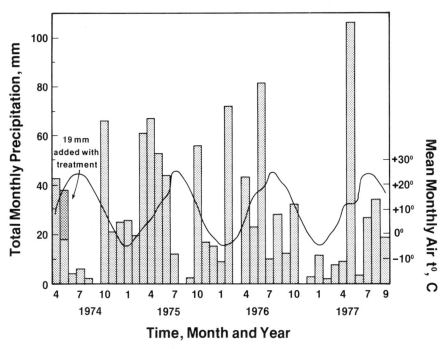

Figure 46-1. Monthly mean air temperatures and total monthly precipitation at the Bear River Bay Weather Station, 2 km north of research site, 1974–1977 (data from U.S. Weather Bureau).

Sampling

Soils were sampled at the treated and control sites before the treatments in 1974 and at intervals during the four years following treatments until 1977. Three subsamples were taken from the cultivated soil layer from four 1-m^2 quadrants established randomly at the sites. The subsamples from the quadrants were combined, air dried, ground, sieved (2 mm), and stored at 4°C until analyzed.

Analysis for Oil

One-gram soil samples were extracted with 50 ml toluene and filtered on Whatman 5 paper; the absorbance of the extract was determined at 420 nm (Odu, 1972). A standard was prepared by extraction of the oil emulsion, obtained from the waste-oil pond, with toluene, evaporation at 25°C, weighing, redissolving in toluene, and determination of absorbance of a dilution series at 420 nm. The sensitivity of the method was 5 g of oil in 50 ml toluene.

Chemical Analysis of Soils

Soils were analyzed for chemical and physical properties, including NO_3^- and total N content (Kjeldaly digestion and distillation), by the Utah State University Soil-Testing Laboratory. Total NH_4^+ was determined by alkaline steam distillation (Bremner, 1965).

Microbiological Analysis

The dilution plate count method (Dotson et al., 1970) was used to determine the number of microorganisms in soils. The bacterial aerobic-soil normal flora and actinomycete populations were determined by plating on standard soil-extract agar containing 0.1% glucose. Hydrocarbon utilizers were determined on soil-extract agar with 1% sterilized motor oil added instead of glucose (Walker and Colwell, 1976). The number of organisms growing on glucose-free soil-extract agar was subtracted from the number growing on oil-amended agar to obtain the number of hydrocarbon utilizers. Anaerobic and facultatively anaerobic organisms were assayed in Nathioglycollate broth with 1% agar in deep screw-cap tubes. Fungal propagules were determined by using Martin's medium (Martin, 1950).

For the determination of bacterial genera, colonies from soil-extract-glucose plates were systematically isolated and tested with standard biochemical and physiological test media and determined according to keys by Skerman (1967) and *Bergey's Manual* (Buchanan and Gibbons, 1974). Fungal genera were classified on the basis of spore formation on Martin's medium or further inoculation on water agar to induce sporulation. Identification of fungi was based on the keys provided by Gilman (1957), von Arx (1974), and Clements and Shear (1973).

RESULTS AND DISCUSSION

Degradation of Oil

By the end of the first year (November 1974) following the oil application, the mean value of oil degradation of all treatments was 37%. During the second, third, and fourth years, 81%, 84%, and 91%, respectively, of the added oil was degraded. No significant runoff or leaching of hydrocarbons into the soil below the 12-cm cultivated layer was detected during the study period. The water added with treatments (54% in the oil emulsion and the additionally disposed water phase from the pond), however, infiltrated rapidly into the soil and pervious slate bedrock.

Examination of selected treatments (Fig. 46-2) indicated that there was no significant difference in degradation rates between the various treatments

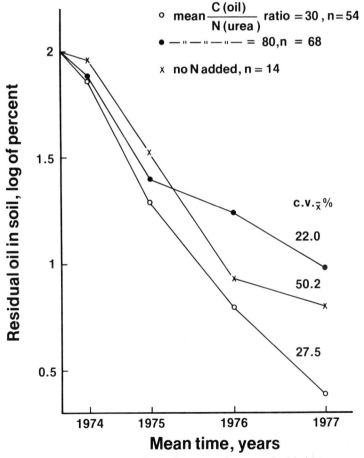

Figure 46-2. Oil content in soil as a function of time and added nitrogen content, expressed as C:N ratio.

with respect to the amount of oil and nitrogen fertilizer (i.e., C:N ratio) added. Although the annual mean values (Fig. 46-2) suggest a lag during the first year and a decreased rate during the third and fourth years as compared to the second year, these variations are not statistically significant due to considerable variability among samples, as reflected by the high coefficients of variation (22.0% to 50.2%). The variation was inherently introduced by the heterogeneous, uneven distribution and mixing of oil with soil by using barber-shank harrow and rototiller as a practical method for oil landfarming purposes, and periodic monthly examination of oil degradation suggested that the maximum rates of degradation took place during the moist and warm spring seasons. Linear regression of exponential annual mean values of degraded oil for the several treatments yielded an equation $y = 1.866 - 0.369t$ (standard deviation $= 0.049$, and $r^2 = 0.78$ for the slope; $t = 1/\text{yr}$).

The added fertilizer nitrogen (i.e., total N of soil, Fig. 46-3) to soil decreased at a rate similar to that of oil degradation (Fig. 46-2). The second fertilizer (urea) addition did not perceptibly increase the total mean annual N amount in soil during 1975. Following urea addition, the mean nitrate

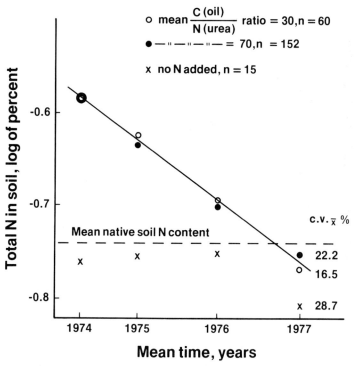

Figure 46-3.   Total nitrogen content in soil as a function of time and C:N ratios of added oil and urea.

content in treated soils increased to 132 μg NO_3^--N/g soil in August 1974, decreased to 11 μg NO_3^--N/g soil during the winter, and increased again to 106 μg/NO_3^--N/g soil in August 1975 following the second year urea addition, decreasing to 15 μg NO_3^--N during the winter 1975-1976. By October 1976 the mean nitrate content in oil- and urea-treated soils decreased to 5 μg NO_3^--N/g soil. The mean value of nitrate content during the three years in nontreated control soils was 3 μg NO_3^--N/g soil. The nitrate content in oil-amended soils receiving no urea fertilizer was not statistically different from that in the control soils. These results indicate that nitrification was taking place in the oil-amended soils, and they suggest that fertilizer nitrogen, de-rived from urea, may be partially lost by denitrification. It has been pre-viously determined (Skujiņš, McDonald, and Knight, 1983) that about 8% of the degraded oil was changed into humic-matter-type components with a C:N ratio below 10:1.

Although it has been suggested that limited amounts of N and P fertilizer addition to oil wastes would improve biodegradation in soil (Huddleston, 1979), results indicate that the degradation rates at various levels of fertilizer amendments are not statistically different, and they conform with the results of other authors (Odu, 1978; Watts, Corey, and McLeod, 1982). The critical factor for oil degradation in arid and semiarid soils appears to be the moisture availability during periods of elevated temperatures.

Characteristics of Microbial Populations

The hydrocarbon-utilizing bacterial population increased severalfold in the oil-amended soils (Fig. 46-4). A very short adaptation time was required for the development of the hydrocarbon-utilizing population, as numbers above 10^8/g soil were detected within seven days following oil application in June 1974. Vanloocke et al. (1975) reported that microbial adaptation to hydrocar-bon substrate can be a rapid process and that a number of soil microorgan-isms have constitutive enzymes for hydrocarbon degradation. The number of hydrocarbon-utilizing bacteria followed the same trend as oil degradation, approaching numbers present in nontreated soils by October 1976.

The numbers of bacteria growing on soil-extract agar with glucose, representing the soil normal flora, also increased during the first and second years following the oil addition, but they returned to the nontreated-soil levels during the third year. The immediate response of this population to initial amendments was probably aided by the water that was added with the waste material application. The changes in bacterial numbers during the subsequent seasons corresponded to the moisture changes in soil reflecting the rainfall pattern (Fig. 46-1), indicating that the microorganisms in this semiarid region were highly dependent on moisture availability for the utilization of added carbon substrate for cell biomass production. Hunt

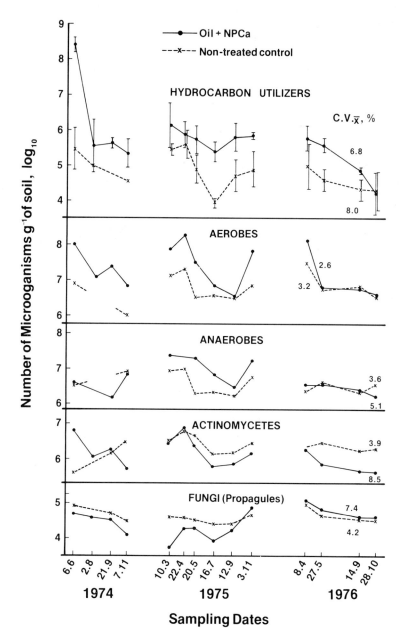

Figure 46-4. Seasonal microbial population changes in oil + NPCa-amended and nonamended control soils during three years following oil application (representative standard deviation ranges are shown for hydrocarbon utilizers).

(1972) reported that numerous microorganisms were able to utilize residues and cell debris of bacteria growing on hydrocarbons in the soil; therefore, a large population of non-hydrocarbon decomposers may accompany the addition of oil to soil.

The numbers of anaerobic and facultative anaerobic bacteria were higher in the oil-amended soils than in control soils during the second year following treatments, indicating an increased anaerobicity of the treated soils during this period. Most investigators agree that there is no evidence for hydrocarbon utilization under anaerobic conditions in nature (Atlas, 1978; de Borger et al., 1978). Microorganisms that utilize nitrate or sulfate as the terminal electron acceptor allowing for anaerobic hydrocarbon utilization, however, have been isolated (Atlas, 1978), and certain oxygen- and sulfur-containing compounds present in petroleum may be utilized by microorganisms under anaerobic conditions (Atlas, 1977, 1978). Considerable amounts of nitrate were present in these oil- and fertilizer-amended soils during 1974 and 1975. It is possible that some of these anaerobic processes were taking place in these soils.

Following the initial increase upon initiation of treatments, the numbers of actinomycetes in oil-treated soils were lower than in the nontreated control soils. As a group, actinomycetes do not compete well with bacterial proliferation upon addition of organic matter and nitrogen to soil.

The participation of certain fungi in oil biodegradation has been emphasized by several investigators (Ahearn and Meyers, 1976; Lehtomäki and Niemelä, 1975; Westlake, Jobson, and Cook, 1978). In our oil-amended soils, the fungal propagule numbers were consistently lower than in the control soils. No significant differences in fungal propagule numbers following oil application were noted by Westlake, Jobson, and Cook (1978). Fungal response to moisture availability in oil-treated soils, however, was notable during the second year of treatment.

It is recognized that the plate count method produces a bias in favor of fungi that produce large numbers of spores. Many soil fungi produce few or no spores, and, therefore, their response to added oil substrate would hardly be detected by dilution plating (Westlake, Jobson, and Cook, 1978). Also, many fungal structures are bound to soil particles and are not released during the dilution process. The fungal biomass determination by hyphal-length measurement was severely hampered by the presence of excessive amounts of hydrophobic substances (oil) in the soil.

There have been numerous studies on the taxonomy of hydrocarbon-utilizing microorganisms, especially from marine environments (Atlas, 1981). The present study examined annual successional changes of soil-normal flora—that is, aerobic bacteria growing on soil-extract agar and fungi developing on Martin's medium in a semiarid soil upon addition of oil (Table 46-1).

Table 46-1. Percent Relative Distribution of Dominant Microbial Populations
(n = Sampling Time × Sampling Site)

Microbial Component	Nontreated Control 3-yr mean	Treatment and Year Oil + NPCa			Oil Without Amendments		
	n = 13	1st	2nd	3rd	1st	2nd	3rd
		6	12	10	3	6	5
Bacteria (100% basis)							
Coryneforms	45	89	59	49	69	62	53
G+ rods (mostly *Bacillus*)	13	8	5	4	17	9	1
G+ cocci (mostly *Micrococcus*)	1		2	4		3	2
G− cocci (mostly *Acinetobacter*)	19	1	10	13		7	9
G− rods:							
pseudomonads	9	1	6	19	7	12	25
Alcaligenes	4		7	5		1	3
others (mostly *Flavobacterium*, *Proteus*, *Serratia*)	9	1	11	6	7	6	7
Fungal genera (100% basis)							
Absidia						0.5	

Genus	1	2	3	4	5	6	7
Alternaria	2	8	4	21	19	2	13
Aphanocladium	2						13
Aspergillus	54	25	14	15		4	
Botrytis	5	3				3	
Chaetomium			3	1			
Choanephora				2			0.5
Chrysosporium				1			
Cladosporium			1	4			
Cunninghamella	2	32	10		20	6	
Drechslera	3		5	1			
Fusarium	0.5		6	11			1
Gliocladium						1	
Monilia		8	2				
Mortierella	2		1	2			
Mucor	1		3	1		1.5	0.5
Paecilomyces	1			1			
Penicillium	23	23	36	26	60	80	66
Peyronellaea				1			
Rhizopus	3						
Syncephalastrum	0.5		1				
Verticillium				5		2	3
Nonsporing, unidentified	1	1	14	8	1		

The relative distribution values indicated that in oil-amended soils coryneforms were by far the most dominant organisms, followed by *Bacillus,* pseudomonads, and other gram-negative rods, especially *Flavobacterium.* As the diversity increased during the four years, the prominence of pseudomonads increased, and that of coryneforms decreased. It is notable that during the first year the coryneforms were especially dominating in soils having oil and fertilizer amendments (89% of population) as compared to soils having oil without fertilizer amendments (69% of population). These results are comparable to those of Jensen (1975), who found that the most important genera of oil degraders in oily waste-amended soils were *Arthrobacter* and *Pseudomonas.*

The fungal genera dominating the soils the first year following oil treatments were *Alternaria, Aspergillus, Cunninghamella, Monilia,* and *Penicillium.* During the second and third years following the treatments, *Chaetomium, Chrysosporium, Cladosporium Fusarium, Syncephalastrum,* and *Verticillium* appeared as a part of dominating mycoflora but were not significantly present in control soils. It has been demonstrated that representative species of most of the listed genera have the capability of utilizing hydrocarbons (Atlas, 1981).

Conclusions

The feasibility of waste-oil biodegradation by landfarming methodology in a semiarid environment has been demonstrated. The biodegradation followed favorable climatic conditions, such as increased precipitation and warmer air temperatures. Consistent with the results of other investigators, addition of nitrogen fertilizers did not increase oil degradation significantly.

ACKNOWLEDGEMENTS

This investigation was supported by a contract from the U.S. Environmental Protection Agency. Coordination and fieldwork of this project were directed by H. J. Snyder and G. B. Rice, EPA.

REFERENCES

Ahearn, D. G., and S. P. Meyers, 1976, Fungal degradation of oil in marine environments, in *Recent Advances in Aquatic Mycology,* E. B. Jones, ed., John Wiley & Sons, New York, pp. 125-133.

Atlas, R. M., 1977, Stimulated petroleum biodegradation, *CRC Crit. Rev. Microbiol.* **5:**371-386.

Atlas, R. M., 1978, Microorganisms and petroleum pollutants, *Bioscience* **28:**387-391.

Atlas, R. M., 1981, Microbial degradation of petroleum hydrocarbons: An environmental perspective, *Microbiol. Rev.* **45:**180-209.

Bremner, J. M., 1965, Inorganic forms of nitrogen, in *Methods of Soil Analysis,* Part 2, C. A. Black, ed., American Society of Agronomy, Madison, Wisconsin, pp. 1179-1237.

Buchanan, R. E., and N. E. Gibbons, eds., 1974, *Bergey's Manual of Determinative Bacteriology,* Williams and Wilkins Co., Baltimore.

Clements, F. E., and C. L. Shear, 1973, The genera of fungi, *Can. J. Microbiol.* **25:**146-156.

de Borger, R., R. Vanloocke, A. Verlinde, and W. Verstraete, 1878, Microbial degradation of oil in surface soil horizons, *Rev. Ecol. Biol. Sol.* **15:**445-452.

Dotson, G. K., R. B. Dean, B. A. Kennar, and W. B. Cooke, 1970, Land spreading, a conserving and nonpolluting method of disposing of oily water, in *Advances in Water Pollution Research,* S. A. Jenkins, ed., Pergamon Press, New York, pp. 1-15.

Gilman, J. C., 1957, *A Manual of Soil Fungi,* Iowa State University Press, Ames.

Huddleston, R. L., 1979, Solid-waste disposal: Landfarming, *Chem. Eng.* **86:**119-124.

Hunt, P. G., 1972, *The microbiology of terrestrial crude oil degradation,* Special Report 168, U.S. Army Corps of Engineers Cold Regions Research and Engineering Laboratory, Hanover, New Hampshire.

Jensen, V., 1975, Bacterial flora of soil after application of oily waste, *Oikos* **26:**152-158.

Lehtomäki, M., and S. Niemelä, 1975, Improving microbial degradation of oil in soil, *Ambio* **4:**126-129.

Martin, J. P., 1950, Use of acid, rose Bengal and streptomycin in a plate method for estimating soil fungi, *Soil Sci.* **69:**215-232.

Odu, C. T. I., 1972, Microbiology of soils contaminated with petroleum. II. Extent of contamination and some soils and microbial properties after contamination, *Inst. Pet. J.* **58:**201-208.

Odu, C. T. I., 1978, The effect of nutrient application and aeration on oil degradation in soil, *Environ. Pollut.* **15:**235-240.

Skerman, V. B. D., 1967, *A Guide to Identification of the Genera of Bacteria,* Williams and Wilkins Co., Baltimore.

Skujiņš, J., S. O. McDonald, and W. G. Knight, 1983, Metal ion availability during biodegradation of waste oil in semiarid soils, *Ecol. Bull.* (Stockholm) **35:**341-350.

Snyder, H. J., G. B. Rice, and J. Skujiņš, 1976, Disposal of waste oil re-refining residues by land farming, in *Proceedings, Hazardous Waste Research Symposium,* EPA Report 600/9-76-015, U. S. Environmental Protection Agency, Washington, D.C., pp. 195-205.

Vanloocke, R., R. de Borger, J. P. Voets, and W. Verstraete, 1975, Soil and ground water contamination by oil spills: Problems and remedies, *Intl. J. Environ. Studies.*

von Arx, J. A., 1974, *The Genera of Fungi Sporulating in Pure Culture,* J. Cramer— Gantner Verlag, Vaduz.

Walker, J. D., and R. R. Colwell, 1976, Enumeration of petroleum-degrading microorganisms, *Appl. Environ. Microbiol.* **31:**198-207.

Watts, J. R., J. C. Corey, and K. W. McLeod, 1982, Land application studies of industrial waste oils, *Environ. Pollut.* **28:**165-175.

Westlake, D. W. S., A. M. Jobson, and F. D. Cook, 1978, *In situ* degradation of oil in a soil of the boreal region of the Northwest Territories, *J. Microbiol.* **24:**254-260.

47

Proteolytic Hydrolysis of Humic Acid by Pronase*

Yiu-Kwok Chan, Peter R. Marshall, and Morris Schnitzer

Research Branch, Agriculture Canada

INTRODUCTION

Proteases have been used to demonstrate the presence of peptides in humic substances (Scharpenseel and Krausse, 1962; Ladd and Brisbane, 1967; Sowden, 1970). One of these, pronase, is a nonspecific proteolytic enzyme prepared from *Streptomyces griseus*. It has been shown to partially release amino acids from humic acids and soil fractions (Ladd and Brisbane, 1967; Sowden, 1970; Brisbane, Amato, and Ladd, 1972). Enzymatic hydrolysis of humic substances is attractive for its mild reaction conditions. Conventionally, prolonged hydrolysis with strong hydrochloric acid (HCl) under reflux is used to release amino-nitrogen (N) and ammonia (NH_3) from soil organic matter. This procedure could induce changes in the humic acid structure, such as increases in condensation or aromaticity. In the interest of maintaining the native forms of the humic components, tests were conducted on the feasibility of using pronase hydrolysis as an alternative method to hydrolyze a humic acid that has been under study. To stabilize the enzyme activity and to reduce autolysis in large-scale hydrolysis, pronase was immobilized on a commercially prepared affinity gel. Griffith and Thomas (1979) showed that pronase immobilized on carboxymethyl cellulose remained much more active than the soluble enzyme in the presence of clay. This paper reports a

---

*Contribution No. 1405 from the Chemistry and Biology Research Institute, Agriculture Canada, Ottawa, Canada.

high efficiency of pronase hydrolysis of a humic acid (HA) compared to acid hydrolysis and the feasibility of using the immobilized pronase to scale up its hydrolysis.

MATERIALS AND METHODS

Source of HA

The HA used in this study was extracted from the surface horizon (10-15 cm) of Bainsville clay loam (Orthic Humic Gleysol) as described previously (Schnitzer and Hindle, 1981). The extraction procedure outlined by Schnitzer (1982) was followed. The preparation was dialyzed in Spectrapor 3 membrane (Spectrum Medical Industries, Inc., Los Angeles, California) that had a molecular weight cutoff of about 3,500 daltons. The Bainsville HA thus prepared contained 2.5% total N on an ash-free dry weight basis, 33.6% of which was proteinaceous in nature. Other analytical data were: 55.3% carbon (C), 5.4% hydrogen (H), 0.9% sulfur (S), 35.9% oxygen (O), 3.9 meq/g carboxyl (CO_2H), and 2.5 meq/g phenolic hydroxyl (OH) groups. A stock solution containing about 2 mg HA/ml was prepared by dissolving it in 0.1 M sodium hydroxide (NaOH) and adjusting the pH to 7.5 with HCl.

Acid Hydrolysis

Fifteen-mg samples of HA were hydrolyzed separately by refluxing with 50 ml of 6 M HCl for 24 hr. Insoluble residues were separated by centrifugation, and the supernatants were retained for amino acid analysis.

Pronase Assay in the Presence of HA

The effect of HA on pronase activity was investigated by the casein-digestion method measuring the 275-nm absorbance of the tyrosine released (Narahashi, 1970). Individual zero-time controls were set up for various concentrations of HA to correct for the absorbance due to HA. An HA-to-pronase ratio from 0 to 200 was used. Results are reported as means of duplicate assays.

Hydrolysis with Soluble Pronase

Pronase containing about 5 units/mg specific activity was obtained from Sigma Chemical Co. (St. Louis, Missouri). The procedure used for hydrolysis was based on that described by Sowden (1970), with the following modifications. Incubation was carried out in a borate-HCl buffer containing 5 mM calcium chloride ($CaCl_2$) at pH 7.5 in accordance with Narahashi (1970), and samples were freeze-dried for amino acid analysis. Methanol was omitted in the incubation mix. One mg pronase was added at zero time for the hydrolysis of 2 mg HA. Controls with enzyme alone or HA alone provided estimates of background amino acids. The hydrolysis was set up in duplicate tubes, and net mean results were converted on a basis of 15 mg HA for comparison with HCl hydrolysis.

Hydrolysis with Immobilized Pronase

Pronase was coupled to Affi-Gel 10 (Bio-Rad Laboratories, Richmond, Calif.) by gently shaking 25 ml of the washed gel overnight with a 50-ml borate buffer solution (0.1 M, pH 7.5) containing 1.25 g pronase at 4°C. The coupled gel was washed free of adsorbed enzyme with the same buffer and was found to have a specific activity of 2.5 units/ml, using the casein assay for pronase mentioned above. The Affi-Gel 10-pronase was diluted 16X with Sephadex G-200. A 200-ml column was packed with the Affi-Gel 10-pronase and Sephadex G-200 and equilibrated with a 1-mM $CaCl_2$ solution at pH 7.5. Ten ml of HA solution (pH 7.5, 1.5 mg/ml) was applied to the column and eluted with the $CaCl_2$ solution at 23°C using anti-gravity flow. The eluate was monitored at 280 nm. Two fractions were collected. Each was concentrated by rotoevaporation at 45°C, acidified to remove HA, and freeze-dried for amino acid analysis.

Amino Acid Analysis

Samples of HA hydrolysates were held at pH 2.2 in 1 ml of sodium citrate, then chromatographed on a Dionex DC-4A column containing sulfonated divinyl benzene and developed with three serial sodium (Na) buffers with increasing pH from 3.28 to 4.95. Duplicate determinations were done for all the amino acids except asparagine, cystine, glutamine, and tryptophan.

RESULTS AND DISCUSSION

Humic Acid Hydrolysis by Soluble Pronase

Before using pronase to hydrolyze the Bainsville HA, two feasibility tests were carried out. First, the extent of autolysis of the enzyme was investigated by incubating 1.5 mg of the enzyme alone in the assay buffer for various periods up to 16 hr. Table 47-1 shows that the enzyme itself contained some free amino acids. Assuming that the pronase contained about 10% N, less than 7% of its N content existed as free amino acids, indicating that it was not a recrystallized preparation. Autolysis increased as a function of incubation time up to 36% in 16 hr. Incubation for 3 hr resulted in less than 20% of the total N released as free amino acids. To minimize the background

Table 47-1. Autolysis of Soluble Pronase during Incubation in the Buffered Assay System

Incubation Time (hr)	Total Free Amino Acids Released	
	$\mu g\,N$	% of Total N
0	10.2	6.8
3	29.2	19.5
6	33.2	22.1
16	54.5	36.3

Table 47-2. Effect of Bainsville HA on Pronase Hydrolysis of Casein

Humic Acid Concentration (mg/ml)	Weight HA / Weight Pronase	Pronase Activity (%)
0	0	100
0.1	40	93.4
0.2	80	96.5
0.3	120	90.7
0.4	160	91.1
0.5	200	98.5

amino acids, a standard 3-hr incubation was then adopted for the subsequent enzyme hydrolysis of the HA. Under such conditions and with an HA substrate-to-pronase ratio of 2:1, a maximum release of 80% of amino acid-N from the HA was estimated over that due to pronase autolysis. This relatively short incubation time was desirable compared to prolonged incubation used earlier, where the degree of pronase autolysis was not reported (Ladd and Brisbane, 1967; Sowden, 1970).

The possibility that pronase could be inhibited by the HA was examined in the second feasibility test. Using casein as substrate, pronase activity was assayed spectrophotometrically in the presence of various concentrations of HA (Table 47-2). The percent inhibition of pronase was small, ranging from 1.5% to 9.3% with an HA-to-pronase ratio of 40:200. Since the degree of inhibition was not related to the amount of HA present, the effect of the Bainsville HA on pronase activity was insignificant. This finding is in contrast to the reported inactivation of pronase by incubation with some HA's (Butler and Ladd, 1971).

The pronase hydrolysis pattern of the Bainsville HA is shown in Table 47-3, where the amino acids released are expressed as a percentage of those released by hot-acid hydrolysis. These results are net amounts of amino acids

Table 47-3. Comparison of Pronase and HCl Hydrolysis of Bainsville HA

	Amino Acid or NH$_3$ Released (μg N)		Amino Acid-N (Pronase) / Amino Acid-N (6 M HCl)
	Pronase	6 M HCl	(%)
Asp	4.3	6.7	64.2
Thr	3.4	2.3	147.8
Ser	3.4	2.4	141.7
Glu	1.7	5.7	29.8
Pro	—[a]	—	
Gly	7.7	7.1	108.5
Ala	6.8	4.5	151.1
CysH	—	—	
Val	8.5	5.8	146.6
Met	—	—	
Ile	5.1	3.7	137.8
Leu	7.7	5.2	148.1
Tyr	1.7	0.9	188.9
Phe	5.1	3.4	150.0
Lys	13.6	14.6	93.2
His	10.2	6.9	147.8
Arg	11.9	7.8	152.6
NH$_3$	1.7	169.9 (150.1)[b]	1.0 (1.1)
Total amino acid-N	91.1	77.0 (96.8)	118.2 (94.1)
Total amino acid + NH$_3$-N	92.8	246.9	

[a]Not detected.
[b]Corrected for acid degradation of amino-N to NH$_3$.

released from the HA after subtracting controls containing enzyme alone and HA alone. Free amino acids were not detected in the HA preparation itself except for small quantities of lysine and histidine. Four salient features of the amino acid pattern are discernible:

1. Three amino acids (proline, cysteine, and methionine) were not released by pronase or by acid hydrolysis. It may be concluded that these amino acids are insignificant components of peptides associated with the Bainsville HA.
2. Pronase release of the dicarboxylic amino acids (aspartic and glutamic acids), their amides (asparagine and glutamine), or both, was incomplete compared to HCl hydrolysis. This finding is similar to that of Brisbane, Amato, and Ladd (1972) and may reflect the specificity of pronase.

3. The release of glycine and lysine was comparable by either method of hydrolysis.
4. An apparent over-release of the rest of the amino acids was obtained by pronase hydrolysis.

However, it was estimated that about 8% of the total amino acid-N and NH_3-N could represent amino-N degradation to NH_3 during acid hydrolysis (Y.-K. Chan, R. R. Marshall, and M. Schnitzer, unpub. results). In this case, it amounts to 19.8 μg N. The total apparent overrelease of amino acids by pronase calculated from Table 47-3 is 20.9 μg N. Hence, pronase did not really release more amino acids than hot HCl. In fact, pronase was shown to be highly efficient (94% after correcting for acid degradation of amino-N) in hydrolyzing peptides and proteins from the Bainsville HA. Ladd and Brisbane (1967) reported that proteolytic hydrolysis of HAs yielded less than 40% of the amino acids released by acid hydrolysis. It is recognized that the susceptibility of HAs to proteolysis by pronase depends on the source and method of their extraction. The absence of pronase-inhibiting components in the Bainsville HA is considered to be a major factor in contributing to the high efficiency of the enzymatic hydrolysis. The amino acids released by pronase were also reported to be inversely related to the aromaticity of the HA (Ladd and Brisbane, 1967). Conversely, a high efficiency of pronase hydrolysis suggests that the peptides and proteins are associated with a low concentration of aromatic rings in the HA structure. This is in agreement with information obtained from ^{13}C NMR spectroscopy that shows an aromaticity of only 33% for the Bainsville HA (Preston and Schnitzer, in press). In addition, it has been suggested that peptides are loosely attached to humus (Haworth, 1971; Kowalenko, 1978).

Pronase released little NH_3 from the Bainsville HA. On the other hand, about 40% of the total N was released by hot HCl as NH_3. This high percentage of NH_3 could be partly attributed to the preparation of the HA in NaOH solution and partly to the hydrolysis of the Unknown N (Schnitzer and Hindle, 1981). Since there was no overrelease of amino acids or NH_3 by pronase, this enzyme was unlikely to have acted on the Unknown N. This observation is consistent with the suggestion that the Unknown N is not proteinaceous (Schnitzer and Hindel, 1981).

Humic Acid Hydrolysis by Immobilized Pronase

The Bainsville HA was progressively hydrolyzed and separated into two fractions (I and II) when passing through the immobilized pronase dispersed throughout the Sephadex gel column. Free amino acids were found in both fractions. More amino acids eluted with fraction I than fraction II, indicating

Table 47–4. Comparison of Immobilized Pronase and HCl Hydrolysis of Bainsville HA

	Amino Acid or NH₃ Released (μg N)				Amino-Acid N (Pronase) / Amino-Acid N (6 M HCl) (%)
	Immobilized Pronase			6 M HCl	
	I	II	Total		
Asp	0.9	1.0	1.9	10.2	18.6
Thr	UDª	UD	UD	3.4	
Ser	UD	UD	UD	3.4	
Glu	UD	UD	UD	8.5	
Pro	—ᵇ	—	—	—	
Gly	8.5	3.7	12.2	11.1	109.9
Ala	8.0	4.4	12.4	6.8	182.3
CysH	—	—	—	—	
Val	7.1	7.5	14.6	8.5	171.2
Met	1.3	1.4	2.7	—	
Ile	2.9	2.4	5.3	5.1	103.9
Leu	4.6	3.9	8.5	7.7	110.4
Tyr	4.4	2.2	6.6	1.7	388.2
Phe	2.1	0.6	2.7	5.1	52.9
Lys	4.3	4.5	8.8	22.2	39.6
His	8.9	4.4	13.3	10.2	130.4
Arg	15.8	—	15.8	11.9	132.8
NH₃	0.5	1.7	2.2	155.6 (133.9)ᶜ	1.4
Total amino acid-N	68.8	36.0	104.8	115.8 (137.5)	90.5 (76.2)
Total amino acid + NH₃-N	69.3	37.7	107.0	271.4	

ªUndetermined due to nonresolution.
ᵇNot detected.
ᶜCorrected for acid degradation of amino-N to NH₃.

that more proteinaceous substances associated with the larger-molecular-size fraction. The distribution of amino acids in the hydrolysate is summarized in Table 47-4. First, similar to soluble pronase, the immobilized enzyme did not release proline and cysteine from the Bainsville HA. However, a small amount of methionine was released. Sulfur-containing amino acids are unstable in acid and, therefore, could not have been detected in acid hydrolysates. Second, immobilized pronase was less efficient for releasing aspartic acid, phenylalanine, and lysine than the soluble enzyme, while glycine, isoleucine, and leucine were released in amounts comparable to those in the acid hydrolysate. The difference in the activity of the two forms of pronase was probably caused by the presence of steric constraints in the immobilized enzyme. For the same reason, incomplete hydrolysis of peptides and proteins was suspected. The short peptides released could have interfered with the separation of threonine, serine, and glutamic acid during amino acid analysis. These amino acids co-eluted and, consequently, were not quantified (Table 47-4). Finally, the apparent overrelease of the rest of the amino acids (23.6 μg N) was equivalent to the estimated degradation of amino-N to NH_3 (21.7 μg N) by hot HCl. As in the case of the soluble pronase, the immobilized pronase did not release more amino acids from the Bainsville HA than did acid hydrolysis. There was no evidence that the HA caused the release of pronase from the gel matrix. Because threonine, serine, and glutamic acids were not quantified, the overall efficiency for releasing amino acids by immobilized pronase was only 76.2% compared to hot HCl after correcting for acid degradation of amino-N. Assuming that these amino acids were released in amounts similar to those released by the soluble enzyme, the efficiency would be raised to about 80%.

CONCLUSION

Pronase hydrolysis has obvious advantages over acid hydrolysis for deproteinating soil fractions. The most important of these is that it reacts under relatively mild conditions and thus avoids possible modification of the humic structure. The successful proteolytic hydrolysis of the Bainsville HA is particularly useful for preparing humic fraction rich in the Unknown N and would thus aid in its characterization. Bainsville soil was chosen for such studies because of its relatively high HA content. Proteolytic hydrolysis should be tested for its applicability to other soil types and soil materials. It should also be extended to include other proteases. This is the first report on the use of affinity gel-immobilized pronase for hydrolyzing HA and its efficiency compared to that of the soluble enzyme. In spite of its somewhat lower efficiency, immobilized pronase is indispensable for scaling up the hydrolysis reaction with little loss of enzyme activity and little interference from autolysis. Moreover, separation of pronase from the deproteinated HA after hydrolysis is not required since the enzyme remains on the reaction column. This aspect is particularly useful in studies of soil peptides and proteins. Further improvements in the use of immobilized pronase in order to increase its hydrolytic efficiency are possible.

ACKNOWLEDGEMENTS

We thank Derek Tutte of the Analytical Chemistry Services for the amino acid analyses.

REFERENCES

Brisbane, P. G., M. Amato, and J. N. Ladd, 1972, Gas chromatographic analysis of amino acids from the action of proteolytic enzymes on soil humic acids, *Soil Biol. Biochem.* **4:**51-61.

Butler, J. H. A., and J. N. Ladd, 1971, Importance of the molecular weight of humic and fulvic acids in determining their effects on protease activity, *Soil Biol. Biochem.* **3:**249-257.

Griffith, S. M., and R. L. Thomas, 1979, Activity of immobilized pronase in the presence of montmorillonite, *Soil Sci. Soc. Am. J.* **43:**1138-1140.

Haworth, R. D., 1971, The chemical nature of humic acid, *Soil Sci.* **11:**71-79.

Kowalenko, C. G., 1978, Organic nitrogen, phosphorus and sulfur in soils, in *Soil Organic matter,* M. Schnitzer and S. U. Khan, eds., Elsevier Scientific Publishing Co., New York, pp. 95-136.

Ladd, J. N., and P. G. Brisbane, 1967, Release of amino acids from soil humic acids by proteolytic enzymes, *Aust. J. Soil Res.* **5:**161-171.

Narahashi, Y., 1970, Pronase, *Methods in Enzymology,* Vol. 19, *Proteolytic Enzymes,* G. E. Perlmann and K. Lorand, eds., Academic Press, New York, pp. 651-664.

Preston, C., and M. Schnitzer, in press, Effects of chemical modifications and extractants on the ^{13}C NMR spectra of humic materials, *Soil Sci. Soc. Am. J.*

Scharpenseel, H. W., and R. Krausse, 1962, Aminosäureuntersuchungen an verschiedenen organischen Sedimenten, besonders Grau-und Braunhuminisäurefraktionen verschiedener Bodentypen (einschliesslich ^{14}C markierter Huminsäuren), *Z. PflErnähr. Düng. Bodenk.* **96:**11-34.

Schnitzer, M., 1982, Organic matter characterization, in *Methods of Soil Analysis,* Part 2, *Chemical and Microbiological Properties,* 2nd ed., A. L. Page, et al., eds., American Society of Agronomy, Madison, Wisconsin, pp. 581-594.

Schnitzer, M., and D. A. Hindle, 1981, Effects of different methods of acid hydrolysis on the nitrogen distribution in two soils, *Plant and Soil* **60:**237-243.

Sowden, F. J., 1970, Action of proteolytic enzymes on soil organic matter, *Can. J. Soil Sci.* **50:**233-241.

48

Effects of Erosional Processes on Nutrient Cycling in Semiarid Landscapes

**David S. Schimel, Eugene F. Kelly,
Caroline Yonker, Richard Aguilar,
and Robert D. Heil**
Colorado State University

INTRODUCTION

The influence of topography on distribution of chemical, physical, and biological soil properties has long been recognized. Different topographic positions not only have different microclimatic conditions but also receive material from and lose material to other topographic positions. This results in sequences, or catenas, of soils along hillslopes. The *catena* was first defined as "a grouping of soils which, while they fall wide apart in a natural system of classification, are yet linked by conditions of topography and are repeated whenever the same conditions are met with" (Milne, 1935). The concept of the catena has been fundamental to pedological research since that first definition. The purpose of this paper is to relate the pedologist's idea of the catena to studies of ecosystem processes. The catena, and the geomorphic processes that the concept implies, are essential to understanding nutrient and organic-matter dynamics in semiarid landscapes.

Catenas contain a sequence of sites whose distribution reflects the influence over time of geomorphological, pedological, and biological processes (Gerrard, 1981). The soils of a catena reflect both the redistribution of material and different intensities of processes at different sites within the catena. Texture, solum depth, and organic-matter content are key variables in understanding changes in primary productivity and nutrient cycling along catenas. These variables control nutrient storage and availability, aggregation and erosion resistance, and soil-water dynamics. Biomass and species composition of grassland vegetation often vary with topographic position (Barnes, Tieszen, and Ode, 1983; Ayyad and Dix, 1964; Dix and Smiens, 1967; Brown, 1971; Ralston and Dix, 1966; Bazzaz and Parish, 1982). Variations in the characteristics of vegetation along hillslopes have often been attributed to parallel variations in water and nutrient availability (reviewed in Bazzaz and Parish, 1982).

The structure of this paper is a synthesis of recent research with information from the literature. The paper addresses three topics: (1) the erosional processes that generate diversity in soils and topography; (2) the effects of erosion on the spatial distribution of nutrients and organic matter; and (3) variations in controls of nutrient cycling along catenas.

DISCUSSION

Mechanisms of Erosion

The processes of wind and water erosion have long been identified and discussed in the literature. Wind-erosion processes include suspension, where soil particles are lifted into the atmosphere and suspended for prolonged periods of time; saltation, where soil movement proceeds as a series of low bounces across the surface, initiated by the impact of other particles; and creep, where the largest soil particles are essentially pushed over the soil surface. Water-erosion processes include raindrop impact and splash, rill and gully erosion, and sheet surface flow or wash erosion. Detachment and transport of soil particles by the impact and splash process is initiated by the kinetic energy of falling raindrops as they strike the soil surface and disperse. Rill and gully erosion occur where there is sufficient precipitation to generate runoff; the runoff is concentrated in microchannels, rills, or gullies. Sheet erosion, implying fairly uniform laminar flow, occurs when runoff is not concentrated in distinct channels.

The magnitude of the effect of these processes is not uniform across a variety of environments; their relative importance and interaction is largely governed by climate (Gerrard, 1981). For example, the process of sheet erosion predominates in dry, sparsely vegetated ecosystems but is not as significant in more humid ecosystems where greater annual precipitation results in increased vegetative cover. Because of the protective effects of increased cover, other erosional processes dominate. A simple conceptual model relates water erosion to climate, using mean annual temperature (MAT) and mean annual precipitation (MAP) (Schumm, 1977). For any given MAT, sediment yield as a function of MAP increases to a maximum and then drops. The maximum sediment yield occurs at higher MAPs as MAT increases. The model thus implicitly partitions precipitation between runoff and evapotranspiration as a function of temperature. The decrease in sediment yield occurs when effective precipitation is high enough to support vegetation sufficient to reduce erosion.

The above model is useful for predicting the relative importance of water erosion under defined conditions. It does not predict actual erosion rates and does not address storage of sediment within landscapes. To interpret the mosaic of soils within a landscape, more complex models are required, and these models must include some detail about landforms and geology.

Erosional Movement of Nutrients and Organic Matter

Landscapes consist of curved, sloping surfaces that are formed by geomorphic processes, including mass-wasting processes (e.g., soil creep, debris flow, slump), wind, and water erosion. Although water movement is most often the driving force in landscape evolution, aeolian processes are also important.

Typical slope profiles have an upper convex segment, a lower concave segment, and a straight segment between these two. The features of some landscapes are more complex; nine slope segments are recognized in one recent classification (Dalrymple, Blong, and Conacher, 1968). Distinct geomorphic and pedogenic processes are associated with each of the different landscape components and are controlled by the geometric configuration (slope shape, length, and gradient) at the portion of the landscape. Ruhe and Walker (1968) classified hillslope profile components into five categories: summits, shoulders, backslopes, footslopes, and toeslopes (Fig. 48-1). In the absence of wind erosion, summit components generally contain stable and strongly developed soils, although they may be shallower than depositional soils. The shoulder—the convex rounded component between the summit and the linear backslope—is the most highly eroded portion of the landscape and contains shallow and poorly developed soils. The backslope area is also characterized by little pedogenic development; it is usually a transportational zone where inputs and outputs balance each other with no net change in soil profile depth. Footslope and toeslope components are characterized by continual inputs of material and solutes from the upslope areas. Soil development in these areas is controlled by the deposition of sediments. The above describes a simple landform, and real landscapes are often more complex. Many of the variations in landforms with climate, parent material, and other factors are reviewed in Gerrard (1981).

In closed drainage systems, debris eroded from the higher elevations will be trapped within the system (Walker and Ruhe, 1968). In open drainage systems, a certain amount of surficial sediment will be transported to toeslope positions from downstream areas or deposited from upstream areas.

Soil morphology varies by landscape position and provides important information about the intensity of ongoing processes. These changes are the result of variations in the vertical and lateral movements of water and mineral and organic material along the catena. Muhaimeed (1981), for example, found that the thickness of the A and B horizons, total solum, and the ratio of the thickness of B/A increased from the upslope position (summit, shoulder) to the lower slope position (footslope). Ruhe and Walker (1968) found the greatest solum thickness on the linear summit and the shallowest solum on the more convex and steeper shoulder landscape position. These differences can be attributed to erosional and pedogenic activity.

Changes in soil physical properties as a result of erosional and depositional activity can be observed in the changes in particle size along soil catenas. Malo et al. (1974) used particle size and organic carbon data to identify erosional and depositional landscapes. In general, maximum erosional activity at the shoulder landscape position is reflected in coarse-textured material, since the fines are selectively removed. Fine-textured material, rich

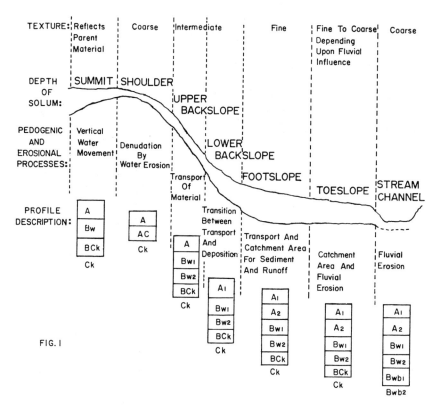

Figure 48–1. Structural features and pedology of an idealized soil catena. *(Source: Horizon designations from Soil Survey Staff, 1981.)*

in organic matter, accumulates in the depositional footslopes position (Kleiss, 1970; Malo et al., 1974).

Removal of the finer mineral material and organic matter from slope shoulders results in increased infiltration rates and reduced water-holding capacity. The redeposition of fine mineral material and organic matter in depositional sites causes increased water-holding capacities and decreased infiltration rates. This leads to markedly different water relations in the erosional and depositional sites of a catena, which in turn leads to changes in primary production, weathering, and leaching along the catena (Gerrard, 1981; Kelly, 1983). Erosional sites do not store water effectively, and this is often reflected in low primary production. Depositional sites store more water and are less leachable for a given amount of water. If they receive significant runon, however, they may be leached more deeply. Depositional soils, because they are wet longer, are more intensely weathered.

Changes in the quantity of organic matter as a function of landscape position can be attributed to some extent to erosional and depositional processes (Malo et al., 1974) along with variations in production and decomposition. Because of its concentration in the surface soil and its low density, organic matter is one of the first soil constituents to be removed through erosion (Slater, 1942). A recent study demonstrates both the concentration of carbon in depositional sites and the rapidity with which organic matter can be redistributed. Three paired cultivated and uncultivated catenas on three parent materials in North Dakota were sampled in order to determine loss and the redistribution of carbon simultaneously (Table 48-1). All three uncultivated catenas show increasing carbon content in lower slope positions. The three parent materials differed in their response to disturbance, however. The sandstone and shale catenas each lost 35% of total landscape carbon after 50 years of cultivation. The patterns of redistribution were somewhat different. In the shale-derived catena, both the summit and shoulder lost a higher proportion of carbon than did the entire catena on average, while both lower slope positions lost less. Thus the lower sites may have received carbon in sediment. In the sandstone-derived catena, the two upper positions lost more than the average for that catena. The upper backslope may have served primarily for transport since it declined by about the same proportion as did the average of the catena. The lower backslope may have received deposition. Its large area and relatively coarse texture prevented much sediment from reaching the footslope, which lost a large proportion of its carbon.

The siltstone catena lost 46% of total landscape carbon, more than either of the others. The pattern of proportional loss on the sandstone- and shale-derived catenas suggests that downslope movement of sediment in overland flow occurred. The pattern of proportional losses is less clear for the siltstone site and was not consistent with water erosion. Silt is a highly

Table 48-1. Redistribution and Loss of Carbon due to 44 Years of Cultivation in Three North Dakota Catenas

	Native C Solum Depth $(kg \cdot m^{-2})$*	Cultivated C Solum Depth $(kg \cdot m^{-2})$*	Change due to Cultivation (%)
Shale-derived catena (6596 m²)			
Summit	12.43	6.70	−45
Shoulder	12.77	5.92	−53
Backslope	12.10	9.08	−24
Footslope	20.17	15.20	−24
Total for catena*	98.58*	64.09*	−35
Average, weighted by area	14.48	9.42	−35
Siltstone-derived catena (10,834 m²)			
Summit	10.01	5.55	−45
Shoulder	10.51	5.79	−49
U Backslope	11.30	6.93	−39
L Backslope	15.52	7.95	−49
Footslope	14.21	8.94	−37
Total for catena*	131.05*	70.38*	−46
Average, weighted by area	11.49	6.17	−46
Sandstone-derived catena (11,006 m²)			
Summit	7.70	3.47	−54
Shoulder	5.99	3.59	−40
U backslope	8.54	5.69	−56
L backslope	11.61	9.49	−18
Footslope	13.51	8.92	−34
Total for catena*	115.70*	70.42*	−34
Average, weighted by area	10.12	6.65	−34

*Or $Mg \cdot catena^{-1}$.
Source: Adapted from Kelly (1983).

Table 48-2. Carbon, Nitrogen and Phosphorus Contents of Erosional and Depositional Sites in a Shortgrass Steppe

Landscape Position	Percent of Total Area	N Solum Depth $(kg \cdot ha^{-1})$	P Solum Depth $(kg \cdot ha^{-1})$	C Solum Depth $(kg \cdot ha^{-1})$
Upslopes (eroded)	70	1,235	650	10,725
Swales (depositional)	30	6,060	1,030	41,140

Source: Adapted from Schimel (1982) and Senft (1983).

wind-erodible particle size fraction, as discussed above. The high loss and anomalous pattern of proportional loss on the siltstone catena may have resulted from severe wind erosion.

Nitrogen (N) and phosphorus (P) are also redistributed by erosion. Phosphorus often increases downslope (Smeck, 1973, Table 2). This has a significant impact on the genesis of soils along catenas since P is a control over inorganic and biological carbon (C) and N dynamics (McGill and Cole, 1981; Cole and Heil, 1981; Smeck, 1973). Phosphorus moved downslope may move largely as organic P forms, although this is not proven. The nitrogen content of soils often increases downslope (Aandahl, 1948; Voroney, Van Veen, and Paul, 1981). It is not clear whether this is due primarily to movement of N downslope, differential N_2 fixation, or both (Schimel, 1982; Aandahl, 1948). This pattern is clear in areas of rolling or stream-dissected topography within the shortgrass steppe (Table 48-2). Much of the increase in N in depositional sites may be due to transport. A first approximation of the amount of N moved can be made using the following assumptions: (1) N and P move together as organic matter, and (2) the transported organic matter has an N:P ratio of 10:1. Gaseous and leaching losses of N are not common in this system (Mosier et al., 1981; Woodmansee, 1978). Under the assumed conditions, 3800 kg/ha of N would have moved to the depositional sites, as compared with the actual increase of 4825 kg/ha. The amount of carbon moved cannot be estimated quantitatively because of respiration losses over time, but matching C to N, even with narrow C:N ratios, yields estimates of carbon transported equal to the total carbon in depositional sites. Much of the increment of carbon may result from deposition.

Variations in Nutrient Availability and Turnover

The movement of nutrients is only one of the effects that erosion has on nutrient cycles. The concentration of clay particles and organic matter in depositional sites changes a number of controls over nutrient-cycling processes. These changes include increased aggregation because of increased clay and organic-matter contents, increased cation-exchange capacity, and increased water-holding capacity. The effects of clay absorption and aggregation on turnover of organic matter are still unclear. However, many authors have suggested that both act to slow down the turnover of organic matter through chemical or physical protection, which should result in increased soil carbon contents (Tisdale and Oades, 1982; Van Veen and Paul, 1981; Cheshire, Sparling, and Mundie, 1983). The turnover of nutrients may be slowed in depositional sites by burial and protection within aggregates. The concentration of P in lower slope positions may result in C and N accumulation above that due to deposition since those three elements tend to equilibrate (Cole and Heil, 1981). Depositional sites also tend to receive additional water as runoff. Primary production is often much higher in these sites because of favorable nutrient and water relations. Thus, the carbon input from primary production is increased, leading to a positive feedback that tends to further increase soil organic matter. Increased plant cover also increases the retention of sediment and organic matter entering depositional sites. In dissected regions of the shortgrass steppe, depositional sites constitute 20% to 30% of the total area but may contain 60% of the organic carbon (Table 48-2). Nitrogen and phosphorous are similarly concentrated.

CONCLUSIONS

The depositional sites of semiarid environments have favorable nutrient and water relations for plant growth. Nutrient mineralization rates are high because of the large total amounts of nutrients and the more favorable water relations. High soil organic-matter levels thus reflect both deposition and relatively high levels of primary productivity. Nutrient turnover rates (mineralization divided by pool size) may be low (Schimel, 1982). This may result from several factors. Much of the carbon may be deep in alluvial soils, where temperature and moisture relations are unfavorable. Increased levels of clay and organic matter will result in physical protection or organic matter within aggregates, resulting in the formation of an inactive fraction of soil organic matter. A small portion of the organic matter may be highly active and contribute a high proportion of total mineralization (Van Veen and Paul, 1981). Depositional sites are sinks for carbon and nutrients because of burial and reduced decomposition due to protection. Agricultural erosion further concentrates nutrients in depositional sites, reducing the proportion of total nutrients that are actively cycling and reducing fertility below the level that would be predicted from the disappearance of nutrients.

ACKNOWLEDGEMENTS

Preparation of this paper was supported by National Science Foundation Grants Nos. DEB 7906009, DEB 8114822, DEB 8105281, DEB 8207015.

REFERENCES

Aandahl, A. R., 1948, The characterization of slope positions and their influence on total nitrogen content of a few virgin soils of western Iowa, *Soil Sci. Soc. Am. Proc.* **13**:449-454.

Ayyad, M. A. G., and R. L. Dix, 1964, Vegetation microenvironment of prairie slopes in Saskatchewan, *Ecol. Monog.* **34**:421-442.

Barnes, P. W., L. L. Tieszen, and D. J. Ode, 1983, Distribution, production, and diversity of C$_3$- and C$_4$-dominated communities in a mixed prairie, *Can. J. Bot.* **61**:741-751.

Bazzaz, F. A., and J. A. D. Parish, 1982, Organization of grassland communities, in *Grasses and Grasslands: Systematics and Ecology,* J. R. Estes, R. J. Tyrl, and J. N. Brunken, eds., University of Oklahoma Press, Norman, Oklahoma, pp. 233-254.

Brown, R. W., 1971, Distribution of plant communities in southeastern Montana badlands, *Am. Midl. Nat.* **85**:458-477.

Cheshire, M. V., G. P. Sparling, and C. M. Mundie, 1983, Effect of periodate treatment of soil on carbohydrate constituents and soil aggregation, *J. Soil Sci.* **34**:105-112.

Cole, C. V., and R. D. Heil, 1981, Phosphorus effects on terrestrial nitrogen cycling, in *Terrestrial Nitrogen Cycles,* F. E. Clark and T. Rosswall, eds., *Ecol. Bull.* (Stockholm) **33**:363-374.

Dalrymple, J. B., R. J. Blong, and A. J. Conacher, 1968, A hypothetical nine unit landsurface model, *Z. fur Geomorphol.* **12**:60-76.

Dix, R. L., and F. E. Smeins, 1967, The prairie, meadow, and marsh vegetation of Nelson County, North Dakota, *Can. J. Bot.* **45**:21-58.

Gerrard, A. J., 1981, *Soils and Landforms: An Integration of Geomorphology and Pedology,* George Allen and Unwin, London.

Kelly, E. F., 1983, *Evaluating Erosion along Catenary Sequences of the Northern Great Plains,* M. S. thesis, Colorado State University, Fort Collins.

Kleiss, H. J., 1970, Hillslope sedimentation and soil formation in northeastern Iowa, *Soil Sci. Soc. Am. Proc.* **34**:287-290.

Malo, D. D., B. K. Worchester, D. K. Cassel, and K. D. Matzdorf, 1974, Soil-landscape relationships in a closed drainage system, *Soil Sci. Soc. Am. Proc.* **38**:813-818.

McGill, W. B., and C. V. Cole, 1981, Comparative aspects of cycling of organic C, N. S and P through soil organic matter, *Geoderma* **26**:267-286.

Milne, G., 1935, Some suggested units of classification and mapping particularly for East African soils, *Soil Res.* **4**:(3).

Mosier, A. R., M. Stillwell, W. J. Parton, and R. G. Woodmansee, 1981, Nitrous oxide emissions from a native shortgrass prairie, *Soil Sci. Soc. Am. J.* **45**:617-619.

Muhaimeed, A. A., 1981, *Soil Property Relationships on Selected Landscape Segments under Cultivated vs. Rangeland Conditions,* Ph.D. dissertation, Colorado State University, Fort Collins.

Ralston, R. D., and R. L. Dix, 1966, Green herbage production of native grasslands in the Red River Valley—1965, *N. Dak. Acad. Sci. Proc.* **XX**:57-66.

Ruhe, R. V., and P. H. Walker, 1968, Hillslope models and soil formation. I. Open systems, *9th Int. Cong. Soil Sci. Trans.* (Adelaide, Australia) **IV**:551-560.

Schimel, D. S., 1982, *Nutrient and Organic Matter Dynamics in Grasslands: Effects of Fire and Erosion,* Ph.D. dissertation, Colorado State University, Fort Collins.

Schumm, S. A., 1977, *The Fluvial System,* Wiley-Interscience, New York.

Senft, R., 1983, *The Redistribution of Nitrogen by Cattle,* Ph.D. dissertation, Colorado State University, Fort Collins.

Slater, C. S., 1942, Variability of eroded material, *J. Agri. Res.* **65**:209-219.

Smeck, N. E., 1973, Phosphorus: An indication of pedogenetic weathering processes, *Soil Sci.* **115**:119-206.

Soil Survey Staff, 1981, *Soil Survey Manual,* USDA Handbook 18, U.S. Government Printing Office, Washington, D.C., Chap. 4.

Tisdall, J. M., and J. M. Oades, 1982, Organic matter and water-stable aggregates in soils, *Soil Sci.* **33**:141-163.

Van Veen, J. A., and E. A. Paul, 1981, Organic carbon dynamics in grassland soils. 1. Background information and computer simulation, *Can. J. Soil. Sci.* **61:**185-201.

Voroney, R. P., J. A. Van Veen, and E. A. Paul, 1981, Organic C dynamics in grassland soils. 2. Model validation and simulation of the long-term effects of cultivation and rainfall erosions, *Can. J. Soil Sci.* **61:**211-224.

Walker, P. H., and R. V. Ruhe, 1968, Hillslope models and soil formation. II. Closed systems, *9th Int. Cong. Soil Sci. Trans.* (Adelaide, Australia) **IV:**561-568.

Woodmansee, R. G., 1978, Additions and losses of nitrogen in grassland ecosystems, *Bioscience* **28:**488-453.

Index

Acid precipitation, 500, 501, 506, 510, 516, 518-519
Acridine Orange (AO). *See* Ground water, techniques for measuring microorganisms in
Actinomycetes. *See also Thermoactinomyces vulgaris*
   for oil degradation, 549, 552, 555-557, 558, 560
Agroecosystems, 575, 576
Algal mats. *See* Mats, microbial
Aluminum. *See* Soil, aluminum in
Amino acids, analysis of, 564, 566-567, 569
Apatite. *See* Phosphorous
Aquifer. *See* Ground water
Archaebacteria, thermophilic, from submarine hydrothermal areas, 320-330
Arctic fjord system, Greenland, 137, 138-139, 142-143, 148-153
Arsenic
   biogeochemical cycle of, 169-170, 173-176

biotransformation of, 169-170, 173-176
leaching of
   from oil shale by microorganisms, 169-176
   from soil by microorganisms, 169-176
volatilization by microorganisms, 169-176

Bacillus, distribution of, over sulfur occurrences, 311, 313-318
Bacteria
   iron oxidation by. *See* Iron-oxidizing bacteria
   manganese oxidation by. *See* Manganese-oxidizing bacteria
   thermophilic. *See* Thermophiles
Belemnite, 54-60
Bermuda, 383-396
Big Soda Lake, Nevada, 81-87
Biodegradation, of oil, 549-560. *See also* Hydrocarbons
Biodenitrification. *See* Denitrification

About the Editors

DOUGLAS E. CALDWELL received both the B.S. and Ph.D. degrees from Michigan State University, where he studied under James Tiedje. Dr. Caldwell was assistant professor of biology and curator of microorganisms at the University of New Mexico, and presently is associate professor of biology at the University of Saskatchewan. His field of research is thermophilic micro-organisms—the kinetics of their growth and their attachment to surfaces. He is currently investigating the use of thermophiles in cellulose degradation.

JAMES A. BRIERLEY began research on thermophilic microorganisms in metals leaching while preparing to receive the Ph.D. degree from Montana State University. Subsequently, he joined the Biology Department at New Mexico Institute of Mining and Technology, where he was named depart-ment chairman in 1979. During his 17-year association with the college, Dr. Brierley developed research on the use of microorganisms for the recovery of metals in the mining industry. In 1983, he was named to his current position as research director for Advanced Mineral Technologies, Inc., where he continues research on microorganisms in metals recovery.

CORALE L. BRIERLEY is president of Advanced Mineral Technologies, Inc., a company she founded in 1982, soon after receiving the Ph.D. degree from The University of Texas at Dallas. Previously, Dr. Brierley worked with the New Mexico Bureau of Mines and Mineral Resources as chemical microbiologist. At present, her efforts are devoted to developing processes for use in the mining and metals industries.